MW01114887

Cytokines

Series on

Modern Insights into Disease from Molecules to Man

Series Editor

Victor R. Preedy
Professor of Nutritional Biochemistry
School of Biomedical & Health Sciences
King's College London
and
Professor of Clinical Biochemistry
King's College Hospital
UK

Books in this Series

- Adhesion Molecules
- Apoptosis
- Cytokines
- Adipokines

Cytokines

Editors

Victor R. Preedy
Professor of Nutritional Biochemistry
School of Biomedical & Health Sciences
King's College London
and
Professor of Clinical Biochemistry
King's College Hospital
UK

Ross J. Hunter *MD*
Cardiology Research Fellow
St Bartholomew's Hospital
London
UK

Science Publishers
Jersey, British Isles
Enfield, New Hampshire

Published by Science Publishers, an imprint of Edenbridge Ltd.
- St. Helier, Jersey, British Channel Islands
- P.O. Box 699, Enfield, NH 03748, USA

E-mail: *info@scipub.net* Website: *www.scipub.net*

Marketed and distributed by:

6000 Broken Sound Parkway, NW
Suite 300, Boca Raton, FL 33487
270 Madison Avenue
New York, NY 10016
2 Park Square, Milton Park
Abingdon, Oxon OX14 4RN, UK

Copyright reserved © 2011

ISBN: 978-1-57808-690-0

Cover illustrations: Reproduced by kind courtesy of the undermentioned authors:
Courtney C. Kurtz *et al.* (Figure No. 4 from Chapter 7)
Elieser Gorelik *et al.* (Figure No. 1 from Chapter 15)
Kate L. Graham and Helen E. Thomas (Figure No. 1 from Chapter 18)
Patrizia D'Amelio (Figure No. 2 from Chapter 25)

Library of Congress Cataloging-in-Publication Data

Cytokines / editors, Victor R. Preedy, Ross Hunter.
p. ; cm. -- (Series on modern insights into disease from molecules to man)
Includes bibliographical references and index.
ISBN 978-1-57808-690-0 (hardcover)
1. Cytokines. I. Preedy, Victor R. II. Hunter, Ross, 1977- III.
Series: Series on modern insights into disease from molecules to man.
[DNLM: 1. Cytokines. QW 568]
QR185.8.C95C9825 2011
616.07'9--dc22

2010051277

The views expressed in this book are those of the author(s) and the publisher does not assume responsibility for the authenticity of the findings/conclusions drawn by the author(s). No responsibility is assumed by the publishers for any injury and/or damage to persons or property as a matter of products liability, negligence or otherwise, or from any use operation of any methods, products, instructions or ideas contained in the material herein. Because of rapid advances in the medical sciences, in particular, independent verification of diagnoses and drug dosages should be made.

All rights reserved. No part of this publication may be reproduced, stored in a retrieval system, or transmitted in any form or by any means, electronic, mechanical, photocopying or otherwise, without the prior permission of the publisher, in writing. The exception to this is when a reasonable part of the text is quoted for purpose of book review, abstracting etc.

This book is sold subject to the condition that it shall not, by way of trade or otherwise be lent, re-sold, hired out, or otherwise circulated without the publisher's prior consent in any form of binding or cover other than that in which it is published and without a similar condition including this condition being imposed on the subsequent purchaser.

Printed in the United States of America

Preface

The cytokines are generally considered to be a group of peptides secreted by cells of the immune system such as macrophages, lymphocytes and T cells, although the same peptides may also be secreted by non-immune cells such as neurological tissues and adipocytes. The term *cytokine* is, however, simplistic and in fact they are divided into functional families such as the interferons and colony-stimulating factors. In normative situations they act to control the immune milieu but in some circumstances they can be harmful. That is, they can be anti- and pro-inflammatory. They can be "good" and "bad". However, the actions of the cytokines are not confined to the immune or inflammatory response. They can respond to metabolic dysregulation such as oxidative stress and even alter in psychiatric conditions. In other words they have a diverse range of responses in numerous pathophysiological and lifestyle conditions. It is conceivable that pathways and points of therapeutic intervention in one tissue may be applicable to other tissues or conditions. In *Cytokines* the Editor has invited distinguished scholars to write chapters which together will provide readers with a holistic knowledge-base to enable them to understand cytokines in specific or broad details. There are six sections as follows:

(1) General and cellular aspects
(2) Lifestyle factors and life events
(3) Immunology and infections
(4) Cancer
(5) Cardiovascular and metabolic disease
(6) Organs and tissue systems

The book has a mixture of preclinical and clinical information and includes chapters on microRNAs, tight junctions, cytokine networks and MAPKs, natural killer cells, T helper subsets, viral infection, pulmonary tuberculosis, HIV, exercise, tissue repair, alcoholism, the elderly, lung cancer, breast cancer, cancer immunotherapy, chemotherapeutic drugs, cardiovascular risk, angiogenic cytokine therapy, diabetes, hypo- and hyperglycemia, metabolic syndrome, skeletal muscle insulin resistance, Alzheimer's disease, Parkinson's disease, cerebrospinal fluid cytokines, encephalitis, bone and rheumatoid arthritis. The book is designed for a broad scientific readership including health scientists, doctors, physiologists, immunologists, biochemists, college and university teachers and lecturers, undergraduates and graduates. The chapters are written either by national or international experts or specialists in their field.

Professor Victor R. Preedy
Dr Ross J. Hunter, MD

Contents

Section I
General and Cellular Aspects

CHAPTER 1

MicroRNAs and Cytokines

Xiangde Liu
985910 Nebraska Medical Center, Omaha, Nebraska 68198-5910, USA

ABSTRACT

MicroRNAs (miRNAs) regulate posttranscriptional gene expression by inhibiting protein translation or by destabilizing target mRNA. While most miRNAs are widely expressed in various types of organs or tissues, a few are expressed only in limited developmental stages or in specific tissues or cells. Alteration of miRNA expression in human organs or tissues may contribute to the development of many diseases including cancer and chronic inflammation. Cytokines are also involved in the development of cancer and many other chronic inflammatory diseases such as chronic obstructive pulmonary disease (COPD). MiRNAs can regulate cytokine expression either by directly binding to its target sequence or by indirectly regulating a cluster of adenine and uridine-rich element binding proteins (ARE-BPs). Vice versa, cytokines, especially pro-inflammatory cytokines (IL-1ß and TNF-α), can also regulate expression of miRNAs such as miR-146a and miR-155. Expression of miR-146a is dramatically increased in response to the stimulation of inflammatory cytokines in many human cells including human bronchial epithelial cells (HBECs). Transient transfection of miR-146a mimic into HBECs not only protects the cells from death, but also promotes cell proliferation, suggesting miR-146a may provide a link between chronic inflammation and lung cancer or peri-bronchial fibrosis via extending cell survival and stimulating cell proliferation.

INTRODUCTION

MicroRNAs (miRNAs) are small, non-protein-coding RNAs that mediate posttranscriptional gene repression by inhibiting protein translation or by destabilizing target gene transcripts. Unlike small interfering RNA (siRNA) that requires perfect sequence match with targeted mRNA, miRNAs bind to the 3'-untranslated region (UTR) of target gene(s) through imperfect base-pairing, with one or more mismatches in sequence complementary seed region. While we are making rapid progress in understanding biosynthesis and molecular mechanism of miRNA biological process, the precise biological functions of these small (22 nucleotides) miRNAs is less clear, and it is estimated that approximately one third (33%) of human genomes are regulated by miRNAs.

The first non-coding RNA, named small temporal RNA (stRNA) at that time, *lin-4*, was discovered by Ambros and colleagues from *C. elegans* in 1993 (Lee *et al.* 1993), but it was only in 2000 that the homologs of let-7, the second tiny RNA discovered in *C. elegans*, was identified in humans and other animals (Pasquinelli *et al.* 2000, Reinhart *et al.* 2000, Slack *et al.* 2000). Afterwards, approximately 30 in human, 60 in worms, and 20 in Drosophila were identified in less than one year (Lagos-Quintana *et al.* 2001). Thus, the term microRNA was then used to refer to the stRNAs and all other tiny RNAs with similar features but unknown functions (Lagos-Quintana *et al.* 2001, Lau *et al.* 2001). In 2004, a registry for microRNAs (miRNAs) was set up to catalog the miRNAs and facilitate the naming of newly identified genes (Griffiths-Jones 2004), and the October 2007 release of the miRNA Registry records 5922 distinct mature miRNA sequences from 58 species, including 555 distinct mature miRNA sequences from human (Saini *et al.* 2007). To date, miRNA Registry records 10,883 entries and 721 miRNAs from human (www.mirbas.org,_Release 14, September 2009).

In the animal cells, functionally mature miRNAs 22 nucleotides long are excised from a long endogenous RNA strand by the sequential cleavage of endonucleases (Drosha and Dicer, Fig. 1). For example, in a normal alveolar epithelial cell, a strand of primary miR-200c (pri-miR-200c) is transcribed from independent gene by the activity of RNA polymerase II (Park *et al.* 2008). The pri-miR-200c is processed by the RNase III Drosha into 50- to 80-base hairpin-like premature miR-200c (Pre-miR-200c). These hairpin-like pre-miR-200cs are then exported from the nucleus to the cytoplasm by the Exportin-5, where a second RNase III-like nuclease, Dicer, processes them further into duplexes, each duplex usually yields a single functionally mature miR-200c through unwinding and being loaded with RNA-induced silencing complex (RISC). This functioning *miR-200c/RISC* complex will imperfectly bind to the 3'-UTR of the target mRNAs, that is, transcription factor 8 (TF8) and MMP mRNAs. Recognition of 3'-UTR of the target mRNA by partial sequence complementary to the miR-200c results in translation repression of TF8, a factor that inhibits E-cadherin expression (Park *et al.* 2008), and the MMPs that can cleave the E-cadherin. When the cells are undergoing epithelial-mesenchymal transition (EMT) in response to cytokine stimulation (Fig. 2), transforming growth factor ß (TGF-ß) inhibits synthesis of the pri-miR200c through the Snail family proteins, which results in restoring inhibitory function of TF8 on E-cadherin

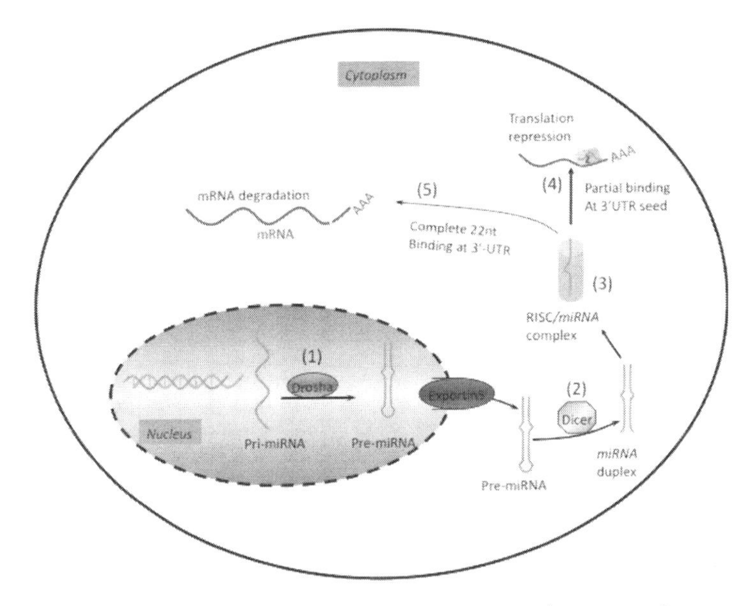

Fig. 1 MiRNA biogenesis and function in a cell. Primary miRNAs (pri-miRNAs) are transcribed from independent genes by the RNA polymerase II. (1) The pri-miRNAs are further cleaved into 50- to 80-base premature miRNAs (pre-miRNAs) by Drosha. (2) The pre-miRNAs are translocated into cytoplasm by the GTP-dependent factor exportin-5, where the nuclease Dicer processes further into the miRNA duplexes. (3) Functional mature miRNAs are loaded with the RNA-induced silencing complex (RISC), and (4) represses translation of the target mRNAs by partial binding to the 3'UTR (7-8 mer) or (5) degrades mRNA by complete complementary binding to target (22nt).

Color image of this figure appears in the color plate section at the end of the book.

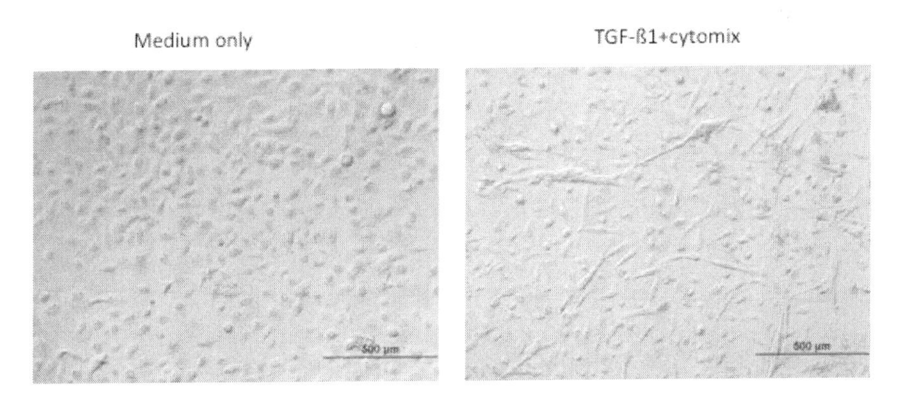

Fig. 2 Human bronchial epithelial cells (HBEC) undergo epithelial-mesenchymal transition (EMT). HBECs were treated with or without TGF-ß1(2 ng/ml) plus cytomix (a mixture of 2.5 ng/ml IL-1ß + 5 ng/ml TNF-α + 5 ng/ml IFN-γ) for 3 d. Scale bar: 500 µm.

expression. The effect of TGF-ß inhibition on E-cadherin expression is further augmented by IL-1ß and TNF-α through up-regulating TGF-ß Receptor type I (TßR-I) (Liu 2008). In addition, production of MMPs (most likely, MMP-1, -3, -9) is increased in EMT cells by the stimulation of TGF-ß1, IL-1ß and TNF-α, which results in the cleavage of E-cadherin.

MiRNAs may contribute to the pathogenesis of many diseases mainly in the following three ways: (1) Loss or down-regulation of the miRNA expression. MiRNAs such as let-7 families function as tumor suppressor and down-regulation of this kind of miRNAs is a common phenomenon in solid tumors. (2) Over-expression of miRNA. A number of miRNAs function as oncogenes. In this regard, over-expression of oncogenic miRNAs such as miR-17-92 cluster results in extended cell survival and blockade of apoptosis and thus may contribute to the development of cancer. (3) Mutation or epigenetic alteration in the 3' UTR of mRNA or seed region of miRNA. This would result in over-expression of the target gene. For example, a single-nucleotide polymorphism (SNP) of miR-146a can turn the pro-apoptotic function of this miRNA into an anti-apoptotic (Jazdzewski *et al.* 2009). A SNP in the 3'UTR of HLA-G, a known asthma-susceptibility gene, disrupts binding of corresponding miR-148 and thus leads to high susceptibility to asthma (Tan *et al.* 2007).

Most miRNAs are widely expressed in various types of organs or tissues. A few, however, are expressed only in limited developmental stages or in specific tissues or cells. For instance, miR-195 and miR-200c are expressed specifically in the lung, miR-203 is specifically and highly expressed in the skin among the 21 human organs or tissues, and miR-1 is predominantly expressed in the heart (Zhao *et al.* 2005, Sonkoly *et al.* 2007, Wang *et al.* 2007). Alteration of miRNA expression in human organs or tissues plays an important role in many diseases such as cancer, developmental abnormalities, cardiovascular disorders, and the pathogenesis of chronic inflammatory diseases. Like miRNAs, cytokines are also involved in the development of cancer and many other chronic inflammatory diseases such as chronic obstructive pulmonary disease (COPD).This review will specifically discuss miRNA expression in response to inflammatory cytokine stimulation and its potential role in modulating cellular functions.

MICRORNA AND CYTOKINES

Although the number of distinct mature miRNA in the Registry database (miRBase) continues to grow rapidly (Saini *et al.* 2007), only a few miRNAs are directly associated with regulating cytokine production in that the 3' untranslated regions (3'UTRs) of many cytokines lack direct binding sites for a miRNA. Cytokines that contain miRNA target site at their 3'UTRs include IL-1A, IL-4, IL-7, IL-8, IL-10, IL-11, IL-12A, IL-12B, IL-13, IL-15, IL-16, IL-17A/D/F, IL-18, IL-22, IL-23A, IL-24, IL-25, IL-29, IL-33 and IL-34 (Table 1). Unlike plant miRNA, in which miRNA are completely (22nt) complementary to their targets, the majority of animal miRNAs are only partially complementary (7-8 mer) to their targets at seed region and thus, in most cases, miRNAs can repress protein translation rather than degrade mRNA.

Besides the direct binding of miRNA to its target 3'UTR seed region, production of many cytokines is also indirectly controlled by miRNAs through a mechanism of regulating a cluster of adenine and uridine-rich elements (AREs). The basic unit of the ARE is a pentamer of AUUUA or nonamer of UUAUUAUU. The ARE recruits several different ARE-binding proteins (ARE-BPs) that can positively or negatively regulate cytokine mRNA stability and/or translation (Anderson 2008). ARE-BPs such as Tristetraprolin (TTP), AU-rich binding factor 1 (AUF1), and members of Hu protein R (HUR) family are regulated by miRNAs. Currently, the ARE registry lists over 4000 genes as potential targets of post-transcriptional regulation (Bakheet *et al.* 2006) and many cytokines contain ARE sites at their 3'UTR region. Cytokines that contain ARE sites are IL-1A, IL-1B, IL-2, IL-3, IL-4, IL-5, IL-6, IL-7, IL-8, IL-10, IL-11, IL-12A, IL-12B, IL-17, IL-20 and IL-27(Asirvatham *et al.* 2009).

Table 1 Cytokines that contain conserved target for miRNA(s)

Cytokine	*Access#*	*miRNAs*
IL-1A	NM 000575	miR-181
IL-4	NM 000589	miR-410; 340
IL-7	NM-000880	miR-768; 432; 892a; 656; 548; 559
IL-8	NM 000584	miR-302; 372; 302; 372; 520; 373
IL-10	NM_000572	let-7; 98
IL-11	NM_000641	miR-124; 506
IL-12A	NM_000882	miR-21; 590-5P
IL-12B	NM_002187	miR-23
IL-13	NM_002188	let-7; 98
IL-15	NM_000585	miR-203
IL-16	NM_004513	miR-125
IL-17A	NM_002190	miR-1266
IL-17D	NM_138284	miR-607; 938
IL-17F	NM_052872	miR-590-3p
IL-18	NM_001562	miR-590-3p
IL-22	NM_020525	miR-944
IL-23A	NM_016584	miR-211; 204
IL-24	NM_006850	miR-203
IL-25	NM_022789	miR-93; 20; 17; 106; 519
IL-29	NM_172140	miR-548n
IL-33	NM_033439	miR-890; 125a-3p; 563; 584; 1244
IL-34	NM_152456	miR-31

*Based on TargetScan. org

While the cytokine expression can be regulated by miRNAs through the mechanisms of direct binding to the target or indirect controlling ARE-BPs, a number of cytokines can also regulate miRNA synthesis (Table 2). For instance, IL-1ß and TNF-α are potent stimulators for induction of miR-146a and miR-155 in a variety of cells. MiR-146a is an inflammation-responsive miRNA and plays a role in regulating immune response, chronic inflammation, and cell proliferation and differentiation. Here I will focus on the role of miR-146a in regulating chronic inflammation and potential connection to the development of cancer.

Table 2 miRNA expression in response to cytokine stimulation

Cytokines	*miRNAs*	*Cell or tissue*
IL-1β/TNF-α	miR-146a	HBEC, A549;
IL-1β/TNF	miR-146a/155	Synoval fibroblasts
IL-1β	miRNA-7	Intestinal epithelium
IL-1β/TNF	miRNA-9	Monocytes, Neutrophil
TNF-α	miRNA-155	B cell
TGF-β1	miRNA-155	Mammary epithelium
IFN-β	miRNA-155	Macrophage
IFN-β	miRNA-122	Liver specific cells
TGF-β1	miRNA-200C	Epithelial cell
TGF-β1	miRNA-24	Skeletal fibroblast
IL-6	miRNA-21	Embroys
IL-6	miRNA-370	Cholanglocarcinoma
IL-6/STAT3	Let-7	Cholanglocarcinoma
IL-6/STAT3	miR-17/92	HPAEC
IL-6/STAR3	miRNA-21	Myeloma cell

HBEC: human bronchial epithelial cells
HPAEC: human pumonary artery endothelial cells

MIR-146A AND CHRONIC INFLAMMATORY DISORDERS

To date, 721 human miRNAs have been identified [Release 14, September 2009. http://www.mirbase.org], but only a few miRNAs have been reported that might be associated with chronic inflammation (Table 3). Among these miRNAs associated with chronic inflammation, miRNA-146a is one of the most extensively investigated. In this regard, basal expression of miR-146a is elevated in the tissues of psoriasis and rheumatoid arthritis (Sonkoly *et al.* 2007, Nakasa *et al.* 2008), diseases that are associated with chronic inflammation. The expression of miR-146a in rheumatoid arthritis synovial fibroblasts was higher than that in the fibroblasts from osteoarthritis (Stanczyk *et al.* 2008), and miR-146a was found to localize at the CD68+ macrophages, CD3+ T-cells and CD79a+ B-cells at the superficial and sub-lining layers of the synovial tissue (Nakasa *et al.* 2008). These findings suggest that miR-146a is associated with not only chronic inflammation, but also innate

Table 3 miRNAs and chronic inflammatory disorders

miRNA	*Disorders*	*Alteration*	*Target genes*
miR-146a	RA, psoriasis	Up	TRAF6, IRAK1
miR-155	RA, VI	Up	IKK, MMP-3
miR-125b	Psoriasis, eczema	Down	N/A
miR-21	Psoriasis, eczema	Up	N/A
miR-203	Psoriasis	Up	SOCS-3, p63
miR-17-92 cluster	SLE, ITP, PBC	Down	PTEN, BIM
miR-296	ITP	Down	N/A
miR-198	SLE	Up	N/A
miR-342	SLE, ITP	Up	N/A
miR-383	SLE, ITP	Down	N/A
let-7b	PBC	Down	IL-6
miR-346	PBC	Down	N/A
miR-20a	PBC	Down	N/A
miR-451	PBC	Up	N/A
miR-129	PBC	Up	N/A
miR-126	VI	Down	VCAM-1

RA, rheumatoid arthritis; VI, vascular inflammation; SLE, systemic lupus erythematosus; ITP, idiopathic thrombocytopenic purpura; PBC, primary biliary cirrhosis; TRAF-6, TNF receptor associated factor 6; IRAK1, IL-1 receptor associated kinase.

immunity. *In vitro* studies also demonstrated that miR-146a expression was rapidly increased in response to IL-1ß and TNF-α in various types of cells including A549 cells, BEAS2B cells, primary human airway epithelial cells, primary human airway smooth muscle cells, and human rheumatoid arthritis synovial fibroblasts (Nakasa *et al.* 2008, Perry *et al.* 2008, Liu *et al.* 2009). Consistent with these reports, we have demonstrated that expression of miR-146a in response to inflammatory cytokine stimulation was significantly increased in primary human airway epithelial cells by microarray assay (Fig. 3A) as well as by real time RT-PCR (Fig. 3B). Interestingly, however, miR-146a was not increased in tissues obtained from patients with other chronic inflammatory disorders such as lung biopsies from mild asthma (Williams *et al.* 2009) or the skin from atopic eczema (Sonkoly *et al.* 2007).

The miR-146a family is composed of two members, that is, miR-146a and miR-146b that are located on chromosomes 5 and 10, respectively. In addition to the inflammatory cytokines (IL-1ß and TNF-α), lipopolysaccharide (LPS) is also a potent stimulator of miR-146a through TLR-4 activation (Taganov *et al.* 2006). Studies have shown that NF-κB mediates miR-146a expression driven by either LPS or inflammatory cytokines. While a negative feedback loop exists through targeting TNF-α receptor associated factor 6 (TRAF6) and IL-1ß receptor associated kinase 1 (IRAK1) by miR-146a, little is known regarding miR-146a biological function and its mechanism. Perry *et al.* showed that the increase of miR-146a expression in human alveolar epithelial cells (A549 cell) negatively

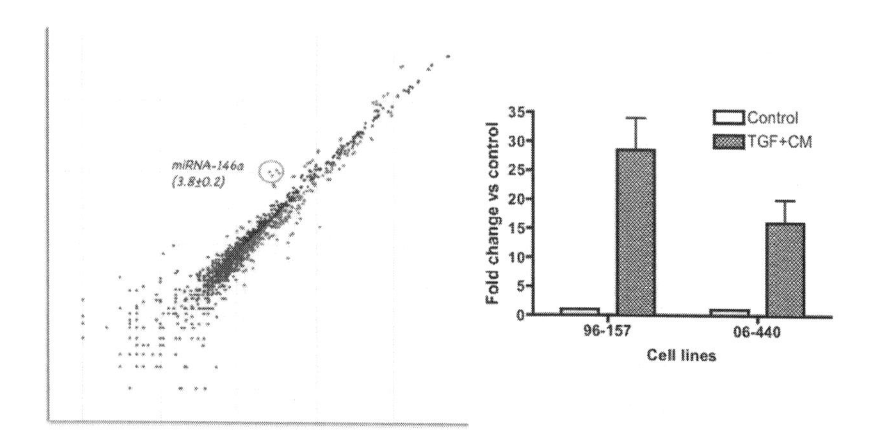

Fig. 3 MicroRNA-146a was up-regulated in HBECs in response to the cytokine stimulation. HBECs were treated with TGF-ß1 (2 ng/ml) plus cytomix (2.5 ng/ml IL-1ß + 5 ng/ml TNF-α + 5 ng/ml IFN-γ) for 24 h. MicroRNA expression was assessed by miRNA microarray and further confirmed by real time RT-PCR. (A) miRNA expression by microRNA microarray. Plotted data shown is one representative strain from 3 HBEC strains. Four dots in the circle indicate up-regulated miRNA-146a, an average of 3.8±0.2 fold increase in all 3 strains tested. Vertical axis: cells treated with TGF-ß1 plus cytomix. Horizontal axis: cells cultured in medium only. (B) Quantitative assay of miRNA-146a expression by real time RT-PCR. Data presented was from two strains of the HBECs (96-157 and 06-440). Vertical axis: fold change of miRNA-146a expression compared to control. Horizontal axis: cell strains. Control: medium only. TCM: TGF-ß1 plus cytomix.

Color image of this figure appears in the color plate section at the end of the book.

regulates the IL-1ß-induced release of IL-8 (CXCL8) and Regulated on Activation Normal T Cell Expressed and Secreted (RANTES, also called CCL5) (Perry *et al.* 2008). They also found that down-regulation of TRAF6 and/or IRAK1 by the over-expressed miR-146a appears to have nothing to do with suppression of the chemokine release following IL-1ß stimulation, suggesting miR-146a is involved in indirect targeting of IL-8 and RANTES translation.

MIR-146A AND CANCER

Although the role of miR-146a in cancer development, prognosis and response to the therapy has been actively studied, it is yet unknown whether changes in miR-146aexpression are causally linked to the development of cancer. However, it is now well established that chronic inflammation and the activation of NF-κB is associated with development of multiple cancers (Kundu and Surh 2008), and that inflammatory cytokines are potent stimulators of miR-146a through NF-κB pathway (Perry *et al.* 2008, Liu *et al.* 2009). It might therefore be speculated that up-regulated miR-146a expression through activated NF-κB signaling is involved in the development of cancers. In support of this concept, it has been reported that expression of miR-146a is increased in various kinds of malignancies including papillary thyroid carcinoma (Jazdzewski *et al.* 2008), cervical cancer (Wang *et al.* 2008), pancreatic cancer (Shen *et al.* 2008), and breast cancer (Shen *et al.* 2008). In addition, latent membrane protein 1(LMP1) of Epstein-Barr virus (EBV), a known virus implicated in the development of a number of tumors such as Burkitt's lymphoma, Hodgkin's lymphoma and nasopharyngeal carcinoma, has been shown to induce miR-146a (Motsch *et al.* 2007, Sonkoly *et al.* 2007). *In vitro* studies indicated that transfection of miR-146a precursor resulted in cell proliferation (Wang *et al.* 2008). Consistently, we have also found that transfection of miR-146a mimic into human bronchial epithelial cells resulted in cell proliferation (Fig. 4A), which seems to occur through suppression of p16 expression (Fig. 4B). Furthermore, transient transfection of a miRNA-146a mimic into human bronchial epithelial cells can protect the cells from apoptosis in response to inflammatory cytokines plus TGF-ß1 (Fig. 5), which seems mediated by STAT3 signaling pathway and its downstream anti-apoptotic protein Bcl-XL (Fig. 6), suggesting miR-146a may link the chronic inflammation and cancer through regulation of IL-6/STAT3 signaling.

Elevated IL-6 and constitutive activation of signal transducer and activator of transcription 3 (STAT3) were observed in various kinds of tumors including non-small cell lung cancer, colorectal cancer, renal cell carcinoma, and breast and ovarian cancer. Besides persistent elevation of IL-6, decreases of suppressor of cytokine signaling-3 (SOCS-3) also contribute to the constitutive activation of STAT3. Although a direct binding site for miR-146a at the 3'UTR sequence of SOCS-3 was not found (TargetScan, Release 5.1, April 2009), both miR-146a and miR-203 were up-regulated in psoriasis, and over-expressed miR-203 suppresses SOCS-3 protein translation (Sonkoly *et al.* 2007). In addition, we have found that miR-146a mimic transfection into human bronchial epithelial cells prevents STAT3 de-phosphorylation induced by inflammatory cytokines (Fig. 6), suggesting that miR-146a may be involved in the development of cancer through increasing cell survival and stimulating cell proliferation as illustrated in Fig. 7.

Table 4 Key features of miRNAs and cytokines

- MicroRNAs (miRNAs) are small, non-protein-coding RNAs that mediate posttranscriptional gene repression by inhibiting protein translation or by destabilizing target gene transcripts.
- Unlike small interfering RNA (siRNA) that requires perfect sequence match with targeted mRNA, miRNAs bind to the 3'UTR seed region of target gene(s) through imperfect base-pairing.
- While most miRNAs are widely expressed in various types of organs or tissues, a few are expressed only in limited developmental stages or in specific tissues or cells.
- miR-146a is one of the inflammation-responsive miRNAs and it is mainly involved in regulating innate immunity, chronic inflammation and cancer development.

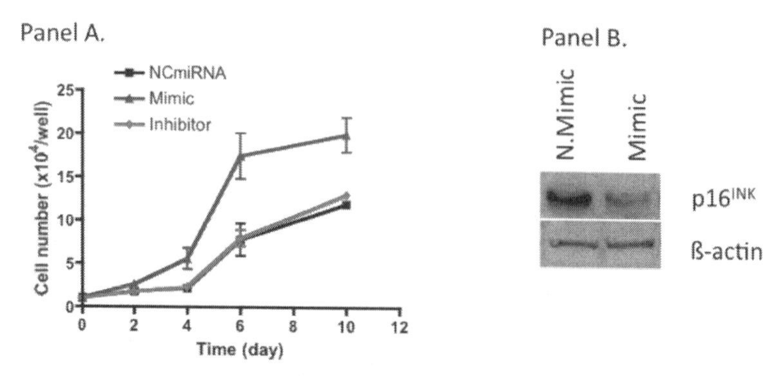

Fig. 4 miR-146a promotes HBEC proliferation and partially inhibits $p16^{INK4A}$. (A) Effect on cell growth. A negative control miRNA or a miRNA-146a mimic were transfected into HBECs using Lipofectamine 2000. Cells were then allowed to grow in complete medium for 10 d and medium was changed every 2 d. Cell number was counted by a Coulter Counter. Vertical axis: cell number per well. Horizontal axis: time (day). (B) Effect on $p16^{INK4A}$. HBECs were transfected with either a negative control miRNA (NC miRNA) or a mimic of miRNA-146a. After 4 d culture in complete medium, cell lysate was immunoblotted for $p16^{INK4A}$ and ß-actin (as loading control).

Color image of this figure appears in the color plate section at the end of the book.

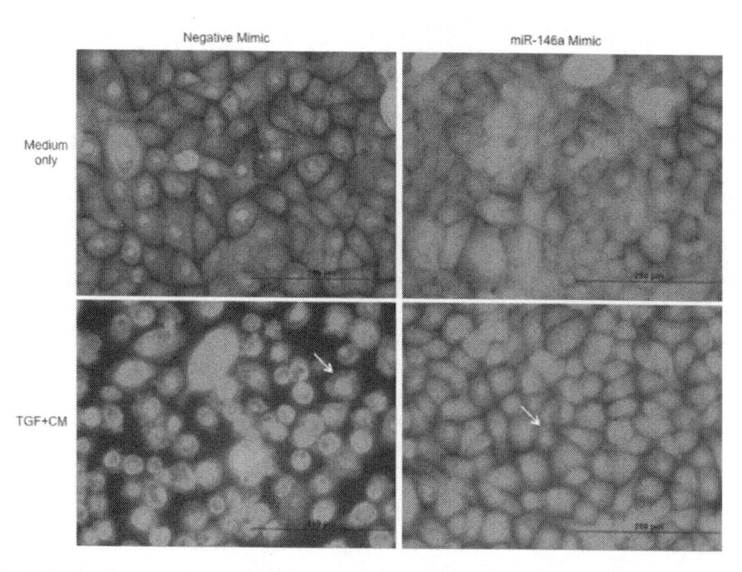

Fig. 5 Introduction of a miR-146a mimic protected HBECs from apoptosis. A negative control miRNA and a miR-146a mimic (100nM for both) were transfected into HBECs using Lipofectamine 2000. Cells were then treated with or without TGF-ß1 (2 ng/ml) plus cytomix (2 ng/ml IL-1ß, 4 ng/ml TNF-α and 4 ng/ml IFN-γ). After 48 h, cell morphology and viability was assessed by LIVE/DEAD staining. Arrows indicate examples of apoptotic cells with intact membrane but cell shrinkage with blebs, membrane asymmetry and condensed nuclei.

Color image of this figure appears in the color plate section at the end of the book.

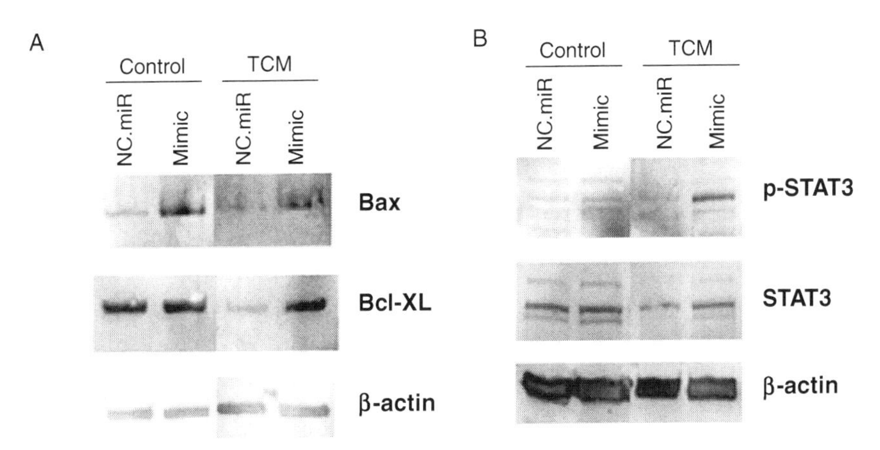

Fig. 6 Alteration of Bcl-XL expression and STAT3 phosphorylation in HBECs transfected with a miRNA-146a mimic. HBECs were transfected with negative control miRNA or a miRNA-146a mimic. Cells were then treated with TGF-ß1 plus cytomix for 24 h. Cell lysates were subjected to immunoblotting as described in methods. (A) Expression of pro-apoptotic protein (Bax) and anti-apoptotic protein (Bcl-XL). (B) STAT3 expression (total STAT3) and activation (tyrosine 705 phosphorylated STAT3). Control: cells were not treated with cytokines. TCM: cells were treated with TGF-ß1 plus cytomix.

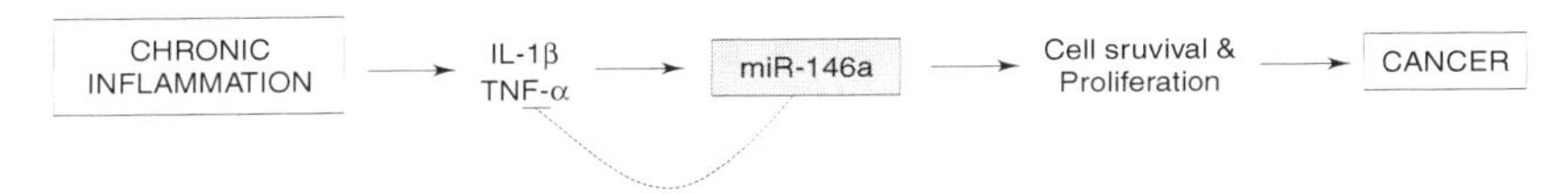

Fig. 7 miR-146a may provide a link between chronic inflammation and cancer. IL-1ß and TNF-α, released in the milieu of chronic inflammation, stimulate miR-146a synthesis. Up-regulated miR-146a can not only negatively block IL-1ß and TNF-α signaling by targeting IRAK1 and TRAF6, but also stimulate cell proliferation and extent cell survival. By this mechanism, miR-146a may provide a link between chronic inflammation and cancer formation.

Summary Points

- The first non-coding RNA, named small temporal RNA (stRNA) at that time, *lin-4*, was discovered by Ambros and colleagues from *C. elegans* in 1993.
- The term *microRNA* was used in 2001 to refer to the stRNAs and all other small RNAs with similar features but unknown functions.
- To date, miRNA Registry records 10,883 entries and 721 miRNAs from human (www.mirbase.org, Release 14, September 2009).
- Regulation of cytokine expression by a miRNA or miRNAs is mediated through either direct binding of miRNA to its target 3'UTR seed region, or indirect modulating ARE-BPs.

- Cytokines, especially pro-inflammatory cytokines (IL-1ß and TNF-α), can stimulate several miRNAs including miR-146a and miR-155.
- Interaction of miRNAs and cytokines is a complicated regulatory network, which might be the link between chronic inflammation and various diseases including cancer and chronic obstructive pulmonary diseases (COPD).
- Targeting miRNAs may be a radical new therapeutic strategy for the treatment of chronic inflammatory diseases such as COPD as well as malignant diseases in the next decades.

Abbreviations

ARE	:	adenine and uridine-rich element
ARE-BP	:	ARE binding proteins
COPD	:	chronic obstructive pulmonary disease
CCL	:	CC chemokine ligand
CXCL	:	chemokine (C-X-C motif) ligand
EMT	:	epithelial-mesenchymal transition
HBEC	:	human bronchial epithelial cells
LPS	:	lipopolysaccharide
miR	:	microRNA
MMP	:	matrix metalloproteinase
RANTES	:	regulated on activation normal T cell expressed and secreted
RISC	:	RNA-induced silencing complex
STAT3	:	signal transducer and activator of transcription 3
SNP	:	single nucleotide polymorphism
SOCS	:	suppressor of cytokine signaling
TF8	:	transcription factor
TGF-ß	:	transforming growth factor ß
TLR	:	toll-like receptor
UTR	:	un-translated region

Definition of Terms

Adenine and uridine-rich elements (AREs): a cluster of sequences found in the 3' UTR regions of cytokines. The basic unit of the ARE is a pentamer of AUUUA or nonamer of UUAUUAUU.

ARE-binding proteins (ARE-BPs): proteins that can positively or negatively regulate cytokine mRNA stability and/or translation through binding to ARE sequences.

miRNA: a small non-coding RNA, short in sequence (~22 nucleotide), imperfectly binds to complementary sequences of 3' UTR.

References

Anderson, P. 2008. Post-transcriptional control of cytokine production. Nat. Immunol. 9(4): 353-359.

Asirvatham, A.J. and W.J. Magner, *et al.* 2009. miRNA regulation of cytokine genes. Cytokine 45(2): 58-69.

Bakheet, T. and B.R. Williams, *et al.* 2006. ARED 3.0: the large and diverse AU-rich transcriptome. Nucleic Acids Res. 34(Database issue): D111-114.

Griffiths-Jones, S. 2004. The microRNA Registry. Nucleic Acids Res. 32(Database issue): D109-111.

Jazdzewski, K. and S. Liyanarachchi, *et al.* 2009. Polymorphic mature microRNAs from passenger strand of pre-miR-146a contribute to thyroid cancer. Proc. Natl. Acad. Sci. USA 106(5): 1502-1505.

Jazdzewski, K. and E.L. Murray, *et al.* 2008. Common SNP in pre-miR-146a decreases mature miR expression and predisposes to papillary thyroid carcinoma. Proc. Natl. Acad. Sci. USA 105(20): 7269-7274.

Kundu, J.K. and Y.J. Surh. 2008. Inflammation: gearing the journey to cancer. Mutat. Res. 659(1-2): 15-30.

Lagos-Quintana, M. and R. Rauhut, *et al.* 2001. Identification of novel genes coding for small expressed RNAs. Science 294(5543): 853-858.

Lau, N.C. and L.P. Lim, *et al.* 2001. An abundant class of tiny RNAs with probable regulatory roles in Caenorhabditis elegans. Science 294(5543): 858-862.

Lee, R.C. and R.L. Feinbaum, *et al.* 1993. The *C. elegans* heterochronic gene lin-4 encodes small RNAs with antisense complementarity to lin-14. Cell 75(5): 843-854.

Liu, X. 2008. Inflammatory cytokines augments TGF-beta1-induced epithelial-mesenchymal transition in A549 cells by up-regulating TbetaR-I. Cell. Motil. Cytoskeleton 65(12): 935-944.

Liu, X. and A. Nelson, *et al.* 2009. MicroRNA-146a modulates human bronchial epithelial cell survival in response to the cytokine-induced apoptosis. Biochem. Biophys. Res. Commun. 380(1): 177-182.

Motsch, N. and T. Pfuhl, *et al.* 2007. Epstein-Barr virus-encoded latent membrane protein 1 (LMP1) induces the expression of the cellular microRNA miR-146a. RNA Biol. 4(3): 131-137.

Nakasa, T. and S. Miyaki, *et al.* 2008. Expression of microRNA-146 in rheumatoid arthritis synovial tissue. Arthritis Rheum. 58(5): 1284-1292.

Park, S.M. and A.B. Gaur, *et al.* 2008. The miR-200 family determines the epithelial phenotype of cancer cells by targeting the E-cadherin repressors ZEB1 and ZEB2. Genes Dev. 22(7): 894-907.

Pasquinelli, A.E. and B.J. Reinhart, *et al.* 2000. Conservation of the sequence and temporal expression of let-7 heterochronic regulatory RNA. Nature 408(6808): 86-89.

Perry, M.M. and S.A. Moschos, *et al.* 2008. Rapid changes in microRNA-146a expression negatively regulate the IL-1beta-induced inflammatory response in human lung alveolar epithelial cells. J. Immunol. 180(8): 5689-5698.

Reinhart, B.J. and F.J. Slack, *et al.* 2000. The 21-nucleotide let-7 RNA regulates developmental timing in *Caenorhabditis elegans*. Nature 403(6772): 901-906.

Saini, H.K. and S. Griffiths-Jones, *et al.* 2007. Genomic analysis of human microRNA transcripts. Proc. Natl. Acad. Sci. USA 104(45): 17719-17724.

Shen, J. and C.B. Ambrosone, *et al.* 2008. A functional polymorphism in the miR-146a gene and age of familial breast/ovarian cancer diagnosis. Carcinogenesis 29(10): 1963-1966.

Slack, F.J. and M. Basson, *et al.* 2000. The lin-41 RBCC gene acts in the *C. elegans* heterochronic pathway between the let-7 regulatory RNA and the LIN-29 transcription factor. Mol. Cell 5(4): 659-669.

Sonkoly, E. and T. Wei, *et al.* 2007. MicroRNAs: novel regulators involved in the pathogenesis of psoriasis? PLoS One 2(7): e610.

Stanczyk, J. and D.M. Pedrioli, *et al.* 2008. Altered expression of MicroRNA in synovial fibroblasts and synovial tissue in rheumatoid arthritis. Arthritis Rheum. 58(4): 1001-1009.

Taganov, K.D. and M.P. Boldin, *et al.* 2006. NF-kappaB-dependent induction of microRNA miR-146, an inhibitor targeted to signaling proteins of innate immune responses. Proc. Natl. Acad. Sci. USA 103(33): 12481-12486.

Tan, Z. and G. Randall, *et al.* 2007. Allele-specific targeting of microRNAs to HLA-G and risk of asthma. Am. J. Hum. Genet. 81(4): 829-834.

Wang, X. and S. Tang, *et al.* 2008. Aberrant expression of oncogenic and tumor-suppressive microRNAs in cervical cancer is required for cancer cell growth. PLoS One 3(7): e2557.

Wang, Y. and T. Weng, *et al.* 2007. Identification of rat lung-specific microRNAs by micoRNA microarray: valuable discoveries for the facilitation of lung research. BMC Genomics 8: 29.

Williams, A.E. and H. Larner-Svensson, *et al.* 2009. MicroRNA expression profiling in mild asthmatic human airways and effect of corticosteroid therapy. PLoS One 4(6): e5889.

Zhao, Y. and E. Samal, *et al.* 2005. Serum response factor regulates a muscle-specific microRNA that targets Hand2 during cardiogenesis. Nature 436(7048): 214-220.

CHAPTER 2

Cytokines and Tight Junctions

Christopher T. Capaldo and Asma Nusrat*
Emory University, Department of Pathology and Laboratory Medicine
Whitehead Bld, Rm 105b, Atlanta, Georgia 30322, USA

ABSTRACT

Epithelial and endothelial tight junctions (TJs) act as a physical barrier against the diffusion of material between adjacent cells. Cell-cell barrier failure is associated with the loss of cellular differentiation and increased antigen exposure. These events play an important role in the pathogenesis of a number of diseases, including inflammatory bowel diseases, asthma, cystic fibrosis, and cancer. At the molecular level, the TJ barrier can be attributed to a complex array of proteins, including claudins, a diverse family of transmembrane proteins that form ion and size-selective aqueous pores. The stability of claudin-based junctions is maintained by an array of protein scaffolds, with linkages to the intercellular actin network. Recent studies have determined the detrimental effects of proinflammatory cytokines on TJ-dependent barrier properties. Proinflammatory cytokines, such as interferon gamma and tumor necrosis factor alpha, and interleukins, remodel TJs through the regulation of both claudin and TJ scaffold proteins. Additionally, proinflammatory cytokines regulate actin cytoskeletal dynamics, through the myosin light chain intercellular signaling cascade, resulting in structural and functional changes in TJs. Conversely, anti-inflammatory cytokines such as IL-10 and transforming growth factor beta attenuate pro-inflammatory stimuli. While it remains unclear whether TJ dysfunction is causative or symptomatic of detrimental inflammatory stimuli, investigations into the molecular mechanisms involved have increased our understanding of TJ regulation. This chapter will review recent advances in our understanding of the mechanisms of cytokine modulation of TJ structure and barrier function.

*Corresponding author

INTRODUCTION

Cytokine-mediated changes in TJ structure contribute to diverse physiological processes involving the modulation of epithelial/endothelial barrier properties including spermatocyte transmigration across the blood testis barrier and mammary epithelial cell differentiation (Khaled *et al.* 2007, Lui and Cheng 2007). Additionally, cytokines mediate changes in TJ function that contribute to the pathogenesis of a number of diseases, such as inflammatory bowel diseases (IBD), asthma, cystic fibrosis, psoriasis, and diseases involving the perturbation of the blood brain and blood retinal barrier. As a case in point, the inflammatory bowel diseases ulcerative colitis and Crohn's disease are conditions of chronic inflammation of the intestinal mucosa. IBD patients suffer from chronic colitis with relapsing acute inflammation, ulceration of the intestinal lining, elevated levels of proinflammatory cytokines, and gross changes in the structure and composition of TJs (Mankertz and Schulzke 2007, Nakamura *et al.* 1992).

Epithelial and endothelial tissues perform vital functions by forming regulated barriers between the body and its environment, between tissues within the body, and within tissue compartments. Examples include the protective function of the skin, the vascular network, and the tubule system within the kidney. For didactic purposes we will discuss the intestinal epithelium, where the individual cells that line the intestine form a single-cell monolayer and have a distinctive polarized architecture. Intestinal epithelial cells have plasma membrane surfaces that are specialized for either absorptive/secretory or adhesive properties. The apical domain, which faces the gut lumen, contains proteinaceous channels and transporters necessary for the absorptive properties of the gut, as well as apical brush border structures that increase the absorptive area of each cell. The basal surface, which contacts the extracellular matrix, is specialized for adhesion. The lateral domains consist of plasma membrane in contact with adjacent cells. At the cellular level, epithelial cell polarity and tissue integrity requires a series of interconnected, lateral membrane cell-cell contacts referred to as junctions. These include interdigitated junctions called tight junctions, adherens junctions, and desmosomes, which form contacts between cells and cytoskeletal components. This chapter will focus on the tight junction (TJ) because of its two primary functions: (1) maintaining plasma membrane polarity by preventing the intermixing of the lipids and proteins of the apical surface with those of the lateral domains and (2) forming a seal between cells, thereby creating a barrier to the penetration of luminal contents (Fig. 1B).

TIGHT JUNCTION STRUCTURE

Under electron microscope, TJs appear as areas of apparent membrane fusion at the luminal side of the lateral membrane and consist of tetraspan proteins, occludin

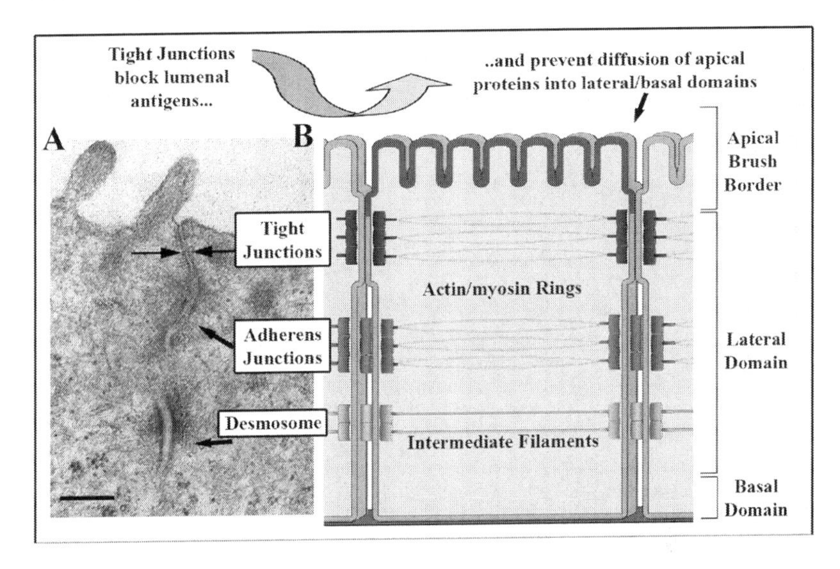

Fig. 1 Structure and function of tight junctions. **A**. Using electron microscopy, cell contacts are seen as electron-dense structures. TJs are the apical-most structure, nearest the lumen. **B**. Epithelial junction structure. Both TJ and adherens junction structures consist of membrane spanning complexes that are linked to the actin/myosin cytoskeletal ring. Desmosomal adhesions are linked to intermediate filaments and provide tensile strength. Intact TJs provide two vital functions: (1) they prevent penetration of lumenal contents into the basal environment and (2) they maintain cell polarity by preventing apical surface proteins from diffusing into the lateral membranes (the apical surface is shown in dark grey). Scale bar = 0.2 μm. Figure 1A is reproduced with permission (James L. Madara).

and claudins, and the single pass transmembrane proteins of the immunoglobulin family, junctional adhesion molecules (JAMs) and the Coxackie adenovirus receptor (CAR) (Figs. 1A and 2A). Transmembrane TJ proteins interact inside the cell with a variety of cytoplasmic scaffold proteins. The first TJ scaffold protein to be discovered was Zonula Occludens 1 (ZO-1), which interacts with claudins, occludin, other plaque proteins, and the actin cytoskeleton (for a detailed review see Gonzalez-Mariscal *et al.* 2003). Scaffold proteins like ZO-1 also serve as a platform for the sequestration of signaling proteins such as kinases, GTPase exchange factors, and transcription factors. Importantly, scaffold and signaling proteins of the TJ associate with the actin cytoskeletal network. Because of the promiscuity of scaffold protein interactions, the meshwork of interacting proteins is collectively referred to as the TJ scaffold plaque. A generalized interaction map is shown in Fig. 2A. Transmembrane TJ components bind to specific scaffold components, which in a reciprocal fashion confer stability on transmembrane proteins. Scaffolding proteins self-associate, bind to actin, and sequester regulatory proteins. Indeed, each class of TJ protein has the potential to regulate other TJ constituents, either directly or indirectly through scaffold protein mediators.

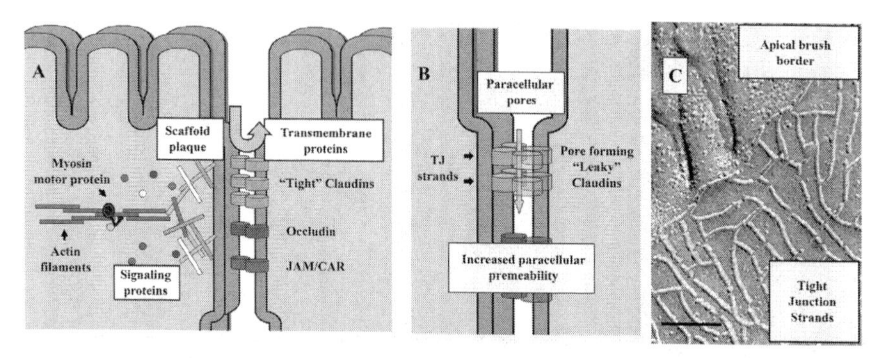

Fig. 2 Paracellular permeability is regulated through claudin-based tight junctions. **A.** A schematic of epithelial TJs. TJs consist of three classes of proteins: (1) transmembrane proteins that form contacts which seal adjacent cells together, (2) scaffolding proteins that provide structure and link the junction to the actin network, and (3) signaling and regulatory molecules that function to modulate actin dynamics, stabilize scaffold and transmembrane proteins, and communicate the status of the junctions to the nucleus. Claudin transmembrane proteins form the TJ seal. Also shown are non-sealing TJ components occludin, junctional adhesion molecules (JAMs), and Coxackie adenovirus receptor (CAR); these proteins regulate claudins. **B.** The "tightness" of the TJs can be modified by exchanging or inserting different combinations of claudin family transmembrane proteins. Note that the figure depicts the insertion of pore-forming claudins, as well as fewer claudin strands within the junction. **C.** A freeze fracture micrograph showing TJs as multiple strands within the membrane. Scale bar = 50 nm. Figure 2C is reproduced with permission (James L. Madara).

TJ CLAUDIN PROTEINS DETERMINE PARACELLULAR PERMEABILITY PROPERTIES

Claudin transmembrane proteins regulate TJ barrier properties. The claudin family, with 24 genes in mammals, hetero- and homo-oligomerize to form ion and size-selective aqueous pores. Although most claudin-pair interactions remain uncharacterized, a review of recent data grouped claudins into two types: pore-forming claudins (2, 7, 10, 15, and 16) and sealing claudins (1, 4, 5, 8, 11, 14 and 19) (Krause *et al.* 2007). For example, mice lacking the claudin 16 gene phenocopy the human disease hypomagnesemia because of a lack of Mg+ conductance in the kidney. Conversely, the sealing claudin, claudin-5, is required for endothelial tightness, as exhibited by the leaky blood brain barrier found in claudin-5-defective mice (reviewed in Furuse 2009). Subsequently, the combination of claudin gene expression within a specific tissue determines the character of the paracellular pores, contributing to the leakiness or tightness of the tissue to ions and solutes.

Alteration of the claudin protein composition is thought to regulate TJ function through three different mechanisms. As discussed above, differential claudin expression can enhance paracellular permeability, for example, by increasing the number of "leaky" pore-forming claudins or decreasing "tight" sealing claudins (see Fig. 2). This is due to the differential ion selectivity that is conferred by the amino acid composition of extracellular loops, which protrude from the membrane into the extracellular space (mechanisms of claudin-based barrier function are reviewed in Anderson and Van Itallie 2009). Co-culture of stable cells shows that not all claudin hetero-dimers are compatible. Therefore, a change in the protein expression level one claudin has the potential to alter the stability of other claudins within the junction. Additionally, the relative levels of pore-forming claudins reflect the number of pores. Indeed, claudin-2 expression is sufficient to confer increased Na+ conductance, and over-expression of claudin-2, but not claudin-4, alters small molecule flux by increasing pore number. The third mechanism involves the insertion of additional TJ strands within the tight junction (Claude and Goodenough, 1973). Specific claudin isoforms lead to differential strand number and architecture, altering paracellular permeability. In summary, TJ properties can be altered by three different methods, exchanging claudins, changing the relative amounts of claudins within the junction, and/or altering TJ strand number or complexity.

METHODS FOR THE STUDY OF CYTOKINE MODULATION OF TJS

Through the use of laboratory cell culture systems it has been shown that a broad array of cytokines perturb epithelial and endothelial barrier function. TJ function is measured in these cultures after they have been grown on semi-permeable membranes, as shown in Fig. 3A. TJ barrier function is then assessed by the measurement of transepithelial (or endothelial) electrical resistance (TER), and/or the ability of TJs to restrict the passage of small molecules such as inulin, mannitol, or dextran through the paracellular space. In this manner barrier integrity can be compared between cells cultured in the presence or the absence of exogenous cytokine in the growth media. *In vivo*, permeability across the epithelium of a mucosa can be similarly determined by measurement of TER and paracellular flux of solutes. Another common method is to visualize TJ integrity by immunofluorescence labeling and confocal microscopy. TJs can then be viewed "top down", where claudins can be seen forming a continuous belt around the cell (Fig. 3B). Cells cultured with proinflammatory cytokines are then analyzed for TJ integrity. As shown in Fig. 3B, cytokine-treated cells exhibit increased TJ disorder (white arrow) and rings of claudin inside the cell (*, macropinocytosis) (Bruewer *et al.* 2005).

Table 1 Key features of tight junctions

- Tight junctions are vital in order to (A) seal adjacent cells by restricting the paracellular pathway and (B) maintain cellular polarity.
- Tight junctions are composed of four classes of proteins: transmembrane proteins, scaffold proteins, regulatory molecules, and actin/myosin cytoskeletal components.
- Tight junction "tightness" is determined by the Claudin family of transmembrane proteins.
- Claudin protein expression can alter tissue permeability due to the ability of specific Claudin pairs to form aqueous paracellular pores between cells.
- Tight junction barrier properties can be modulated by exchanging Claudins, altering the amount of Claudins, or changing TJ strand architecture.
- Tight junction proteins are mutually dependent, such that any class of TJ protein has the potential to regulate the others.

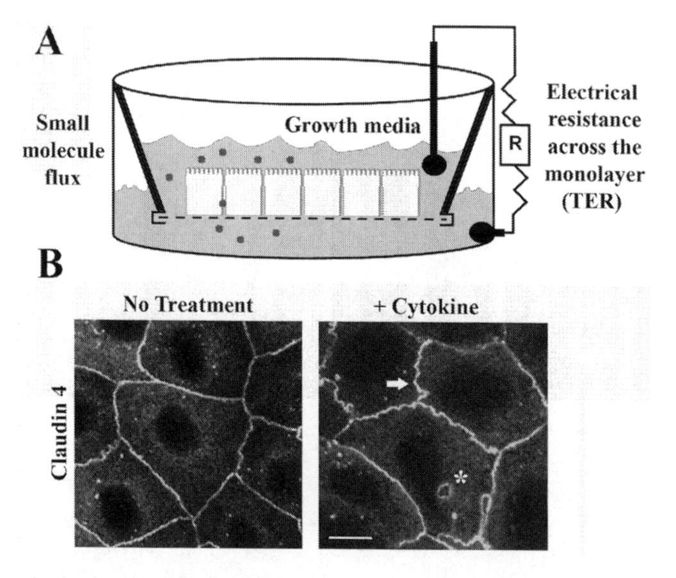

Fig. 3 Methodologies for the measurement of cytokine-induced changes in paracellular permeability. A. Transwell culture system for the measurement of TJ function. Cells are grown on a semi-permeable support matrix, with growth media above and below the cells. Cellular barrier function is then assessed by measuring the electrical resistance of the monolayer (TER), or by adding small molecules to the top chamber and assessing penetration into the lower chamber over time. B. Claudin protein as imaged by immunofluorescence microscopy in cultured human colonic epithelial cells (CaCo-2). Cells were either untreated (left panel) or treated with IFN-γ and TNFα for 48 h. Note that cytokine-treated cells exhibit disordered TJs (white arrow) and internalized TJ proteins (*). Scale = 10 μm.

IFN-γ and TNF-α

Interferon gamma (IFN-γ) and tumor necrosis factor alpha (TNF-α) are pro-inflammatory cytokines that, in addition to their immunomodulatory role during inflammation, act to modify epithelial and endothelial barrier function (reviewed in Capaldo and Nusrat 2009). In model cell culture systems of inflammation, direct treatment with either cytokine increases the paracellular permeability of endothelial and epithelial monolayers. First observed in endothelial cell cultures by Stolpen *et al.*, treatment with recombinant IFN-γ causes actin rearrangement into stress fibers and is suggestive of alteration in the contraction of cortical actin fibers (Stolpen *et al.* 1986). Similarly, treatment of epithelial and endothelial cells with TNF-α induces actin restructuring, with a more pronounced effect observed with INF-γ co-stimulation (Goldblum *et al.* 1993, Wang *et al.* 2006). Recent studies have explored in greater depth the molecular mechanism behind cytokine-mediated actin contractility and barrier function defects. IFN-γ exposure activates the small GTPase RhoA and increases the expression of rho-associated kinase (ROCK), which in turn phosphorylates and activates myosin light chain (MLC) (Utech *et al.* 2005). RhoA is a powerful regulator of actin dynamics whose activity is associated with increased actin fiber formation (reviewed in Hall 1998)). ROCK then directly enhances the activity of the myosin motor proteins at the TJ, thereby enhancing the tension on the actin fibers. These data help to explain two consistently observed cellular responses to IFN-γ treatment, actin restructuring and increased peripheral acto-myosin contractility. Increased tension on the actin cables, via increased myosin motor activity, increases internalization of TJ proteins through macropinocytosis: the budding-off of plasma membrane into the cell's interior (Bruewer *et al.* 2005, Utech *et al.* 2005). Indeed, both IFN-γ and TNF-α treatments result in the displacement of TJ transmembrane proteins claudin, occludin, and JAM as well as the scaffold protein ZO-1 (Capaldo and Nusrat 2009). TNF-α has been reported to influence TJ proteins by an additional mechanism. TNF-α treatment increases protein levels of myosin light chain kinase (MLCK) through the activation of the transcription factor nuclear factor kappa B (NF-κB) (reviewed in Shen *et al.* 2009). Similar to ROCK activation seen with IFN-γ treatment, increases in MLCK protein levels are then consistent with MLC hyperphosphorylation and enhanced acto-myosin contractility.

Together, the above findings are suggestive of a common mechanism of IFN-γ and TNF-α activity through the regulation of acto-myosin contractility and increased internalization of TJ structures. Interestingly, mucosal biopsies from patients with actively inflamed ulcerative colitis show internalized sub-apical vesicles that contain TJ transmembrane proteins. These are similar to those found in cytokine-treated epithelial cells in culture, indicating that vesicle internalization of TJ proteins is an *in vivo* mechanism involved in permeability changes (Bruewer *et al.* 2005). In summary, intercellular signaling after IFN-γ and TNF-α exposure appears to be propagated through parallel pathways; IFN-γ signals through ROCK, and TNF-α signals through NF-κB. Either pathway ultimately converges to enhance actin-myosin contractility and TJ internalization.

INTERLEUKINS

Interleukins are a large family of cytokines that are under increasing scrutiny for their effects on epithelial and endothelial paracellular permeability. IL-1, 2, 4, 6, 8, 10 and 13 have been found to have variable effects on permeability, depending on the tissue involved and additional cytokines present (Al-Sadi *et al.* 2009, Capaldo and Nusrat 2009). Interleukin-1 is a pro-inflammatory cytokine that is elevated in the intestinal mucosa of patients with IBD and in the bronchoalveolar lavage fluid from asthma patients (Broide *et al.* 1992, Ligumsky *et al.* 1990). In both epithelial and endothelial *in vitro* cell culture systems, IL-1 addition increases paracellular permeability and coincides with lower occludin mRNA levels (Al-Sadi and Ma 2007). This is consistent with previous findings in astrocytes, where IL-1β treatment suppressed occludin protein levels (Duffy *et al.* 2000). The mechanisms through which occludin regulates permeability are unclear, but exogenous occludin over-expression in canine kidney epithelial cells increased TJ strand complexity/number and decreased cell permeability (McCarthy *et al.* 1996). As with IL-1, IL-4 treatment causes increased epithelial permeability and corresponds with a decrease in the proteins levels of ZO-1 and occludin, and increased claudin-2 levels (Prasad *et al.* 2005, Wisner *et al.* 2008). As mentioned above, claudin-2 over-expression is sufficient to impair TJ barrier properties. In IL-6 knockout mice, increased intestinal permeability correlates with ZO-1 instability (Yang *et al.* 2003). Consistently with this finding, IL-6 treatment of endothelial cells in culture increases permeability and induces ZO-1 mislocalization, filamentous actin remodeling, and increased actin contractility (Desai *et al.* 2002). Both IL-2 and IL-10 oppose the influence of pro-inflammatory cytokines such as IFN- γ on barrier properties of the epithelium or endothelium (Al-Sadi *et al.* 2009). IL-2 and IL-10 knockout mice exhibit spontaneous colitis, and IL-10 knockout mice have increased levels of mucosal proinflammatory cytokines TNF-α, IL-1, and IL-6 (Mazzon *et al.* 2002, Sadlack *et al.* 1993). IL-10 knockout correlates with ZO-1 and claudin-1 mislocalization away from TJs in intestinal epithelial cells and may reflect the action of increased mucosal pro-inflammatory cytokines (Mazzon *et al.* 2002). Interestingly, IL-10 knockout mice do not develop colitis when reared in a bacteria-free environment, lending credence to the hypothesis that disease results from an unregulated immune response to enteric bacteria. Conversely, exposure of airway epithelial cells to IL-13 induces increased epithelial paracellular permeability that is associated with decreased TER, enhanced mannitol flux, and lowered ZO-1 protein levels (Al-Sadi *et al.* 2009). IL-13-mediated barrier dysfunction in colonic epithelial cells also correlates with increased claudin-2 protein levels (Prasad *et al.* 2005).

The above data indicate that interleukins act as modulators of TJ protein components, controlling their subcellular localization and total protein levels, resulting in leaky TJs. Alternatively, interleukins 2 and 10 act to antagonize the action of pro-inflammatory cytokines on TJ permeability, although the mechanisms involved are poorly understood (for review see Al-Sadi *et al.* 2009).

GROWTH FACTORS

Growth factors influence permeability depending on the cell environment (Capaldo and Nusrat 2009). Transforming growth factor beta (TGF-β) is a multifunctional cytokine that has been shown to enhance epithelial barrier properties *in vitro*. Indeed, pretreatment of intestinal epithelial cells with TGF-β blocks the TJ protein disruption associated with enterohemorrhagic *Escherichia coli* exposure (Howe *et al.* 2005). However, like many cytokines, TGF-β exhibits pleiotropic effects; for example, it increases permeability in uterine epithelial cells (Grant-Tschudy and Wira 2005). One mechanism for this function is the disruption of RhoA signaling to the actin network. Indeed, TGF-β induced intercellular signaling functions to ubiquitinate and degrade RhoA (Ozdamar *et al.* 2005). Additionally, in rat hepatocytes, TGF-β exposure decreases claudin-1 and increases claudin-2 protein expression (Kojima *et al.* 2007). This may reflect multiple functions for TGF-β depending on the biological context. Indeed, TGF-β treatment is a model for the study of cell-cell contact disruption observed during epithelial to mesenchymal transition, a process occurring during both early development and cancer, where epithelial cells lose adhesive properties and become motile, fibroblast-like cells. Variable effects of cell barrier function are also seen with hepatocyte growth factor/scatter factor, heparin binding epidermal growth factor, and platelet derived growth factor (reviewed in Capaldo and Nusrat 2009). Although varied in their effects on barrier properties, growth factors act by mechanisms similar to cytokines, including displacement or down-regulation of TJ protein components, and regulation of the actin network through RhoA.

SUMMARY

Modulation of TJ properties by cytokines occurs through two distinct processes, the wholesale restructuring of TJ and actin networks (Fig. 4A) or the remodeling of TJs by selectively removing or introducing TJ components, most importantly, claudins family proteins (Fig. 4B). However, cytokine regulation of cell barrier properties is not limited to the modulation of TJ structures. Most notably, pro-inflammatory cytokine exposure has been shown to cause cell death by apoptosis, thereby disrupting the epithelial barrier. During *in vivo* inflammatory events, the cytokine environment experienced by the cell is quite complex, with multiple cytokines contributing to the ultimate biological response to stimulation. Additionally, cell barrier modulations due to cytokine exposure exhibit cell-type-specific, pleiotropic, temporal, and dose-dependent effects. It is as yet unclear how cytokines dictate specific cellular responses, such as TJ remodeling, TJ restructuring,

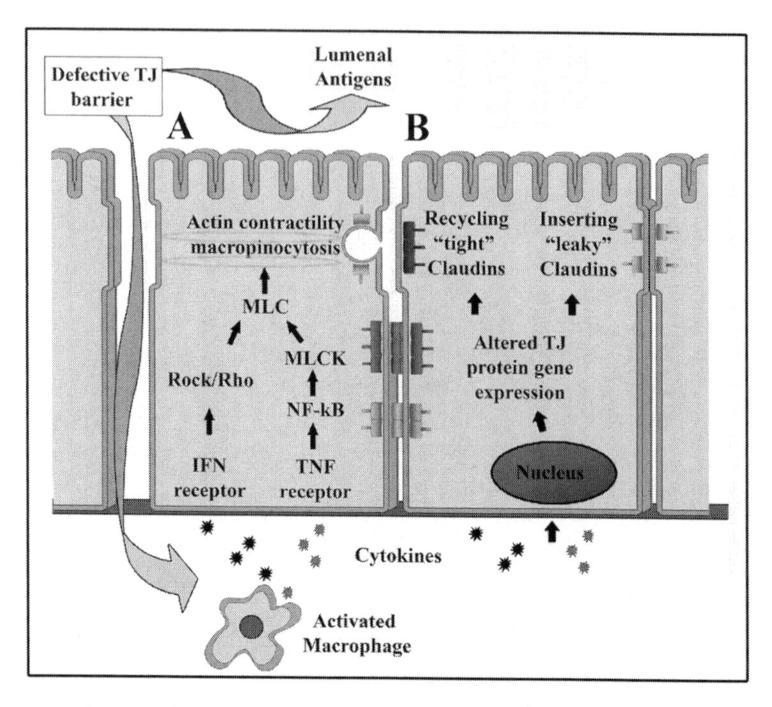

Fig. 4 Mechanisms of cytokine-mediated TJ regulation. A. Tight junction (TJ) barrier defects lead to enhanced cytokine production. Cytokine stimulation proceeds through cell surface receptors, thereby activating intracellular signaling, and promoting myosin motor activation (MLC) and actin contractility. This results in internalization of TJ proteins. B. Cytokine remodeling of TJs by stimulating the recycling of "tight" claudin proteins and inserting "leaky" claudin proteins.

or cell death, under such complex conditions. However, our understanding of the molecular mechanisms involved in the cell's response to cytokine stimulation is expanding rapidly through the use of reductionistic cell culture systems. This approach, in combination with *in vivo* models of human disease, will aid in the development of new therapeutic strategies aimed at diminishing the detrimental effects of enhanced paracellular permeability due to cytokine stimulation.

Summary Points

- Tight junction structures seal adjacent cells and regulate the passage of substances between cells.
- The claudin family of TJ proteins form contacts with adjacent cells and provide the intercellular seal. They are stabilized within the junction by protein scaffolds, signaling molecules and the actin cytoskeleton.

- Pro-inflammatory cytokines promote the remodeling and/or removal of TJ proteins, thereby increasing paracellular permeability.
- Cytokines act to remodel TJs by (1) changing TJ protein gene expression and (2) regulating the localization of TJ proteins, including transmembrane, scaffolding and signaling molecules.
- Pro-inflammatory cytokines remove TJ proteins from the lateral membranes through an intercellular signaling cascade that enhances actin-myosin contractility, resulting in internalization of TJs.
- Anti-inflammatory cytokines and growth factors act to enhance or antagonize pro-inflammatory signals in a context-dependent manner.

Abbreviations

CAR	:	coxackie adenovirus receptor
IFN-γ	:	interferon gamma
IL	:	interleukin
JAM	:	junctional adhesion molecules
MLC	:	myosin light chain
MLCK	:	myosin light chain kinase
NF-κB	:	nuclear factor kappa B
ROCK	:	rho-associated kinase
TGF-β	:	transforming growth factor beta
TJ	:	tight junctions
TNF-α	:	tumor necrosis factor alpha
ZO-1	:	zonula occludens 1

Definition of Terms

Actin-myosin ring: Cytoskeletal actin networks associated with TJs through interactions with the TJ scaffold form a ring around the circumference of the cell. This ring can be tightened, like the drawstring of a bag, by myosin motor protein activation.

Barrier function: The ability of a group of cells, connected by TJs, to restrict the flow of ions and solutes.

Cell polarity: The ability of cells to maintain functionally distinct plasma membrane surfaces. Loss of cell polarity is a defining characteristic of carcinogenesis.

Claudins: Claudins are a diverse family of transmembrane proteins that provide the sealing function of the TJ. Claudins span the extracellular space and interact with claudins from neighboring cells.

Macropinocytosis: The budding-off of plasma membrane into the cell's interior. Also referred to as a process of endocytosis.

Tight junctions: Multi-protein complexes that form contacts between adjacent cells. TJs function both to maintain cell polarity and to provide a regulatable barrier to paracellular flux.

References

Al-Sadi, R. and M. Boivin, and T. Ma. 2009. Mechanism of cytokine modulation of epithelial tight junction barrier. Front. Biosci. 14: 2765-2778.

Al-Sadi, R.M. and T.Y. Ma. 2007. IL-1beta causes an increase in intestinal epithelial tight junction permeability. J. Immunol. 178: 4641-4649.

Anderson, J.M. and C.M. Van Itallie. 2009. Physiology and function of the tight junction. Cold Spring Harb. Perspect. Biol. 1: a002584.

Broide, D.H. and M. Lotz, A.J. Cuomo, D.A. Coburn, E.C. Federman, and S.I. Wasserman. 1992. Cytokines in symptomatic asthma airways. J. Allergy Clin. Immunol. 89: 958-967.

Bruewer, M. and M. Utech, A.I. Ivanov, A.M. Hopkins, C.A. Parkos, and A. Nusrat. 2005. Interferon-gamma induces internalization of epithelial tight junction proteins via a macropinocytosis-like process. FASEB J. 19: 923-933.

Capaldo, C.T. and A. Nusrat. 2009. Cytokine regulation of tight junctions. Biochim. Biophys. Acta. 1788: 864-871.

Claude, P. and D.A. Goodenough. 1973. Fracture faces of zonulae occludentes from "tight" and "leaky" epithelia. J. Cell Biol. 58: 390-400.

Desai, T.R. and N.J. Leeper, K.L. Hynes, and B.L. Gewertz. 2002. Interleukin-6 causes endothelial barrier dysfunction via the protein kinase C pathway. J. Surg. Res. 104: 118-123.

Duffy, H.S. and G.R. John, S.C. Lee, C.F. Brosnan, and D.C. Spray. 2000. Reciprocal regulation of the junctional proteins claudin-1 and connexin43 by interleukin-1beta in primary human fetal astrocytes. J. Neurosci. 20: RC114.

Furuse, M. 2009. Knockout animals and natural mutations as experimental and diagnostic tool for studying tight junction functions *in vivo*. Biochim. Biophys. Acta. 1788: 813-819.

Goldblum, S.E. and X. Ding, and J. Campbell-Washington. 1993. TNF-alpha induces endothelial cell F-actin depolymerization, new actin synthesis, and barrier dysfunction. Am. J. Physiol. 264: C894-905.

Gonzalez-Mariscal, L. and A. Betanzos, P. Nava, and B.E. Jaramillo. 2003. Tight junction proteins. Prog. Biophys. Mol. Biol. 81: 1-44.

Grant-Tschudy, K.S. and C.R. Wira. 2005. Paracrine mediators of mouse uterine epithelial cell transepithelial resistance in culture. J. Reprod. Immunol. 67: 1-12.

Hall, A. 1998. Rho GTPases and the actin cytoskeleton. Science. 279: 509-514.

Howe, K.L. and C. Reardon, A. Wang, A. Nazli, and D.M. McKay. 2005. Transforming growth factor-beta regulation of epithelial tight junction proteins enhances barrier function and blocks enterohemorrhagic Escherichia coli O157:H7-induced increased permeability. Am. J. Pathol. 167: 1587-1597.

Khaled, W.T. and E.K. Read, S.E. Nicholson, F.O. Baxter, A.J. Brennan, P.J. Came, N. Sprigg, A.N. McKenzie, and C.J. Watson. 2007. The IL-4/IL-13/Stat6 signalling pathway promotes luminal mammary epithelial cell development. Development 134: 2739-2750.

Kojima, T. and K.I. Takano, T. Yamamoto, M. Murata, S. Son, M. Imamura, H. Yamaguchi, M. Osanai, H. Chiba, T. Himi, and N. Sawada. 2007. Transforming growth factor-beta induces epithelial to mesenchymal transition by down-regulation of claudin-1 expression and the fence function in adult rat hepatocytes. Liver Int. 28(4): 432-434.

Krause, G. and L. Winkler, S.L. Mueller, R.F. Haseloff, J. Piontek, and I.E. Blasig. 2007. Structure and function of claudins. Biochim. Biophys. Acta. 1778: 631-645.

Ligumsky, M. and P.L. Simon, F. Karmeli, and D. Rachmilewitz. 1990. Role of interleukin 1 in inflammatory bowel disease--enhanced production during active disease. Gut 31: 686-689.

Lui, W.Y. and C.Y. Cheng. 2007. Regulation of cell junction dynamics by cytokines in the testis: a molecular and biochemical perspective. Cytokine Growth Factor Rev. 18: 299-311.

Mankertz, J. and J.D. Schulzke. 2007. Altered permeability in inflammatory bowel disease: pathophysiology and clinical implications. Curr. Opin. Gastroenterol. 23: 379-383.

Mazzon, E. and D. Puzzolo, A.P. Caputi, and S. Cuzzocrea. 2002. Role of IL-10 in hepatocyte tight junction alteration in mouse model of experimental colitis. Mol. Med. 8: 353-366.

McCarthy, K.M. and I.B. Skare, M.C. Stankewich, M. Furuse, S. Tsukita, R.A. Rogers, R.D. Lynch, and E.E. Schneeberger. 1996. Occludin is a functional component of the tight junction. J. Cell Sci. 109 (Pt 9): 2287-2298.

Nakamura, M. and H. Saito, J. Kasanuki, Y. Tamura, and S. Yoshida. 1992. Cytokine production in patients with inflammatory bowel disease. Gut 33: 933-937.

Ozdamar, B. and R. Bose, M. Barrios-Rodiles, H.R. Wang, Y. Zhang, and J.L. Wrana. 2005. Regulation of the polarity protein Par6 by TGFbeta receptors controls epithelial cell plasticity. Science. 307: 1603-1609.

Prasad, S. and R. Mingrino, K. Kaukinen, K.L. Hayes, R.M. Powell, T.T. MacDonald, and J.E. Collins. 2005. Inflammatory processes have differential effects on claudins 2, 3 and 4 in colonic epithelial cells. Lab. Invest. 85: 1139-1162.

Sadlack, B. and H. Merz, H. Schorle, A. Schimpl, A.C. Feller, and I. Horak. 1993. Ulcerative colitis-like disease in mice with a disrupted interleukin-2 gene. Cell. 75: 253-261.

Shen, L. and L. Su, and J.R. Turner. 2009. Mechanisms and functional implications of intestinal barrier defects. Dig. Dis. 27: 443-449.

Stolpen, A.H. and E.C. Guinan, W. Fiers, and J.S. Pober. 1986. Recombinant tumor necrosis factor and immune interferon act singly and in combination to reorganize human vascular endothelial cell monolayers. Am. J. Pathol. 123: 16-24.

Utech, M. and A.I. Ivanov, S.N. Samarin, M. Bruewer, J.R. Turner, R.J. Mrsny, C.A. Parkos, and A. Nusrat. 2005. Mechanism of IFN-gamma-induced endocytosis of tight junction proteins: myosin II-dependent vacuolarization of the apical plasma membrane. Mol. Biol. Cell. 16: 5040-5052.

Wang, F. and B.T. Schwarz, W.V. Graham, Y. Wang, L. Su, D.R. Clayburgh, C. Abraham, and J.R. Turner. 2006. IFN-gamma-induced TNFR2 expression is required for TNF-dependent intestinal epithelial barrier dysfunction. Gastroenterology 131: 1153-1163.

Wisner, D.M. and L.R. Harris, 3rd, C.L. Green, and L.S. Poritz. 2008. Opposing regulation of the tight junction protein claudin-2 by interferon-gamma and interleukin-4. J. Surg. Res. 144: 1-7.

Yang, R. and X. Han, T. Uchiyama, S.K. Watkins, A. Yaguchi, R.L. Delude, and M.P. Fink. 2003. IL-6 is essential for development of gut barrier dysfunction after hemorrhagic shock and resuscitation in mice. Am. J. Physiol. Gastrointest. Liver Physiol. 285: G621-629.

CHAPTER 3

Cytokine Networks and MAPKs—Toward New Therapeutic Targets for Rheumatoid Arthritis

Zafar Rasheed, Nahid Akhtar, and Tariq M. Haqqi*
Department of Medicine, Division of Rheumatology,
MetroHealth Medical Center, 2500 MetroHealth Drive,
Cleveland, Ohio 44109, USA

ABSTRACT

In contrast to traditional disease-modifying antirheumatic drugs, biologics are novel agents that are developed specifically to target key pathophysiologic pathways in autoimmune diseases, such as rheumatoid arthritis (RA), with high specificity. Success achieved so far in the blockade of tumor necrosis factor (TNF)-α in RA exemplifies the feasibility and potential therapeutic application of antagonizing cytokine signaling. Identification of additional pro-inflammatory factors and an understanding of their effector function now offer major possibilities for the generation of additional novel biological therapeutics to address unmet clinical needs. Such interventions will ideally fulfill several of the following criteria: control of inflammation, modulation of underlying immune dysfunction by promoting the re-establishment of immune tolerance, protection of targeted tissues such as bone and cartilage, and preservation of host immune capability to avoid profound immune suppression and amelioration of co-morbidity associated with underlying RA. A number of preclinical development programs are ongoing to target a variety of cytokines that are central to immune regulation and tissue-matrix destruction in RA. Considerable evidence has been presented to indicate that pro-inflammatory cytokine network involving IL-6, IL-15 and IL-17 are suitable therapeutic targets for effective amelioration of inflammation and bone destruction. In addition,

*Corresponding author

IL-7, IL-18, IL-19, IL-20, IL-22 and IL-32 are fascinating novel pro-inflammatory cytokines, which should be targeted and offer new therapeutic options for RA. The quest for chemically amenable targets has recently led to the identification and characterization of the intracellular signaling pathways associated with these inflammatory cytokines. In particular, the mitogen-activated protein kinase (MAPK) pathways, which provide the essential link between receptor signals and nuclear transcription, offer several potential therapeutic opportunities for RA. This review provides an update on cytokine activities, MAPK signaling pathways activated, and novel strategies that represent, in our opinion, optimal utility as future therapeutic targets.

INTRODUCTION

Cytokines targeting is now established as a feasible method for treating rheumatoid arthritis (RA), which may occur because of the imbalance of pro- and anti-inflammatory cytokines and leads to the development of chronic synovial inflammation and joint destruction. Although the precise cause of RA remains unknown, the importance of inflammatory cytokines in the pathogenesis of RA has been documented in a number of studies (Tarner *et al.* 2007). Inflammatory cytokines, which include tumor necrosis factor-α (TNF-α) interleukin-1 (IL-1), IL-6, IL-17 and the downstream inflammatory mediators produced by activated cells in the arthritic joints, are an essential component of the milieu that drives cartilage and bone destruction. The effect of cytokine neutralization as a driving force for inhibition of inflammation was first exemplified by the use of anti-TNF-α as the first biologic therapy to be licensed for RA treatment (Taylor and Feldmann 2009). This approach has been subsequently extended to include IL-1 and IL-1 receptor (IL-1R) targeting. Now these agents are also used to treat a variety of autoimmune and chronic inflammatory disorders (Senolt *et al.* 2009). In RA, TNF blockade in the clinic is clearly associated with reduced synovial joint inflammation as well as cartilage and subchondral bone damage with improvement in quality of life measures. A variety of *in vitro* and *in vivo* models used in the elucidation of TNF-dependent networks in RA synovitis are now being employed to validate the therapeutic potential of other inflammatory cytokines that have also been implicated in the disease pathogenesis (Taylor and Feldmann 2009). The success of anti-TNF biologics on one hand, and their shortcomings on the other, triggered an enormous research effort to identify additional novel potential targets. Other cytokine targets such as IL-6, IL-17 and IL-15 have also been validated and are in the process of being tested. However, there remains a significant unmet clinical need for new therapies as existing regimens are effective in only a proportion of patients and crucially leave the host at significantly higher risk of overwhelming infection. Moreover, the high cost, short half-life and i.v. application of these biologics have made it necessary to consider alternative treatments such as orally active small molecule inhibitors of signaling cascades relating to these important inflammatory mediators. Much interest has focused on inhibitors of the MAPKs primarily because they have been implicated as key regulators of pro-inflammatory cytokines such as TNF-α, IL-1, IL-6, IL-17 production as well as playing crucial roles in signaling via receptors for TNF-α, IL-1, IL-17, B cell antigen receptor (BCR) and toll-like receptor (TLR) (Thalhamer *et al.* 2008). Indeed, p38 MAPK was initially discovered as the target of pyridinyl imidazoles, a class of compounds that inhibits the production of TNF-α and IL-1 by activated human monocytes, and p38-MAPK inhibitors are currently undergoing phase II trials for RA (www.sciosinc.com/scios/pr104637 6591; www.vpharm.com/Pressreleases2006/pr030806.html). However, some of the current inhibitors under trial are likely to induce side effects due to cross inhibition of subclasses of p38-MAPKs (α,β,γ,δ) in different arthritic models (Thalhamer *et al.* 2008). These data highlight the complexity of MAPK signaling in arthritis and provide a basis for the design of novel strategies to treat human RA. Of note, inhibition of the p38α isoform seems to be particularly effective with regard to cartilage and bone destruction, since a selective p38α inhibitor reduced bone loss, numbers of osteoclasts and serum levels of cartilage breakdown metabolites in arthritic mice (Tarner *et al.* 2007). Of these, strategies that target specific subclasses of p38-, ERK- and JNK-MAPKs, in our opinion, offer several potential therapeutic opportunities for RA.

This article reviews the highlight of recent research activities in the clinical development of novel cytokines, including IL-7, IL-18, IL-19, IL-20, IL-23 and IL-32, in the pathogenesis of RA. It also focuses on the MAPKs signaling pathways, emphasizing their role in inflammation and damage to articular tissues and their modulation with therapeutic agents currently in use, and potential future strategies.

INTERLEUKIN-7

Recent data from several groups demonstrate high levels of IL-7 in the joints of patients with rheumatoid and juvenile idiopathic arthritis (Churchman and Ponchel 2008). In contrast, circulating levels of IL-7 in RA remain a point of debate. IL-7 has many roles in T cell, dendritic cell and bone biology in humans. Reduced levels of circulating IL-7 probably underlie a number of the dysfunctions associated with circulating T cells in RA and may provide a mechanism for some of the unexplained systemic manifestations of the disease (reviewed by Churchman and Pounchel 2008). In joint disease such as RA and osteoarthritis, IL-7, via immune activation, can induce joint destruction. Now, it has also been demonstrated that activated human articular chondrocytes from osteoarthritis patients produce IL-7 (Von Roon and Lafeber 2008). IL-7 stimulates production of proteases by IL-7 receptor expressing RASFs, synovial macrophages and chondrocytes, thereby enhancing cartilage matrix degradation (Fig. 1) (von Roon and Lafeber 2008).

Besides this, IL-7 induces cytokines produced by arthritogenic T cells (for example, interferon c (IFNc), IL-17), T cell differentiating factors (e.g., IL-12), chemokines capable of attracting inflammatory cells (e.g., macrophage-induced gene (MIG), macrophage inflammatory protein (MIP)-1a) as well as molecules involved in cell adhesion, migration, and costimulation (e.g., lymphocyte function associated antigen (LFA)-1, CD40, CD80). In addition, IL-7 induces bone loss by stimulating osteoclastogenesis that is dependent on receptor activator of nuclear factor kB ligand (RANKL) (reviewed by Churchman and Ponchel 2008). IL-7 induces TNF-α secretion by T cells and by macrophages after T cell dependent synovial monocyte/macrophage activation suggesting that IL-7 is an important cytokine in RA, as it is capable of inducing inflammation and immunopathology. To date no clinical trial has been conducted on anti-IL-7 agent in humans; however, a vaccination strategy has been developed to enhance patients' anti-tumor activity against melanoma (von Roon and Lafeber 2008). The foregoing discussion indicates that it is not clear to what extent IL-7 mediates the unique pro-inflammatory activity in RA pathogenesis and its utility as a monotherapy in RA, but data clearly suggest its role in RA and support further studies.

INTERLEUKIN-18

IL-18 is a pleiotropic cytokine involved in the regulation of innate and acquired immune response, playing a key role in autoimmune and inflammatory diseases (Gracie *et al.* 2003). The importance of IL-18 in the induction and perpetuation of chronic inflammation during experimental and clinical rheumatoid synovitis has been demonstrated, and it may also be involved in development of RA (Gracie *et al.* 1999). There seems to be a correlation between IL-18 levels in damaged joints, serum and disease severity. IL-18 is released by synovial cells of RA patients without extrinsic stimulation (Gracie *et al.* 1999). Activation of RASFs by IL-18 produces TNF-α, IL-1, chemokines and adhesion molecules, which thereby induce inflammation and cartilage destruction (Fig. 1). IL-18 also activates memory T cell stimulation in damaged joints to produce various pro-inflammatory cytokines, such as IFN-γ and TNF-α (Fig. 1) (Gracie *et al.* 2003). Besides this, IL-18 also acts on various other cell types such as keratinocytes, osteoclasts and chondrocytes (Gracie *et al.* 2003) but often in synergy rather than independently; therefore, some of its activities remain unclear. In synovial compartment, IL-18 functions with numerous cytokines including IL-12 and IL-15 and as such it serves to amplify ongoing inflammatory responses (Petrovii-Rackov and Pejnovic 2006). Numerous *in vivo* studies using both IL-18-gene-targeted mice and neutralizing agents such as anti-IL-18 antibody or IL-18 binding protein implicate IL-18 in components of host defense and in responses in autoimmune models of disease (Dinarello

---

Fig. 1 Inflammatory cytokines are the potential intercellular targets for biologic therapy of RA. Cytokines are important elements of the intercellular communication systems within the synovium that contribute to synovial inflammation and cartilage and bone destruction. IL-6 has

Fig. 1 Contd...

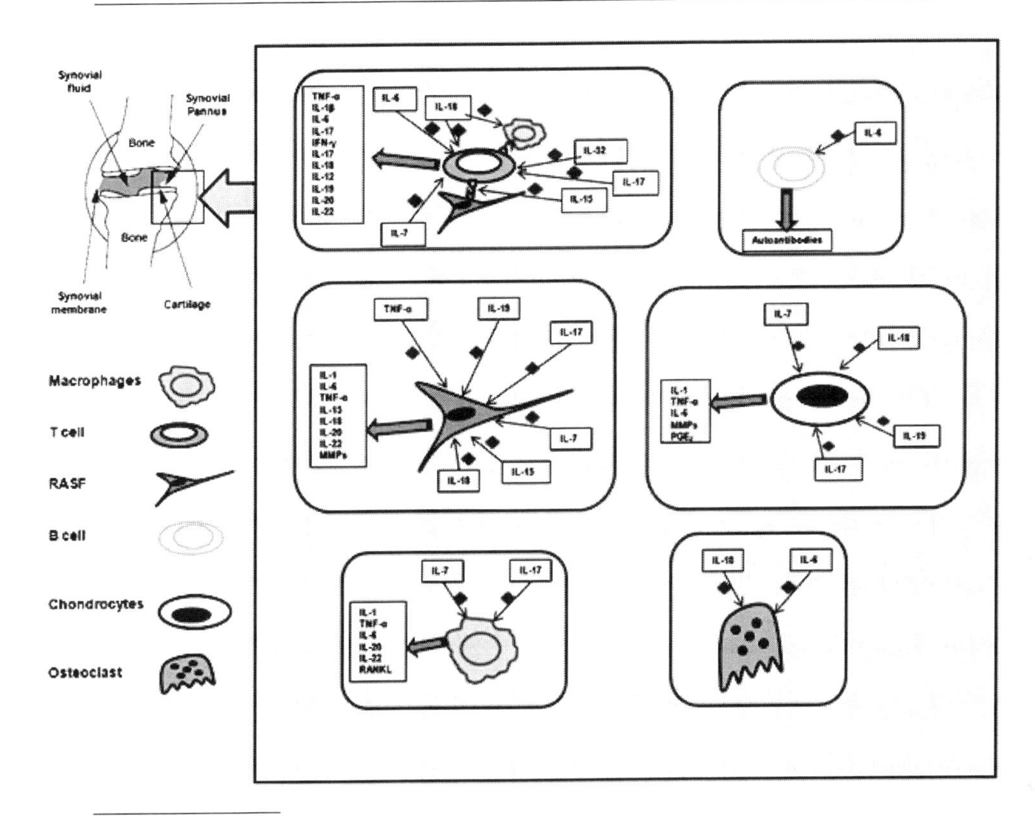

Fig. 1 Contd...

numerous pro-inflammatory functions, including production of TNF-α, IL-1, MMPs etc. by T cells or RASFs activation, induction of B cell differentiation into plasma cells, inhibition of matrix synthesis by articular chondrocytes, and induction of osteoclast differentiation. Use of anti-IL-6 agents has shown great promise for the treatment of RA. IL-7, a novel cytokine, has multiple effects that include mediation of interactions between T cells, chondrocytes, synovial macrophages and RASFs, which results in up-regulation of TNF-α and MMPs and enhances synovial inflammation and joint destruction. IL-15 also has multiple effects that include interactions with T cells and RASFs, which induce the production of TNF-α, IFNγ and IL-17 as shown here. IL-17 induces TNF-α, IL-1 and IL-6 production by RASFs and RA macrophages. Cartilage destruction is also promoted by IL-17-mediated inhibition of chondrocytes metabolism and matrix synthesis and by induction of matrix breakdown by the release of MMPs by chondrocytes and RASFs. IL-18 also promotes the production of TNF-α, IL-1 and IFN-γ by RASFs and T cells. IL-18 also promotes osteaoclast differentiation. Cartilage destruction is also promoted by IL-17- or IL-18-mediated inhibition of chondrocytes metabolism and matrix synthesis and by induction of matrix breakdown by the release of MMPs by chondrocytes and RASFs. IL-10 family members IL-19, IL-20 and IL-22 are produced by activated synovial cells as shown here. IL-19 promotes the induction of TNF-α, IL-6 and IFN-γ by RASFs and also promotes cartilage destruction by activated chondrocytes. Inhibition of these cytokines seems to be anti-inflammatory and chondroprotective. Thin arrows indicate activation, thick arrows indicate production of cellular products, and black squares denote therapeutic targets. IL, interleukin; IFN, interferon; RASF, RA synovial fibroblast; PGE_2, prostaglandin E_2; TNF, tumor necrosis factor. (Unpublished figure by authors.)

2004), increasing interest in it as a therapeutic target. Generation of anti-IL-18 monoclonal neutralizing antibodies represents an attractive approach, although at this time clinical studies have not yet commenced. Directly targeting the IL-18 receptor, e.g., via antibody or specific antagonist, is also of potential interest, although shared utilization with other IL-1 cytokine, e.g., IL-1F7, may reduce the specificity of such an approach. The small molecules approach includes inhibitors of caspase-1 and generation of inhibitors to components of the IL-18 receptor signaling pathway. The latter approaches will provide limited specificity for IL-18, since generation and release of other IL-1 superfamily members may also be inhibited. Whether this offers therapeutic disadvantage in RA, however, is unclear and need not be assumed. This also clearly indicates that IL-18 offers potential as a therapeutic target in RA and should be investigated further.

INTERLEUKIN-10 FAMILY MEMBERS

IL-19, IL-20 and IL-22 are members of the IL-10 family, which may play an inflammatory role in RA. IL-10, in contrast, is a potent anti-inflammatory cytokine (Mazza *et al.* 2009) that has been shown to regulate endogenous pro-inflammatory cytokine production in RA synovial tissue (Mitchell *et al.* 2008). IL-19 is a novel cytokine from the IL-10 family and was first originally cloned by searching EST database for IL-10 homologue. Liao *et al.* (2002) were the first to report pro-inflammatory nature of IL-19 in monocytes as it up-regulated the production of potent inflammatory cytokines TNF-α and IL-6. Recently, Sakurai *et al.* (2008) demonstrated that IL-19 produced by synovial cells, promotes joint inflammation in RA by inducing IL-6 secretion and decrease synovial cell apoptosis. It is also reported that activation of monocytes with IL-6, TNF-α, IFN-γ, LPS or GM-CSF resulted in induction of IL-19 (Hsing *et al.* 2006). In summary, however, the broad functional activity and expression of IL-19 in RA, together with the elegant work thus far performed to elucidate its activities, render it an interesting potential target. IL-20 and all three of its receptors are expressed in synovial membranes and RASFs derived from the synovial tissue of RA patients and in rats with CIA (Hsu *et al.* 2006). IL-20 was expressed at significantly higher levels in synovial fluid of RA patients and it was demonstrated that RASFs and synovial macrophages produce IL-20, which acts on RASFs using an autocrine pathway, inducing the production of MCP-1 and IL-8 causing more severe inflammation (Hsu *et al.* 2006). Furthermore, it was also demonstrated in CIA that administration of soluble IL-20 receptor type 1 significantly reduced disease, indicative of *in vivo* blockade of disease (Hsu *et al.* 2006). A separate study examined the other IL-10 family member; IL-22 was expressed in RA synovial tissues and in RA synovial fluid mononuclear cells (Ikeuchi *et al.* 2005). High levels of IL-22 were expressed in both the lining and sublining layers of RA synovial tissues, predominantly in RASFs and synovial macrophages. IL-22R1 was similarly expressed, but restricted to the RASFs. In short, IL-22 produced by RASFs and synovial macrophages promotes inflammatory responses in RA synovial tissue by inducing the proliferation and chemokine production (Ikeuchi *et al.* 2005). These studies therefore indicate a novel role of this pro-inflammatory member of the IL-10 family in RA and further indicate that some of the effects would be insensitive to IL-10 immunoregulation. Interestingly, IL-22 expression has been identified in murine T cells, with substantially higher expression levels in Th17 cells, as compared with Th1 and Th2 counterparts (Liang *et al.* 2006). It is also reported that this cytokine acts synergistically with IL-17 to induce genes associated with innate immunity in primary keratinocytes and that may have implications for defining their roles in RA (Liang *et al.* 2006). These IL-10 family members may have potential as therapeutic target in RA, but the clinical efficacy of targeting these cytokines in humans remains to be determined.

INTERLEUKIN-32

IL-32 is a recently described cytokine produced by T lymphocytes, natural killer cells, epithelial cells and blood monocytes (Kim *et al.* 2007, Netea *et al.* 2005). Of particular importance, IL-32 is prominently induced by IFN-γ in epithelial cells and monocytes (Kim *et al.* 2007). Human recombinant IL-32 exhibits several properties typical of a pro-inflammatory cytokine (Kim *et al.* 2007). For example, the cytokine induces other pro-inflammatory cytokines and chemokines such as TNF-α, IL-1β, IL-6, and IL-8 by means of the activation of NF-κB and p38-MAPK (Netea *et al.* 2005). An unexpected property of IL-32 is its ability to augment by 10-fold production of IL-1β and IL-6 induced by muramyl dipeptides by means of the nucleotide oligomerization domains 1 and 2 (NOD1 and NOD2) through a caspase-1-dependent mechanism (Netea *et al.* 2005). A single mutation in NOD2 plays a role in a subgroup of patients with Cohn's inflammatory bowel disease. Together, these studies suggest that IL-32 has an important role in inflammation, both during host defenses against microorganisms and in autoimmune diseases. Joosten *et al.* (2006) used immunohistochemistry to describe high expression levels of IL-32 in RA synovial tissue biopsies, but not osteoarthritis synovium. Injection of human IL-32 into the knee joints of mice resulted in joint swelling, infiltration and cartilage damage (Joosten *et al.* 2006). In TNF-α-deficient mice, IL-32-driven joint swelling was absent, and cell influx was markedly reduced, but loss of proteoglycan was unaffected, suggesting that at least some IL-32 activity is, in part, TNF-α dependent (Joosten *et al.* 2006). Another study using transgenic mice overexpressing human IL-32 (Shoda *et al.* 2006) demonstrated that splenocytes from these mice produced more TNF-α, IL-1 and IL-6 in response to LPS stimulation and contained more TNF-α in serum. These mice also showed exacerbation of CIA that could be negated by TNF-α blockade, adding further support to the proposed TNF-α dependence of IL-32 activity. Taken together these data suggest that IL-32 may represent an attractive therapeutic target in RA.

MITOGEN-ACTIVATED PROTEIN KINASES

MAPKs have been implicated as playing key regulatory roles in the production of inflammatory cytokines (TNF-α, IL-1, IL-6, IL-17 etc.) and downstream signaling events leading to inflammation and joint destruction by providing the essential link between receptor signals and nuclear gene transcription (Tarner *et al.* 2009). MAPKs comprise a family of highly conserved serine/threonine protein kinases that have been implicated in the regulation of key cellular processes including gene induction, cell survival/apoptosis, proliferation and differentiation as well as cellular stress and inflammatory responses (Thalhamer *et al.* 2008). There are three major classes of MAPKs in mammals, p38-MAPK, c-jun N-terminal kinase (JNK) and the extracellular signal-regulated kinases (ERKs). MAPKs are activated via a signaling cascade as stimulation of MAPKs requires the upstream activation of a MAPK kinase (MAPKK, MEK or MKK) and an MAPK kinase kinase (MAPKKK, MEKK or MKKK) (Fig. 2). Clearly, MAPKs play important roles in transducing inflammation and joint destruction and therefore receive considerable attention as potential therapeutic targets in RA. However, there are multiple isoforms of these kinases (p38α,β,γ and δ, JNK1–3, ERK-1–8) that have been implicated in the regulation of a plethora of essential cellular responses (Thalhamer *et al.* 2008), dictating that inhibitors that simply ablate p38-, JNK- or ERK-MAPKs activities are likely to have serious side effects. Moreover, most of the current inhibitor compounds under trial are competitors for the ATP-binding site of p38-MAPK and hence are likely to induce side effects due to cross-inhibition of other classes of kinases (Thalhamer *et al.* 2008). Interestingly, the ERK/MEK inhibitors are not ATP competitive but rather act to prevent the MAPKKK, Raf1 from activating MEK and thus their observed relative lack of side effects probably reflects that they do not non-specifically inhibit other kinases *in vivo* (Thalhamer *et al.* 2008). Similarly, peptide-based approaches to disrupting JNK signaling by targeting scaffold protein interactions show promise but as yet have not been tested in *in vivo* models of inflammation (Thalhamer *et al.* 2008). Of note, inhibition of the p38α-MAPK isoform seems to be particularly effective with regard to cartilage and bone destruction, since a selective p38α-MAPK inhibitor reduced bone loss, numbers of osteoclasts and serum levels of cartilage breakdown metabolites in arthritic model of mice (Tarner, *et al.* 2007). Therefore, in our opinion, future strategies should target the development of inhibitors that target inflammation-restricted MAPK signaling and, in particular, the aberrant inflammation related to chronic disease while leaving intact the 'healthy' inflammation that is essential for fighting infection. The recent explosion in identifying isoforms of the major components (MAPKs, MAPKKs, MAPKKKs) of MAPK cascades (Thalhamer *et al.* 2008) and the delineation of their specific roles in coupling individual receptors to particular p38-, JNK- and ERK- MAPK signals and their functional outcomes will ultimately

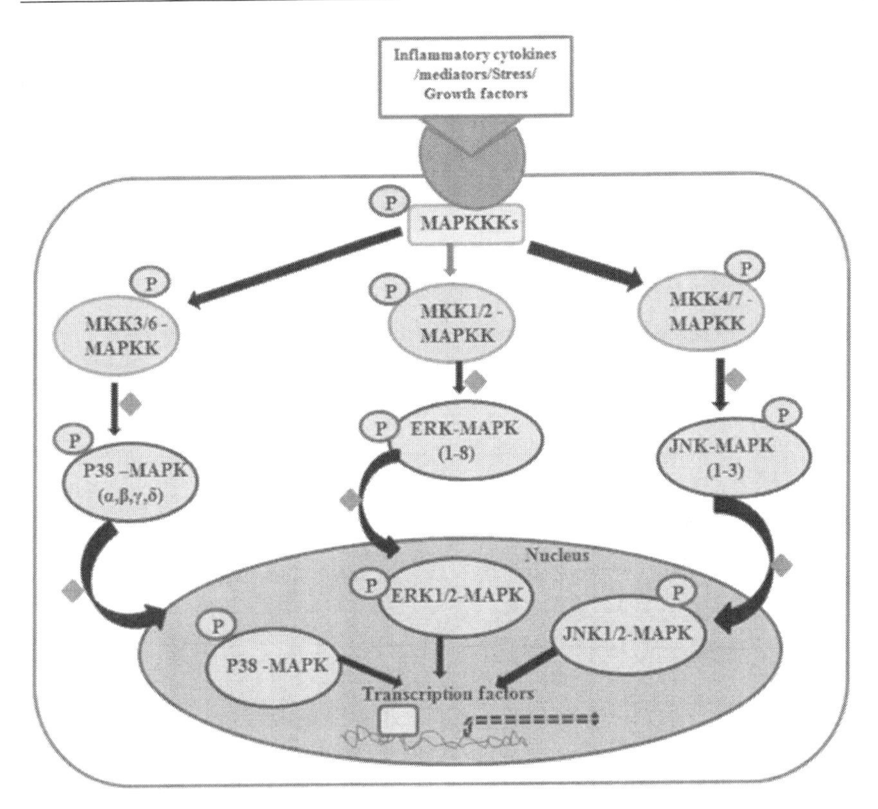

Fig. 2 Mitogen-activated protein kinases are potential intracellular targets for biologic therapy of RA. Intracellular signaling pathways that have been targeted experimentally for the treatment of RA include MAPK pathway in particular. The MAPKs are a group of signaling enzymes that include many different members (or submembers); of these, p38-MAPKs (α,β,γ,δ), ERK-MAPKs (1-8) and JNK-MAPKs (1-3) have received particular attention. The MAPKs relay pro-inflammatory signals to the nucleus by sequential phosphorylation. Specific small molecule inhibitors have been developed to intercept signal transduction by these kinases. However, owing to overlapping functions of the different MAPKs and the multitude of diverse physiologic processes in which they are involved, therapeutic compounds need to be highly selective in order to achieve the desired effect and minimize potential adverse effects. Black arrows indicate activation, broken arrow indicates initiation of transcription in the nucleus, and gray squares denote therapeutic targets. MAPKKKs, MAPK kinase kinase; MAPKK (MKK), MAPK kinase; p38-MAPKs, p38 mitogen-activated protein kinases; JNK, c-Jun-N-terminal kinase; ERK, extracellular signal-related kinase. (Unpublished figure by authors.)

lead to the unraveling of their role in disease pathogenesis and the development of specific inhibitors that will provide novel, safe small molecule therapeutics for rheumatoid arthritis.

Table 1 Key features of Mitogen Activated Protein Kinases

- MAPKs are a family of serine/threonine protein kinases widely conserved among eukaryotes and are involved in many cellular programs such as cell proliferation, cell differentiation, cell movement and cell death.
- MAPKs are intracellular signaling pathways which provide the link between receptor signals and nuclear gene transcription. MAPK signaling cascades are organized hierarchically into three-tiered modules. MAPKs are phosphorylated and activated by MAPK-kinases (MAPKKs), which in turn are phosphorylated and activated by MAPKK-kinases (MAPKKKs). The MAPKKKs are in turn activated by interaction with the family of small GTPases and/or other protein kinases, connecting the MAPK module to cell surface receptors or external stimuli.
- Aberrant or inappropriate functions of MAPKs have now been identified in diseases ranging from cancer to inflammatory disease to obesity and diabetes.
- MAPK subtypes, p38MAPK (also known as MAPK14), c-Jun-N-terminal kinase (also known as MAPK8) and extracellular signal-related kinase (also known as MAPK1) have received considerable attention as potential therapeutic targets.
- MAPKs play key roles in the molecular and cellular events underpinning the pathogenesis of RA.
- MAPKs are rational targets of drug design for novel therapies in RA.
- MAPKs: mitogen-activated protein kinases, GTP: guanosine triphosphate, RA: rheumatoid arthritis

Summary Points

Recent advances in our understanding of cellular and molecular mechanisms in RA and the blockade of TNF-α exemplify the feasibility and potential therapeutic application of antagonizing cytokine signaling. Despite these advances, there remains a considerable unmet clinical need in this field.

Several interesting, novel pro-inflammatory cytokines have emerged in the recent literature as contributing to the pathogenesis of RA. Much of this evidence has been gathered from animal studies, however, and it remains to be seen whether any of these cytokines become validated targets in clinical trials in human participants. It is currently unclear how to design appropriate predictive clinical trials for testing novel cytokines, or how to optimize the development of targeting agents that offer therapeutic synergy with TNF blockers or act in a manner sufficiently distinct to capture TNF blockade failures.

Intracellular signaling and transcription pathways represent attractive therapeutic targets, provided that potential therapeutic modulators are carefully selected and have high molecular specificity.

Preliminary clinical data on inhibition of specific isoforms of p38-, ERK- or JNK-MAPKs seem promising so far.

The therapeutic goals in RA are now remission induction and the management of immune dysregulation and autoimmunity. Optimizing cytokine or intracellular signaling modulation in order to achieve such targets remains a considerable challenge.

Abbreviations

CIA	:	collagen-induced arthritis
ERK	:	extracellular signal-regulated kinase
GM-CSF	:	granulocytes/macrophage colony stimulating factor
IFN	:	interferon
IL	:	interleukin
JNK	:	c-jun N-terminal kinase
MAPK	:	mitogen-activated protein kinase
MIG	:	macrophage-induced gene
MIP	:	macrophages inflammatory protein
MMPs	:	matrix metalloproteinases
p38-MAPKs	:	p38-mitogen-activated protein kinases
RA	:	rheumatoid arthritis
RASF	:	RA synovial fibroblast
TNF	:	tumor necrosis factor

Definition of Terms

Autoimmune diseases: diseases that arise from an overactive immune response of the body against substances and tissues normally present in the body.

Intracellular signaling pathways: pathways that provide the essential link between receptor signals and nuclear gene transcription.

Mitogen activated protein kinases: are serine/threonine-specific protein kinases that respond to extracellular stimuli (mitogens, osmotic stress, heat shock and pro-inflammatory cytokines) and regulate various cellular activities, such as gene expression, mitosis, differentiation, proliferation, and cell survival/apoptosis.

Pro-inflammatory cytokines: cytokines produced predominantly by activated immune cells, which promote systemic inflammation.

Rheumatoid arthritis: is a chronic, systemic autoimmune inflammatory disorder that may affect many tissues and organs, but principally attacks the joints producing an inflammatory synovitis that often progresses to destruction of the articular cartilage and ankylosis of the joints.

Tumor necrosis factor (formally known as TNF-α): cytokine involved in systemic inflammation and is a member of a group of inflammatory cytokines that stimulate the acute phase reaction.

Acknowledgements

This work was supported in part by NIH/NCCAM grants, RO1 AT-003267, RO1 AT-005520, and R21 AT504615 and funds from the MetroHealth Medical Center, Cleveland, Ohio.

References

Churchman, S.M. and F. Ponchel. 2008. Interleukin-7 in rheumatoid arthritis. Rheumatology (Oxford). 47: 753-759.

Dinarello, C.A. 2004. Interleukin-18 and the treatment of rheumatoid arthritis. Rheum. Dis. Clin. North Am. 30: 417-434.

Gracie, J.A. and R.J. Forsey, W.L. Chan, A. Gilmour, B.P. Leung, M.R. Greer, K. Kennedy, R. Carter, X.Q. Wei, D. Xu, M. Field, A. Foulis, F.Y. Liew, and I.B. McInnes. 1999. A proinflammatory role for IL-18 in rheumatoid arthritis. J. Clin. Invest. 104: 1393-1401.

Gracie, J.A. and S.E. Robertson, and I.B. McInnes. 2003. Interleukin-18. J. Leukoc. Biol. 73: 213-224.

Hsing, C.H. and M.Y. Hsieh, W.Y. Chen, E. Cheung So, B.C. Cheng, and M.S. Chang. 2006. Induction of interleukin-19 and interleukin-22 after cardiac surgery with cardiopulmonary bypass. Ann. Thorac. Surg. 81: 2196-2201.

Hsu, Y.H. and H.H. Li, M.Y. Hsieh, M.F. Liu, K.Y. Huang, L.S. Chin, P.C. Chen, H.H. Cheng, and M.S. Chang. 2006. Function of interleukin-20 as a proinflammatory molecule in rheumatoid and experimental arthritis. Arthritis Rheum. 54: 2722-2733.

Ikeuchi, H. and T. Kuroiwa, N. Hiramatsu, Y. Kaneko, K. Hiromura, K. Ueki, and Y. Nojima. 2005. Expression of interleukin-22 in rheumatoid arthritis: potential role as a proinflammatory cytokine. Arthritis Rheum. 52: 1037-1046.

Joosten, L.A. and M.G. Netea, S.H. Kim, D.Y. Yoon, B. Oppers-Walgreen, T.R. Radstake, P. Barrera, F.A. van de Loo, C.A. Dinarello, and W.B. van den Berg. 2006. IL-32, a proinflammatory cytokine in rheumatoid arthritis. Proc. Natl. Acad. Sci. USA 103: 3298-3303.

Kim, H.R. and M.L. Cho, K.W. Kim, J.Y. John, S.Y. Hwang, C.H. Yoon, S.H. Park, S.H. Lee, and H.Y. Kim. 2007. Up-regulation of IL-23p19 expression in rheumatoid arthritis synovial fibroblasts by IL-17 through PI3-kinase-, NF-kappaB- and p38 MAPK-dependent signaling pathways. Rheumatology (Oxford) 46: 57-64.

Liang, S.C. and X.Y. Tan, D.P. Luxenberg, R. Karim, K. Dunussi-Joannopoulos, M. Collins, and L.A. Fouser. 2006. Interleukin (IL)-22 and IL-17 are coexpressed by Th17 cells and cooperatively enhance expression of antimicrobial peptides. J. Exp. Med. 203: 2271-2279.

Liao, Y.C. and W.G. Liang, F.W. Chen, J.H. Hsu, J.J. Yang, and M.S. Chang. 2002. IL-19 induces production of IL-6 and TNF-alpha and results in cell apoptosis through TNF-alpha. J. Immunol. 169: 4288-4297.

Mazza, G. and C.A. Sabatos-Peyton, R.E. Protheroe, and D.C. Wraith. 2010. Isolation and characterization of human interleukin (IL)-10-secreting T cells from peripheral blood. Hum. Immunol. 71: 225-234.

Mitchell, A. and C. Rentero, Y. Endoh, K. Hsu, K. Gaus, C. Geczy, H.P. McNeil, L. Borges, and N. Tedla. 2008. LILRA5 is expressed by synovial tissue macrophages in rheumatoid arthritis, selectively induces pro-inflammatory cytokines and IL-10 and is regulated by TNF-alpha, IL-10 and IFN-gamma. Eur. J. Immunol. 38: 3459-3473.

Netea, M.G. and T. Azam, G. Ferwerda, S.E. Girardin, M. Walsh, J.S. Park, E. Abraham, J.M. Kim, D.Y. Yoon, C.A. Dinarello, and S.H. Kim. 2005. IL-32 synergizes with nucleotide oligomerization domain (NOD) 1 and NOD2 ligands for IL-1beta and IL-6 production through a caspase 1-dependent mechanism. Proc. Natl. Acad. Sci. USA 102: 16309-16314.

Petrovic-Rackov, L. and N. Pejnovic. 2006. Clinical significance of IL-18, IL-15, IL-12 and TNF-alpha measurement in rheumatoid arthritis. Clin. Rheumatol. 25: 448-452.

Sakurai, N. and T. Kuroiwa, H. Ikeuchi, N. Hiramatsu, A. Maeshima, Y. Kaneko, K. Hiromura, and Y. Nojima. 2008. Expression of IL-19 and its receptors in RA: potential role for synovial hyperplasia formation. Rheumatology (Oxford). 47: 815-820.

Senolt, L. and J. Vencovsky, K. Pavelka, C. Ospelt, and S. Gay. 2009. Prospective new biological therapies for rheumatoid arthritis. Autoimmun. Rev. 9: 102-107.

Shoda, H. and K. Fujio, Y. Yamaguchi, A. Okamoto, T. Sawada, Y. Kochi, and K. Yamamoto. 2006. Interactions between IL-32 and tumor necrosis factor alpha contribute to the exacerbation of immune-inflammatory diseases. Arthritis Res. Ther. 8: R166.

Tarner, I.H. and U. Muller-Ladner, and S. Gay. 2007. Emerging targets of biologic therapies for rheumatoid arthritis. Nat. Clin. Pract. Rheumatol. 3: 336-345.

Taylor, P.C. and M. Feldmann. 2009. Anti-TNF biologic agents: still the therapy of choice for rheumatoid arthritis. Nat. Rev. Rheumatol. 5: 578-582.

Thalhamer, T. and M.A. McGrath, and M.M. Harnett. 2008. MAPKs and their relevance to arthritis and inflammation. Rheumatology 47: 409-414.

van Roon, J.A. and F.P. Lafeber. 2008. Role of interleukin-7 in degenerative and inflammatory joint diseases. Arthritis Res. Ther. 10: 107.

Section II
Life Style Factors and Life Events

CHAPTER 4

Exercise, Cytokines and Tissue Repair

Chunming Dong, Liyong Wang, and Pascal J. Goldschmidt-Clermont*

The University of Miami Miller School of Medicine, Miami, Florida 33136, USA

ABSTRACT

Physical activity and exercise not only provide health benefits that contribute to the quality of life in healthy subjects, but also mitigate many aspects of disease or chronic conditions, particularly cardiovascular disease (CVD), by reducing risk and controlling disease progress. Atherosclerosis, the underlying pathological process for CVD, occurs as the result of imbalance between endothelial injury resulting from chronic inflammation and production of reactive oxygen species (ROS) and vascular repair induced at least in part by circulating endothelial progenitor cells (EPCs). Importantly, pro-inflammatory cytokines stimulate the production of ROS that in turn activate various intracellular signaling pathways leading to further increase in ROS production, creating a positive feedback loop. Considerable evidence suggests that exercise can disrupt the positive feedback loop, thus suppressing inflammation and ROS production. Exercise also mobilizes EPCs from the bone marrow and improves their self-renewal potential and differentiation capability, although the mechanisms for such effects are less well understood. Acute exercise can, however, trigger a cardiovascular event. Thus, exercise appears to be a double-edge sword, which in the long run can maintain the equilibrium between vascular injury and repair. As with pharmacological approaches, individuals respond to exercise differently. The Genetics, Exercise and Research (GEAR) program at the University of Miami uses genomic and epigenomic approaches to decipher the molecular mechanisms underlying EPC mobilization and the spectrum of responses to various forms of exercise, in an attempt to provide the scientific basis for optimized personalized training programs.

Corresponding author

INTRODUCTION

Chronic systemic inflammation is an underlying cause of many age-associated illnesses, including cardiovascular disease (CVD), diabetes mellitus and cancer, as it can inflict devastating insults throughout the body. On the other hand, the capability of tissue repair declines with aging. Indeed, endothelial dysfunction and frank atherosclerosis arise from an imbalance between vascular injury and repair (Dong and Goldschmidt-Clermont 2007). Although pharmacological interventions are mandatory in many circumstances to reduce inflammation, potentially noxious drugs are often not necessary for the control of chronic inflammation. Instead, regular exercise has gained widespread advocacy as an effective preventive measure. Exercise training has been shown to reduce chronic inflammation by decreasing pro-inflammatory cytokines and increasing anti-inflammatory cytokines (Strohacker and McFarlin 2010). Exercise training also improves the beneficial effects of circulating endothelial progenitor cells (EPCs). In this chapter, we provide evidence supporting the impact of exercise on health and on disease states, particularly CVD. We also review the effects of exercise in controlling inflammatory cytokine profile and EPC changes. The potential of cytokines and growth factors as the link between exercise and EPC changes will be explored. Finally, the prospects that genomic approaches may play a significant role in elucidating the mechanisms underlying the health benefits of exercise will be elaborated.

EXERCISE AND HEALTH

Exercise is an important part of preventive strategies for many chronic conditions, including osteoporosis, diabetes, CVD, and mental health problems (Erikssen *et al.* 1998). Aerobic exercise training produces favorable cardiovascular responses, including increased stroke volume, improved skeletal muscle oxidative capacity, and widening of the systemic arteriovenous oxygen difference. Exercise is associated with remodeling of the vessel wall, including arteriogenesis and angiogenesis. Arteriogenesis refers to the enlargement of existing arterial vessels in both diameter and wall thickness. Instructively, the size of major vessels is proportional to the level of exercise and/or pattern of use with conduit vessel size greater in specific muscle groups recruited for activity, e.g., subclavian arteries for ping-pong players and femoral arteries for cyclists. Although the molecular mechanisms underlying arteriogenesis remain elusive, mechanotransduction of elevated shear stress by the endothelium into a vessel enlargement signal seems to play a central role. This process is dependent on endothelial nitric oxide synthase (eNOS) activity and extracellular matrix (ECM) remodeling by matrix metalloproteinases (MMPs). Recruitment of monocytes follows the up-regulation of monocyte chemotactic protein 1 (MCP-1), leading to release of cytokines, such as vascular endothelial growth factor (VEGF) and basic fibroblast growth factor (FGF-2). Angiogenesis is regulated by a myriad factors, including VEGF and MMPs (Brown and Hudlicka 2003). Increasing evidence has shown that arteriogenesis and angiogenesis depend not only on cytokines released from cells within the vessel wall but also on circulating EPCs derived from the bone marrow (Critser and Yoder 2010), which is discussed in detail later.

EXERCISE AND CARDIOVASCULAR DISEASE

Although physical exertion is a precipitating factor for myocardial infarction and sudden cardiac death, this adverse outcome usually occurs in sedentary individuals who suddenly engage in strenuous physical activity. On the other hand, daily aerobic exercise training reduces the risk of heart disease by affecting various aggravating factors, such as chronic inflammation, diabetes mellitus, blood pressure and EPCs (Yung *et al.* 2009). Multiple exercise modalities, including aerobic exercise, yoga or resistance training, reduce chronic inflammation. Chronic exercise slows the progression of hypertension and improves cardiac function in spontaneous hypertensive rats. Physical fitness and exercise prevent the onset of non–insulin-dependent diabetes mellitus (DM type II) and improve blood glucose regulation in diabetes patients. Exercise also increases the responsiveness of coronary arteries: the arteries of the runners expanded twice as much as the non-runners in response to nitroglycerin (Sattelmair *et al.* 2009). Exercise also increases circulating EPCs and EPC mobilizing factors. With greater coronary artery responsiveness, lower blood pressure, controlled inflammation, and enhanced arterial repair capability, chronic exercise slows the heart disease process and enables the cardiac muscle to handle more stress and therefore lower the threshold for myocardial ischemia, infarction and sudden cardiac death.

MOLECULAR AND CELLULAR MECHANISMS FOR EXERCISE-INDUCED EFFECTS

Considerable efforts have been committed to understanding the mechanisms that mediate exercise-induced effects at the molecular and cellular levels. We will review the evidence on inflammatory cytokines and EPC functions.

Exercise and Inflammation

The pathogenesis of the atherosclerosis is characterized as a chronic inflammatory response that promotes accumulation of monocyte-derived macrophages in arterial plaques. Activated macrophages release cytokines that result in arterial cell death and tissue damage (Libby 2007). The evidence that exercise training suppresses inflammatory cytokines has been overwhelming, from animal studies to healthy subjects and to an array of cardiovascular disorders. Swimming increases the anti-inflammatory cytokine IL-10, decreases muscle lipid peroxidation and improves left ventricular function in chronic heart failure rats (Nunes *et al.* 2008). We have demonstrated that regular exercise leads to a potent suppression of circulating pro-inflammatory cytokines in both young and old mice (Ajijola *et al.* 2009). However, reduction in atherosclerotic lesions was seen only in young mice, not old mice. Instructively, the reduction in systemic inflammation was associated with increased lifespan, in spite of the atherosclerosis burden, suggesting that inflammation may trigger cardiac death and that strategies aiming at reducing inflammation may protect individuals from sudden death.

In healthy young adults, a 12 week high-intensity aerobic training down-regulates tumor necrosis factor (TNF) production (Sloan *et al.* 2007). In middle-aged men, 12 week moderate exercise training results in a substantial decrease in serum interleukin-6 (IL-6), which is reversible by only 2 week of detraining (Thompson *et al.* 2009). In elderly women, 10 wk moderate to high-intensity resistance training markedly reduces serum levels of TNF-α. It also reduces lipopolysaccharide-stimulated monocyte production of IL-6, IL-1β and TNF-α (Phillips *et al.* 2010). Indeed, two thirds of about 40 observational studies report an inverse correlation between inflammatory factors and fitness after adjustment for obesity/overweight.

The impact of 2 year regular exercise training versus percutaneous coronary intervention (PCI) on chronic inflammation and cardiovascular events was studied in patients with coronary artery disease (CAD) in a prospective, randomized trial. Exercise but not PCI leads to a significant reduction in serum CRP and IL-6 levels. Both exercise and PCR reduced cardiac events, with exercise having a more pronounced effect (Hambrecht *et al.* 2004). Moderate exercise in the early period after myocardial infarction decreases acute phase reactants, CRP, fibrinogen, and soluble TNF-α receptor-1. Yoga practice in patients with chronic heart failure improved exercise tolerance, decreased IL-6 and CRP and increased antioxidant enzyme extracellular superoxide dismutase (Pullen *et al.* 2008). Remarkably, exercise, but not diet-induced weight loss, suppresses skeletal muscle inflammation gene expression in frail, obese elderly (Lambert *et al.* 2008).

Importantly, pro-inflammatory cytokines and reactive oxygen species (ROS) are intricately connected. Indeed, pro-inflammatory cytokines activate the production of ROS, which, in turn, can activate various intracellular signaling pathways, including nuclear factor-kappaB (NF-κB). Activation of NF-κB induces transcription of multiple pro-inflammatory cytokine genes, which, in turn, leads to further increase in ROS production, sustaining a positive feedback loop that leads to worsening inflammation. In spontaneously hypertensive rats, chronic exercise is associated with reduction in pro-inflammatory cytokines and inducible nitric oxide (NO) synthase, down-regulation of superoxide and derived ROS production, and increase in antioxidant levels. Thus, exercise may represent one of the few approaches at our disposal that can break the positive feedback loop between pro-inflammatory cytokines and ROS, which may prove to be extremely important in CVD prevention, rehabilitation and therapy.

Multiple studies have attempted to identify the molecular events underlying the effects of regular exercise on inflammation suppression. The gene encoding peroxisome-proliferator-activated receptor-γ coactivator 1α (PGC1α) has gained attention. PGC1α is rapidly induced after a single bout of endurance exercise but normalizes when physical activity stops. Importantly, PGC1α exerts anti-inflammatory effects both *in vivo* and *in vitro* (Handschin *et al.* 2007). Muscle-specific *PGC1a*$^{-/-}$ mice show up-regulation in the muscle of many genes involved in local or systemic inflammation, particularly those encoding IL-6 and TNF-α, and suppressor of cytokine signaling 1 (SOCS1), SOCS3 and CD68. *PGC1a*$^{+/-}$ mice showed a smaller, but significant, increase in the expression of many of these genes as compared with their wild-type counterparts. Primary culture and myotube formation of *PGC1a*$^{-/-}$ muscle cells resulted in up-regulation of *TNF*-α and *Il-6* mRNA and more secretion of IL-6 than did wild-type myotubes, suggesting that circulating inflammatory cytokines associated with inflammatory state can originate from the skeletal muscle cells, and that PGC1α suppresses the production of these inflammatory mediators. Instructively, the reduction in *PGC1a* mRNA in *PGC1a*$^{+/-}$ mice relative to wild-type animals is comparable to the down-regulation of this gene in muscles of sedentary mice compared with exercising mice and to the down-regulation of this gene in the muscle of humans with type II diabetes relative to healthy volunteers. Remarkably, people with type II diabetes have increased expression of IL-6 and TNF-α in skeletal muscle, as well as an increased concentration of IL-6 in the serum. The reduction of *PGC1a* mRNA in the skeletal muscle of people with type II diabetes and sedentary individuals is likely to be linked to the chronic, low-grade inflammation in these individuals. Exactly how PGC1α mediates inflammatory gene expression in muscle remains to be determined. Multiple lines of evidence support ROS as mediators. Indeed, PGC1α is a powerful suppressor of ROS production as it is required for the induction of many ROS-detoxifying enzymes and proteins that attenuate ROS production. Hence, *PGC1A*$^{-/-}$ mice have elevated ROS, which contributes to the chronic inflammatory state (Handschin and Spiegelman 2008). Therefore, exercise training would be expected to induce *PGC1A* expression and suppress chronic inflammation.

Other studies have suggested a role for IL-6 as an anti-inflammatory cytokine in mediating exercise-induced anti-inflammation (Pedersen 2007). It is hypothesized that the long-term effects of exercise are due to increased IL-6 production. Recombinant human IL-6 infusion and exercise inhibit the endotoxin-induced increase in circulating levels of TNFα in healthy humans. Furthermore, levels of TNFα are markedly elevated in anti-IL-6-treated mice and in IL-6-

knockout mice. Physiological concentrations of IL-6 stimulate the production of anti-inflammatory cytokines IL-1α and IL-10. Other signaling factors involved mediating exercise-induced anti-inflammatory effects include Signal Transducer and Activator of Transcription 3 (STAT3).

In contrast to the suppression of chronic inflammation by regular, moderate exercise, prolonged, high-intensity training may elicit elevated systemic inflammation. It is well known that marathon runners experience transient immunosuppression. A decline of the total lymphocyte concentration is characteristic after exercise of prolonged duration and/or high intensity. The mechanisms underlying exercise-induced lymphocytopenia remain elusive. It is noteworthy that exercise tolerance decreases with certain conditions, such that mild exercise training could elevate inflammation in certain patients. It remains to be determined whether the inflammatory response in these circumstances is at all unhealthy. For example, angiogenesis is stimulated in patients with intermittent claudication by exercise but it usually coincides with increased inflammation.

We have addressed in prior reviews (Dong and Goldschmidt-Clermont 2007, Goldschmidt-Clermont *et al.* 2005) the issue of inflammation as a beneficial response versus inflammation as an insult to tissues and organs. Even ROS display a dual role. We have shown that endothelium disrupted by mechanical injury undergoes rapid endothelial-to-mesodermal transformation of the endothelial cells adjacent to the wound. Such transformation is required for the repair of the endothelial layer and depends on ROS (Moldovan *et al.* 2000). Cells adjacent to the wound display heightened levels of ROS, up to several cell rows from the wound edge (Fig. 1). Blocking ROS production resulted in no endothelial transformation and cell migration to close the wound. Hence, ROS are not just noxious molecules that contribute to inflammation. Instead, they are essential for tissue repair and, in this case, their primary effect may be the inhibition of phosphatases at a time when kinases, and in particular receptor tyrosine kinases, are particularly critical. If ROS production is an important aspect of the response to exercise, it is therefore no surprise that such response can either contribute to tissue repair and healing, or to tissue injury and inflammation. The difference between the two opposite effects may be related to the anti-oxidant capacity of the cells in which ROS are produced as well as the cellular location of ROS production (Fig. 2). An important element of arterial repair was discovered just a few years ago and requires the participation of a somatic stem cell: the endothelial progenitor cell (EPC).

Exercise and EPCs

The progressive obsolescence of tissues and organs during aging is in part due to the senescence of somatic stem cells, and consequently the diminished repair capability. For example, EPCs continuously repair the arterial wall to maintain endothelial integrity. When successful, this process halts the inflammation

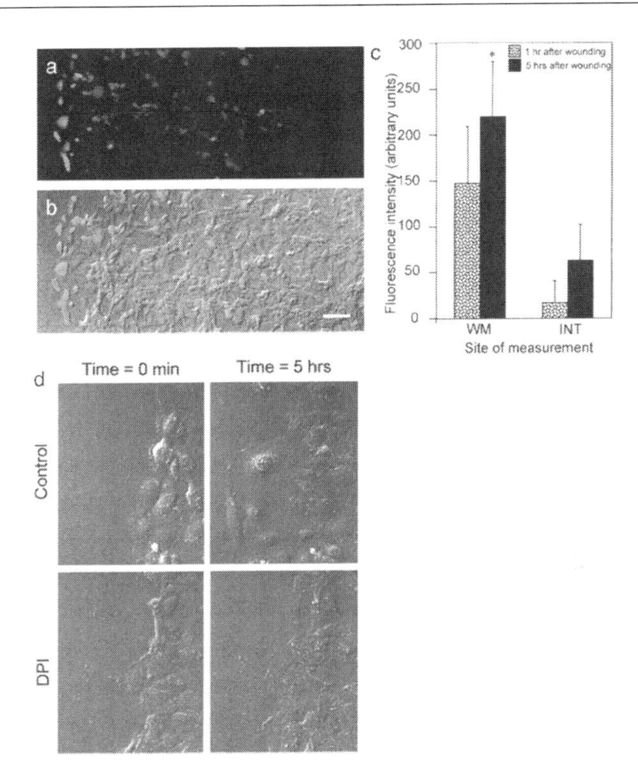

Fig. 1 ROS are produced and necessary for successful repair of a wounded endothelium. Increased production of ROS at the wound margin. a and b, CM-DCF-DA fluorescence (a) and overlay of the fluorescence and Differential Interference Contrast (DIC) Light Microscopy images of the same field (b). Bar = 100 μm. c, Measurement of CM-DCF-DA fluorescence in mouse aortic endothelial cells, 1 and 5 hr after wounding, at the wound margin (WM) and distant from the wound, in the intact monolayer (INT). Data correspond to mean ± SEM. *$P<0.05$ for WM vs INT at 5 hr. ROS are required for endothelial wound healing. d, DIC images of cells at the beginning (0 min) and end (5 hr) of representative experiments for each condition (control and 10 μm DPI). Bar = 50 μm. (With permission from Circ. Res. 2000: 86: 549-557.)

Color image of this figure appears in the color plate section at the end of the book.

response and the development of atherosclerosis. Age-associated senescence of EPCs may play a significant role in age-related cardiovascular dysfunction, leading to increased cardiovascular risk. Indeed, EPCs from old mice fail to repair arteries in atherosclerosis models (Fig. 3) (Rauscher *et al.* 2003). Physical inactivity is an integral part of the aging process and it is a recognized risk factor for CVD. Hence, it is conceivable that EPCs may mediate, at least in part, the beneficial effects of exercise on CVD.

The effects of exercise on EPCs have been studied in both healthy subjects and in patients with CVD. A bout of modified Bruce treadmill protocol or a 1 hr spinning exercise routine results in significant increase in circulating EPCs.

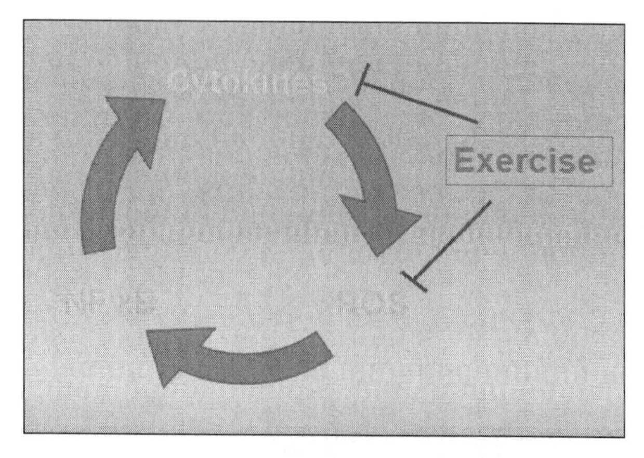

Fig. 2 Exercise disrupts the positive feedback loop between pro-inflammatory cytokines and ROS. Pro-inflammatory cytokines activate ROS, which activates NF-κB. Activation of NF-κB induces gene transcription of multiple pro-inflammatory cytokines, which leads to further increase in ROS production, cultivating a positive feedback mechanism leading to worsening inflammation. Exercise disrupts the positive feedback loop by inhibiting both cytokines and ROS production. Unpublished.

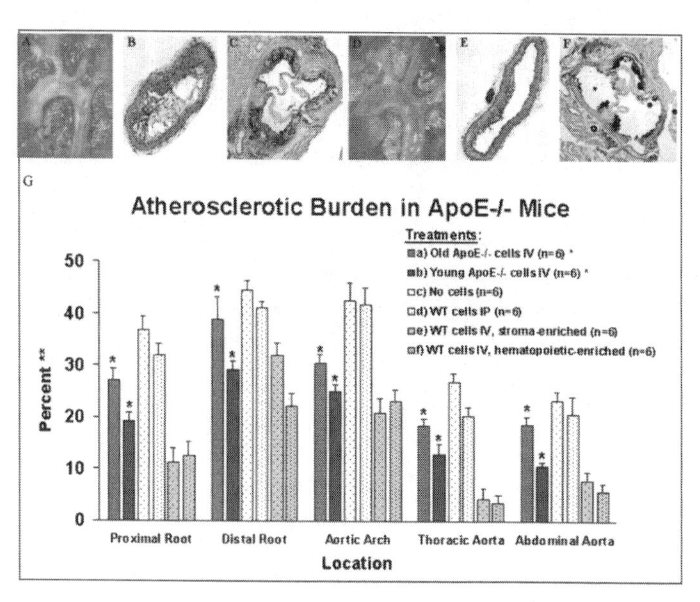

Fig. 3 Bone marrow–derived EPCs from young but not aged mice suppress atherosclerosis. Atherosclerosis assessment in mice receiving no cell treatment (**A-C**) or EPCs from age-matched wild-type mice (**D-F**). The relative efficacy of EPCs from young and aged apoE-/- mice as well as wild-type mice in different parts of the aortic wall is depicted in **G**. (With permission from Circulation 2003: 108: 457-463.)

Color image of this figure appears in the color plate section at the end of the book.

Such response may be mediated through NO production or VEGF up-regulation. On the other hand, a marathon race significantly decreased CD34^{+} cells, cKit^{+} cells, and CD133^{+} cells—cell populations enriched for EPCs—in older runners. This finding coincides with a significant down-regulation of VEGF (Adams *et al.* 2008). In contrast, a 246 km-"Spartathlon" ultradistance race increases EPCs by nearly 11-fold in a younger population. This increase is associated with a markedly increased inflammatory cytokine release (Goussetis *et al.* 2009). Together, these results suggest that a single bout of prolonged, strenuous exercise can stimulate the mobilization of bone marrow cells. Yet, the specific type of released cells, inflammatory cells versus EPCs, may depend on the age of the subjects and the relative cellular composition and reserve integrity of the bone marrow (Fig. 4). For young individuals whose supply of EPCs is large, the cells that are released will be EPCs, whereas in older individuals whose supply of EPCs is diminished, inflammatory cells may be mobilized instead. This is particularly true for older individuals who do not exercise regularly, or even for fit older individuals who experience a particularly strenuous type of exercise (marathon running, for example) (Fig. 4).

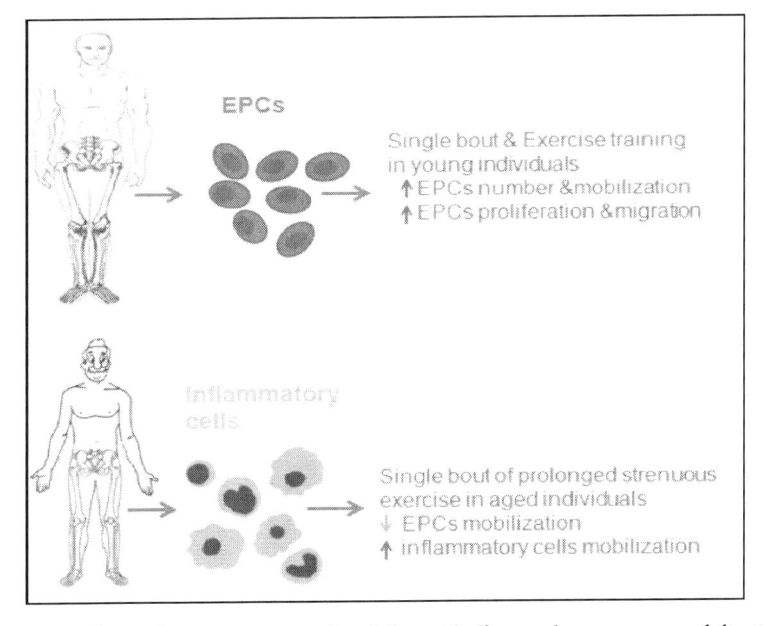

Fig. 4 A single bout of strenuous exercise has differential effects on bone marrow mobilization in young and aged individuals. In young individuals, a single bout of exercise stimulates the release of EPCs and improves EPC functions, whereas in aged subjects where EPCs are exhausted, a bout of strenuous exercise triggers the release of inflammatory cells. Unpublished.

Extensive studies in animal models and human subjects have demonstrated that exercise training can alter the number and functional characteristics of EPCs. The beneficial effects of exercise training on the mobilization of EPCs have been observed in patients with a spectrum of cardiovascular-related diseases, including chest pain, heart failure, CAD, peripheral artery disease (PAD), and hemodialysis, or following cardiac surgery or stent implantation. Healthy subjects, ranging from sixth graders, high school students and sedentary working adults to elderly subjects, show improved EPC markers in response to exercise training. Furthermore, improved functional capacity of EPCs, evidenced by an improved EPC colony-forming and migratory capacity, in patients with PAD and CAD has been demonstrated after walking training. Thus, exercise may exert its beneficial effects not only by affecting the abundance of EPCs, but also by improving their function.

Importantly, studies in both mice and humans have provided insights into the signaling mediators for EPC mobilization. Indeed, eNOS-knockout mice and wild-type mice treated with the NO synthase inhibitor or hydrogen peroxide show lower EPC numbers at baseline and an attenuated increase of EPCs in response to exercise training, whereas in wild-type mice without treatment, exercise training causes approximately a three-fold increase in the number of EPCs (Laufs *et al.* 2004). Red wine antioxidants have synergistic effects with exercise training on increasing EPCs in mice. In older men, the rise and fall of EPC levels with long-term training and short-term detraining, respectively, correlate with total anti-oxidant capacity. Furthermore, exercise training in active and inactive men has differential effects on the suppression of NADPH oxidase subunits and enhancement of NO production, and these effects correlate with increased EPC mobilization (Witkowski *et al.* 2009). Multiple studies have suggested the role of growth factors and cytokines in mediating EPC release. The most implicated ones are VEGF, IL-6 and IL-10. Overall, however, there is limited systematic understanding for the factors that mediate exercise-induced EPC mobilization.

It has been shown that a three-month cardiac rehabilitation program increases EPC numbers by two-fold, and colony-forming units, a marker for the replicative potential of stem cells, by three-fold (Paul *et al.* 2007). This effect is associated with increased blood nitrite concentrations and reduced EPC apoptosis. Remarkably, the study identifies a minority of patients who do not respond to the rehabilitation program with an increase in blood nitrites or EPC numbers. Indeed, several studies showed no effect of exercise training on EPCs, which may indicate heterogeneity of response and that some patients are non-responders. This is important if we attempt to use exercise as a therapeutic and preventive strategy. We need to be able to identify responders versus non-responders and the underlying mechanisms for the difference, such that specific exercise regimens can be prescribed to best suit a given individual. To fully characterize the genomic and epigenomic basis for the beneficial effects, or lack thereof, of exercise training and to understand the molecular mechanisms that mediate exercise-induced EPC mobilization and potentiation, we have initiated a substantial research program named **G**enetics, **E**xercise and **R**esearch (GEAR) at the University of Miami Miller School of Medicine's Hussman Institute for Human Genomics. This study employs unbiased genomic approaches that include global gene expression mapping, genome-wide microRNA and DNA methylation profiling to study the effects of various forms of exercise, including combined endurance and resistance Western training as well as Eastern Tai Chi in large cohorts of participants focusing on responders versus non-responders and mobilization of EPCs.

Replacement of senescent and damaged cells contributes to the maintenance of tissue functions and longevity of the organism. The presence of competent tissue-specific progenitor or stem cells throughout the body reservoirs serves as the "fountain of youth" to replenish obsolete and damaged cells. Thus, maintaining adequate supply of stem cells, including EPCs, is of fundamental scientific and clinical importance in the design of optimal protocols for the prevention and treatment of age-related degenerative diseases. Although this may be achieved with the administration of growth factors or autologous stem cell transplantation, exercise training may prove to be the most physiological, cost-effective and useful strategy (in combination with pharmacological interventions) for promoting the proliferation and potentiation of these regenerative cells. Further studies, including the GEAR program, will allow us to better understand the underlying molecular mechanisms for EPC mobilization and for individual responsiveness to exercise and to design personalized training programs.

Table 1 Key points about exercise

- **Exercise and health** Exercise produces favorable changes in many organ systems, including the cardiovascular system, skeletal muscle and mental health.
- **Exercise and cardiovascular disease** Exercise improves outcomes in CAD, hypertension, hyperlipidemia, congestive heart failure and diabetes mellitus.
- **Molecular and cellular mechanisms for exercise-induced effects** Exercise suppresses inflammation by regulating pro-inflammatory cytokines, ROS and PGC1. Exercise improves arterial repair by stimulating EPC mobilization and enhancing their functions.

Summary Points

- Chronic diseases, including cardiovascular disease, involve tissue injury mediated by inflammation and tissue repair mediated by stem/progenitor cells.
- Inflammatory cytokines, NFκB and ROS regulate inflammation and EPC mobilization and function.
- Regular exercise suppresses pro-inflammatory cytokins and ROS production.

- Regular exercise increases the mobilization and function of EPCs, enhancing vascular repair.
- Genomic and epigenomic approaches focusing on responders versus non-responders help develop personalized exercise regimen aiming to maximize the beneficial effects of exercise.

Abbreviations

CAD	:	coronary artery disease
CM-DCF-DA	:	chloromethyl-dichlorodihydrofluorescein-diacetate
CRP	:	C-reactive protein
CVD	:	cardiovascular disease
ECM	:	extracellular matrix
eNOS	:	endothelial nitric oxide synthase
EPC	:	endothelial progenitor cell
FGF	:	fibroblast growth factor
GEAR	:	Genetics, Exercise, and Research program
IL	:	interleukin
MCP-1	:	monocyte chemotactic protein 1
MMPs	:	matrix metalloproteinases
NF-κB	:	nuclear factor-kappaB
NO	:	nitric oxide
PAD	:	peripheral arterial disease
PCI	:	percutaneous coronary intervention
PGC1α	:	peroxisome-proliferator-activated receptor-γ coactivator 1α
ROS	:	reactive oxygen species
SOCS1	:	suppressor of cytokine signaling 1
STAT3	:	Signal Transducer and Activator of Transcription 3
TNF	:	tumor necrosis factor
VEGF	:	vascular endothelial growth factor

Definitions of Terms

Cardiovascular disease: The class of diseases that affect the heart and blood vessels, mainly arteries. The term usually refers to arterial diseases related to atherosclerosis. These conditions share commonalities in causes, mechanisms, and treatments.

Cytokines: A group of proteins, peptides and glycoproteins secreted by immune cells and a variety of other cells. They function as signaling molecules in cellular communications.

Endothelial progenitor cells (EPCs): A group of precursor cells that are able to differentiate into mature endothelial cells that line the inner lumen of the blood vessels. They are derived from the bone marrow and exist in the peripheral blood.

Inflammation: The complex biological response of the organism to injurious stimuli, such as damaged cells, pathogens or irritants. Inflammation is a protective attempt by the organism to fight off the harmful stimuli and to initiate the healing process for the tissue.

Reactive oxygen species (ROS): Small molecules that contain oxygen atom, including oxygen ions and peroxides. They are highly reactive and are byproducts of the normal metabolism of oxygen. They serve as signaling molecules in the cells. Abnormal ROS levels can result in significant damage to cell structures—oxidative stress.

References

Adams, V. and A. Linke, F. Breuckmann, K. Leineweber, S. Erbs, N. Krankel, M. Brocker-Preuss, F. Woitek, R. Erbel, G. Heusch, R. Hambrecht, G. Schuler, and S. Möhlenkamp. 2008. Circulating progenitor cells decrease immediately after marathon race in advanced-age marathon runners. Eur. J. Cardiovasc. Prev. Rehabil. 15: 602-607.

Ajijola, O.A. and C. Dong, E.E. Herderick, Q. Ma, P.J. Goldschmidt-Clermont, and Z. Yan. 2009. Voluntary running suppresses proinflammatory cytokines and bone marrow endothelial progenitor cell levels in apolipoprotein-E-deficient mice. Antioxid. Redox Signal. 11: 15-23.

Brown, M.D. and O. Hudlicka. 2003. Modulation of physiological angiogenesis in skeletal muscle by mechanical forces: involvement of VEGF and metalloproteinases. Angiogenesis 6: 1-14.

Critser, P.J. and M.C. Yoder. 2010. Endothelial colony-forming cell role in neoangiogenesis and tissue repair. Curr. Opin. Organ. Transplant. 15: 68-72.

Dong, C. and P.J. Goldschmidt-Clermont. 2007. Endothelial progenitor cells: a promising therapeutic alternative for cardiovascular disease. J. Interv. Cardiol. 20: 93-99.

Erikssen, G. and K. Liestol, J. Bjornholt, E. Thaulow, L. Sandvik, and J. Erikssen. 1998. Changes in physical fitness and changes in mortality. Lancet 352: 759-762.

Goldschmidt-Clermont, P.J. and M.A. Creager, D.W. Losordo, G.K. Lam, M. Wassef, and V.J. Dzau. 2005. Atherosclerosis 2005: recent discoveries and novel hypotheses. Circulation 112: 3348-3353.

Goussetis, E. and A. Spiropoulos, M. Tsironi, K. Skenderi, A. Margeli, S. Graphakos, P. Baltopoulos, and I. Papassotiriou. 2009. Spartathlon, a 246 kilometer foot race: effects of acute inflammation induced by prolonged exercise on circulating progenitor reparative cells. Blood Cells Mol. Dis. 42: 294-299.

Hambrecht, R. and C. Walther, S. Mobius-Winkler, S. Gielen, A. Linke, K. Conradi, S. Erbs, R. Kluge, K. Kendziorra, O. Sabri, P. Sick, and G. Schuler. 2004. Percutaneous coronary angioplasty compared with exercise training in patients with stable coronary artery disease: a randomized trial. Circulation 109: 1371-1378.

Handschin, C. and B.M. Spiegelman. 2008. The role of exercise and PGC1alpha in inflammation and chronic disease. Nature 454: 463-469.

Lambert, C.P. and N.R. Wright, B.N. Finck, and D.T. Villareal. 2008. Exercise but not diet-induced weight loss decreases skeletal muscle inflammatory gene expression in frail obese elderly persons. J. Appl. Physiol. 105: 473-478.

Laufs, U. and N. Werner, A. Link, M. Endres, S. Wassmann, K. Jurgens, E. Miche, M. Bohm, and G. Nickenig. 2004. Physical training increases endothelial progenitor cells, inhibits neointima formation, and enhances angiogenesis. Circulation 109: 220-226.

Libby, P. 2007. Inflammatory mechanisms: the molecular basis of inflammation and disease. Nutr. Rev. 65: S140-146.

Moldovan, L. and N.I. Moldovan, R.H. Sohn, S.A. Parikh, and P.J. Goldschmidt-Clermont. 2000. Redox changes of cultured endothelial cells and actin dynamics. Circ. Res. 86: 549-557.

Nunes, R.B. and M. Tonetto, N. Machado, M. Chazan, T.G. Heck, A.B. Veiga, and P. Dall'Ago. 2008. Physical exercise improves plasmatic levels of IL-10, left ventricular end-diastolic pressure, and muscle lipid peroxidation in chronic heart failure rats. J. Appl. Physiol. 104: 1641-1647.

Paul, J.D. and T.M. Powell, M. Thompson, M. Benjamin, M. Rodrigo, A. Carlow, V. Annavajjhala, S. Shiva, A. Dejam, M.T. Gladwin, J.P. McCoy, G. Zalos, B. Press, M. Murphy, J.M. Hill, G. Csako, M.A. Waclawiw, and R.O. Cannon 3rd. 2007. Endothelial progenitor cell mobilization and increased intravascular nitric oxide in patients undergoing cardiac rehabilitation. J. Cardiopulm. Rehabil. Prev. 27: 65-73.

Pedersen, B.K. 2007. IL-6 signalling in exercise and disease. Biochem. Soc. Trans. 35: 1295-1297.

Phillips, M.D. and M.G. Flynn, B.K. McFarlin, L.K. Stewart, and K.L. Timmerman. 2010. Resistance training at eight-repetition maximum reduces the inflammatory milieu in elderly women. Med. Sci. Sports Exerc. 42: 314-325.

Pullen, P.R. and S.H. Nagamia, P.K. Mehta, W.R. Thompson, D. Benardot, R. Hammoud, J.M. Parrott, S. Sola, and B.V. Khan. 2008. Effects of yoga on inflammation and exercise capacity in patients with chronic heart failure. J. Card. Fail. 14: 407-413.

Rauscher, F.M. and P.J. Goldschmidt-Clermont, B.H. Davis, T. Wang, D. Gregg, P. Ramaswami, A.M. Pippen, B.H. Annex, C. Dong, and D.A. Taylor. 2003. Aging, progenitor cell exhaustion, and atherosclerosis. Circulation 108: 457-463.

Sattelmair, J.R. and J.H. Pertman, and D.E. Forman. 2009. Effects of physical activity on cardiovascular and noncardiovascular outcomes in older adults. Clin. Geriatr. Med. 25: 677-702, viii-ix.

Sloan, R.P. and P.A. Shapiro, R.E. Demeersman, P.S. McKinley, K.J. Tracey, I. Slavov, Y. Fang, and P.D. Flood. 2007. Aerobic exercise attenuates inducible TNF production in humans. J. Appl. Physiol. 103: 1007-1011.

Strohacker, K. and B.K. McFarlin. 2010. Influence of obesity, physical inactivity, and weight cycling on chronic inflammation. Front. Biosci. (Elite Ed) 2: 98-104.

Thompson, D. and D. Markovitch, J.A. Betts, D.J. Mazzatti, J. Turner, and R.M. Tyrrell. 2010. Time-course of changes in inflammatory markers during a 6-month exercise intervention in sedentary middle-aged men: A randomized-controlled trial. J. Appl. Physiol. 108: 769-779.

Witkowski, S. and M.M. Lockard, N.T. Jenkins, T.O. Obisesan, E.E. Spangenburg, and J.M. Hagberg. 2009. Relation of circulating progenitor cells to vascular function and oxidative stress with long term training and short term detraining in older men. Clin. Sci. (Lond). 118: 303-811.

Yung, L.M. and I. Laher, X. Yao, Z.Y. Chen, Y. Huang, and F.P. Leung. 2009. Exercise, vascular wall and cardiovascular diseases: an update (part 2). Sports Med. 39: 45-63.

CHAPTER 5

Cytokines in Alcoholics

E. González-Reimers* **and F. Santolaria-Fernández**
Servicio de Medicina Interna, Hospital Universitario de Canarias
Ofra s/n 38320, La Laguna, Tenerife, Canary Islands, Spain

ABSTRACT

Cytokines constitute a broad family of inflammatory and regulatory mediators that play outstanding roles in organic complications of alcoholism, contributing to the protean manifestations of this disease. Alcohol increases gut permeability to endotoxin, leading to Kupffer cell activation and pro-inflammatory cytokine secretion. This initiates a cascade of events characterized by inflammation, lipid peroxidation, neutrophil recruitment, and immune activation, closing a positive feedback loop, ultimately leading to liver cell necrosis and apoptosis. Therefore, alcoholic hepatitis can be considered as a TNF-α-mediated disease; in more advanced stages of the disease, TGF-β plays a more important role, promoting fibrosis both in the liver (leading to cirrhosis) and in the pancreas (leading to chronic pancreatitis). Increased TNF-α favours muscle atrophy, is probably involved in neurodegeneration and brain atrophy, and contributes to alcoholic cardio-myopathy and bone alterations, together with other cytokines, especially IL-6, a well-known activator of osteoclasts. Moreover, altered cytokine secretion in alcoholics may predispose them to sepsis and severe pneumonia. However, despite the proven action of TNF-α in some organs, such as the liver, anti-TNF treatment trials in alcoholic hepatitis have led to disappointing results, suggesting that vigorous research is still needed to fully elucidate the role of these inflammatory modulators in alcohol-related organic dysfunction.

*Corresponding author

INTRODUCTION

Cytokines constitute a broad family of inflammatory and regulatory mediators that play key roles in several features of alcoholism, contributing to the protean manifestations of this disease. They exert their multiple actions in an autocrine, paracrine and endocrine manner (Balkwill and Burke 1989).

Although cytokines are mainly produced by macropahges and immune cells, virtually all nucleated cells are able to secrete them under certain circumstances. Adipocytes have emerged as an important source of cytokines including leptin, TNF-α, adiponectin, resistin, visfatin, apelin, IL-6, IL-8, IL-10, TGF-β, and nerve growth factor (Trayhurn and Wood 2004). Recently, even muscle has been implicated in cytokine (myokine) production (Pedersen and Febbraio 2008).

ALCOHOL-MEDIATED LIVER DISEASE

Increased cytokine levels in alcoholic hepatitis and liver cirrhosis were described 20 years ago (Bird *et al.* 1990). They are responsible for most of the systemic effects and histological alterations described in alcoholics with liver disease. These cytokines mainly derive from hepatocytes and Kupffer cells. Currently, clinical and experimental data suggest that raised Th-1 cytokine levels are important in the early stages of alcoholic liver disease, whereas a shift to a Th-2 response may be required for liver fibrosis and cirrhosis to develop. During this late stage, cytokines like IL-13 promote collagen deposition, both by a direct effect on liver fibroblasts and by an indirect effect via TGF-β, which activates stellate cells. In accordance with this, in a series of 70 alcoholic patients (unpublished data), we found significantly higher IL-13 levels among cirrhotics than among non-cirrhotics and controls ($p=0.001$), associated with worse prognosis in a subset of 64 patients followed for a median period of 60 months: those with IL-13 values above the median showed shorter survival (Log-rank, $p=0.05$, Fig. 1).

Pathophysiology: Endotoxin and Pro-inflammatory Cytokines

The first step is Kupffer cell activation by gut-derived endotoxin (LPS) reaching the liver via the portal system. Increased intestinal permeability is a well-documented feature of both liver cirrhosis and ethanol (Cariello *et al.* 2009). Ethanol and acetaldehyde disrupt the epithelial tight junctions and increase the permeability of the paracellular space to macromolecules, including bacteria. LPS levels increase after ethanol intake, trigger cytokine release by liver macrophages, and induce oxidative stress, which, in turn, promote secretion of additional TNF-α. In the study by Cariello *et al.* (2009), patients with altered intestinal permeability showed

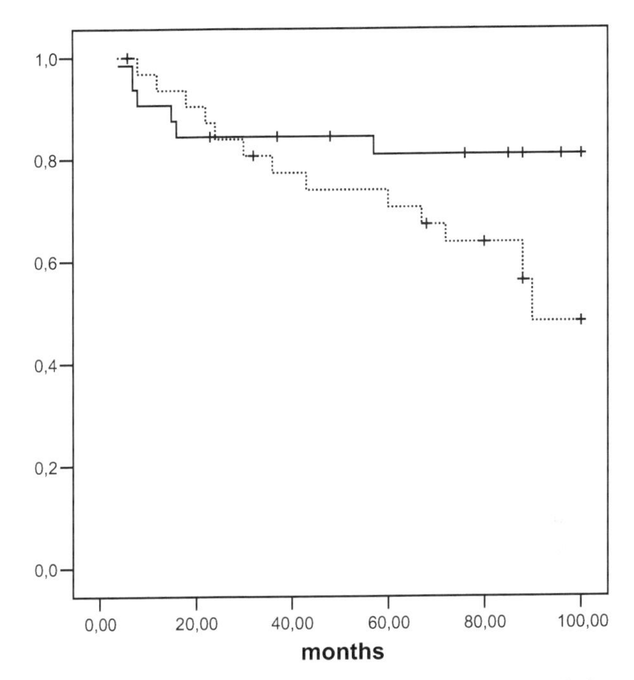

Fig. 1 Interleukin-13 (IL-13) and survival in 64 alcoholic patients, classified according to IL-13 values above (broken line) or below (solid line) the median. Patients were followed during a median of 60 months (some of them even 100 months), so that these data represent long-term survival in relation with IL-13 values at admission and fit well with the raised levels of IL-13 in cirrhotics. Log rank p=0.05 (unpublished results of personal observations).

raised IL-6 and TNF-α alpha levels. These cytokines are produced by Kupffer cells, which become activated by LPS binding to TLRs, especially TLR-4. This effect is potentiated by ethanol, through activation of NADPH oxidase and increased production of ROS, which in turn increase the expression of TLR-4. Activation of this receptor by LPS modulates the activity of transcription factors, such as egr-1, which binds to the TNF-α promoter and increases TNF-α synthesis and secretion (Pritchard and Nagyi 2005). Adiponectin counteracts these effects, decreasing DNA-binding activity of egr-1, and also exerts an inhibitory effect on NADPH oxidase activity, reducing ROS production. Adiponectin also activates AMPK, an effect counteracted by ethanol, a key step in ethanol-mediated liver disease, since inhibition of AMPK by ethanol promotes enhanced fatty acid synthesis and decreased oxidation, thus leading to liver steatosis (Sozio and Crabb 2008). However, ethanol inhibits adiponectin production by adipose tissue and adiponectin effects on the liver, by blocking adiponectin receptors.

TNF-α and IL-6 are directly involved in the pathogenesis of alcoholic hepatitis and, at least in the case of TNF-α, in haemodynamic alterations: treatment with infliximab leads to a marked decrease in portal hypertension (Mookerjee *et al.* 2003). In addition, serum levels of pro-inflammatory cytokines may have prognostic value (Fukui 2005). They attract neutrophils and macrophages, which contribute to the inflammatory response and also synthesize more TNF-α and IL-8. Moreover, ethanol also induces hepatocytes to secrete IL-8, a potent chemoattractant for neutrophils and monocytes. Infiltration by neutrophils is further enhanced by LPS- and ethanol-mediated increased liver synthesis of osteopontin, another chemoattractant, and by IL-17, levels of which are significantly higher in patients with alcoholic liver disease, especially in the more severe cases (Lemmers *et al.* 2009). This is accompanied by up-regulation of adhesion molecules, such as E-selectin, ICAM-1 and VCAM-1 (Sacanella and Estruch 2003). Increased free-radical production derived from phagocytic activity aggravates the scenario, enhancing cytokine production and creating a positive feedback loop. Free radicals lead to lipid peroxidation, including cell membrane components, and form adducts with acetaldehyde and proteins, leading to increased hepatocyte damage and triggering an immune response. Antioxidant response is therefore essential in the defence against ethanol-induced liver injury, although treatment of alcoholic hepatitis with antioxidants has, as yet, not proved successful.

Regulatory, Th-1 and Th-2 Cytokines

Hepatocytes and Kupffer cells are able to secrete IL-10, an anti-inflammatory cytokine, levels of which are increased in alcoholic hepatitis, although not enough to counteract the pro-inflammatory effects of TNF-α , as shown by Naveau *et al.* (2005) in a study on 83 patients, 41 of whom had severe disease. We have also observed raised IL-10 levels in cirrhotics compared with non-cirrhotics ($Z=2.41$, $p=0.006$) and controls ($KW=6.24$, $p=0.044$, Fig. 2). IL-6 may also exert some protective effects, since it may increase PPAR-α activity, which increases the export of VLDL, up-regulating MTP, leading to decreased fat accumulation. In addition, IL-6 inhibits apoptosis, although ethanol renders the liver less sensitive to the anti-apoptotic action of IL-6. Whether the net effects of IL-6 are protective or not for the liver, they are higher in cirrhotics (Fig. 3) and related with prognosis, as we observed in a study of 98 alcoholic patients followed during a median period of 60 months (Fig. 4).

In addition to TNF-α, IL-6, IL-8 and IL-10, Th-1 cytokines are also secreted, at least in early stages. Possibly, Kupffer cell stimulation by endotoxin induces the secretion of IL-12, which promotes a shift of the immune system to a Th-1-type cytokine secretion pattern. It is unclear why, later in the disease, a Th-2 cytokine pattern develops. Some data indicate that IL-12 secretion decreases with time, so that in more advanced stages of alcoholic liver disease an increase of cytokines such as IL-13, IL-15 and TGF-β1 ensues, whose effects on collagen synthesis were commented on before.

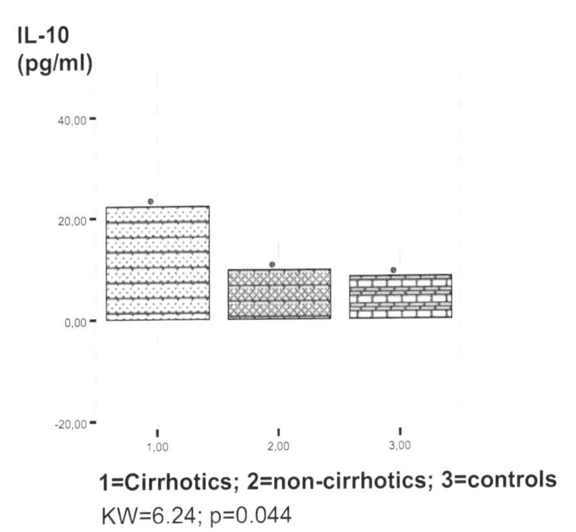

Fig. 2 IL-10 in cirrhotics, non-cirrhotics and controls. Interleukin-10 is a regulatory, anti-inflammatory cytokine, whose values (mean and standard error) were considerably higher in cirrhotics (46 cases) than in non-cirrhotics (64 cases) and controls (24 cases). Despite its anti-inflammatory action, elevated IL-10 in cirrhotics is not enough to counteract the effects of the pro-inflammatory cytokines (unpublished results of personal observations). KW=Kruskall-Wallis.

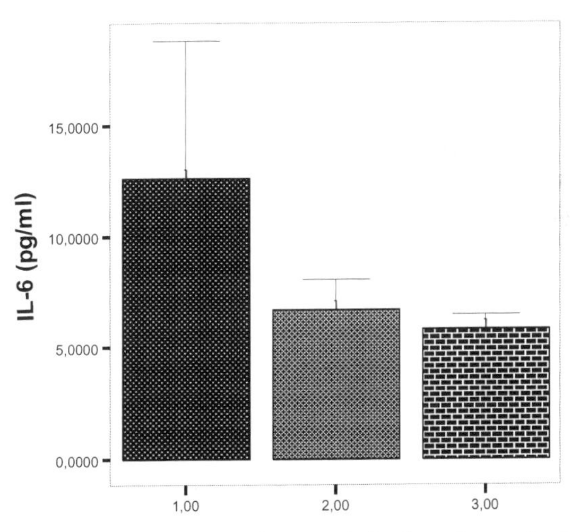

Fig. 3 IL-6 in cirrhotics, non-cirrhotics and controls. Intereleukin-6 is a cytokine, with both pro-inflammatory and liver-protective effects. This figure shows considerably higher (mean and standard error) values in cirrhotics (46 cases) than in non-cirrhotics (64 cases) and controls (24 cases) (unpublished results of personal observations). KW=Kruskall-Wallis.

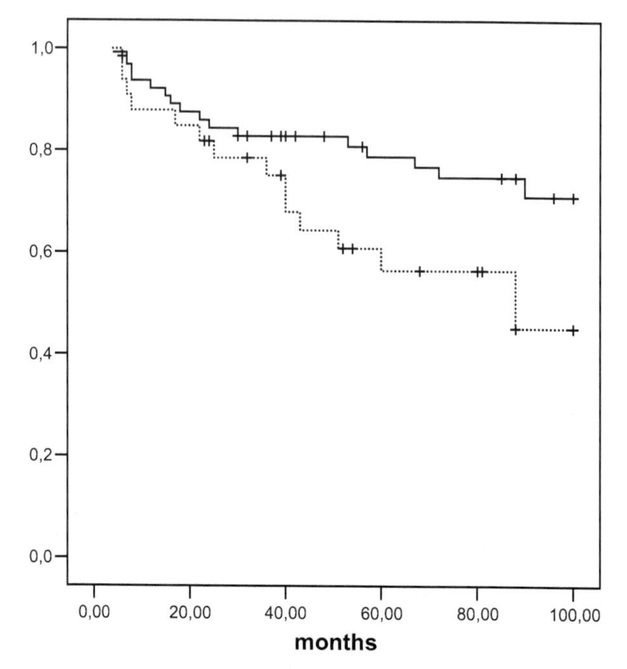

Fig. 4 IL-6 and survival in 98 alcoholic patients, classified according to IL-6 values above (broken line) or below (solid line) the median. Log rank p = 0.021. Whichever the predominant effect (pro-inflammatory or protective), this figure clearly shows that higher values are associated with worse prognosis (unpublished results of personal observations).

Leptin and Other Adipokines

Leptin regulates fat mass by decreasing food intake and increasing resting energy expenditure, but also exerts a pro-inflammatory effect and may increase the intensity of liver damage in alcoholics. Leptin promotes Th-1 immune response, increases macrophage production of IL-6 and TNF- α, and activates neutrophils (Guzik *et al.* 2006). In a study of 79 alcoholics, serum leptin levels were decreased, especially among those with anorexia and weight loss, even after adjusting for the amount of fat, and showed a close correlation with total fat (Santolaria *et al.* 2003). In addition, especially in non-alcoholic steatohepatitis, leptin exerts a pro-fibrogenic effect, which highlights the hazards of the co-existence of obesity and underlying liver disease (Marra and Bertolani 2009). Increased levels of another adipokine, resistin, were found in severe alcoholic hepatitis, as well as in other forms of liver disease; resistin exerts pro-inflammatory effects, increasing the expression of IL-8 and MCP-1. Although some studies suggest a role of the newly described adipokines visfatin and apelin in non-alcoholic steatohepatitis, their effects in alcoholic hepatitis are still unclear.

Cytokines in Stable Alcoholics

Most studies showing increased cytokine levels and/or increased production by the liver were performed in patients with acute alcoholic hepatitis, but their role in stable alcoholics has been less studied. In the recovery phase of severe alcoholic hepatitis, IL-6 and IL-8 tend to decrease (Fukui 2005) in survivors. Ethanol may contribute to cytokine production by other tissues including liver, lung, and brain, which may contribute to long-term organ dysfunction (Crews *et al.* 2006), and, probably, to deranged immune response and susceptibility to infection, a situation that may persist for months after abstinence. Protein malnutrition, a frequent condition in alcoholics, is associated with higher pro-inflammatory cytokine levels, such as TNF-α, IL-1 and IL-6, but adipose tissue may also serve as a continuous source of inflammatory cytokines, thus contributing to progression of liver damage and to muscle atrophy. Interleukin-6, an adipokine, was found to be increased in 77 male alcoholics with stable disease, showing a correlation with body fat stores (González-Reimers *et al.* 2007).

ALTERED IMMUNE RESPONSE

Monocytes respond in opposite ways to acute and chronic ethanol consumption, leading to decreased or enhanced cytokine production, respectively (Crews *et al.* 2006). However, after co-stimulation of different toll-like receptors (i.e., TLR-2 and TLR-4), ethanol led to an increase in TNF-α secretion by monocytes. Moreover, acute administration of moderate doses of ethanol led to a decrease in the secretion of pro-inflammatory cytokines, but also to an increase in IL-10 secretion. These alterations may have consequences in the defence against bacteria. Impaired cytokine regulation renders the lung of the alcoholic more susceptible to acute respiratory distress, a condition for which alcoholism carries a relative risk of 3.7 (1.83-7.71; Moss *et al.* 2003). Normally, alcohol consumption alone would not alter alveolar permeability, but it does increase the expression of TGF-β in rats (Bechara *et al.* 2004), without secretion into the alveolar space. However, after an endotoxemia challenge in that study, there was a five-fold increase of TGF-β into the alveolar space of alcoholized rats, inducing oedema. The release of TGF-β could be inhibited by the addition of glutathione. Furthermore, TGF-β inhibits both Th-1 and Th-2 subsets, and may therefore hamper the defensive immune response. In addition, ethanol may also inhibit GM-CSF expression in the lung of experimental animals, a factor that is essential for activating and priming alveolar macrophage function, thus leading to a blunted defensive response.

PANCREATITIS

Controversy exists regarding a specific effect of ethanol in altered cytokine pattern associated with pancreatitis. Increased secretion of IL-6 and TNF-α plays a key role in the pathophysiology of acute pancreatitis, and high serum levels of both are observed in acute alcoholic pancreatitis, but no higher than those observed in gallstone-associated pancreatitis, thus suggesting that alcohol does not specifically alter the cytokine pattern in acute pancreatitis (except for IL-8). Recent research suggests that progressive fibrous tissue deposit in chronic pancreatitis may be due to a direct effect of ethanol on pancreatic stellate cells, but also to the activation of pancreatic myofibroblasts by cytokines, especially TGF-β1. In a study of 109 patients, Yasuda *et al.* (2008) reported significantly increased serum TGF β-1 and fractalkine—a soluble chemoattractant able to recruit monocytes, but not MCP-1, in the 52 patients with chronic alcoholic pancreatitis compared with the 57 non-alcoholic patients. Fractalkine levels increased early in the course of the disease and were proposed as a diagnostic tool for early stage chronic pancreatitis.

BRAIN

Brain damage may be also related to cytokine secretion. Peripheral endotoxemia results in brain inflammation and neuronal activation, which may be due to increased cytokine secretion. Endotoxin receptor signalling molecules such as CD14 and TLR4 are within the brain and could explain a direct action of LPS, but LPS also induces secretion of many growth factors and cytokines, such as TNF-α, IL-1 β, IL-10, MCP-1, IL-6, and IL-12. The different behaviour of TNF-α after alcohol exposure, with rapid decrease in liver and serum but persistence in brain, may be related to local synthesis in the brain of TNF-α, or to the absence of regulatory T cells and anti-inflammatory cytokines in the brain, a feature that allows TNF-α to persist for up to 10 months after an LPS challenge, inducing neuroinflammation and demyelination. These effects could be especially important in binge drinkers (Crews *et al.* 2006). TNF-α potentiates glutamate excitotoxicity, linked to excessive glutamate activation of N-methyl-D-aspartate (NMDA) receptor. TNF-α reduces glial glutamate transporter activity and thus may also play a role in neurodegeneration (Zhou and Crews 2005), but the effects may also extend to other aspects: increased glutamate is related to an increased desire to consume ethanol. Therefore, increased TNF-α would be related not only to brain damage, but also to alcohol dependence.

Alcohol-induced neurodegeneration is due to both neuronal death and impaired neuronal regeneration. Both mechanisms have been proposed to explain disruption of executive frontal cortical function, thus leading to impulsive behaviour and loss of control, leading to more alcohol consumption, creating a hyperglutamatergic state and closing a positive feedback loop.

MUSCLE

Chronic alcoholic myopathy is a frequent alteration in alcoholics. Muscle is a source of myokines, but it is also a target for the action of other cytokines. Increased TNF-α levels impair skeletal and cardiac muscle protein synthesis (Lang *et al.* 2002) and also increase protein catabolism via the ubiquitin/proteasome system. However, the effects of TNF-α are complex and modulated by other cytokines: IL-6, IFN-γ, and IL-1β all facilitate TNF-α-induced cachexia, whereas IL-15 and leptin inhibit this effect. Increased apoptosis has also been described (Fernández-Solá *et al.* 2003). TNF-α exerts its action by binding to both kinds of receptors, TNF receptor 1 (p55) and TNF receptor 2 (p75), activating the nuclear factor κB pathway in muscle, an effect strongly potentiated by IFN-γ but also influenced by type 1 interferons. In a certain way, the effect of TNF-α and other interleukins is opposite to that of growth factors, such as insulin, IGF-1 and leptin, promoters of protein synthesis and muscle cell growth and differentiation. Oxidative stress blunts the response to these growth factors. The role of pro-inflammatory cytokines on muscle atrophy is supported by the finding of an inverse association between TNF-α and IL-6 and muscle mass in elderly individuals (Kanapuru *et al.* 2009). Regarding non-cirrhotic alcoholics, in a study of 55 patients, TNF-α was inversely related to lean mass (Fig. 5), especially at the legs. As expected, close correlations were observed between lean mass and IGF-1 levels in alcoholics (Fig. 6).

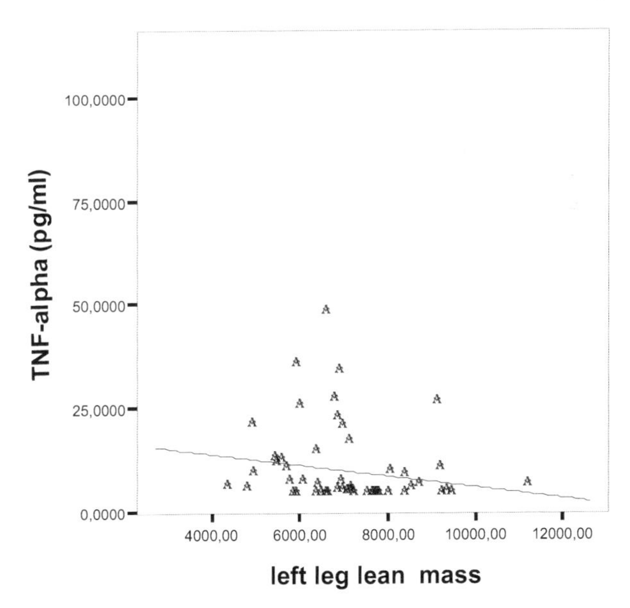

Fig. 5 Correlation between left leg lean mass and serum TNF-α levels (rho = -0.03, p = 0.015). Tumor necrosis factor (TNF)-α (initially termed cachectin) exerts an important effect on muscle protein synthesis and breakdown and also induces apoptosis of muscle fibres. The negative relation between left leg lean mass and TNF-α in 55 non-cirrhotic alcoholics is shown in this figure (unpublished results of personal observations).

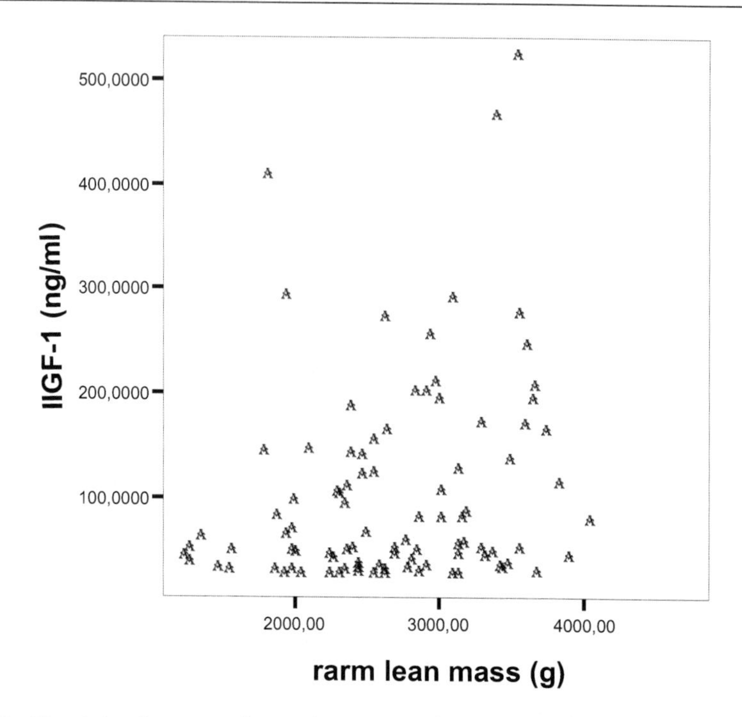

Fig. 6 Correlation between right arm lean mass and IGF-1 levels ($r = 0.25$, $p = 0.01$). This figure shows a positive, significant relationship between insulin-like growth factor type 1 (IGF-1) and right arm lean mass. Unlike TNF-α, IGF-1 promotes growth of muscle fibres; therefore, the significant relationship shown in this figure (106 alcoholics) is in accordance with the physiological effects of this hormone (unpublished results of personal observations).

BONE

In bone there is also a complex interplay between pro-inflammatory and anti-inflammatory cytokines and bone remodeling. Osteoprotegerin, RANK and RANK-L are involved in the regulation of osteoclastogenesis. Activated RANK promotes osteoclastogenesis, but binding of osteoprotegerin to RANK-L prevents osteoclast activation and preserves bone.

Several pro-inflammatory cytokines and hormones modulate bone homeostasis acting on the osteoprotegerin/RANK-L system, and they become altered in alcoholism. Estrogens lead to increased osteoprotegerin levels, but cause no change in RANK-L levels; *in vitro*, osteoprotegerin production is up-regulated by estrogens. An opposite effect is observed for androgens. Vitamin D promotes osteoclastogenesis by reciprocally up-regulating the expression of RANK-L and down-regulating that of osteoprotegerin. Cytokines, such as TNF-α and IL-6, which regulate the osteoprotegerin/RANK-L system (Yamada *et al.* 2002), also become altered in alcoholics. Ethanol stimulates IL-6 production, and IL-6 activates osteoclastogenesis via induction of RANK-L (Dai *et al.* 2000). Thus, in addition to hormone alterations, this is another way whereby ethanol may be associated with osteoporosis. It is therefore not surprising that osteoprotegerin levels increased in 77 alcoholic subjects, especially in cirrhotics, showing a strong relation with TNF-α and IL-6. In accordance with other observations, this suggests that osteoprotegerin levels increase in order to compensate for excessive bone loss (Gracía-Valdecasas-Campelo *et al.* 2006). The effects of other cytokines, such as IL-4, IL-8, IL-13, IL-15, IL-17 and IL-10 on ethanol-mediated bone alterations are less consistent.

CONCLUSION

Cytokines play outstanding roles in several organic complications of the alcoholic, some of which are reviewed in this chapter. Despite the considerable bulk of knowledge about the effect of cytokines on some organs, such as the liver, anti-cytokine treatment trials in alcoholic hepatitis have had disappointing results, suggesting that intensive research is still needed to fully elucidate the role of these inflammatory modulators in alcohol-related organic dysfunction.

APPLICATIONS IN OTHER AREAS OF HEALTH AND DISEASE

Ethanol abuse affects virtually any organ. Recently, considerable research has highlighted the paramount importance of cytokines in the pathogenesis of most alterations in the alcoholic patient. Especially remarkable are the effects of increased TNF-α on neurocognitive impairment and organic brain damage, considering that it may particularly affect adolescents and young adults. Also important is the relation between adipokines and liver damage, which may also affect obese, non-alcoholic individuals. In addition, cytokine-mediated muscle atrophy, a common and sometimes devastating complication in the alcoholic, may share some features with other forms of muscle wasting. Although our knowledge is still incomplete, a relation may exist between cytokine-mediated alterations in atherosclerosis and heavy chronic alcohol consumption. Despite these facts, the results of trials using anti-cytokine products are inconclusive. Therefore, further research is needed in this area, which constitutes a serious health problem.

Table 1 Key features of alcoholism

- Alcoholism is any condition that results in the continued consumption of alcoholic beveraged, despite health problems and negative social consequences.
- Weekly amount of ethanol potentially leading to organic problems: 21 drinks for men and 14 for women (approximately 40 g and 25 g/day, respectively).
- Alcohol metabolism produces acetaldehyde, a highly reactive toxic compound, via two main pathways: alcohol-dehydrogenase and microsomal ethanol oxidizing system, inducible by chronic alcohol consumption.
- Tolerance is a state in which a subject needs progressively larger doses to achieve the same effects.
- Physical dependence is a state in which the patient needs ethanol to feel well, so that symptoms of withdrawal result from abrupt discontinuation or dosage reduction.
- Addiction refers to the compulsive psychological necessity to consume alcohol.
- Organic complications of alcoholism affect virtually any organ, especially the liver (cirrhosis), brain (brain atrophy, neurocognitive impairment), heart (dilated cardiomyopathy, arrhythmia), muscle (muscle atrophy, rhabdomyolysis), pancreas (acute and chronic pancreatitis), gonads (hyopogonadism) and bone (osteoporosis) and also increase the frequency of certain kinds of cancer.

Table 2 Main effects of cytokines on selected organs

Organ	*Proposed Main Mechanism*	*Main Cytokine Involved*	*Alteration*
	Increased intestinal permeability		
	Increased circulating endotoxin		
Liver	Stimulation of Kupffer cells via toll-like receptors	TNF-α	
	Increased cytokine production by Kupffer cells	IL-6	ALCOHOLIC
	Increased lipid peroxidation		HEPATITIS
	Activation of Th-1 immune cells Neutrophils recruitment		
	Increased proinflammatory cytokine secretion	Th-1 cytokines (IFN-γ)	
	With time, activation of Th-2 and regulatory T cells	TGF-β	LIVER FIBROSIS AND CIRRHOSIS
Muscle	Increased apoptosis	TNF-α	MUSCLE
	Impaired protein synthesis		WASTING
	Increased catabolism		
Bone	Increased resorption	TNF-α	OSTEOPENIA
	Impaired synthesis	Direct effect of ethanol. Possible effect of cytokines (IL-10, IL-13)	
Brain	Demyelination	Direct effect of ethanol	BRAIN ATROPHY
	Neurodegeneration		COGNITIVE IMPAIRMENT
	Impaired regeneration	Cytokines (especially TNF-α)	
	Functional alteration		DEPENDENCE?

IL, interleukin; Th, T-helper; IFN, interferon; TNF, tumor necrosis factor.

Summary Points

- Ethanol-mediated increased intestinal permeability causes endotoxemia, which stimulates Kupffer cells to produce pro-inflammatory cytokines and initiates a pro-inflammatory cascade of events including lipid peroxidation, neutrophils and immune cells recruitment and activation, liver steatosis, hepatocyte necrosis and apoptosis, and, ultimately, liver fibrosis.
- Increased TNF secretion may be responsible for distant organ damage, such as in bone, muscle or the brain, but ethanol and/or acetaldehyde may independently cause end-organ damage and induce further cytokine secretion.
- Alcoholic hepatitis is considered a TNF-α-mediated disease. It represents the earliest stage of alcohol-mediated liver injury, although in the heavy drinker it may be additional to liver cirrhosis, being associated with high mortality.
- Liver cirrhosis is the chronic form of alcoholic liver disease, characterized by fibrosis and architectural disorganization of the liver, leading to portal hypertension and liver failure. TGF-β plays a major role in progressive fibrous tissue deposition in the liver.
- Chronic brain damage associated with excessive ethanol consumption includes cerebellar degeneration, central pontine myelinolysis, cortical and subcortical atrophy, toxic amblyopia, pellagra, Wernicke-Korsakoff encephalopathy, Marchiafawa-Bignami disease, and increased prevalence of stroke and cerebral trauma. TNF-α may play a main role in these alterations
- TNF-α is related to decreased muscle protein synthesis, increased breakdown and apoptosis of muscle cells, leading to muscle atrophy.
- Normal bone mass is the result of an equilibrium between bone synthesis and bone resorption. Ethanol directly affects bone formation, but cytokines, especially TNF-α and IL-6, increase bone resorption, leading to decreased bone mass.

Abbreviations

AMP : adenosine monophosphate
AMPK : AMP-activated protein kinase
CD : cluster of differentiation
DNA : deoxyribonucleic acid
Egr-1 : early growth response factor -1
GM-CSF : granulocyte/monocyte colony stimulating factor

ICAM	:	intercellular adhesion molecules
IGF	:	insulin-like growth factor
IFN	:	interferon
IL	:	interleukin
LPS	:	lipopolysaccharide
MCP	:	monocyte chemoattractant protein
MTP	:	microsomal triglyceride transfer protein
NADPH	:	reduced nicotinamide adenine dinucleotide phosphate
NMDA	:	N-methyl-D-aspartate
PPAR	:	peroxisome proliferator-activated receptors
RANK	:	receptor activator of nuclear factor κ B
RANKL	:	receptor activator of nuclear factor κ B ligand
ROS	:	reactive oxygen species
TGF	:	transforming growth factor
Th	:	T-helper
TLR	:	toll-like receptor
TNF	:	tumor necrosis factor
VCAM	:	vascular cell adhesion molecule
VLDL	:	very low density lipoprotein

Definition of Terms

Antioxidants: Enzymes or cofactors of enzymatic pathways that transform reactive oxidant species into less reactive ones.

Apoptosis: Also known as prpgrammed cell death, a series of biochemical events that lead to changes in the cell membrane, cell shrinkage, and nuclear and DNA fragmentation, without inflammatory reaction, and disposal of cellular debris that does not damage the organism.

Innate immune system: Non-specific, first step defence against micro-organisms. It includes skin and mucosal barriers with secretion of many substances including mucus and the so-called natural peptide antibiotics, ciliated cells of the mucosal surfaces, phagocytes, and proteins such as acute phase reactants, complement and non-γ interferons (or type-1 interferons).

Oxidative damage: De-naturalization of different molecules (DNA, lipids, enzymes) by reaction with highly reactive oxygen species. These are powerful triggers of cytokine secretion, alter cell structure and function, and may be carcinogenic or immunogenic.

Th-1, Th-2 and regulatory T cells: CD4+ lymphocytes that exert their action by secreting different patterns of cytokines. Interferon-γ (a macrophage activator) is the main cytokine of Th-1 cells, IL-4 (involved in allergic reactions) is the main cytokine secreted by Th-2 cells, whereas IL-10 and TGF-β are the main cytokines secreted by regulatory T cells.

Toll-like receptors: Cell surface receptors abundantly expressed especially in macrophages and antigen-presenting cells. They recognize and become activated by pathogen-associated molecular patterns of different micro-organisms and play a key role in the so-called innate immune response.

References

Balkwill, F.R. and F. Burke. 1989. The cytokine network. Immunol. Today 10: 299-304.

Bechara, R.I. and L.A. Brown, J. Roman, P.C. Joshi, and D.M. Guidot. 2004. Transforming growth factor beta 1 expression and activation is increased in the alcoholic rat lung. Am. J. Resp. Crit. Care Med. 170: 188-194.

Bird, G.L. and N. Sheron, A.K. Goka, G.J. Alexander, and R.S. Williams.1990. Increased plasma tumor necrosis factor in severe alcoholic hepatitis. Ann. Intern. Med. 112: 917-920.

Cariello, R. and A. Federico, A. Sapone, C. Tuccillo, V.R. Scialdone, A. Tiso, A. Miranda, P. Portincasa, V. Carbonara, G. Palasciano, L. Martorelli, P. Esposito, M. Carteni, C. del Vecchio Blanco, and C. Loguercio. 2009. Intestinal permeability in patients with chronic liver disease: its relationship with the aetiology and the entity of liver damage. Digest. Liver Dis. Doi: 10.106/j.dld.2009.05.001.

Crews, F.T. and R. Bechara, L.A. Brown, D.M. Guidot, P. Mandrecker, S. Oak, L. Qin, G. Szabo, M. Wheeler, and J. Zou. 2006. Cytokines and alcohol. Alcohol. Clin. Exp. Res. 30: 720-730.

Dai, J. and D. Lin, J. Zhang, P. Habib, P. Smith, J. Murtha, Z. Fu, Z. Yao, Y. Qi, and E.T. Keller. 2000. Chronic alcohol ingestion induces osteoclastogenesis and bone loss through IL-6 in mice. J. Clin. Invest. 106: 887-895.

Fernández-Solà, J. and J.M. Nicolás, F. Fatjó, G. García, E. Sacanella, R. Estruch, E. Tobías, E. Badia, and A. Urbano-Márquez. 2003. Evidence of apoptosis in chronic alcoholic skeletal myopathy. Hum. Pathol. 34: 1247-1252.

Fukui, H. 2005. Relation of endotoxin, endotoxin binding proteins and macrophages to severe alcoholic liver injury and multiple organ failure. Alcohol. Clin. Exp. Res. 29(11 Suppl): 172S-179S.

García-Valdecasas-Campelo, E. and E. González-Reimers, F. Santolaria, M.J. de la Vega-Prieto, A. Milena-Abril, M.J. Sánchez-Pérez, A. Martínez-Riera, and M.A. Gómez-Rodríguez. 2006. Serum Osteoprotegerin and RANKL levels in chronic alcoholic liver disease. Alcohol Alcohol. 41: 261-266.

González-Reimers, E. and E. García-Valdecasas, F. Santolaria-Fernández, M.J. de la Vega-Prieto, R. Ros Vilamajó, A. Martínez-Riera, and M. Rodríguez-Gaspar. 2007. Pro-inflammatory cytokines in stable chronic alcoholics: relationships with lean and fat mass. Food Chem. Toxicol. 45: 904-909.

Guzik, T.J. and D. Mangalat, and R. Korbut. 2006. Adipocytokines novel link between inflammation and vascular function? J. Physiol. Pharmacol. 57: 505-528.

Kanapuru, B. and W.B. Ershler. 2009. Inflammation, coagulation and the pathway to frailty. Am. J. Med. 122: 605-613.

Lang, C.H. and R.A. Frost, A.C. Nairn, D.A. McLean, and T.C. Vary. 2002. TNF-alpha impairs heart and skeletal muscle protein synthesis by altering translation initiation. Am. J. Physiol. Endocrinol. Metab. 282: E336-347.

Lemmers, A. and C. Moreno, T. Gustot, R. Maréchal, D. Degré, P. Demetter, P. de Nadai, A. Geerts, E. Quertinmont, V. Vercruyss, O. Le Moine, and J. Devière. 2009. The interleukin-17 pathway is involved in human alcoholic liver disease. Hepatology 49: 646-657.

Marra, F. and C. Bertolani. 2009. Adipokines in liver diseases. Hepatology 50: 957-969.

Mookerjee, R.P. and S. Sen, N.A. Davies, S.J. Hogdges, R. Williams, and R. Jalan. 2003. Tumour necrosis factor-alpha is an important mediator of portal and systemic haemodynamic derangements in alcoholic hepatitis. Gut 52: 1182-1187.

Moss, M. and E.L. Burnham. 2003. Chronic alcohol abuse, acute respiratory distress syndrome, and multiple organ dysfunction. Crit. Care 31: S207-212.

Naveau, S. and A. Balian, F. Capron, B. Raynard, D. Fallik, H. Agostini, L. Grangeot-Keros, A. Portier, P. Galanaud, J.C. Chaput, and D. Emilie. 2005. Balance between CT pro- and anti-inflammatory cytokines in patients with acute alcoholic hepatitis. Gastroenterol. Clin. Biol. 29: 269-274.

Pedersen, B.K. and M.A. Febbraio. 2008. Muscle as an endocrine organ: focus on muscle-derived interleukin-6. Physiol. Rev. 88: 1379-1406.

Pritchard, M.Y. and L.E. Nagy. 2005. Ethanol-induced liver injury: potential roles for egr-1. Alcohol. Clin. Exp. Res. 29(11 Supplement): 146S-150S.

Sacanella, E. and R. Estruch. 2003. The effect of alcohol consumption on endothelial adhesion molecule expression. Addict. Biol. 8: 371-378.

Santolaria, F. and A. Pérez-Cejas, M.R. Alemán, E. González Reimers, A. Milena, M.J. de la Vega, A. Martínez Riera, and M.A. Gómez Rodríguez. 2003. Low serum leptin levels and malnutrition in chronic alcohol misusers hospitalized by somatic complications. Alcohol Alcohol 38: 60-66.

Sozio, M. and D.W. Crabb. 2008. Alcohol and lipid metabolism. Am. J. Physiol. Endocrinol. Metab. 295: E10-E16.

Trayhurn, P. and I.S. Wood. 2004. Adipokines: inflammation and pleiotropic role of white adipose tissue. Br. J. Nutr. 92: 347-355.

Yamada, N. and S. Niwa, T. Tsujimura, T. Iwasaki, A. Sugihara, H. Futani, S. Hayashi, H. Okamura, H. Akedo, and N. Terada. 2002. Interleukin 18 and interleukin 12 synergistically inhibit osteoclastic bone–resorbing activity. Bone 30: 901-908.

Yasuda, M. and T. Ito, T. Oono, K. Kawabe, T. Kaku, H. Igarashi, T. Nakamura, and R. Takayanagi. 2008. Fractalkine and TGF-β1 reflect the severity of chronic pancreatitis in humans. World J. Gastroenterol. 14: 6488-6495.

Zhou, J.Y. and F.T. Crews. 2005. TNF-α potentiates glutamate neurotoxicity by inhibiting glutamate uptake in organotypic brain slice cultures: neuroprotection by NF-κB inhibition. Brain Res. 1034: 11-24.

CHAPTER 6

Cytokines in the Elderly

Dimitrios Adamis
Research and Academic Institute of Athens
27 Themistokleous Street & Akadimias, Athens, 106 77, Greece
Health Services Research, Institute of Psychiatry, King's College, London, UK

ABSTRACT

There are changes in the immune system and the responsiveness of the system slows with advancing age. It is suggested that dysregulation of cytokines plays a significant role in this alteration of the immune response. Here the role of some cytokines in mortality, longevity, and in two clinical conditions with high prevalence in elderly population, frailty and delirium, is presented and discussed.

The levels of IL-6 and possibly of TNF-α increase with age, while the levels of IL-2 decrease. Similarly, high levels of IL-6 and perhaps of TNF-α are associated with mortality. However, genetic studies have produced conflicting results regarding longevity and cytokines.

Furthermore, there is evidence that frailty, which is characterized by weight loss, low activity, functional decline, slow motor performance and cognitive decline, is associated with high levels of IL-1α, IL-6 and TNF-α.

Finally, in delirium, which is often caused by infections and frequently is a side effect of therapeutic use of cytokines, a close relationship is expected with cytokines. However, only few studies have investigated this relationship and the results are contradictory.

Despite the conflicting results, it seems that levels of cytokines change with age, elevated levels of pro-inflammatory cytokines are associated with frailty in older adults, and they possibly have a direct effect in cognitive and functional decline, while their relationship with delirium needs to be further evaluated. The mechanisms of the above effects are not fully understood and further studies, preferably longitudinal and in groups of cytokines, are needed to elucidate those mechanisms.

INTRODUCTION

As the population ages, the prevalence of age-related disease is becoming higher. Similarly, the immune system changes in the elderly, and the immune responsiveness slows with age. It is suggested that dysregulation of cytokines plays a significant role in this alteration of the immune response. The clinical consequences of this alteration may include increased susceptibility to infectious disease and increased morbidity and mortality. Given that the most frequent disease in the elderly population is dementia, more often of Alzheimer's type, there is a body of evidence that inflammatory mechanisms are linked to AD. Similarly, as cognitive and functional decline are age-related, current research is focused on exploring the relation of cognitive and functional decline (frailty) with inflammation. Finally, in delirium, a very common clinical condition often caused by infections, the inflammatory mechanisms underlying this condition began to be investigated only recently.

In this chapter we will review the current evidence connecting cytokines with mortality and longevity and two frequent clinical conditions in the elderly, frailty and delirium. The relation of cytokines with AD and degenerative disease (Parkinson's disease) are covered in other chapters of this book (Velluto *et al.* and Iarlori *et al.*)

LEVELS OF CYTOKINES IN THE ELDERLY AND THEIR RELATION WITH MORTALITY AND LONGEVITY

In normal situations cytokines are not detectable or they exist in very low levels; their elevation is an indication of inflammation or disease progress. However, the secretion of some cytokines seems to be age-related. Below the most often investigated age-related cytokines are discussed.

Interleukin 6 (IL-6)

It is observed that healthy older people have higher serum levels of IL-6 compared to young adults and those levels are positively correlated with age. Harris *et al.* (1999) reported an increased level of IL-6 of 0.016 pg/ml per year of life. Also Stowe *et al.* (2009) in a large population-based study found the same results though not confirmed by all studies. It is possible that the relationship is with undetected disease rather than age. However, Daynes *et al.* (1993) found that lymphoid cells isolated from older adult humans and mice spontaneously produce higher levels of IL-6 than those isolated from younger individuals. This finding suggests that increases in serum IL-6 with age are not just due to undiagnosed disease(s).

A number of studies investigated the association of IL-6 levels with mortality in elderly people. For example, Bruunsgaard *et al.* (2003a) found in a relatively healthy cohort of 80-year-olds that serum levels of IL-6 at baseline were strong predictors of all-cause mortality during the following 6 yr. IL-6 was also a strong predictor of mortality in another study of relatively healthy non-disabled elderly population after a follow-up of an average 4.6 yr (Harris *et al.* 1999). Similarly, in our study (Adamis *et al.* 2007a) we found that elevated IL-6 levels were a predictor of mortality in hospitalized elderly patients. The association of elevated IL-6 with mortality, independently of other risk factors and morbidity, suggests that IL-6 has some more specific biological activities that are not yet known.

On the other hand, the sIL6R increases progressively up to an individual's seventh decade and declines thereafter, reaching, in centenarians, levels equivalent to those of young controls (Sansoni *et al.* 2008). The reduction of sIL6R in centenarians may be a longevity factor, perhaps by modulating the pro-inflammatory activity of IL-6 (Sansoni *et al.* 2008).

Genetic studies that investigated the effect of IL-6 polymorphisms on human longevity have shown rather conflicting results. Increased IL-6 levels occur in GG genotype; thus, if increased IL-6 levels are associated with mortality, then there may be less IL-6 GG homozygotes in older survivors in a population. However, the results of the relevant studies are contradictory. In a recent meta-analysis by Di Bona *et al.* (2009) of eight case-control studies, high heterogeneity and inconclusive results were found among the studies, while in sub-analysis of three studies in an Italian population, an association of longevity and non-GG genotype male carriers was found.

Thus, although clinical studies show that IL-6 levels increase with age and are associated with mortality, genetic studies are not conclusive about the role of IL-6 genotype in longevity. Possibly other environmental factors also play a role in longevity, apart from genotype (Di Bona *et al.* 2009).

Tumor Necrosis Factor alpha (TNF-α)

Levels of TNF-α increase with advanced age (e.g., Bruunsgaard *et al.* 2003a, b). Similarly, Stowe *et al.* (2009) found a strong association of levels of TNF-r1 (a surrogate marker for TNF-α) and advance age. However, not all studies have confirmed those results and the alternative hypothesis is that the slightly elevated levels of TNF-α in the elderly are due to underlying disease(s) like atherosclerosis, osteoporosis and age-associated pathological events and are not the consequences of an independent age-related mechanism.

Like IL-6, TNF-α has been found to be an independent predictor of mortality (Bruunsgaard *et al.* 2003b), but studies that investigated the TNF-α polymorphisms to assess its association with longevity showed negative results (Rea *et al.* 2006).

Thus, there are not yet unambiguous indications that TNF-α levels are elevated in advanced age or that TNF-α is a predictive factor for mortality or longevity.

Interleukins IL-1 (family), IL-2 and IL-10

Both IL-1α and IL-1β are presumed to be pro-inflammatory cytokines and so is IL-2, but IL-1Ra and IL-10 are anti-inflammatory and they counteract pro-inflammatory cytokines.

A number of studies show that the production of IL-1 is elevated in healthy aged subjects as well as in centenarians but again there are studies that found the opposite results. For instance, Cavallone *et al.* (2003) found that IL-1Ra plasma levels showed an age-related increase, whereas IL-1β plasma levels did not show association with age, but Stowe *et al.* (2009) reported that IL-1ra, and IL-10 did not significantly increase with age. Apart from methodological problems it seems that underlying undiagnosed disease(s) may contribute to those discrepancies in the literature; for example, Ferrucci *et al.* (2005) found no correlation between age and IL-1Ra after adjusting for cardiovascular risk factors.

In addition, studies that investigated genotype of the IL-1 family found no evidence of change in IL-1 family gene frequencies as a result of age (e.g., Cavallone *et al.* 2003), suggesting that IL-1 gene does not affect longevity and that the increased levels that were found in some studies are not genetically regulated.

In contrast, most of the studies that examine the IL-2 cytokine report a decrease in IL-2 production that is associated with age. However, two studies (Ross *et al.* 2003, Scola *et al.* 2005) that investigated the IL-2 polymorphism did not find any change in the frequencies of IL-2 genotypes as a function of advanced age.

Regarding the IL-10, few studies have measured circulating levels of this cytokine, but there have been reports that indicate no association of the IL-10 levels with ageing (Stowe *et al.* 2009). On the contrary, the IL-10-1082GG allele (which is associated with high IL-10 production) has been reported to be a male-specific marker for longevity in genetic studies of an Italian population but this finding has not been confirmed by other studies in different populations (Rea *et al.* 2006).

From the above it seems that with few exceptions (IL-6, IL-2) there is no firm evidence that the levels of the cytokines change with advancing age. Although some of the discrepancies of the studies reported above are most likely related to different methodologies, to small sample sizes or to variations of age of the selected subjects, it needs to be noticed that no cytokine works in isolation. For example, TNF-α stimulates IL-6 production, while IL-6 influences the synthesis of IL-1 and TNF-α by inducing the increase of TNF receptors and IL-1Ra, which can subsequently bind and down-regulate the function of TNF-α. Thus, there is a close relationship among cytokines and there is a need for studies to look at the balance between cytokines or groups of cytokines with respect to ageing.

Furthermore, the increased or altered production of some cytokines is perhaps relevant to age-associated pathological events like atherosclerosis, osteoporosis, or undiagnosed disease(s) as has been proposed, but the fact that the same phenomenon is observed in healthy centenarians, who have avoided these pathologies, indicates that the age-related increase in these cytokines could be due to basic abnormalities of the cells or to changes in the microenvironment to which the cells are exposed and not necessarily to systematic diseases. Therefore, no matter what the involved mechanisms are, the abnormal secretion observed in old subjects may be relevant to a number of abnormalities associated with ageing.

Similarly, the predictive value of some cytokines for mortality is perhaps an epiphenomenon of underlying mechanism(s) that lead to death. Although the mechanisms are not known yet, this has clinical and prognostic significance, if fully investigated and further validated. Unfortunately, genetics studies in longevity, which are more powerful and more independent of clinical conditions or environmental influences, have not yet given conclusive results (Tables 1 to 3).

Table 1 Theories of ageing

- Organ- and system-based theories: Failure of one major organ or system contributes to ageing process.
- Genetic base theories: The ageing process is encoded to genes, or is the result of damage to genes, or represents a genetic failure of information flow to produce functional proteins.
- Stochastic theories: Minor random adverse changes occur during the lifetime and the accumulation of those unfavourable changes overwhelms the capacity of an organism to survive.

Table 2 Key points about cytokine levels in ageing

- Serum IL-6 levels increase with age.
- It seems that levels of TNF-α increase with age but not all studies agree on that.
- Studies disagree on whether IL-1 levels change with age.
- IL-2 production decreases with age.
- There are indications of no association of IL-10 levels with ageing.
- Underlying disease(s), changes in the cell environment and genetically determined mechanisms have been proposed as potential causes for alteration of cytokine levels with advancing age, but none of them is fully supported by the findings.

Table 3 Key points of the role of cytokines in mortality and longevity

- High levels of IL-6 are associated with mortality.
- The above holds for TNF-α but not all studies confirm it.
- There is interaction between cytokines, so investigation of an individual cytokine does not necessarily lead to reliable conclusions.
- Although the underlying mechanisms of the association of elevated IL-6 levels with mortality are not known, levels of IL-6 could be used as predictors if this is confirmed by further studies.
- Genetic studies have produced conflicting results regarding longevity and cytokines.

CYTOKINES AND FRAILTY

With advancing age, functional capacity in performing complicated or even self-care activities of daily living is becoming increasingly limited.

Interleukin-1

Studies on inflammatory myopathies have reported that IL-1α promotes muscle proteolysis and is associated with muscle degradation resulting in muscle loss and reduced strength in cases of illness (Winkelman 2004).

IL-1 reduces food intake, and increased IL-1 levels occur in patients with cachexia, even when infection and cancer are not present. Similarly, IL-1 has a direct link in synaptic transmission in the hippocampus, plays a part in the memory process and in high levels appears to have a detrimental role in cognition.

On the other hand, it has been suggested that IL-1α can be an autocrine growth factor, and at moderate levels can induce NGF and therefore play a neuroprotective role (Basu *et al.* 2004). In a study (Adamis 2007) it was found that IL-1α levels have a positive association with functional capacity (those with low IL-1α levels have a better function).

Thus, although IL-1α may have an association with frailty (weight loss, sarcopenia, anorexia, cognitive and functional decline), low levels of IL-1α may have the opposite effects. As there is a "window" for reversibility of frailty, administration of low dose of IL-1 may have therapeutic effects. But before that, specific studies are necessary to further evaluate its role on frailty.

IL-6 and TNF-α

Both IL-6 and TNF-α have been intensively investigated in functional, cognitive, and other frailty-related symptoms. TNF-α causes significant anorexia, increases leptin- and corticotrophin-releasing hormone levels, and contributes to lipolysis, muscle protein breakdown, and nitrogen loss. In humans exogenous administration of IL-6 stimulates the HPA and causes ACTH release.

Most of the studies have found an adverse association between levels of both IL-6 and TNF-α and functional ability (e.g., Bruunsgaard *et al.* 2003a), although the pathophysiological mechanism is not yet established. One suggestion is the effect of occult inflammatory disease but the long follow-up periods make this unlikely. Ageing-associated trends to higher cytokine levels, independent of identifiable disease, may have detrimental effects on the neuroendocrine system, skeletal muscle, CNS, and other systems, possibly through increased production of oxygen free radicals with subsequent organ damage and systemic functional decline. It has been suggested that IL-6 may be a surrogate marker of TNF-α and not directly involved in the functional decline (Krabbe *et al.* 2004).

Furthermore, it was found by Adamis (2007) that high LIF levels (a less often investigated cytokine in frailty) have a positive effect on functional status. Although LIF and IL-6 use the same receptor group and so may have overlapping biological effects, in this study, the relationship of LIF with functional status was the opposite of that of IL-6. This may be because LIF has a dual anti-inflammatory and pro-inflammatory effect.

Although the pathophysiological mechanisms of the frailty process are not yet fully understood, inflammatory mechanisms, and the disturbances of the balance of pro and anti-inflammatory cytokines, could be considered as among the most important physiological correlates of the frailty syndrome (Table 4).

Table 4 Key points of the role of cytokines in frailty

- High levels of IL-1α may have an association with frailty, but low levels of IL-1α may have the opposite effects.
- High levels of IL-6 and TNF-α are associated with functional decline.
- Pathophysiological mechanism(s) of the association of high levels of IL-1α, IL-6 and TNF-α with frailty have not yet been established.

CYTOKINES AND DELIRIUM

Delirium affects 20% or more of the elderly medically ill patients on hospital admission, with another 10% or more developing delirium within the next few days.

Delirium has been seen often in connection to infection and the use of cytokines for treatment of medical conditions frequently causes delirium. Although there is a direct connection between delirium and cytokines, surprisingly few studies have investigated this relationship. In this section studies that investigated this relationship are reported and the possible pathological mechanisms are discussed.

De Rooij *et al.* (2007) found an association of delirium in elderly medical inpatients with increased levels of IL-6. Similar results for the IL-6 and IL-8 were reported from the same group (van Munster *et al.* 2008) in elderly patients with hip fractures, but no association was reported for the cytokines TNF-α, IL-1β, IL-10 and IL-12 with delirium. Lemstra *et al.* (2008) found no relationship between levels of IL-6 in pre-operative and post-operative delirious and non-delirious elderly patients with hip fracture. In addition Adamis *et al.* (2007b) found no differences in the levels of the cytokines IFN-γ, IL-1α, IL-1β, IL-1RA, IL-6, TNF-α and LIF between those elderly medical inpatients with prevalent delirium and those without. However, from the same study, when the data were analysed longitudinally (including both incident and prevalent cases and data from all the assessments) it was found that those with delirium had lower IL-1Ra levels and that IFN-γ levels were associated with severity of delirium (Adamis *et al.* 2009). Although the studies are too few to arrive at firm conclusions about the role of cytokines in delirium, most of the above reported discrepancies in the findings are due to methodological problems (longitudinal vs. cross-sectional) and to the different analyses of the data.

A working hypothesis for the latter finding (that those with delirium have lower IL-1Ra levels) is that deficits in the immunoreactivity of the brain may be associated with delirium. It is possible that high levels or persistence of pro-inflammatory cytokines are not requirements for delirium if there are deficits in the immunoreactivity of the brain (low cerebral reserve). This would explain the observation that delirium occurs more commonly in children and older adults after relatively minor precipitating illnesses but not in other age groups. This hypothesis can be supported from studies on stroke and AD. The production of a small amount of IL-1β but a large amount of IL-1Ra appears to be a natural response, for example, in stroke. The decreased levels of IL-1Ra observed in these studies may suggest impaired balance between the activities of IL-1β and IL-1Ra in the brains of the patients and this imbalance may contribute to decreased inhibition of the intracerebrally produced IL-1β, which leads to harmful effects of this cytokine (Tarkowski *et al.* 2003).

Another hypothesis is that, given that cognitive deficits are a strong predictor for delirium, systemic inflammation act as a stressor that can initiate an acute exacerbation of underlying dementia (MacLullich *et al.* 2008).

Both of the above hypotheses assume a direct effect of cytokines in the brain. The first assumes deficits in the anti-inflammatory cytokines in the brain, while the second assumes over-expression of pro-inflammatory cytokines in the brain.

However, it is possible that the relationship between cytokines and delirium is indirect through another mechanism(s). A further hypothesis is that of activation of HPA axis. As some cytokines (IL-1β, IL-6, IL-1, TNF-α, IFN) can activate the HPA axis, this results in secretion of glucocorticoids, occupation of the low-affinity glucocorticoid receptor, and down-regulation of the inflammatory responses (MacLullich *et al.* 2008).

The studies which investigated the relation of delirium and cytokines are very few to confirm or disconfirm any of the above suggested hypotheses and thus this field needs to be further investigated before we arrive at any conclusions.

Table 5 Key points of the role of cytokines in delirium

- Therapeutic use of cytokines can cause delirium.
- Studies that investigate the role of cytokines in delirium are few and the results are contradictory.
- Methodological problems (cross-sectional studies) and different statistical analyses may account for those discrepancies.
- Speculative pathophysiological mechanisms are a direct effect of high levels of pro-inflammatory cytokines in the brain or a low production of ant-inflammatory cytokines or an indirect effect via other systems such as the HPA axis.

Table 6 Key features of the elderly

- Most developed countries use the chronological age of 65 years as a definition of "elderly".
- This definition is arbitrary and associated with the age at which one can begin to receive pension benefits.
- Chronological and biological age are not necessarily synonymous.
- Biological changes underlie the ageing process but the experience of ageing is strongly influenced by cultural attitudes to older people.
- As life expectancy has increased consistently over the last decades the number of older people is growing.
- There are age-related clinical conditions such as dementia and AD. They also have an impact on health services.

Table 7 Key facts about cytokines in the elderly

- Certain cytokines are involved in mortality and longevity together with environmental factors.
- Cytokines are implicated in age-related disease although there is no evidence of cause-effect mechanisms.
- Therapeutic use of cytokines can cause delirium and certain cytokines are altered in delirium. However, there is no evidence yet to show what is the mechanism behind it.
- Cytokines are potential drugs for alteration of the progress of age–related disease.

Summary Points

- The levels of some cytokines change with age.
- The mechanisms for those changes in the levels of cytokines are not well understood. Underlying age-related or undiagnosed diseases, genetic determination or change in the microenvironment of the cells are some of the proposed explanations.

- IL-6 and perhaps TNF-α are predictors for mortality. Although the mechanisms are not identified, if this is finally confirmed, it may have significant clinical implications.
- High levels of IL-1α, IL-6 and TNF-α are implicated in the frailty of older adults.
- There is a close relationship between cytokines and delirium but there are surprisingly few studies that investigate this relationship.
- It is not yet clear whether this relationship is due to a direct effect of the cytokines in the brain or is an indirect effect through other pathways.

Abbreviations

ACTH : adrenocorticotropic hormone
AD : Alzheimer's disease
CNS : central nervous system
CRP : C-reactive protein
CSF : cerebrospinal fluid
HPA : hypothalamic-pituitary-adrenal axis
IL : interleukin
IL-1Ra : interleukin-1 receptor antagonist
IL-1α : interleukin-1 alpha
IL-1β : interleukin-1 beta
IFN-γ : interferon gamma
LIF : leukaemia inhibitor factor
NGF : nerve growth factor
sIL6R : soluble IL-6 receptor
TNF-α : tumour necrosis factor alpha

Definitions of Terms

Cytokines: Low molecular weight proteins produced by immune-response as well as in other cells. Their elevation is an indication of inflammation or disease progress. Cytokines associated with acute disease are different from those in chronic disease; also, some cytokines may be present in the early stages of a disease while others appear later in the course of disease.

Delirium: A disorder of acute onset, over hours to days, followed by a course of fluctuation in the level of consciousness, attention and cognition, sometimes accompanied by delusions or hallucinations.

Dementia: The loss of mental ability in a previously unimpaired person, severe enough to interfere with normal activities of daily living, lasting more than six months, and not associated with alteration of consciousness.

Frailty: The clinical syndrome characterized by multiple pathologies such as weight loss, fatigue, weakness, low activity, slow motor performance, balance and gait abnormalities and potential cognitive decline.

Longevity: Duration of an individual life beyond the norm for the species.

Acknowledgements

I would like to thank Dr K. Karagianni for helpful comments and correction of the manuscript.

References

Adamis, D. 2007. APOE and cytokines as biological markers of recovery of delirium in elderly medical inpatients. MD Thesis, University of London.

Adamis, D. and M. Lunn, F.C. Martin, A. Treloar, N. Gregson, G. Hamilton, and A.J. Macdonald. 2009. Cytokines and IGF-I in delirious and non-delirious acutely ill older medical inpatients. Age Ageing 38: 326-332.

Adamis, D. and A. Treloar, F.Z. Darwiche, N. Gregson, A.J. Macdonald, and F.C. Martin. 2007a. Associations of delirium with in-hospital and in 6-months mortality in elderly medical inpatients. Age Ageing 36: 644-649.

Adamis, D. and A. Treloar, F.C. Martin, N. Gregson, G. Hamilton, and A.J. Macdonald. 2007b. APOE and cytokines as biological markers for recovery of prevalent delirium in elderly medical inpatients. Int. J. Geriatr. Psychiatry 22: 688-694.

Basu, A. and J.K. Krady, and S.W. Levison. 2004. Interleukin-1: a master regulator of neuroinflammation. J. Neurosci. Res. 78: 151-156.

Bruunsgaard, H. and S. Ladelund, A.N. Pedersen, M. Schroll, T. Jorgensen, and B.K. Pedersen. 2003a. Predicting death from tumour necrosis factor-alpha and interleukin-6 in 80-year-old people. Clin. Exp. Immunol. 132: 24-31.

Bruunsgaard, H. and K. Andersen-Ranberg, J.B. Hjelmborg, B.K. Pedersen, and B. Jeune. 2003b. Elevated levels of tumor necrosis factor alpha and mortality in centenarians. Am. J. Med. 115: 278-283.

Cavallone, L. and M. Bonafe, F. Olivieri, *et al.* 2003. The role of IL-1 gene cluster in longevity: a study in Italian population. Mech. Ageing Dev. 124: 533-538.

Daynes, R.A. and B.A. Araneo, W.B. Ershler, C. Maloney, G.Z. Li, and S.Y. Ryu. 1993. Altered regulation of IL-6 production with normal aging. Possible linkage to the age-associated decline in dehydroepiandrosterone and its sulfated derivative. J. Immunol. 150: 5219-5230.

de Rooij, S.E. and B.C. van Munster, J.C. Korevaar, and M. Levi. 2007. Cytokines and acute phase response in delirium. J. Psychosom. Res. 62: 521-525.

Di Bona, D. and S. Vasto, C. Capurso, *et al.* 2009a. Effect of interleukin-6 polymorphisms on human longevity: a systematic review and meta-analysis. Ageing Res. Rev. 8: 36-42.

Ferrucci, L. and A. Corsi, F. Lauretani, *et al.* 2005. The origins of age-related proinflammatory state. Blood 105: 2294-2299.

Harris, T.B. and L. Ferrucci, R.P. Tracy, *et al.* 1999. Associations of elevated interleukin-6 and C-reactive protein levels with mortality in the elderly. Am. J. Med. 106: 506-512.

Krabbe, K.S. and M. Pedersen, and H. Bruunsgaard. 2004. Inflammatory mediators in the elderly. Exp. Gerontol. 39: 687-699.

Lemstra, A.W. and K.J. Kalisvaart, R. Vreeswijk, W.A. van Gool, and P. Eikelenboom. 2008. Pre-operative inflammatory markers and the risk of postoperative delirium in elderly patients. Int. J. Geriatr. Psychiatry 23: 943-948.

Maclullich, A.M. and K.J. Ferguson, T. Miller, S.E. de Rooij, and C. Cunningham. 2008. Unravelling the pathophysiology of delirium: a focus on the role of aberrant stress responses. J. Psychosom. Res. 65: 229-238.

Rea, I.M. and G. Candore, L. Cavallone, F. Olivieri, *et al.* 2006. Longevity. pp. 379-394. In: K. Vandenbroeck [ed], Cytokine Gene Polymorphisms in Multifactorial Conditions. CRC Taylor & Francis, Boca Raton, Florida.

Ross, O.A. and M.D. Curran, A. Meenagh, *et al.* 2003. Study of age-association with cytokine gene polymorphisms in an aged Irish population. Mech. Ageing Dev. 124: 199-206.

Sansoni, P. and R. Vescovini, F. Fagnoni, *et al.* 2008. The immune system in extreme longevity. Exp. Gerontol. 43: 61-65.

Scola, L. and G. Candore, G. Colonna-Romano, *et al.* 2005. Study of the association with -330T/G IL-2 in a population of centenarians from centre and south Italy. Biogerontology 6: 425-429.

Solerte, S.B. and L. Cravello, E. Ferrari, and M. Fioravanti. 2000. Overproduction of IFN-gamma and TNF-alpha from natural killer (NK) cells is associated with abnormal NK reactivity and cognitive derangement in Alzheimer's disease. Ann. NY Acad. Sci. 917: 331-340.

Stowe, R.P. and M.K. Peek, M.P. Cutchin, and J.S. Goodwin. 2009. Plasma cytokine levels in a population-based study: relation to age and ethnicity. J. Gerontol. A Biol. Sci. Med. Sci. Apr; 65(4): 429-43. Epub 2009 Dec. 16.

Tarkowski, E. and A.M. Liljeroth, L. Minthon, A. Tarkowski, A. Wallin, and K. Blennow. 2003. Cerebral pattern of pro- and anti-inflammatory cytokines in dementias. Brain Res. Bull. 61: 255-260.

Winkelman, C. 2004. Inactivity and inflammation: selected cytokines as biologic mediators in muscle dysfunction during critical illness. AACN Clin. Issues 15: 74-82.

Section III
Immunology and Infections

CHAPTER 7

Cytokine Production by T Helper Subsets in Response to Infection and Their Role in Health and Disease

Courtney C. Kurtz, Jeffrey M. Wilson, Steven G. Black, and Peter B. Ernst*

Department of Medicine, University of Virginia
Box 801321, MR-4, 409 Lane Road, Charlottesville, VA 22908, USA

ABSTRACT

$CD4^+$ T cells, also called T helper (Th) cells, are an important part of the adaptive immune system. A number of distinct $CD4^+$ T cell subsets have been elucidated that presumably evolved to control a diverse array of microbial pathogens. These subsets are adapted to the tissue in which they reside such that they have unique activities and functions that govern different aspects of the immune system. $CD4^+$ T cells exert their control largely via the regulated production and release of cytokines. Although a number of distinct effector $CD4^+$ T cell subsets have been described, the best characterized to date include Th1, Th2, and Th17 cells. Th1 cells are important in protection from intracellular bacteria, viruses and fungi and secrete the cytokines IL-2, IFN-γ, TNF-α and LT-α. Th2 cells are a critical part of the immune response to helminth infections and produce IL-4, IL-5, IL-13 and IL-25. Th17 cells are a more recently characterized group of Th effectors that play a role in clearing extracellular bacteria and other pathogens and also help maintain epithelial barrier integrity. This is largely achieved by secretion of IL-17A, IL-17F and IL-22. Another group of $CD4^+$ T cells includes regulatory T cells (Treg), which control effector Th cells and suppress other inflammatory pathways often by the production of cytokines IL-10 and TGF-β1. Although $CD4^+$ T cells and cytokines

*Corresponding author

have an important role in host defense, uncontrolled or dysregulated Th responses contribute to a number of autoimmune and inflammatory diseases.

INTRODUCTION

$CD4^+$ T cells are lymphocytes of the adaptive immune system that play an important role in host defense. Notable for helping to coordinate and activate other elements of the immune system, especially B cells and $CD8^+$ T cells, $CD4^+$ T cells are also aptly called T helper (Th) cells. $CD4^+$ T cells are derived from pluripotent bone marrow cells that undergo additional maturation and selection in the thymus before emigrating to peripheral tissues. As a function of the biological process known as somatic recombination each of these antigen-inexperienced "naïve" $CD4^+$ T cells possesses a unique T cell receptor. As they circulate throughout the body those cells that recognize antigen in an appropriate stimulatory context, requiring an antigen-presenting cell (APC), undergo further activation and proliferation. Depending on a variety of factors, particularly the cohort of cytokines produced by innate cells, activated $CD4^+$ T cells differentiate into distinct lineages with unique functions and cytokine expression profiles.

The major $CD4^+$ T cell subsets include Th1, Th2, and Th17 effector cells, as well as a population of regulatory T cells (Treg) that exert suppressive effects on the immune system. Distinguished on the basis of cytokine expression, intracellular signaling pathways and transcription factors, these lineages evolved to combat the diverse array of microbial pathogens within the unique niches those microbes occupy in vertebrate hosts. Consequently, Th cell subsets display heterogeneity that reflects their unique effector functions. In addition to being an important component of $CD4^+$ T cell effector function, cytokines play a critical role during the generation of these Th lineages from naïve precursors (Fig. 1). For instance, IL-2 is produced early during $CD4^+$ T cell activation and acts in an autocrine fashion to promote proliferation and viability. IL-12 is produced by APC and favors the production of a Th1 phenotype. Th2 differentiation is not well understood, although IL-4 and TSLP within the microenvironment are believed to be important. With the recent recognition of additional Th subsets it is now clear that other cytokines such as IL-6 and TGF-β1 are also instrumental in controlling $CD4^+$ T cell differentiation. Ultimately, it is likely that $CD4^+$ T cell differentiation is determined by the combinatorial actions of these and other cytokines.

TH1 CELLS AND CYTOKINES

Th1 cells are preferentially generated in response to fungal, viral and intracellular bacterial infections. By activating cytotoxic $CD8^+$ T cells and other inflammatory cells, Th1 cells facilitate the clearance of intracellular pathogens. The transcription factor T-bet, which is highly expressed in Th1 cells and counter-regulates

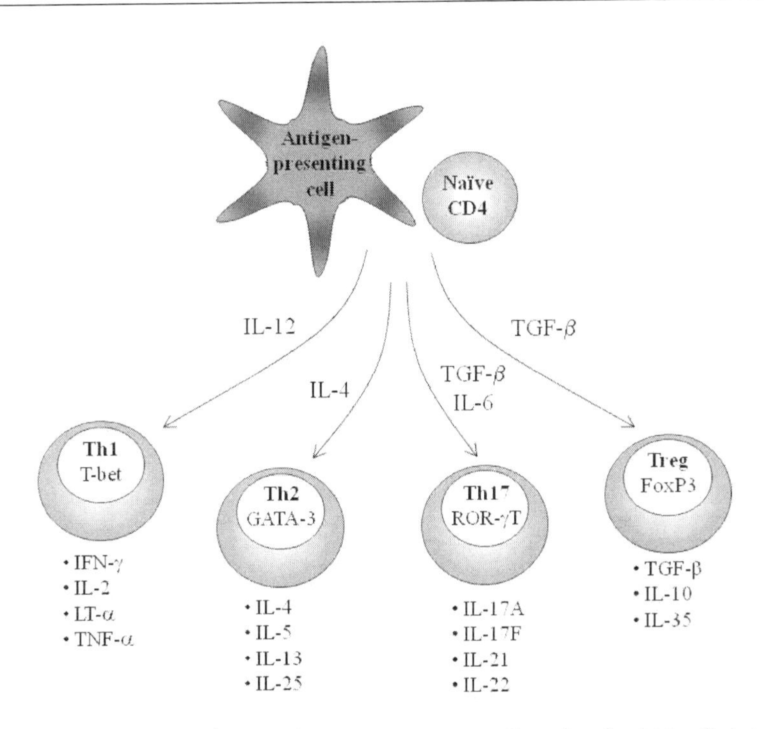

Fig. 1 Explanation of CD4$^+$ T cell subsets. Innate immune cells, such as dendritic cells, interact with naïve CD4$^+$ T (Th) cells. Cytokines, cell surface molecules and other soluble factors derived from antigen-presenting cells (APC) or other innate cells in the microenvironment influence the activation and differentiation of naïve CD4$^+$ T cells into functionally distinct Th subsets. As discussed in the text, cytokines from differentiated Th cells provide a negative feedback on other Th cell subsets while positively selecting for the emerging Th cell phenotype (unpublished diagram by authors).

transcription factors associated with other Th lineages, is often called the "master regulator" of Th1 cells.

The "signature" cytokine produced by Th1 cells is IFN-γ. IFN-γ was initially discovered as an antiviral factor along with IFN-α and IFN-β, although its structure is only slightly similar and its receptor is different from the other IFNs (Billiau and Matthys 2009). Because of these differences, IFN-γ was originally referred to as Type II IFN while IFN-α and -β were referred to as Type I IFNs. Although IFN-γ secretion by Th1 cells does have antiviral capabilities, its ability to activate and recruit leukocytes appears to be more important. IFN-γ production and secretion stimulates a number of pro-inflammatory processes (Fig. 2). These include phagocytosis, recruitment and activation of macrophages, oxidative burst, and upregulation of MHC on APC and other cells (Spellberg and Edwards 2001). These combined effects contribute to the destruction of infected or damaged cells.

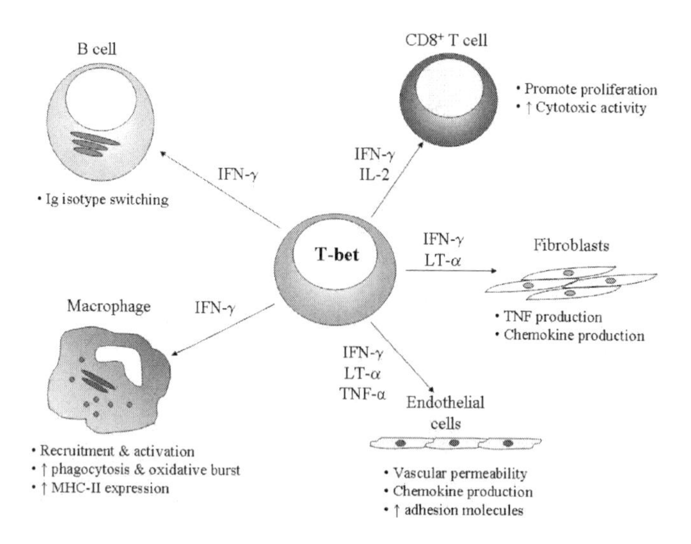

Fig. 2 Th1 cells and cytokines. Th1 cells are characterized by expression of the transcription factor T-bet and the production of the cytokines IFN-γ, TNF-α, LT-α and IL-2. These cytokines promote proliferation and cytotoxic activity of natural killer cells and CD8$^+$ T cells; recruit and activate macrophages; promote isotype switching in B cells; and stimulate production of inflammatory mediators by non-immune cell types such as endothelial cells and fibroblasts. IFN, interferon; TNF, tumor necrosis factor; LT, lymphotoxin; IL, interleukin (unpublished diagram by authors).

In addition, IFN-γ promotes the production of other pro-inflammatory cytokines and chemokines by other non-immune cell types, such as endothelial cells, fibroblasts and keratinocytes (Spellberg and Edwards 2001). IFN-γ plays a helper cell role by directing B cells to switch immunoglobulin isotypes to classes that can activate complement and contribute to cell-mediated immunity. IFN-γ also activates the cytotoxic activity of CD8$^+$ T cells (Billiau and Matthys 2009, Zhang *et al.* 2009). These effects set into motion a cascade of inflammatory events that leads to clearance of the pathogen. Furthermore, production of IFN-γ effectively shuts down the differentiation of naïve precursor cells into Th2 and Th17 cells (Zhu and Paul 2008). This implies a cross-regulation between these subsets of Th cells such that the Th1 subset promotes its own kind at the expense of the others.

In addition to IFN-γ, Th1 cells produce order cytokines such as IL-2, TNF-α, and LT-α (also called TNF-β). The amount of IL-2 produced by Th1 cells is significantly less than that produced by undifferentiated precursor CD4$^+$ T cells. Maintenance of low levels of IL-2 production in Th1 cells is thought to be necessary to develop T cell memory (Zhu and Paul 2008). TNF-α, although not uniquely produced by Th1 cells, is an important pro-inflammatory cytokine that is released early in the inflammatory cascade and plays an important role in recruitment and activation of immune cells (Feldmann and Maini 2001). LT-α stimulates adhesion molecule expression on endothelial cells, allowing attachment and extravasation of a variety of immune cells into the tissue and the subsequent clearance of pathogen (Cuff *et al.* 1998). LT-α also provokes pro-inflammatory cytokines and chemokines from non-immune cells.

Th1 cells are essential for the control of certain infectious diseases, especially intracellular bacteria and viruses. Control of *Mycobacterium* spp. infection, including that of the tuberculosis-causing organism, *Mycobacterium tuberculosis*, has long been attributed to an effective Th1 response (Billiau and Matthys 2009, Roach *et al.* 2001). Similar Th1-mediated control of infection has been noted with *Listeria monocytogenes* and *Salmonella enterica* serovar Typhimurium (Spellberg and Edwards 2001). A Th1 response to parasitic infection with *Leishmania* spp. protozoans, including *Leishmania major* and *Leishmania donovani*, clears the infection quickly. Th1 cells are also effective at ridding the body of fungal infections, including those caused by *Candida* spp., *Cryptococcus* spp. and *Aspergillus* spp. (Spellberg and Edwards 2001). As mentioned earlier, IFN-γ was discovered as an antiviral agent. Th1 cells are therefore effective against virus-infected cells as well. Interestingly, human immunodeficiency virus (HIV) attacks $CD4^+$ T cells and appears to force a shift in Th1/Th2 balance away from the antiviral Th1 cells, effectively promoting infection (Spellberg and Edwards 2001).

TH2 CELLS AND CYTOKINES

A second $CD4^+$ T cell subset known as Th2 cells produce a unique set of cytokines that regulate the function of B cells, eosinophils, basophils, and macrophages and are generally thought to promote defense against extracellular parasites (Anthony *et al.* 2007) (Fig. 3). The transcription factor GATA-3 is expressed in a variety of cell types but has been shown to be especially important in Th2 cell development (Ho *et al.* 2009).

Multiple factors have been reported to promote Th2 development. These include the nature of the stimulus, the inflammatory response within the microenvironment and the selection of APC that favor Th2 cell induction (Perrigoue *et al.* 2008). More recently, basophils have been implicated in the selection of Th2 cells through the production of IL-4. However, it is unknown whether this initiates or perpetuates Th2 responses alone or acts as a complement to traditional APC (Finkelman 2009). Helminth infections typically induce Th2 cells and the molecular structures produced by these microbes contribute to this effect. Once differentiation is initiated, Th2 cells produce IL-4 and IL-25 that reinforce IL-4 expression as well as the production of other Th2 cytokines such as IL-5 and IL-13. At the same time, IL-4 suppresses genes associated with other Th types, including both Th1 and Th17.

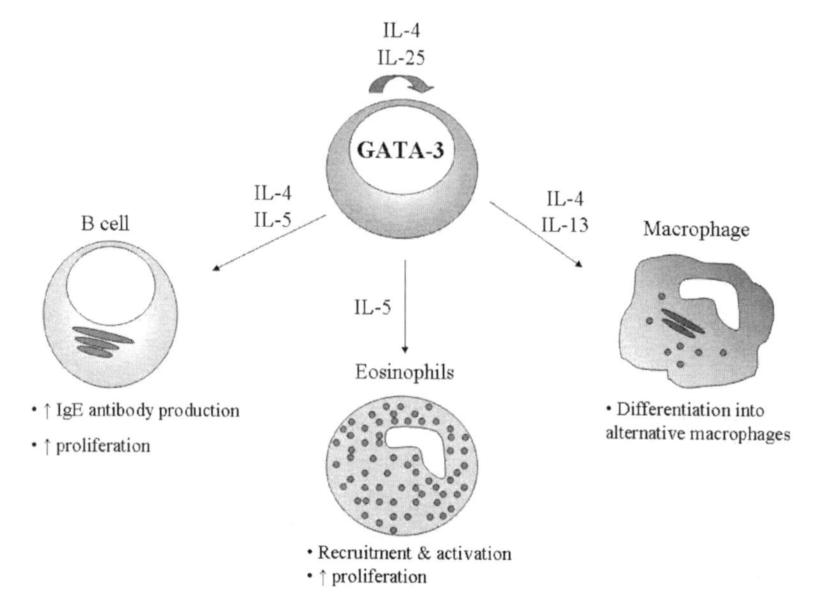

Fig. 3 Th2 cells and cytokines. Th2 cells are characterized by the expression of the transcription factor GATA-3 and the production of the cytokines IL-4, IL-5, IL-13 and IL-25. IL-4, IL-5 and IL-13 act to recruit and activate granulocytes, including eosinophils, and promote IgE production by B cells. IL-4 and IL-25 act in an autocrine manner resulting in increased expression of IL-4, IL-5 and IL-13. Together these responses help to rid the host of parasitic infections, such as those caused by helminths. IL, interleukin (unpublished diagram by authors).

IL-4 plays a major role in B cell differentiation and class switching to IgE, and the expansion of mucosal mast cells making Th2 cells particularly relevant to protection against helminth infection as well as development of allergies and asthma. Another Th2 cytokine, IL-13, has close structural homology to IL-4. Indeed, these two cytokines have been shown to have somewhat overlapping functions. Both play a major role in expulsion of parasites although the relative importance of IL-13 and IL-4 differs depending on the type of infection (Wynn 2003). IL-13, along with IL-4, is capable of directly influencing macrophage activity by inducing what is called an alternatively activated macrophage. As opposed to the IFN-γ-induced classically activated macrophage that is effective in killing intracellular organisms, this cell is thought to play a role in the resolution of wound healing (Mosser and Edwards 2008) and may participate in protection of the host from parasitic infections (Anthony *et al.* 2007).

Other cytokines produced by Th2 cells have important functions in regulating the immune response associated with allergies or parasite infections. For example,

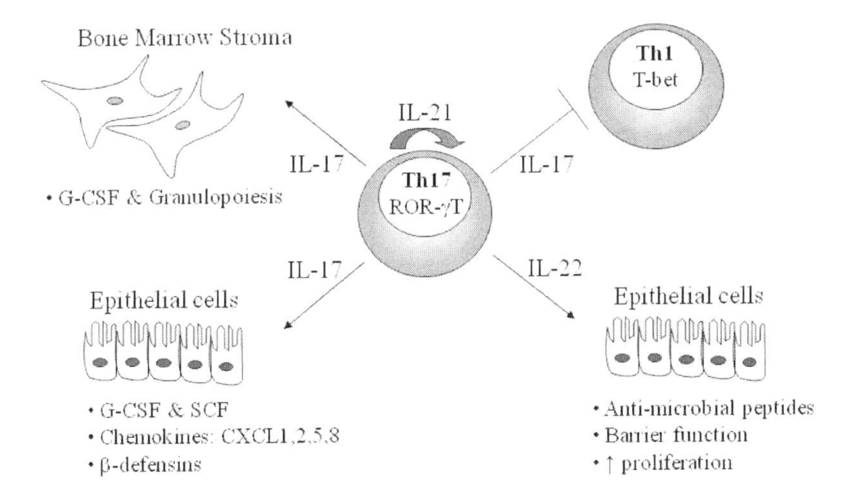

Fig. 4 Th17 cells and cytokines. Th17 cells are characterized by the expression of the transcription factor ROR-γT and produce the cytokines IL-17, IL-21 and IL-22. These cytokines contribute to neutrophil mobilization and epithelial barrier integrity, as well as cross-regulation of competing Th responses and autocrine self-amplification of Th17 cells. IL, interleukin; ROR, retinoic acid receptor-related orphan receptor (unpublished diagram by author).

IL-5 is expressed in response to a variety of parasitic infections and stimulates the recruitment, activation and proliferation of eosinophils. This response contributes to host protection and the resolution of tissue damage. IL-5 also increases the proliferation of B cells (Takatsu and Nakajima 2008). IL-25, another Th2 cytokine, reinforces the Th2 phenotype by promoting the expression of other Th2-associated cytokines including IL-4, IL-5 and IL-13. In some instances, IL-25 expressed by epithelial cells contributes to Th2 response even in the absence of IL-4 (Fort *et al.* 2001).

TH17 CELLS AND CYTOKINES

Only recently defined as a distinct subset of $CD4^+$ T cells, the Th17 lineage is named for its signature cytokine IL-17 (Harrington *et al.* 2005). Other cytokines preferentially secreted by this subset include IL-21 and IL-22. The transcription factor ROR-γT is important in driving many aspects of the Th17 differentiation program. Th17 cells are particularly prevalent at mucosal barriers, such as the intestinal lamina propria, and are thought to be important in defense against extracellular bacteria and other pathogens, as well as in maintaining epithelial barrier integrity.

IL-17, now known as IL-17A, is the founding member of the IL-17 family. This family consists of six distinct cytokines of which both IL-17A and IL-17F are produced by Th17 cells. Typically found as homodimers, IL-17A-F heterodimers can also form and have biological activity (Liang *et al.* 2007). IL-17A has diverse biological functions including neutrophil mobilization and epithelial cell activation. By up-regulating growth factors such as G-CSF and chemokines such as IL-8, IL-17A increases both neutrophil production and recruitment. Deficiencies in IL-17A signaling have been associated with susceptibility to a variety of pathogens such as *Klebsiella pneumoniae*, *Toxoplasma gondii*, and *Candida albicans*, probably relating to defective neutrophil responses (O'Quinn *et al.* 2008). IL-17F has substantial homology to IL-17A and shares many of its biological activities.

IL-21 is another cytokine produced by Th17 cells and acts in an autocrine manner to amplify Th17 responses. By activating ROR-γT, IL-21 can serve in place of, or in addition to, IL-6 in the generation of Th17 cells from naïve precursors (Ouyang *et al.* 2008). The effects of IL-21 are not limited to Th17 cells, however, as it also has been shown to have an impact on other Th cells as well as B cells, $CD8^+$ T cells and epithelial cells.

The Th17 cytokine IL-22 is a member of the IL-10 cytokine family. It is produced by activated $CD4^+$ T cells, particularly Th17 cells, though it can also be expressed by other cell types including dendritic cells (Zheng *et al.* 2008). IL-22 is distinct from IL-17 in that its production does not require a fully differentiated Th17 cell and can be stimulated by IL-6 or IL-23 in the absence of TGF-β1 (Zheng *et al.* 2007). The receptor for IL-22 is widely expressed on non-immune cells such as epithelial cells, and consequently IL-22 elicits strong responses from a variety of epithelial tissues but does not directly impact immune cells (Aggarwal *et al.* 2001). Effects of IL-22 on epithelial cells are broadly related to tissue repair and host defense and include up-regulation of anti-microbial peptides such as β-defensins and S100-family proteins. As many epithelial cells express receptors for both IL-17 and IL-22, the two cytokines together can have additive or synergistic effects in controlling epithelial responses.

REGULATORY T CELLS (TREG)

In addition to the subsets of pro-inflammatory Th cells, a population of $CD4^+$ T cells exists that is intrinsically anti-inflammatory—the Treg. These cells are a topic of intensive study for their ability to control inflammatory responses and prevent autoimmune reactions to self. Although a number of different types of Treg have been characterized, arguably the best studied of these are the natural Treg (nTreg). These cells express high levels of the IL-2 receptor (CD25) and are characterized by expression of the transcription factor FoxP3. For the purposes of this chapter, an extensive discussion of Treg is not possible, although some Treg functions are mediated by anti-inflammatory cytokines such as IL-10, IL-35, and TGF-β1 (Tang and Bluestone 2008) (Fig. 5).

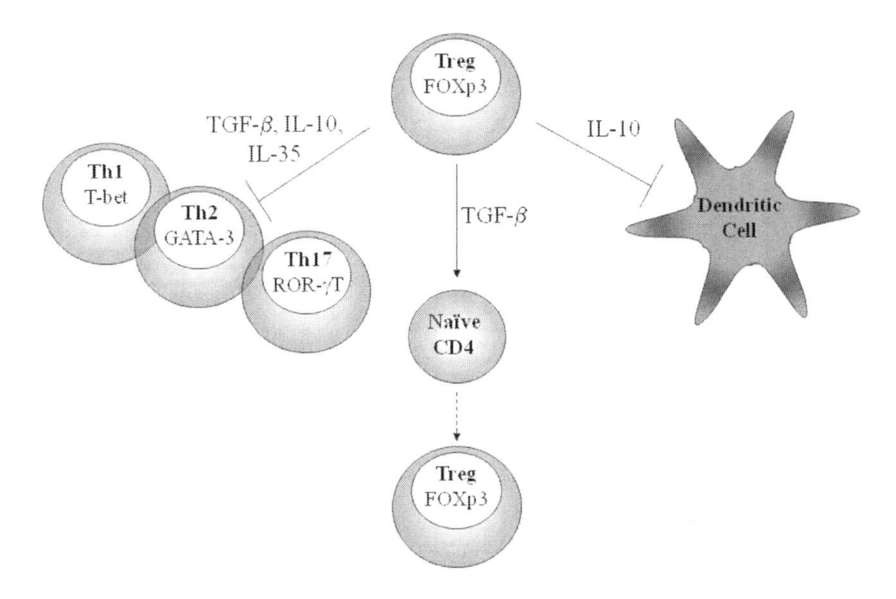

Fig. 5 Regulatory T cells (Treg) and cytokines. Treg-secreted cytokines include TGF-β1, IL-10 and IL-35. Collectively these factors, along with surface molecules and other mediators, suppress the development and function of effector Th cells, via direct effects on the T cells and by modulating innate cells such as dendritic cells. TGF, transforming growth factor; IL, interleukin (unpublished diagram by author).

IL-10 acts on a number of different leukocytes, including effector Th subsets, to suppress inflammatory mechanisms, such as cytokine and chemokine production, oxidative burst, cell differentiation and activation (Moore *et al.* 2001). TGF-β1 is produced my many types of Treg and can simultaneously enhance Treg generation from naïve precursors while hindering the development of effector Th. Although its exact role is unclear, TGF-β1 may exert suppression in membrane-bound form via a cell contact-dependent mechanism (Taylor 2009). Non-cytokine mediators can also affect Treg function, including cellular metabolites like the nucleoside adenosine and cell contact-dependent mechanisms that involve cell surface molecules such as CTLA-4 (Tang and Bluestone 2008).

TH CELLS AND DISEASE

Although $CD4^+$ T cells have presumably evolved to maintain host health in the face of diverse microbial challenges, unchecked Th responses are associated with a number of disease states. Allergies and asthma, both common inflammatory disorders, have long been associated with hyper-responsive Th2 cells. In genetically predisposed individuals exposed to allergen, Th2 cytokines promote IgE production by B cells and subsequent activation of mast cells and basophils. These cells then release factors, such as histamine and leukotrienes, that are important mediators of the allergic reaction (Galli *et al.* 2008).

Over-exuberant Th1 responses have been implicated in the development of gastritis and gastroduodenal ulcers, Crohn's disease, rheumatoid arthritis and type I diabetes. Similarly, Th17 cells have been associated with the pathogenesis of autoimmune diseases such as multiple sclerosis, psoriasis and rheumatoid arthritis (Ouyang *et al.* 2008). Although discreet populations of $CD4^+$ T cells have been well characterized in carefully controlled systems, multiple Th subsets are likely to be generated *in vivo* during inflammation. For example, mixed responses of Th1 and Th17 cells are associated with *Helicobacter pylori*-induced gastritis as well as Crohn's disease.

Th cell heterogeneity also emerges because of their plasticity, as it is increasingly appreciated that conversion between Th subsets is not uncommon. Thus it is not surprising that disease probably results from the combined effects of cytokines associated with multiple Th lineages. Indeed, this seems to be especially true for the Th1 and Th17 lineages where their respective cytokines may act alone or in combination with each other to stimulate chemokine production and lead to the recruitment and activation of myeloid cells (e.g., neutrophils, macrophages, dendritic cells). This, in turn, contributes to chronic inflammation, tissue damage and loss of function associated with the disease state. Moreover, a lack of regulation due to defective or deficient Treg may also contribute to the development of these autoimmune and inflammatory diseases.

Table 1 Key $CD4^+$ T cell cytokines and their functions

Source	*Cytokine*	*Function*
Th1	IL-2	Promotes T cell proliferation
	IFN-γ	Activates neutrophils and macrophages
	LT-α	Stimulates fibroblasts and endothelial cells
	TNF-α	Promotes inflammatory responses
Th2	IL-4	Promotes Th2, B cell isotype switching
	IL-5	Recruits and activates eosinophils
	IL-13	B cell isotype switching
Th17	IL-17A	Recruits neutrophils
	IL-21	Promotes Th17
	IL-22	Induces anti-microbial peptides
Treg	IL-10	General anti-inflammatory effects
	TGF-β1	Promotes Treg and inhibits Th cells

IL, interleukin; IFN, interferon; LT, lymphotoxin; TNF, tumor necrosis factor; TGF, transforming growth factor.

Table 2 Key features of Th cell heterogeneity

- The concept of helper T cell heterogeneity has been appreciated for many years. Tada *et al.* (1978) originally devised the names Th1 and Th2 to describe distinct subsets of Th cells.
- Th1 and Th2 subsets were formally characterized by Mossman et al. (1986). Since then, other $CD4^+$ T cell lineages have been described. These subsets are largely distinguished on the basis of their cytokine expression.
- In addition to the Th17 and Treg discussed here, other subsets that have been described include Th0, Th3, Th9 and Th22 cells.
- The stability and *in vivo* function of many of the Th cell subsets is uncertain and some cells may shift from one phenotype to another based on the physiological conditions.

Summary Points

- $CD4^+$ T cells, also called T helper (Th) cells, can develop into several distinct lineages.
- Th1 cells are characterized by secretion of the cytokines IFN-γ, IL-2, TNF-α and LT-α.
- IFN-γ promotes recruitment of macrophages, phagocytosis and production of other inflammatory cytokines. This is enhanced by other Th1 cytokines.
- Th1 cell responses are important in controlling infections caused by *Leishmania* sp., intracellular bacteria, fungi and viruses.
- Over-exuberant Th1 responses are associated with inflammatory diseases, such as *H. pylori*-induced gastritis and gastroduodenal ulceration, Crohn's disease and diabetes mellitus.
- Th2 cells are characterized by secretion of IL-4, IL-5, IL-13 and IL-25.
- Th2 cytokines play a critical role in protection from many parasitic infections by promoting the recruitment and activation of mast cells basophils and eosinophils and enhancing IgE production by B cells.
- Th2 cells and cytokines are major contributors to allergy and asthma.
- Th17 cells are characterized by production of IL-17, IL-21 and IL-22.
- Th17 cells are common in mucosal tissues, such as the intestine, and the cytokines IL-17A and IL-22 facilitate neutrophil accumulation and maintain mucosal barrier integrity.
- Over-active Th17 cell responses have been linked to some autoimmune diseases, including inflammatory bowel diseases, multiple sclerosis and rheumatoid arthritis.
- Regulatory T cells exert suppressive effects on a number of different leukocytes via secretion of anti-inflammatory cytokines, such as IL-10, IL-35 and TGF-β1.

Abbreviations

APC	:	antigen-presenting cell(s)
CTLA-4	:	cytotoxic T-lymphocyte antigen 4
FoxP3	:	forkhead box P3
GATA	:	guanine adenine thymine adenine sequence in DNA
GATA-3	:	GATA-binding protein 3
G-CSF	:	granulocyte colony-stimulating factor
IFN	:	interferon
IL	:	interleukin
LT	:	lymphotoxin
MHC	:	major histocompatibility complex
ROR-γT	:	retinoic acid receptor-related orphan receptor γT
T-bet	:	T-box expressed in T cells
TGF	:	transforming growth factor
TNF	:	tumor necrosis factor
TSLP	:	thymic stromal lymphopoietin

Definition of Terms

Antigen-presenting cell: A cell, such as a dendritic cell or macrophage, that processes and presents antigen in the context of a MHC to a T cell in order to activate it.

Chemokines: A class of cytokines that chemoattracts, or recruits, specific cells that express the appropriate chemokine receptor.

Extravasation: The process by which immune cells leave the bloodstream and enter a tissue.

Granulopoiesis: The generation of granulocytes (neutrophils, basophils, eosinophils) from stem cells in the bone marrow.

Oxidative burst: A rapid release of reactive oxygen species by some immune cells (e.g., neutrophils) that aids in the destruction of pathogens.

Pluripotent cells: Precursor (stem) cells that are able to differentiate into any number of cell types.

Somatic recombination: Process of generating diversity in antibody and T cell receptors by the combinatorial rearrangement of multiple genes; this allows a wide variety of specificity in B and T cell recognition.

Transcription factor: A protein that binds to particular sequences of genomic DNA and controls the transcription of specific genes.

REFERENCES

Aggarwal, S. and M.H. Xie, M. Maruoka, J. Foster, and A.L. Gurney. 2001. Acinar cells of the pancreas are a target of interleukin-22. J. Interferon Cytokine Res. 21: 1047-1053.

Anthony, R.M. and L.I. Rutitzky, J.F. Urban, Jr., M.J. Stadecker, and W.C. Gause. 2007. Protective immune mechanisms in helminth infection. Nat. Rev. Immunol. 7: 975-987.

Billiau, A. and P. Matthys. 2009. Interferon-gamma: a historical perspective. Cytokine Growth Factor Rev. 20: 97-113.

Cuff, C.A. and J. Schwartz, C.M. Bergman, K.S. Russell, J.R. Bender, and N.H. Ruddle. 1998. Lymphotoxin alpha3 induces chemokines and adhesion molecules: insight into the role of LT alpha in inflammation and lymphoid organ development. J. Immunol. 161: 6853-6860.

Feldmann, M. and R.N. Maini. 2001. Anti-TNF alpha therapy of rheumatoid arthritis: what have we learned? Annu. Rev. Immunol. 19: 163-196.

Finkelman, F.D. 2009. Basophils as Th2-inducing APCs: the dog can sing but is it a diva? Immunol. Cell Biol. 87: 568-570.

Fort, M.M. and J. Cheung, D. Yen, J. Li, S.M. Zurawski, S. Lo, S. Menon, T. Clifford, B. Hunte, R. Lesley, T. Muchamuel, S.D. Hurst, G. Zurawski, M.W. Leach, D.M. Gorman, and D.M. Rennick. 2001. IL-25 induces IL-4, IL-5, and IL-13 and Th2-associated pathologies in vivo. Immunity 15: 985-995.

Galli, S.J. and M. Tsai, and A.M. Piliponsky. 2008. The development of allergic inflammation. Nature 454: 445-454.

Harrington, L.E. and R.D. Hatton, P.R. Mangan, H. Turner, T.L. Murphy, K.M. Murphy, and C.T. Weaver. 2005. Interleukin 17-producing CD4+ effector T cells develop via a lineage distinct from the T helper type 1 and 2 lineages. Nat. Immunol. 6: 1123-1132.

Ho, I.C. and T.S. Tai, and S.Y. Pai. 2009. GATA3 and the T-cell lineage: essential functions before and after T-helper-2-cell differentiation. Nat. Rev. Immunol. 9: 125-135.

Liang, S.C. and A.J. Long, F. Bennett, M.J. Whitters, R. Karim, M. Collins, S.J. Goldman, K. Dunussi-Joannopoulos, C.M. Williams, J.F. Wright, and L.A. Fouser. 2007. An IL-17F/A heterodimer protein is produced by mouse Th17 cells and induces airway neutrophil recruitment. J. Immunol. 179: 7791-7799.

Moore, K.W. and M.R. de Waal, R.L. Coffman, and A. O'Garra. 2001. Interleukin-10 and the interleukin-10 receptor. Annu. Rev. Immunol. 19: 683-765.

Mosmann, T.R. and H. Cherwinski, M.W. Bond, M.A. Giedlin, and R.L. Coffman. 1986. Two types of murine helper T cell clone. I. Definition according to profiles of lymphokine activities and secreted proteins. J. Immunol. 136: 2348-2357.

Mosser, D.M. and J.P. Edwards. 2008. Exploring the full spectrum of macrophage activation. Nat. Rev. Immunol. 8: 958-969.

O'Quinn, D.B. and M.T. Palmer, Y.K. Lee, and C.T. Weaver. 2008. Emergence of the Th17 pathway and its role in host defense. Adv. Immunol. 99: 115-163.

Ouyang, W. and J.K. Kolls, and Y. Zheng. 2008. The biological functions of T helper 17 cell effector cytokines in inflammation. Immunity 28: 454-467.

Perrigoue, J.G. and F.A. Marshall, and D. Artis. 2008. On the hunt for helminths: innate immune cells in the recognition and response to helminth parasites. Cell Microbiol. 10: 1757-1764.

Roach, D.R. and H. Briscoe, B. Saunders, M.P. France, S. Riminton, and W.J. Britton. 2001. Secreted lymphotoxin-alpha is essential for the control of an intracellular bacterial infection. J. Exp. Med. 193: 239-246.

Spellberg, B. and J.E. Edwards. 2001. Type 1/Type 2 immunity in infectious diseases. Clin. Infect. Dis. 32: 76-102.

Tada, T. and T. Takemori, K. Okumura, M. Nonaka, and T. Tokuhisa. 1978. Two distinct types of helper T cells involved in the secondary antibody response: independent and synergistic effects of Ia- and Ia+ helper T cells. J. Exp. Med. 147: 446-458.

Takatsu, K. and H. Nakajima. 2008. IL-5 and eosinophilia. Curr. Opin. Immunol. 20: 288-294.

Tang, Q. and J.A. Bluestone. 2008. The Foxp3+ regulatory T cell: a jack of all trades, master of regulation. Nat. Immunol. 9: 239-244.

Taylor, A.W. 2009. Review of the activation of TGF-beta in immunity. J. Leukoc. Biol. 85: 29-33.

Wynn, T.A. 2003. IL-13 effector functions. Annu. Rev. Immunol. 21: 425-456.

Zhang, S. and H. Zhang, and J. Zhao. 2009. The role of CD4 T cell help for CD8 CTL activation. Biochem. Biophys. Res. Commun. 384: 405-408.

Zheng, Y. and D.M. Danilenko, P. Valdez, I. Kasman, J. Eastham-Anderson, J. Wu, and W. Ouyang. 2007. Interleukin-22, a T(H)17 cytokine, mediates IL-23-induced dermal inflammation and acanthosis. Nature 445: 648-651.

Zheng, Y. and P.A. Valdez, D.M. Danilenko, Y. Hu, S.M. Sa, Q. Gong, A.R. Abbas, Z. Modrusan, N. Ghilardi, F.J. de Sauvage, and W. Ouyang. 2008. Interleukin-22 mediates early host defense against attaching and effacing bacterial pathogens. Nat. Med. 14: 282-289.

Zhu, J. and W.E. Paul. 2008. CD4 T cells: fates, functions, and faults. Blood 112: 1557-1569.

CHAPTER 8

Cytokine Production in Viral Infection

Monica Tomaszewski and Frank J. Jenkins*
Department of Infectious Diseases and Microbiology
University of Pittsburgh, G.17 Hillman Cancer Research Pavilion
5117 Centre Avenue, Pittsburgh, Pennsylvania 15213, USA

ABSTRACT

Cytokines are small proteins excreted by cells with the purpose of intercellular communication. They are known to modulate a host's immune response to infection, particularly viral infections, by recruiting immune cells to the site of infection and signaling defense mechanisms in both infected and un-infected cells. As such, cytokines are part of a strong antiviral defense mechanism by the host and represent a formidable obstacle to the virus surviving in the infected individual. They can also be used by viruses to aid in the replication and persistence of the virus. As a result, while many viruses have evolved different mechanisms to subvert the action of cytokines, others produce homologues of cytokines that assist in their replication and/or pathogenicity. In this chapter, we discuss these different viral responses to cytokines. Hepatitis B and C and influenza viruses are used as examples to demonstrate the up-regulation of cytokine production by virus infection and ways in which the virus combats this response. Herpesviruses are used as examples of how expression of viral homologues of cytokines can influence the immune response, producing a different response than their cellular counterparts. We review how viral cytokines can assist in fighting a host's immune response as well as benefit viral replication.

INTRODUCTION

Cytokines are small, secreted proteins that regulate immunity, inflammation and hematopoiesis. Cytokines share the ability to influence the immune system,

*Corresponding author

whether by recruitment of lymphocytes or by cellular modulation. A major function of cytokines is to regulate the body's immune system in regard to injury, such as viral infection. The purpose of this chapter is to demonstrate some of the ways in which viral infection can manipulate the cytokine response, and how this impacts the ultimate pathogenesis of the disease.

For the purposes of this chapter, we will divide cytokines into two main types involved in viral infection: (1) those produced by the cell in response to infection and (2) viral homologues of cellular cytokines. While the former is the body's obvious response to combating viral infection, the latter is an interesting adaptation of viruses to this response. To demonstrate the first type of cytokines, we will discuss viruses that induce host-derived cytokines, such as the human hepatitis viruses B and C, and influenza viruses, which induce cellular cytokines during infection. These viruses express viral proteins capable of influencing the effects of cytokines, which in turn affect the immune response. To demonstrate the second type of cytokines, we have chosen members of the herpesviruses, which have acquired viral homologues of cytokines and cytokine receptors. These viruses are of particular interest because while the sequence of these virally produced cytokines can be very similar to the human counterpart, not all of the immunomodulatory actions are conserved. This non-conservation of function can lead to targeted actions against the immune response, which allow for survival of the virus while at the same time inhibiting the anti-viral response.

VIRALLY INDUCED CELLULAR CYTOKINES

Hepatitis Viruses B and C

There are several human hepatitis viruses that are named for their ability to cause inflammation of the liver (i.e., hepatitis). The hepatitis viruses belong to several different virus families including Picornavirus, Flavivirus and Hepadnavirus. Hepatitis B (HBV), a member of the Hepadnavirus family, and Hepatitis C, a member of the Flavivirus family, both cause an acute self-limiting disease, as well as a severe chronic disease that leads to liver cirrhosis and hepatocellular carcinoma. The pathology of chronic hepatitis infections caused by HBV and HCV is viewed as a result of immunopathology rather than direct effects by the virus. It is the host's immune response by the adaptive arm of the immune system that has been implicated in the outcomes of chronic disease. Cytokines clearly play a role in this immunopathology.

During HBV/HCV infection, the virus-infected hepatocytes are recognized by cytotoxic T lymphocytes (CTL) and are killed in an effort to combat viral replication and prevent infection of neighboring cells. In a chronic viral infection, ineffective killing of infected hepatocytes, as well as the subsequent, constant production of cytokines (e.g., IFN-γ) by the CTLs, are felt to contribute to a chronic inflammation in the liver leading to cirrhosis and, eventually, hepatocellular carcinoma (Rehermann 1999).

In order for the hepatitis viruses to persist in the body and maintain a productive infection, they would have to compromise immune function. The host's response to virus infection is a multistep process that depends on the recognition of the virus by the infected cell and infiltrating immune cells. Identification of infection by the infected cells results in the production of IFN-α/β, which induces antiviral responses, such as interfering with viral replication. In the case of infiltrating immune cells, recognition of infection is achieved by the binding of pathogen-associated molecular patterns to Toll-like receptors on cells such as DCs and macrophages (reviewed in Bode 2008). This recognition initiates a cascade of intercellular signaling that activates several transcription factors (such as NF-κB) and consequently mediates the production of several anti-viral outcomes, such as production of type 1 interferons (IFN-β). Once viral RNA is recognized within the cell, interferon-gamma (IFN-γ) is produced, which up-regulates the production of MHC:peptide complexes for presentation of foreign proteins and promotes memory T cell proliferation and dendritic/natural killer cell activation. All of these actions are designed to kill infected cells and protect un-infected cells from future infection.

The recruitment and presence of dendritic cells in the liver during infection also affects cytokine production. The hepatitis virus antigens present in the infected liver are engulfed by the immature DCs (iDCs), processed, and presented to immune cells. NK cells mature DCs, which produce pro-inflammatory cytokines including IFN-α, TNF-α, IL-12 and IL-18. The mature DCs present the processed hepatitis antigens to T cells, inducing a Th1 response (cellular immunity) that includes the production of IL-2 and IFN-γ. IL-2 combats some anti-viral recognition by iDCs, by inducing NK cells to lyse them (reviewed in Castello 2009) (Fig. 1).

Taken together, this controlled system effectively alerts the immune system to infection. However, it is hindered by mechanisms the virus has evolved to change this scenario. For instance, the major glycoprotein (E2) of Hepatitis C viruses bind the cell surface tetraspannin CD81 (reviewed in Castello *et al.* 2009). CD81 is present on B- and T-lymphocytes as well as NK cells, and when engaged will inhibit the production of cytokines (IFN-γ) and cytotoxic activity, thus inhibiting the mechanism by which NK cells are involved in anti-viral response. This inhibition of cytokine production exerts its effect on T cell responses by skewing T cell differentiation toward the Th2 response (humoral immunity). The Th1 response involves production of IL-1 and IFN-γ (anti-inflammatory) cytokines, while Th2 promotes chronic inflammation in HCV due to production of IL-4, IL-5, IL-10, and IL-13, which promotes plasma cell differentiation and consequential antibody production.

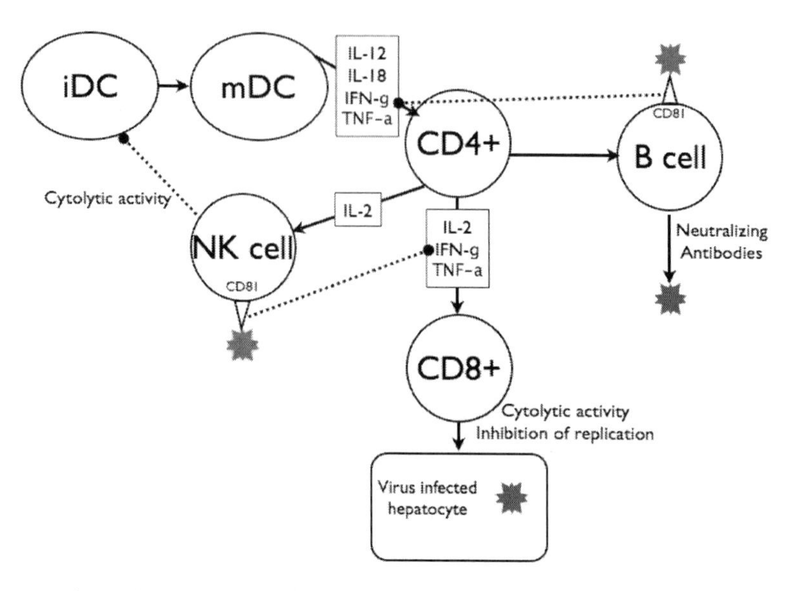

Fig. 1 Infection by hepatitis viruses both induces and blocks cytokine production. Hepatitis-infected hepatocytes induce cytokine production in numerous cell types (black boxes), whereas the interaction of the cell surface protein CD81 (triangle) and the hepatitis virus major glycoprotein can inhibit several cytokines (dotted lines) (unpublished).

Currently, treatment for the HCV is to administer pegelated IFN-γ and ribavirin. The modified molecule of IFN-γ is more stable and, as a whole, increases the amount of IFN-γ in the system, thus "re-skewing" the Th1/Th2 population back to normal (Myrmel *et al.* 2009). HBV can also be treated by pegelated IFN-γ and nucleoside/nucleotide analogues.

Influenza

Influenza viruses belong to the family Orthomyxovirus and are sub-classified into types A, B and C. Influenza A infects respiratory epithelial cells, which produce copious amounts of virus that then infect alveolar macrophages. These cells die by necrosis and apoptosis, respectively, and induce production of multiple cytokines including TNF-α, IFN-α/β, IL-1α/β, IL-6 and IL-8. Other cytokines, such as those involved in T cell regulation (IFN-γ and IL-10), are present transiently during infection, as well as chemokines needed to recruit T cells to the site of infection (MIP-1α and MCP-1) (Wu *et al.* 2009). The production of these cytokines is directly related to pathogenesis as it has been shown that increased disease severity correlates with increased cytokine levels.

The influenza virus has developed mechanisms to thwart the anti-viral effect of cytokines. For example, the anti-viral aspects of interferon in influenza are largely blocked by the presence of a non-structural viral protein (NS1) that specifically targets interferon-dependent transcriptional pathways, thus interfering with the initial antiviral response (Wolff *et al.* 2008). Those subtypes of influenza that are particularly pathogenic can be traced to the structure of NS1, specifically a 4 amino acid C-terminal domain, and this is correlated to up-regulation of cytokine expression (Jackson *et al.* 2008). Additionally, NS proteins deactivate protein kinase R, which is a dsRNA-dependent kinase that phosporylates elf2-α and inhibits virus replication in an interferon-dependent manner (Fig. 2). Lastly, NS1 can inhibit recognition of ssRNA by inhibiting the ubiquitination-induced activation of the RIG-1 complex; this inhibition ultimately stops expression of IFN-α in the antiviral response (reviewed in Wolff *et al.* 2008).

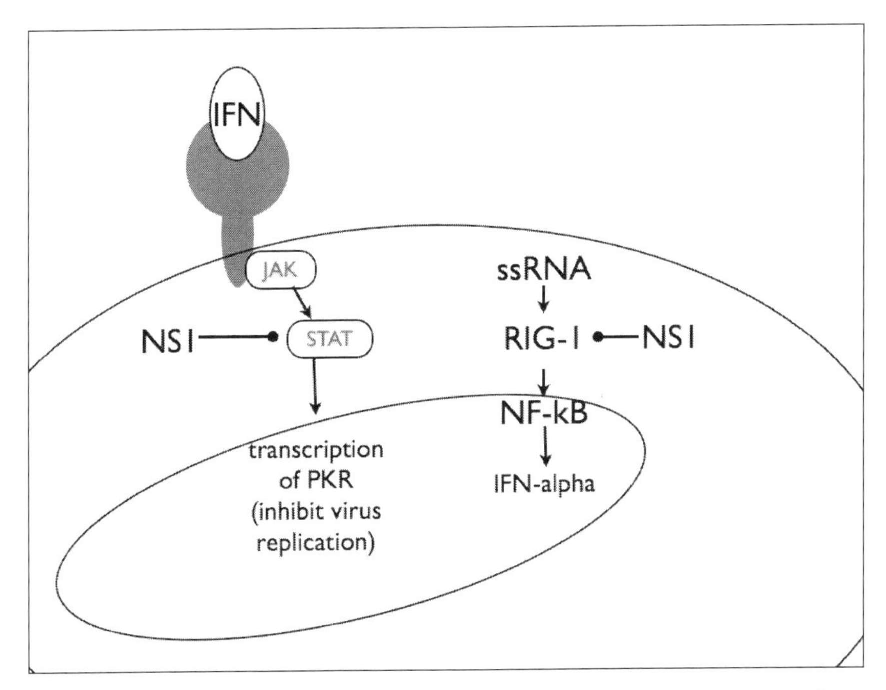

Fig. 2 The non-structural protein of influenza virus directly inhibits the production of interferon-inducible proteins. Non-structural protein 1 (NS1) of influenza virus is responsible for inhibiting the production of type 1 interferon responses. NS1 inhibits the ubiquitination and consequential signaling of RIG1 (retinoic acid inducible gene 1). NS1 also inhibits the downstream signaling post-interferon beta, which would lead to transcription of PKR (protein kinase R) and consequential inhibition of viral DNA synthesis (unpublished).

Table 1 Key facts about the influence of cytokines on viral infection

- Cytokines induced by viral infection are part of the immune response, but viruses have evolved ways to either down-regulate or process these cytokines to evade immune detection.
- In the case of influenza, cytokine response in a healthy individual is responsible for the symptoms of the disease.
- Viruses can encode cytokine and cytokine receptor homologues as a mechanism for immune evasion and to promote survival.
- Virally encoded cytokines and cytokine receptors may not need the requisite receptor or ligand to produce a signal.
- Virus-associated cytokines often do not share all of the functions of the cellular counterpart.

CTL response is also involved in the pathology of influenza. CTLs are a large proportion of the immune infiltrate and contribute to the destruction of infected cells by direct lysis or cytokine production. Part of the cytokine response induced by CTLs involves the recruitment of macrophage and monocytes to the site of infection. This infiltrate can be infected by influenza A virus and while the virus does not replicate in these cells, it does result in the death of the infected cells by apoptosis within 24-48 hr after infection. Prior to death, they are able to produce a sustained level of pro-inflammatory cytokines such as TNF-α, IL-6, and IFN-α/β. This increased level of cytokines recruits more immune cells to the site of infection producing a cycle of infection, cytokine production, apoptosis, and immune cell recruitment.

Multiple studies have indicated that this deregulation of cytokine expression is what contributes to the pathogenesis of influenza. A phenomenon called a "cytokine storm", or hypercytokinesis, can contribute to the virus-induced pathology. Hypercytokinesis occurs when large quantities of cytokines and chemokines are released to promote T cell and macrophage activation, which leads to an overzealous immune response, which, in turn, becomes life threatening. This predominately occurs in the more pathogenic influenza type A subtypes, such as H5N1, in which macrophages can be induced to release more TNF-α than less virulent strains (Salomon *et al.* 2007).

VIRALLY PRODUCED CYTOKINE HOMOLOGUES

Herpesviruses

Herpesviruses are a family of large double-stranded DNA viruses. There are currently eight recognized human herpesviruses.

The herpesvirus family is characterized by its ability to maintain a latent infection. While the site of latency differs among different herpesviruses, the ability to establish latency and remain undetected in the human body represents an efficient immune evasion strategy. Part of this strategy allows for the mimicry of host proteins to maintain a metabolically active cell in light of the viral infection. Some of this mimicry extends to virally produced cytokines (Table 2).

Table 2 Cytokine homologues produced by viruses (unpublished)

Viral Gene	*Cellular Counterpart*	*Degree of Similarity*	*Function in virus pathology*
EBV BCLFI	hIL-10	85%	B-cell proliferation,
CMV ULIIIa	hIL-10	27%	Inhibition of IFN-gamma
Orf Virus ovIL-10	hIL-10	75%	and MHCI antigen presentation
HHV-8 K2	hIL-6	25%	B-cell proliferation, blocks IFN-alpha
HVS 13 vIL-17	hIL-17	57%	Supports T cell proliferation; activates nfkb, promotes IL-6 production

EBV

The Epstein-Barr virus (EBV) is a human gamma-herpesvirus that persists asymptomatically for life. EBV maintains latency in the memory B cell compartment.

Human IL-10 has suppressive effects on the production of pro-inflammatory cytokines by monocytes and neutrophils and on the down-regulation of activation/co-stimulation molecules on monocytes and DCs. It is also an important growth factor for B cells. The EBV BCRF1 gene encodes a protein with ~84% homology to human IL-10 (hIL-10) and because of this is referred to as viral IL-10 (vIL-10) (Fig. 3) (Moore *et al.* 1990). vIL-10 shares some of the functions of human IL-10 such as the ability to modulate expression of proteins involved in antigen presentation and suppression of pro-inflammatory cytokine production (de Waal Malefyt *et al.* 1991a, b). However, as discussed below, vIL-10 is unable to induce B-cell MHC1 expression (Zeidler *et al.* 1997), or co-stimulate thymocyte proliferation (MacNeil *et al.* 1990).

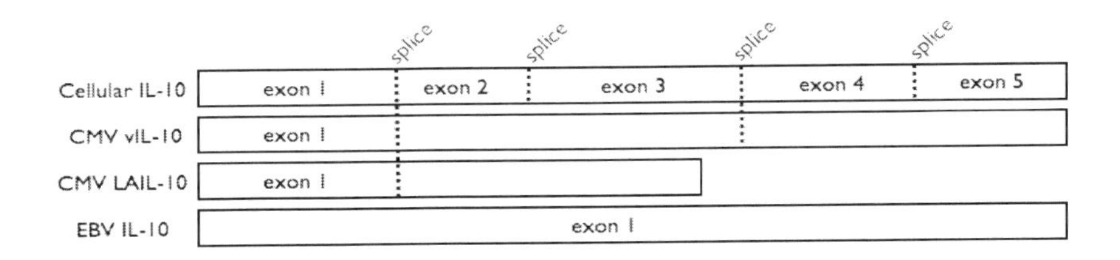

Fig. 3 Genomic structure of cellular and viral IL-10. The human genomic construction of IL-10 consists of five exons and four splice sites, creating a protein of 178 amino acids. The viral IL-10 of Cytomegalovirus (CMV) shares some of the genomic construction, but lacks overall sequence homology. EBV vIL-10 shares the most sequence homology, but the least genomic structure (unpublished).

The distinct role that vIL-10 plays in EBV's immune evasion strategy is also necessary for its viral life cycle. vIL-10 is produced both early in transformation (Miyazaki *et al.* 1993) and during lytic reactivation (Stewart *et al.* 1994). It enhances survival of recently infected cells (Miyazaki *et al.* 1993), which is similar to the function of hIL-10. This particular action also assists in escape from memory T cells during early transformation events (Bejarano and Masucci 1998). vIL-10 also decreases the amount of surface MHC1 via the down-regulation at the mRNA level of TAP1, as well as inhibiting foreign protein processing by inhibiting the LMP2 component of the proteosome (Zeidler *et al.* 1997). In monocytes, vIL-10 down-regulates MHCI, ICAM-1, and B7 expression, which affects monocyte/macrophage presentation of antigens, consequentially acting as a protective mechanism (Salek-Ardakani *et al.* 2002).

Interestingly, because B cells are the main target for EBV, this produces a situation in which the infected cells produce both hIL-10 and vIL-10. These proteins are produced at different times, with vIL-10 produced within 4 hr of infection, which would correspond to the timing associated with presentation of viral antigens. The initial burst of vIL-10 initiates the inhibition of the cascade of events that would lead to CTL and NK response due to this infection. This timing is unlike the production of hIL-10, which is produced 12-18 hr post-infection, and has a more prototypical cellular action.

Pathologically, vIL-10 is thought to have a sustaining effect in several of the EBV-associated diseases. Because of the immunomodulatory effects of the vIL-10 expression, it has been shown that EBV-associated diseases, such as oral hairy leukoplakia, demonstrate an up-regulation of vIL-10 message, but a lack of surrounding immune infiltrate (Hayes *et al.* 1999). In contrast, diseases that have a noticeable immune infiltrate do not express vIL-10, such as NPC (Hayes *et al.* 1999).

Because vIL-10 can act as an anti-inflammatory agent, it is also used to modulate inflammatory cascades in different animal models for possible use as therapy in inflammatory conditions such as arthritis (Zhang *et al.* 2008) and organ transplantation (Zuo *et al.* 2001).

HHV-8

Human herpes virus 8 (HHV-8; also known as Kaposi's Sarcoma Associated Herpesvirus, KSHV) is a lymphotropic virus that infects activated B cells. Like other herpesviruses, HHV-8 can exist in either a lytic or latent state. While HHV-8 is not as prevalent in the immunocompetent population as other herpesviruses, it causes several malignancies, including Kaposi's sarcoma (KS), primary effusion lymphoma (PEL), and some types of multicentric Castleman's disease (MCD).

vIL-6

Human IL-6 is a cytokine that is produced by many different cell types. It plays a role in immune activation, such as production of immunoglobulin, T cell activation, growth and proliferation.

The open reading frame K2 (ORF K2) of HHV-8 encodes the viral homologue of IL-6 (vIL-6). vIL-6 shares only 25% sequence homology, but is structurally similar to hIL-6. The function of vIL-6 extends to support proliferation of B cells, and as a growth factor for primary effusion lymphoma cells. Human IL-6, originally defined as a B-cell differentiation factor, also supports these functions, with the exception of PEL growth. Discordantly, vIL-6 disrupts IFN-α signaling, which promotes survival of virus-infected cells (Asou *et al.* 1998).

Human IL-6 binds to the hIL-6 binding receptor (IL-6Rα/gp80), producing a low-affinity complex that then associates with the IL-6 signaling receptor, gp130. The binding of gp130 (by the IL-6/gp-80 complex) results in the triggering of an intracellular signaling cascade that includes activation of the JAK/STAT3 pathway as well as activation of the mitogen-associated protein kinase pathway and nuclear localization of NF-κB (reviewed in Heinrich *et al.* 2003). Binding of hIL-6 to the gp80 receptor alone does not result in further signaling and the hIL-6 protein does not bind directly to the gp130 receptor. Thus, both the gp80 and gp130 receptors are necessary and important in hIL-6 signaling. Binding of IL-6 to its receptors results in a down-regulation of receptor expression, presumably to insure that over-stimulation does not occur.

The vIL-6 can also bind to gp80; however, it differs significantly from hIL-6 in that it can also bind directly to gp130. As a result, there can be vIL-6:gp130 mediated signaling on cells that do not have gp80; this allows for gp130 signaling on cells not normally receptive to IL-6 induction because of the ubiquitous nature of gp130. vIL-6 is very poorly secreted (Fig. 4) (Chen *et al.* 2009b), with a high percentage of the protein remaining in the ER/Golgi complex, and initializing an autocrine signaling pathway from this location (Chen *et al.* 2009a).

vIL-6 is critical to the pathogenesis of HHV-8–dependent malignancies. In Kaposi's sarcoma lesions, vIL-6 can induce angiogenesis (via VEGF) and hematopoiesis. It has also been demonstrated to be a growth factor for the spindle cells that are a hallmark of KS as well as some lymphocytes. vIL-6 has been shown to be necessary to support the growth of some PEL and it is expressed in some MCD, and may be necessary for the progression of the disease.

CONCLUSION

An integral mechanism in virus persistence is immune evasion. Different families of viruses have developed different mechanisms of evading the immune system,

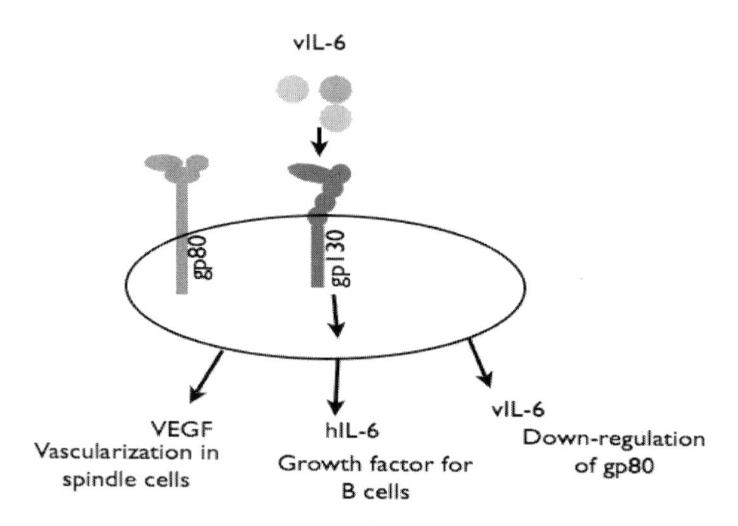

Fig. 4 vIL-6 of HHV-8 is able to augment downstream signaling to induce survival, immune dysregulation and pathology. vIL-6 is able to bind gp130 in the absence of gp80. Once bound, transduced signals through the JAK/STAT and MEK pathways yield pro-virus signals, such as production of VEGF (implicated in the vascular nature KS spindle cells), hIL-6 (for use as a B cell growth factor) and vIL-6 (which down-regulates gp80, which IFN-alpha also uses in its receptor complex) (unpublished).

from producing only a select set of viral proteins in order to stay hidden to a more proactive stance of down-regulating and degrading proteins necessary to mount an immune response. Perhaps the most obvious mechanism would be to subvert the immune system's messaging ability.

In this chapter, two separate mechanisms of virally induced cytokines were discussed. In the first section, hepatitis viruses were used as an example to demonstrate what types of cellular cytokines could be induced by viral infection and how the virus evolved to interfere with the consequential mechanisms of immunity induced by these cytokines. Hepatitis viruses take the more direct approach of disrupting IFN production by inhibition of transcription factors involved in its production. The production of cytokines by influenza viruses generally produces the quick immune response to resolve the infection; however, an overzealous immune system can produce an overwhelming response, creating a situation in which the cytokines are the cause of damage. In the second part of the chapter, we discussed the phenomena of virally produced cytokines. In some cases, it can be inferred from genomic construction of the gene that the cytokine encoding sequences were most likely duplicated from the human genome and eventually modified. This mechanism of adoption provides the virus a functional template for modification of the cytokine function. While it seems that it would be a rather simple solution for viruses to capture the genomic information and then encode anti-inflammatory cytokines without modification, the delineation of what makes something anti-inflammatory may not indicate functions significantly downstream, so modification to encompass the wanted outcomes is of great importance. In fact, herpesvirus-encoded cytokines often provide most but not all of the cellular cytokine actions, indicating that modification is needed for the appropriate actions in the viral context.

Together, this information demonstrates that the mechanisms by which viral infection manipulate cytokines for immunomodulation are varied among viruses and may not be dependent on the ability of the virus to generate a directed inhibitory protein response against the cytokine itself.

Table 3 Basic facts of viral infection

- In viral replication cycles, several key components are common among viruses in infection, protein processing and virus progeny production.
- Viruses infect cell types that have the necessary receptor on the cell surface. EBV, for instance, uses CD21, which is a cell surface protein on B cells. Consequently, EBV is B-cell tropic.
- Viruses require cellular proteins for replication; without the correct proteins, the virus will be unable to replicate its genetic information or produce progeny virus.
- Some types of viruses (e.g., Herpesvirus) are able to persist in the body without apparent clinical symptoms; however, they are able to maintain genomic information. This phenomenon, called latency, is part of the viral mechanism to evade the immune system.
- Immune cells are able to detect virus-infected cells because after infection viruses produce viral proteins and can down-regulate or dysregulate some cellular proteins. These actions are monitored by immune cells and can trigger immune action.
- Production of virus by a cell ultimately causes cell death by using cellular components to physically produce virus.

Summary Points

- Hepatitis virus infection skews T cell response though the ability of the virus to bind and signal through CD81, decreasing production of IFN-γ. By treating hepatitis infections with IFN-γ, the T cell response can be re-established and lead to clearance of the virus.
- The herpesvirus family of viruses encodes both viral cytokines and cytokine receptors. While homology to amino acid sequence varies, viral cytokines can often complement functions of the cellular cytokines in other arenas.
- Viral cytokines can act as an autocrine growth factor mediating the survival of the infected cells. The timing and expression of the virally produced cytokine may be different from those of the cell-produced cytokine, allowing for the appropriate signals to be delivered to the virus-infected cell at the correct time to combat apoptosis or promote cell growth.

- Virally encoded cytokines may not require the entire receptor complex. The vIL-6 protein of HHV-8 requires only the gp130 portion of the gp80:gp130 heterodimer; this decreased requirement allows for the viral cytokine to activate a more diverse selection of cell types.
- An indirect method for inducing an autocrine cytokine response in virus-infected cells is for the virus to produce a constitutively active cytokine receptor. This allows for a specific response in the virus-infected cell without having to secrete a protein.

Abbreviations

BCRF1	:	BamH1 C Right Fragment 1
CTL	:	cytotoxic T-lymphocytes
dsDNA	:	double stranded deoxyribonucleic acid
EBV	:	Epstein-Barr Virus
EIF-2α	:	eukaryotic initiation factor 2 alpha
HHV-8	:	Human Herpesvirus 8
hIL-10	:	human Interleukin 10
HBV	:	Hepatitis B Virus
HCV	:	Hepatitis C Virus
H+SV	:	Herpes simplex Virus
iDCs	:	immature dendritic cells
IL-6	:	Interleukin 6
INF-γ	:	Interferon gamma
KS	:	Kaposi's sarcoma
LAcmvIL-10	:	Latency associated Cytomegalovirus encoded Interleukin 10
LMP2	:	low molecular mass polypeptide-2
MCD	:	Multicentric Castleman's Disease
MHCI	:	major histocompatibility complex 1
MHCII	:	major histocompatibility complex 2
mRNA	:	messenger ribonucleic acid
NK cells	:	natural killer cells
PEL	:	primary effusion lymphoma
PI3K	:	phosphoinositide kinase-3
RIG-1	:	retinoic acid inducible gene 1
SAPK	:	stress-activated protein kinase
TAP1	:	antigen peptide transporter 1

UL111A	:	Unique Long open reading frame 111A of Cytomegalovirus
VEGF	:	vascular endothelial growth factor
vIL-10	:	viral Interleukin 10
vIL-6	:	viral IL-6

Definition of Terms

Cytotoxic T lymphocytes (CTL): Effector T cell that kills virus-infected and tumor cells.

Latent infection: A persistent viral infection without clinical symptoms, virus production, and limited viral protein production.

MHC antigen presentation: Mechanism in which short peptides are presented to effector cells. The presentation is done by a protein called the major histocompatibility complex (MHC), in which the peptide is held in a cleft to facilitate binding and recognition by the requisite T cell receptor.

Plasma cell: Terminally differentiated B-lymphocytes that produce soluble immunoglobulin.

Transformation: Process by which cells are changed by an agent (chemical or protein) to become perpetually proliferating.

References

Asou, H. and J.W. Said, R. Yang, R. Munker, D.J. Park, N. Kamada, and H.P. Koeffler. 1998. Mechanisms of growth control of Kaposi's sarcoma-associated herpesvirus-associated primary effusion lymphoma cells. Blood 91: 2475-2481.

Bode, J.G. and E.D. Brenndorfer, and D. Haussinger. 2008. Hepatitis C virus (HCV) employs multiple strategies to subvert the host innate antiviral response. Biol. Chem. 389: 1283-1298.

Castello, G. and S. Scala, G. Palmieri, S.A. Curley, and F. Izzo. 2009. HCV-related hepatocellular carcinoma: From inflammation to cancer. Clin. Immunol. 134: 237-250.

Chen, D. and G. Sandford, and J. Nicholas. 2009a. Intracellular signaling mechanisms and activities of human herpesvirus 8 interleukin-6. J. Virol. 83: 722-733.

Chen, D. and Y.B. Choi, G. Sandford, and J. Nicholas. 2009b. Determinants of secretion and intracellular localization of human herpesvirus 8 interleukin-6. J. Virol. 83: 6874-6882.

De Waal Malefyt, R. and J. Abrams, B. Bennett, C.G. Figdor and J.E. de Vries. 1991a. Interleukin 10 (IL-10) inhibits cytokine synthesis by human monocytes: an autoregulatory role of IL-10 produced by monocytes. J. Exp. Med. 174: 1209-1220.

De Waal Malefyt, R. and J. Haanen, H. Spits, M.G. Roncarolo, A. te Velde, C. Figdor, K. Johnson, R. Kastelein, H. Yssel, and J.E. de Vries. 1991b. Interleukin 10 (IL-10) and viral IL-10 strongly reduce antigen-specific human T cell proliferation by diminishing the antigen-presenting capacity of monocytes via downregulation of class II major Histocompatibility complex expression. J. Exp. Med. 174: 915-924.

Guidotti, L.G. and F.V. Chisari. 2006. Immunobiology and the pathogenesis of viral hepatitis. Annual Rev. Path. 1: 23: 61.

Hayes, D.P. and A.A. Brink, M.B. Vervoort, J.M. Middeldorp, C.J. Meijer, and A. J. van den Brule. 1999. Expression of Epstein-Barr virus (EBV) transcripts encoding homologues to important human proteins in diverse EBV associated diseases. Mol. Pathol. 53: 97-103

Heinrich, P.C. and I. Behrmann, S. Haan, H.M. Hermanns, G. Muller-Newen, and F. Schaper. 2003. Principles of interleukin (IL)-6-type cytokine signalling and its regulation. Biochem. J. 374: 1-20.

Jackson, D. and M.J. Hossain, D. Hickman, D.R. Perez, and R.A. Lamb. 2008. A new influenza virus virulence determinant: the NS1 protein four C-terminal residues modulate pathogenicity. Proc. Natl. Acad. Sci. USA 105: 4381-4386

La Gruta, N.L. and K. Kedzierska, J. Stambas, and P.C. Doherty. 2007. A question of self-preservation: immunopathology in influenza virus infection. Immunol. Cell Biol. 85: 85-92

MacNeil, I.A. and T. Suda, K.W. Moore, T.R. Mosmann, and A. Zlotnik. 1990. IL-10, a novel growth cofactor for mature and immature T cells. J. Immunol. 145: 4167-4173.

Miyazaki, I. and R.K. Cheung, and H.M. Dosch. 1993. Viral Interleukin 10 is critical for the induction of B cell growth transformation by Epstein-Barr virus. J. Exp. Med. 178: 439-447.

Myrmel, H. and E. Ulvestad, and B. Asjo. 2009. The hepatitis C virus enigma. AMPIS 117: 427-439.

Rehermann, B. 1999. Cellular immune response to hepatitis C virus. J. Viral Hepat. 6: 31-35

Salek-Ardakani, S. and J.R. Arrand, and M. Mackett. 2002. Epstein-Barr virus encoded interleukin-10 inhibits HLA-class1, ICAM-1 and B7 expression on human monocytes; implications for immune evasion by EBV. Virology 304: 341-352.

Salomon, R. and E. Hoffmann, and R.G. Webster. 2007. Inhibition of the cytokine response does not protect against lethal H5N1 influenza infection. Proc. Natl. Acad. Sci. USA 104: 12479-12481.

Spencer, J.V. and J. Cadaoas, P.R. Castillo, V. Saini, and B. Slobedman. 2008. Stimulation of B lymphocytes by cmvIL-10 but not LAcmvIL-10. Virology 374: 164-169.

Stewart, J.P. and F.G. Behm, J.R. Arrand, and C.M. Rooney. 1994. Differential expression of viral and human interleukin-10 (IL-10) by primary B cell tumors and B cell lines. Virology 200: 724-732.

Wolff, T. and F. Zielecki, M. Abt, D. Voss. I. Semmler, and M. Matthael. 2008. Sabotage of antiviral signaling and effectors by influenza viruses. Biol. Chem. 389: 1299-1305.

Wu, W. and J.L. Booth, E.S. Duggan, S.Wu, K.B. Patel, K.M. Coggeshall, and J.P. Metcalf. 2009. Innate immune response to H3N2 and H1N1 influenza virus infection in a human lung organ culture model. Virology 396: 178-188.

Zeidler, R. and G. Eissner, P. Meissner, S. Uebel, R. Tampé, S. Lazis, and W. Hammerschmidt. 2007. Downregulation of TAP1 in B lymphocytes by cellular and Epstein-Barr virus-encoded interleukin-10. Blood 90: 2390-2397.

Zhang, N. and H.D. Cui, and H.X. Xue. 2008. Effect of local viral transfer of interleukin 10 gene on a rabbit arthritis model induced by interleukin 1beta. Chinese Med. J. 121: 435-438.

Zuo, Z. and C. Wang, D. Carpenter, Y. Okada, E. Nicolaidou, M. Toyoda, A. Trento, and S.C. Jordan. 2001. Prolongation of allograft survival with viral IL-10 transfection in a highly histoincompatible model of rat heart allograft rejection. Transplantation 71: 686-691.

CHAPTER 9

Cytokines in Pulmonary Tuberculosis

Katharina Ronacher[1,*], Joel Fleury Djoba-Siawaya[2] and Gerhard Walzl[1]

[1]NRF/DST Centre of Excellence for TB Biomedical Research
Division of Molecular Biology and Human Genetics
Faculty of Health Sciences, Stellenbosch University, South Africa
[2]U 845 Institut National de la Santé et de la Recherche Médicale Faculté de Médecine - Necker 156 rue de Vaugirard 75730, Paris Cedex 15, France

ABSTRACT

Tuberculosis is one of the major health threats worldwide; 2 million people die each year from this disease and about a third of the world's population is infected with *Mycobacterium tuberculosis.* Existing drugs against this bacterium are several decades old and with the alarming increase in multidrug-resistant and extensively drug-resistant strains new drugs are urgently needed. Currently the ultimate success of anti-tuberculosis therapy is the determination of the relapse rate within the first two years after initially successful treatment. However, because of the long duration of clinical trials that rely on this outcome the pharmaceutical industry is reluctant to develop new anti-tuberculosis drugs. Therefore, biomarkers of treatment response and outcome are urgently needed. Certain host markers such as cytokines that are expressed and secreted in response to mycobacteria by macrophages and T cells could possibly be useful biomarkers for tuberculosis treatment response and prediction of treatment outcome.

This chapter describes the host immune response to *Mycobacterium tuberculosis* and in particular the cytokine expression and secretion in humans latently infected with this bacterium as well as in individuals with active tuberculosis disease. Furthermore, the changes in cytokine release during tuberculosis treatment are discussed. Additionally, the usefulness of these cytokines as biomarkers to

*Corresponding author

differentiate between latent infection and active disease and as biomarkers for tuberculosis treatment response and outcome is highlighted.

INTRODUCTION

The leading cause of death due to bacterial infection is active infection with *Mycobacterium tuberculosis* (Mtb). Approximately one third of the world's population harbours latent forms of this infection, resulting in a huge reservoir for future disease. There has been a recent increase in multidrug-resistant disease, which, together with the dangerous interplay between tuberculosis and human immunodeficiency virus (HIV) infection, will pose serious challenges for health care in the years to come (Young *et al.* 2009).

Epidemiology of Tuberculosis

Tuberculosis has evolved with humans for at least thousands of years but became a major public health problem when rapid urbanization and overcrowding promoted its spread at the start of the Industrial Revolution. Currently the TB epidemic is surpassed only by HIV as the global leading cause of death due to an infectious agent (Fitzgerald 2005). The WHO declared tuberculosis a global public health emergency in 1993. In 2007 there were 9.27 million new tuberculosis cases worldwide, with India, China, Indonesia, Nigeria and South Africa being the most severely affected. Global tuberculosis rates peaked in 2004 at 142 cases per 100,000 population (World Health Organisation 2009). There were an estimated 2 million deaths attributed to tuberculosis in 2007. More than one fifth (23%) of these occurred in HIV co-infected individuals, highlighting the burden from HIV-TB co-infection (World Health Organisation 2009). An additional concern is that the HIV epidemic affects the development of multidrug-resistant (MDR) and extensively drug-resistant (XDR) tuberculosis. The large population of HIV-infected susceptible hosts with poor TB treatment success rates, the lack of airborne infection control measures in resource-limited settings where both infections occur most frequently, inadequate drug-sensitivity testing, and overburdened MDR-TB treatment programmes could all lead to a dramatic increase in this dangerous form of TB (Andrews *et al.* 2007).

Pathogenesis

After exposure to infectious droplets, approximately 30% of patients will develop primary Mtb infection. Alveolar macrophages in the terminal airways are the first line of defence and contain the infection in the majority of individuals although a focal pneumonitis does develop. Infected macrophages may subsequently transport viable Mtb bacilli to regional lymph nodes from where widespread haematogenous dissemination can take place in the non-immune host. Seeding to the lungs (most frequent), lymph nodes, kidneys, vertebrae, long bones and meninges may occur (Fitzgerald and Haas 2005). Three to eight weeks after the initial infection, cell-mediated immunity develops that is characterized by a delayed-type hypersensitivity reaction. In the majority of patients the infection is controlled with the development of latent tuberculosis infection, but in 10% active disease develops. This can be in the form of primary progressive tuberculosis (as seen in children or immune-compromised individuals) or post-primary tuberculosis, typically seen in adults. A previously infected patient may also develop active infection due to exogenous re-infection.

PROTECTIVE IMMUNITY AGAINST TB

The accumulation of a variety of immune cells at the site of initial infection leads to formation of a granuloma, characterized by infected phagocytes in the centre surrounded by activated macrophages and an outer cuff of activated lymphocytes (Fig. 1). Most infected individuals are able to contain the infection at this stage leaving only a small scar after disappearance of the granuloma.

Th1-promoting Cytokines

Once inside the lung, Mtb is ingested by dendritic cells and alveolar macrophages, which rapidly induce a pro-inflammatory response. These antigen-presenting cells (APCs) present the Mtb antigens to uncommitted, naïve T cells that subsequently differentiate into T helper 1 (Th1) cells. IL-12 drives natural killer and CD4+T cells to secrete INFγ and together with other cytokines released by the APCs, such as IL-18, IL-23, IL-27 and INFα, causes activation and proliferation of Th1 cells (Fig. 2). These cells produce the bulk of INFγ (and IL-2) as part of the adaptive immune response during latent Mtb infection. Additionally, unconventional T cell subsets (γδ T cells and CD-1 restricted T cells) proliferate considerably after presentation of Mtb lipids and glycoproteins, produce INFγ and thereby amplify the production of IL-12 and IL-18 by APCs. The acute phase cytokines IL-1 and IL-6 secreted by macrophages also play a role in the initial innate immune response and IL-6 knockout mice show increased bacterial burden early during infection (Raja 2004).

An effective T cell-mediated immune response is crucial in containing Mtb infection and protecting the host from developing active TB disease, as has become apparent from genetic mutations in genes involved in the IL-12/IL-23/INFγ system (Ottenhoff *et al.* 2005). The importance of IL-12 in protective immunity to TB was discovered by studying individuals with deficiencies in IL-12p40, who were shown to be predisposed to mycobacterial disease caused by low virulence mycobacteria as well as Mtb (Cooper *et al.* 2007). IL-12 and IL-23 both share the IL-12p40 subunit and the loss of these cytokines is associated with increase in bacterial burden, reduction in INFγ production and increased mortality. IL-12, IL-18 and IL-23 are all potent inducers of INFγ, and IL-23 contributes to protection in the absence of IL-12, but is, like IL-18, not required if IL-12 is functional. Mutations in several other genes implicated in the IL-12/IL-23/INFγ system, e.g., INFγ receptor, IL-12 receptor, STAT1, also result in increased susceptibility to mycobacterial disease (Ottenhoff *et al.* 2005), probably due to decreased INFγ outputs. Although a strong INFγ response is crucial for protective immunity, INFγ alone is not a suitable correlate of protection from infection (Fletcher 2007).

A cytokine of equal importance for protective immunity is TNFα. TNFα triggers chemokine expression by macrophages and is important in controlling primary infection and maintaining latency, as it is required for suppression of bacterial growth as well as the formation and sustainment of granulomas (Stenger 2005). In humans the importance of TNFα has become particularly apparent by the reactivation of TB in individuals treated with TNFα receptor antagonists (Dietrich and Doherty 2009). Additionally, TNFR1 knock-out mice show exacerbated TB disease with impairment of granuloma formation.

Th17-promoting Cytokines

In mycobacterial infections IL-23 drives activation of Th17 cells (Fig. 2) and IL-17 production. IL-23 expression has been shown to increase during Mtb infection; however, both IL-23 and IL-17 are not essential for protection, although IL-17 seems to play a role in proper granuloma formation (Curtis and Way 2009). IL-27 inhibits the actions of IL-17, but is required for regulation of the Th17 response and for prevention of excessive immunopathology (Curtis and Way 2009). Similarly, IL-12 limits the Th17 response; therefore, the balance of IL-12, IL-23 and IL-27 is crucial for protective immunity to TB with minimal tissue destruction. A research group has found that two distinct subsets of IL-17 and IL-22 positive CD4+T cells contribute to the immune response to Mtb in latently infected individuals as well as in TB patients and might play a role in protective immunity (Scriba *et al.* 2008).

Th2-promoting Cytokines

Th2-promoting cytokines, for example TGFβ, which is produced by Mtb-stimulated macrophages as well as regulatory T cells, are thought to contribute to inhibition of bacterial growth, but more importantly the Th2-promoting cytokines are required to balance the Mtb-induced Th1-type response and therefore aid the containment of immunopathology. The immune response of latently infected and recently diseased individuals is predominantly of Th1 type; however, if the infection cannot be contained effectively a shift towards Th2 occurs (see below). Consistent with this, serum IL-10 concentrations are lower in individuals with latent infections than in those with active disease (Jo *et al.* 2003). Healthy individuals who are able to contain the infection also express higher levels of a splice variant of IL-4 (IL-4δ2), which acts as antagonist for the full-length IL-4 (Demissie *et al.* 2004). IL-4 down-regulates microbicidal activity, increases proliferation of antigen-specific regulatory T cells, and promotes TNF toxicity and thereby immunopathology. Inhibiting the actions of IL-4 by increasing expression of IL-4δ2 might therefore have a protective function. Although Th2 cytokines are generally associated with a suppression of the Th1 immune response, IL-4 and IL-13 knockout mice control Mtb infection as well as do wild-type mice (Jo *et al.* 2003), indicating that at least in the particular mouse model used Th2 cytokines do not contribute to suppression of Th1 immunity.

PROGRESSION TO ACTIVE DISEASE

If the host immune system is unable to control bacterial replication the granuloma increases in size and necrosis occurs, eventually leading to cavitation and initiation of the classical TB symptoms (Fig. 1). Figure 3 shows a classical chest X-ray of an individual with active TB disease.

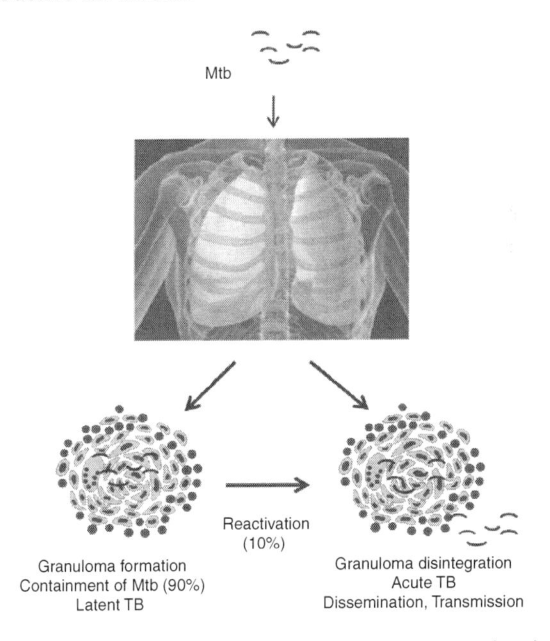

Fig. 1 Latent vs. active tuberculosis (TB). *Mycobacterium tuberculosis* (Mtb) is inhaled in the form of small droplets and taken up by lung macrophages. Most infected individuals contain the infection by formation of granulomas (latent TB). If the infection cannot be contained, about 10% of latently infected persons develop active TB during their lifetime (acute TB). (unpublished).

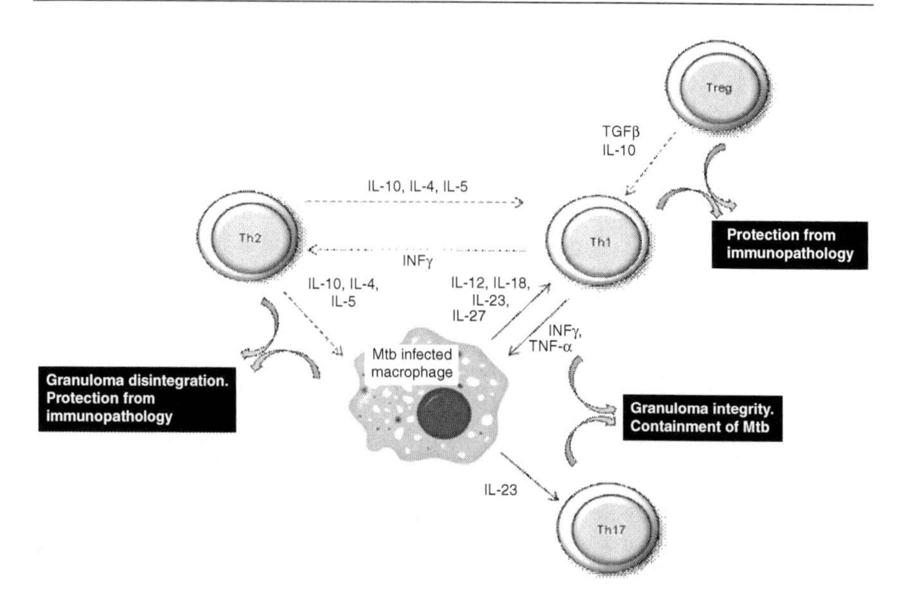

Fig. 2 Illustration of the cytokine interplay between *Mycobacterium tuberculosis* (Mtb) infected macrophages and different T-cell subsets. A balance of the Th1, Th2, Th17 and Treg T cell subsets and cytokine expression is required in order to contain Mtb as well as to protect from excessive immunopathology (unpublished).

As mentioned earlier, it has been suggested that there is a shift from Th1 responses to Th2 responses as latent infection progresses to active disease (Fletcher 2007). The reduced response of peripheral blood T cells to Mtb antigens in TB patients has been attributed to an insufficient Th1 cell activity in the periphery (as opposed to the site of disease), which in turn could be the result of increased regulatory T cell action. Stimulation of diluted whole blood from active TB patients leads to reduced T cell-mediated INFγ production compared with healthy controls, but increased IL-6 and IL-10 production (Jo *et al.* 2003). Also, serum concentrations of IL-10, a cytokine released by regulatory T cells, are increased in active TB patients (Jo *et al.* 2003). Another Th2 cytokine playing a major role in disease progression is IL-4. Latently infected individuals with increased IL-4 mRNA expression develop active disease and individuals with high ratio of INFγ mRNA to IL-4 mRNA remain healthy (Doherty and Arditi 2004). Similarly, the ratio of expression of the splice variant IL-4δ2 to IL-4 increases with treatment and is also indicative of extent of disease in TB patients and of other diseases. A poor prognosis in TB is associated not only with a low INFγ/IL-4 ratio, but also with a low INFγ/IL-10 ratio (Dietrich and Doherty 2009). Interestingly, although TNF-α is associated with protection in TB, high IL-4 levels together with TNF-α seem to cause tissue damage rather than protection (Dietrich and Doherty 2009).

Additionally, cytokines can give an indication as to the severity of pathology. For example, IL-4 levels directly correlate with severity of TB disease and INFγ as well as soluble IL-2Rα concentrations are higher in broncho-alveolar lavage fluid (BALF) from TB patients with severe disease than in those with moderate disease and both parameters decreased during TB therapy (Jo *et al.* 2003).

Finally, different cytokines have been shown to be indicative of different types of active TB disease. For example, patients with pleural TB had elevated levels of markers associated with systemic inflammation (e.g., IL-6, IL-8), whereas individuals with pulmonary TB without effusions had increase levels of cytokines of cell-mediated immunity (e.g., IL-12p40 and sCD40L) (Djoba Siawaya *et al.* 2009b).

BIOMARKERS TO DETECT LATENT INFECTION AND ACTIVE DISEASE

Sputum smear and culture are still the most frequently used tests to detect active TB disease. However, since many active TB patients do not have positive sputum and the results of TB culture take several weeks to obtain, it is of the utmost importance to identify biomarkers that can reliably distinguish between latent TB infection (LTBI) and TB disease. Additionally, it would be of great significance to find markers that can also predict the risk that a latently infected person will progress to active disease, as only approximately 10% of latently infected individuals develop TB during their lifetime (Fig. 1).

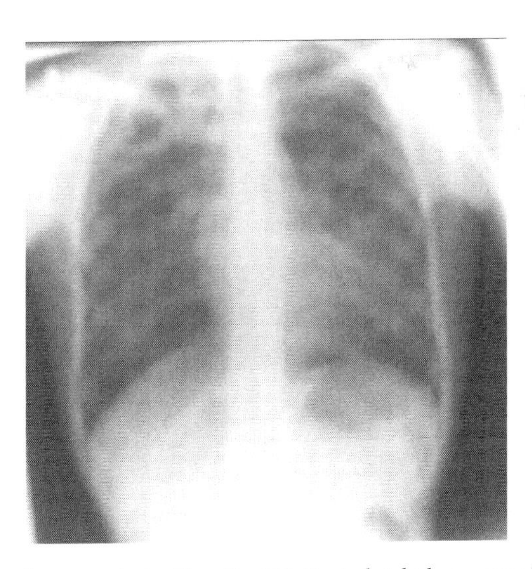

Fig. 3 Chest X-ray from a patient with active TB. Typical right lung upper lobe infiltrate with cavitation as seen with post-primary tuberculosis (unpublished).

Until recently, TST was the only test to detect latent Mtb infection. This test evaluates the skin induration generated by a delayed-type hypersensitivity (DTH) response to mycobacterial antigens 48 h after subcutaneous injection of PPD. Although the size of the skin induration correlates with subsequent development of active TB (Andersen *et al.* 2007), the TST cannot distinguish between LTBI and TB. Because of the importance of INFγ in protective immunity to TB, its potential as a biomarker for Mtb infection and disease status has been investigated by many research groups and has led to the development of INFγ release assays (IGRAs). These assays involve stimulation of memory CD4 T cells with Mtb-specific antigens (ESAT-6, CFP-10 plus additional antigens in different tests) and measurement of secreted INFγ either by ELISA or by ELISPOT assay. Although IGRAs are improved diagnostic tools to detect Mtb infection with higher specificity than the TST, these tests cannot reliably distinguish between LTBI and active TB disease. The magnitude of Mtb antigen-specific secretion of INFγ by peripheral Th1 cells seems to reflect bacterial load in active TB (Doherty *et al.* 2009) and it has been suggested that a strong INFγ response after recent Mtb exposure is indicative of an increased risk of developing clinical TB within the first two years after exposure (Andersen *et al.* 2007). However, INFγ alone is not sufficient as biomarker for detection of LTBI and TB.

Our group extensively studied differential cytokine expression in LTBI individuals and TB patients using the Luminex platform (Fig. 4), which is able to detect multiple different cytokines within one sample. We found that plasma levels of the pro-inflammatory cytokines IL-6 and IP-10 (CXCL10) were significantly lower in latently infected individuals than in TB patients. In contrast, levels of MCP-1 and the chemokine eotaxin were higher in latently infected persons than in TB patients (Djoba Siawaya *et al.* 2009a). Most of these markers reach control levels during or at the end of treatment and have also been used to distinguish fast and slow responders to treatment (see below). Latently infected individuals with increased IL-4 mRNA expression are more likely to develop active disease and individuals with high ratio of INFγ mRNA to IL-4 mRNA more likely to maintain healthy (Doherty *et al.* 2009). The frequency of both IL-17 and IL-22 positive CD4+ cells is lower in TB patients than in healthy TB contacts and BALF of TB patients contains higher IL-22 levels than that of controls (Scriba *et al.* 2008).

CYTOKINE CHANGES DURING ANTI-TUBERCULOSIS TREATMENT AND BIOMARKERS FOR TREATMENT EVALUATION

Directly observed treatment short course (DOTS) is currently the most effective means of treating tuberculosis (TB). The treatment consists of a 6 mon drug regimen divided into an intensive phase of 2 mon with four drugs (isoniazid

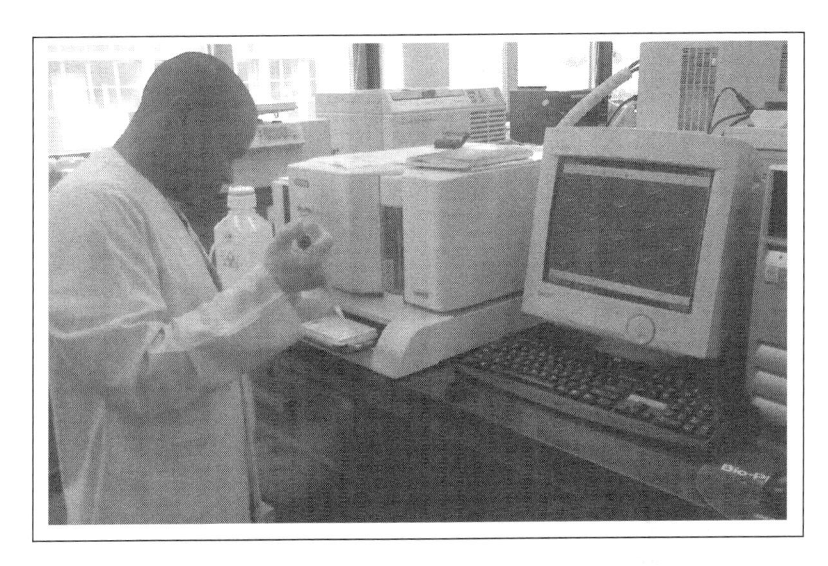

Fig. 4 Staff member performing luminex assays. During a luminex assay a bead linked to a capture antibody binds the cytokine of interest followed by binding of a fluorescently labeled detection antibody. The bead identity and the fluorescence of the detection antibody are determined by lasers. Up to 100 different cytokines can be measured within one sample using this technology.

(INH), ethambutol (EMB), rifampin (RIF) and pyrazinamide (PZA)), followed by a continuation phase of 4 mon with two drugs (INH and RIF). The International Union against Tuberculosis and Lung Disease recommends that sputum smear or culture status after 2 mon of treatment should be used to evaluate early response to chemotherapy. Studies have shown that cytokine expression and secretion during active tuberculosis and modulation thereof during anti-tuberculosis treatment may be useful in establishing and understanding the physio-pathogenic processes as well as clinical phenotypes and may provide an indication of anti-microbial response (Walzl *et al.* 2008, Wallis *et al.* 2009). Therefore, cytokines offer an alternative and valuable measure of Mtb pathogenicity, its related therapeutic interventions, and host immune system recovery after anti-TB treatment.

At the onset of tuberculosis disease, the powerful induction of the inflammatory response by Mtb subsequently leads to the establishment of an anti-inflammatory response, driven by the need of the host to prevent immune-mediated damage. These two events lead to the complex pro- and anti-inflammatory cytokine responses often observed in active TB patients (high levels of EGF, G-CSF, IL-1b, IL-6, CXCL8/IL-8, IL-12p40, CXCL10/IP-10, CCL2/ MCP-1, CCL3/MIP-1a, CCL4/MIP-1b, sCD40L, TNF-a IL-10 and, IL-13) (Djoba Siawaya *et al.* 2009a, Verbon *et al.* 1999, Dlugovitzky *et al.* 1997). Anti-tuberculosis treatment most often leads to the resolution of cytokines to normal levels (Djoba Siawaya *et al.* 2009a, Deveci *et al.* 2005). This normalization suggests a transient induction of regulatory immune mechanisms leading to subsequent restoration of immune homeostasis. If descriptive studies suggest a restoration of immune homeostasis following anti-tuberculosis treatment, functional studies based on cytokine release assays indicate that in some patients the immune system may take longer to regain its protective capacity, leaving a door open for recurrence or relapse to active disease. The persistence of low PPD-induced IFNγ production beyond 6 month after treatment has been observed in some tuberculosis patients (Hirsch *et al.* 1999). There are a number of reports giving ground to the concept that patients' underlying differential response to anti-mycobacterial therapy is reflected in their cytokine secretion or gene expression patterns, which may be used to identify patients with a high risk for a poor treatment outcome (Djoba Siawaya *et al.* 2009a, Wallis *et al.* 2009, Walzl *et al.* 2008). Serum levels of IL-10, IP-10, MCP-1, MIP-1α and TGF-α or cytokine receptors including soluble tumor necrosis factor (TNF) receptors have shown good potential as biomarkers for treatment response and early identification of fast and slow responders to anti-TB treatment when used in multi-variant analysis models (accuracies exceeding 83%) (Djoba Siawaya *et al.* 2008b, 2009a). The studies showed that high levels of IL-10, MCP-1, MIP-1α, TGF-α and IP-10 early after initiation of anti-TB therapy seem to point toward late sputum culture sterilization in patients (Djoba Siawaya *et al.* 2009a). Furthermore, the study of cytokine gene expression showed that fast and slow responders had differential IL-4 and IL-4δ2 gene expression following initiation of treatment (Djoba Siawaya *et al.* 2008a, Siawaya *et al.* 2008). This observation suggests that treatment-induced changes in mycobacterial infection are associated with measurable physiological changes in the host that carry information on treatment effectiveness.

Treatment failure occurs in 1-6% of drug-susceptible patients who completed DOTS. Relapse into active TB after initial cure occurs in 2-7% of patients with drug-susceptible isolates treated with contemporary 6 mon anti-TB therapy. Surrogate or predictive biomarkers that provide an early indication of drug efficacy would significantly decrease the cost of clinical trials and accelerate development of new antibiotics for TB treatment. Additionally, biomarkers may allow stratification of treatment regimens with possibly shortening of treatment in a majority of patients who may not require a 6 mon treatment period and intensified regimens in those with an increased risk for poor response and relapse. A list of cytokines and chemokines that could be potential biomarkers and their association in TB infection is presented in Table 2.

Table 2 Cytokines/chemokines that could be potential host biomarkers

Potential host biomarker (cytokines/chemokines and their receptors)	*Reported association*	Reference
INFγ	Treatment response	
	Bacterial clearance	Walzl *et al.* 2008
IL-4	Progression to active disease	Doherty *et al.* 2009
INFγ/IL-4	Infection status	Doherty *et al.* 2009
IL-4/IL4δ2	Infection status, degree of disease, treatment response	Doherty *et al.* 2009, Djoba Siawaya *et al.* 2008a
IL-10	Infection status, treatment response	Jo *et al.* 2003,
INFγ/IL-10	Infection status, degree of disease	Dietrich and Doherty 2009
TGFα	Infection status	Chegou *et al.* 2009
IL-1α	Infection status	Chegou *et al.* 2009
IL-6	Infection status	Djoba Siawaya *et al.* 2009a
IP-10	Infection status, treatment response	Djoba Siawaya *et al.* 2009a
MCP-1	Infection status, treatment response	Djoba Siawaya *et al.* 2009a
MIP-1α	Infection status, treatment response	Djoba Siawaya *et al.* 2009a
Eotaxin	Infection status	Djoba Siawaya *et al.* 2009a
sIL-2R	Treatment response	Walzl *et al.* 2008
sTNFR1	Sputum conversion	Walzl *et al.* 2008
sTNFR2	Sputum conversion	Walzl *et al.* 2008
	Treatment response	Djoba Siawaya *et al.* 2008b

APPLICATIONS TO OTHER AREAS OF HEALTH AND DISEASE

The potential role of cytokines as biomarkers and clues to pathogenesis and treatment response and effectiveness is not limited to tuberculosis. The critical role of cytokines in the general stimulation of the immune response makes them attractive candidate biomarkers in the assessment, design and development of new effective therapies to infection and disease in general. As current literature shows, there is a great interest in cytokines as biomarkers in almost any biomedical field (Meester and Solis-Soto 2009). Cytokines have shown great potential as biomarkers for prognosis and for assessment of treatment response and toxicity in cancer patients and patients suffering from other pathologies (Meester and Solis-Soto 2009). The fact that the immune response has many redundant mechanisms for specific diseases and that such mechanisms are also involved in responses to multiple types of pathogens makes the use of cytokines as biomarkers challenging (in terms of sensitivity and specificity). It therefore is generally accepted that the model validating the endpoint using cytokines as biomarkers should be designed on a set of markers rather than a single marker.

Table 1 Key facts about tuberculosis

- One third of the world's population is infected with the causative agent of tuberculosis, *Mycobacterium tuberculosis.*
- Every year 1.5 million people worldwide die of tuberculosis.
- Anti-tuberculosis drugs are several decades old and drug-resistant strains of *Mycobacterium tuberculosis* have appeared recently.
- The first and until now only vaccine against tuberculosis BCG (Bacille Calmette Guerin) was developed between 1905 and 1921 at the Pasteur Institute in France, but it has only limited efficacy against the disease.

Summary Points

- Initial tuberculosis infection is characterized by a T helper 1 type immune response.
- If the infection is not contained a gradual shift to T helper 2 occurs.
- Tuberculosis treatment leads to restoration of the T helper 1 cytokine profile.
- Serum cytokines could serve as biomarkers for tuberculosis treatment response and outcome.
- Biomarkers for tuberculosis treatment response are urgently needed to facilitate clinical trials for novel drug development.

Abbreviations

APCs : antigen-presenting cells
BALF : broncho-alveolar lavage fluid
BCG : Bacille Calmette-Guérin
CD : cluster of differentiation
CFP-10 : culture filtrate protein 10
HIV : human immunodeficiency virus
DOTS : directly observed treatment short course
DTH : delayed-type hypersensitivity
EMB : ethambutol
ESAT-6 : early secreted antigenic target 6
ELISA : enzyme-linked immunosorbent assay
IGRA : interferon gamma release assay
IL : interleukin
INH : isoniazid
INFγ : interferon gamma

LTBI	:	latent TB infection
MDR	:	multidrug-resistant
mRNA	:	messenger ribonucleic acid
Mtb	:	*Mycobacterium tuberculosis*
PBMCs	:	peripheral blood mononuclear cells
PPD	:	purified protein derivative
PZA	:	pyrazinamide
RIF	:	rifampin
TB	:	tuberculosis
Th1	:	T helper 1
Th2	:	T helper 2
Th17	:	T helper 17
TNFα	:	tumor necrosis factor alpha
TST	:	tuberculin skin test
XDR	:	extensively drug-resistant

Definition of Terms

Active TB disease: Disease characterized by an Mtb positive sputum smear/culture and clinical or radiological signs of disease. Typical symptoms include chest pain, cough, fever, night sweats, and weight loss.

Biomarker: A biological product that can be objectively measured and that indicates normal biological or pathological processes or a response to drug therapy.

Latent TB infection: Infection characterized by a positive TST or positive IFNγ release assay without bacteriological or radiographic evidence of TB disease.

Sputum conversion: The change from an Mtb positive sputum smear/culture to a negative smear/culture.

References

Andersen, P., and T.M. Doherty, M. Pai, and K. Weldingh. 2007. The prognosis of latent tuberculosis: can disease be predicted? Trends Mol. Med. 13: 175-182.

Andrews, J.R. and N.S. Shah, N. Gandhi, T. Moll, and G. Friedland. 2007. Multidrug-resistant and extensively drug-resistant tuberculosis: implications for the HIV epidemic and antiretroviral therapy rollout in South Africa. J. Infect. Dis. 196 Suppl 3: S482-S490.

Chegou, N.N. and G.F. Black, M. Kidd, P.D. van Helden, and G. Walzl. 2009. Host markers in QuantiFERON supernatants differentiate active TB from latent TB infection: preliminary report. BMC. Pulm. Med. 9: 21.

Cooper, A.M. and A. Solache, and S.A. Khader. 2007. Interleukin-12 and tuberculosis: an old story revisited. Curr. Opin. Immunol. 19: 441-447.

Curtis, M.M. and S.S. Way. 2009. Interleukin-17 in host defence against bacterial, mycobacterial and fungal pathogens. Immunology 126: 177-185.

Demissie, A. and M. Abebe, A. Aseffa, G. Rook, H. Fletcher, A. Zumla, K. Weldingh, I. Brock, P. Andersen, and T.M. Doherty. 2004. Healthy individuals that control a latent infection with Mycobacterium tuberculosis express high levels of Th1 cytokines and the IL-4 antagonist IL-4delta2. J. Immunol. 172: 6938-6943.

Deveci, F. and H.H. Akbulut, T. Turgut, and M.H. Muz. 2005. Changes in serum cytokine levels in active tuberculosis with treatment. Mediators. Inflamm. 2005: 256-262.

Dietrich, J. and T.M. Doherty. 2009. Interaction of Mycobacterium tuberculosis with the host: consequences for vaccine development. APMIS 117: 440-457.

Djoba Siawaya, J.F. and N.B. Bapela, K. Ronacher, N. Beyers, H.P. van, and G. Walzl. 2008a. Differential expression of interleukin-4 (IL-4) and IL-4 delta 2 mRNA, but not transforming growth factor beta (TGF-beta), TGF-beta RII, Foxp3, gamma interferon, T-bet, or GATA-3 mRNA, in patients with fast and slow responses to antituberculosis treatment. Clin. Vaccine Immunol. 15: 1165-1170.

Djoba Siawaya, J.F. and N.B. Bapela, K. Ronacher, H. Veenstra, M. Kidd, R. Gie, N. Beyers, H.P. van, and G. Walzl. 2008b. Immune parameters as markers of tuberculosis extent of disease and early prediction of anti-tuberculosis chemotherapy response. J. Infect. 56: 340-347.

Djoba Siawaya, J.F. and N. Beyers, H.P. van, and G. Walzl. 2009a. Differential cytokine secretion and early treatment response in patients with pulmonary tuberculosis. Clin. Exp. Immunol. 156: 69-77.

Djoba Siawaya, J.F. and N.N. Chegou, M.M. van den Heuvel, A.H. Diacon, N. Beyers, H.P. van, and G. Walzl. 2009b. Differential cytokine/chemokines and KL-6 profiles in patients with different forms of tuberculosis. Cytokine 47: 132-136.

Dlugovitzky, D. and A. Torres-Morales, L. Rateni, M.A. Farroni, C. Largacha, O. Molteni, and O. Bottasso. 1997. Circulating profile of Th1 and Th2 cytokines in tuberculosis patients with different degrees of pulmonary involvement. FEMS Immunol. Med. Microbiol. 18: 203-207.

Doherty, M. and R.S. Wallis, and A. Zumla. 2009. Biomarkers for tuberculosis disease status and diagnosis. Curr. Opin. Pulm. Med. 15: 181-187.

Doherty, T.M. and M. Arditi. 2004. TB, or not TB: that is the question: - does TLR signaling hold the answer? J. Clin. Invest 114: 1699-1703.

Fitzgerald, D. and D.W. Hoos. Mycobacterium tuberculosis. 6, 2852-2886. In: G.L. Mandell, J.E. Bennet and E. Dolin. 2005. Principles and Practice of Infectious Diseases. Elsevier/ Churchill, Livingstone, Philadelphia.

Fletcher, H.A. 2007. Correlates of immune protection from tuberculosis. Curr. Mol. Med. 7: 319-325.

Hirsch, C.S. and Z. Toossi, C. Othieno, J.L. Johnson, S.K. Schwander, S. Robertson, R.S. Wallis, K. Edmonds, A. Okwera, R. Mugerwa, P. Peters, and J.J. Ellner. 1999. Depressed T-cell interferon-gamma responses in pulmonary tuberculosis: analysis of underlying mechanisms and modulation with therapy. J. Infect. Dis. 180: 2069-2073.

Jo, E.K. and J.K. Park, and H.M. Dockrell. 2003. Dynamics of cytokine generation in patients with active pulmonary tuberculosis. Curr. Opin. Infect. Dis. 16: 205-210.

Meester, I. and J.M. Solis-Soto. 2009. Cytokines: Monitors of disease severity for the clinic. Expert. Opin. Med. Diag. 2: 143-155.

Ottenhoff, T.H. and F.A. Verreck, M.A. Hoeve, and V. van der. 2005. Control of human host immunity to mycobacteria. Tuberculosis (Edinb.) 85: 53-64.

Raja, A. 2004. Immunology of tuberculosis. Indian J. Med. Res. 120: 213-232.

Scriba,T.J. and B. Kalsdorf, D.A. Abrahams, F. Isaacs, J. Hofmeister, G. Black, H.Y. Hassan, R.J. Wilkinson, G. Walzl, S.J. Gelderbloem, H. Mahomed, G.D. Hussey, and W.A. Hanekom. 2008. Distinct, specific IL-17- and IL-22-producing CD4+ T cell subsets contribute to the human anti-mycobacterial immune response. J. Immunol. 180: 1962-1970.

Stenger, S. 2005. Immunological control of tuberculosis: role of tumour necrosis factor and more. Ann. Rheum. Dis. 64 Suppl 4: iv24-iv28.

Verbon, A. and N. Juffermans, S.J. van Deventer, P. Speelman, D.H. van, and P.T. van der. 1999. Serum concentrations of cytokines in patients with active tuberculosis (TB) and after treatment. Clin. Exp. Immunol. 115: 110-113.

Wallis, R.S. and T.M. Doherty, P. Onyebujoh, M. Vahedi, H. Laang, O. Olesen, S. Parida, and A. Zumla. 2009. Biomarkers for tuberculosis disease activity, cure, and relapse. Lancet Infect. Dis. 9: 162-172.

Walzl, G. and K. Ronacher, J.F. Djoba Siawaya, and H.M. Dockrell. 2008. Biomarkers for TB treatment response: challenges and future strategies. J. Infect. 57: 103-109.

World Health Organisation. 2009. Global TB control: Surveillance, planning, financing. 2009.

Young, D.B. and H.P. Gideon, and R.J. Wilkinson. 2009. Eliminating latent tuberculosis. Trends Microbiol. 17: 183-188.

CHAPTER 10

HIV and Cytokines: An Immunopatho-Genetic *Liaison*

Luca Cassetta[1], Guido Poli[1,2] and Massimo Alfano[1,*]

[1]AIDS Immunopathogenesis Unit, Department of Immunology
Transplantation and Infectious Diseases
San Raffaele Scientific Institute, Via Olgettina n. 58 - 20132 Milan, Italy
[2]Vita-Salute San Raffaele University, School of Medicine
Via Olgettina n. 58 - 20132 Milan, Italy

ABSTRACT

The human immunodeficiency virus (HIV) is a pathogenic exogenous human retrovirus responsible for the death of millions of people in the world. It infects and spreads mainly in CD4$^+$ cells (mostly T lymphocytes and mononuclear phagocytes), but it can also profoundly interfere with the host innate and adaptive immune responses inducing their profound dysregulation and exhaustion. In this scenario, HIV alters the production of chemokines and cytokines (regulators of the immune system that are necessary for cellular proliferation, trafficking and activation against viral and bacterial pathogens) in order to favor either its own replication or latency. Conversely, cytokines and chemokines regulate most if not all of the crucial steps of HIV replication, including the regulation of its latent state.

This chapter will review these mutual aspects of the interaction between HIV and the cytokine/chemokine network.

[*]*Corresponding author*

INTRODUCTION: THE HIV PANDEMICS

Thirty years after it was described in 1981, the human immunodeficiency virus (HIV), causative agent of the acquired immunodeficiency syndrome (AIDS), remains a major global threat for mankind, having caused an estimated 25 million deaths worldwide and profound demographic changes in the most heavily affected countries. On a global scale, the HIV pandemic has stabilized, although with still high levels of new infections and deaths occurring every day. Globally, 33 million people were estimated to be infected with HIV in 2008 (Fig. 1). The pharmacological treatment of HIV infection with combined anti-retroviral therapy (cART) allows clinical control of disease progression for an indefinite time, although the costs and long-term toxicities of these agents pose new challenges to the scientific and clinical community. Unfortunately, cART still does not reach effectively the most affected regions of the world, such as Sub-Saharan Africa (accounting for almost 70 and 75% of infection and deaths due to AIDS worldwide, respectively). It is indeed estimated that for every person started on cART, two to three others become newly infected (www.unaids.org). In this scenario, at least a partially effective vaccine remains urgently needed; new hope has been introduced following the partially successful RV144 trial conducted in Thailand in 16,000 volunteers (Rerks-Ngarm *et al.* 2009).

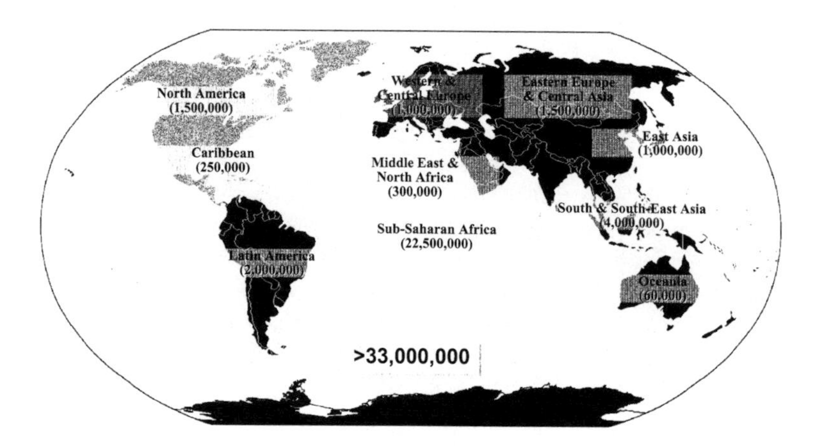

Fig. 1 Adults and children estimated to be living with HIV in 2008. Adapted from http://www.unaids.org.

THE HIV LIFE CYCLE

HIV is a retrovirus belonging to the *Lentiviridae* subfamily, encoding accessory genes in addition to the standard retroviral backbone of structural proteins (Gag, Pol, Env). Like all lentiviruses, HIV is characterized by prolonged periods of silent infection and high mutation rates, due to the lack of proofreading function of its RNA-dependent DNA polymerase (reverse transcriptase, RT) (Coffin 1997). HIV virions are composed of a nucleoprotein core surrounded by an envelope derived from the plasma membrane of the infected cells and exposing surface and transmembrane envelope (Env) glycoproteins gp120 and gp41, respectively (Fig. 2, inset). The core of the virus is constituted by an electron-dense conical capsid (p24 Gag, CA) and contains two copies of genomic viral RNA as well as the viral Nef, Vif and VpR accessory proteins. The viral enzymatic activities (i.e., RT, RNase H, integrase and protease) are encoded by the *pol* gene (Coffin 1997). Two non-structural viral proteins are essential for its replication (Tat and Rev) and have been thus defined as "regulatory", while accessory proteins (Nef, Vif, VpR and VpU) are dispensable for virus replication *in vitro* although they are important for the establishment of the infection *in vivo* where they counteract host restriction factors.

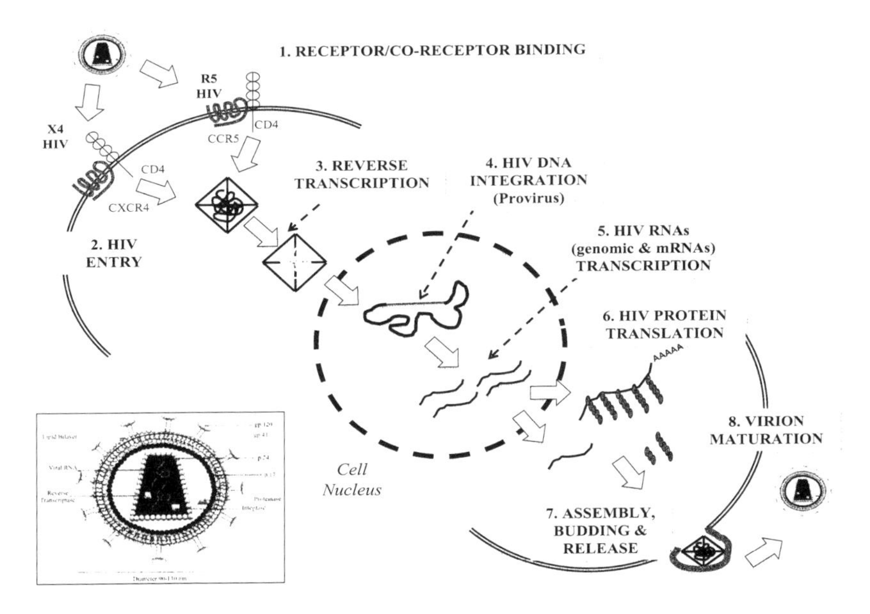

Fig. 2 HIV life cycle. HIV infection of CD4$^+$ cells leading to successful virus production. Inset: viral proteins present in HIV virions. R5 HIV, CCR5-dependent virus; X4 HIV, CXCR4-dependent virus.

Virion Attachment, Binding and Entry

This process is initiated by the binding of cell-free virions to the $\alpha_4\beta_7$ integrin (Arthos *et al.* 2008), followed by the specific interaction of gp120 Env trimers with the CD4 molecule (primary entry receptor) expressed on the surface of a subset of T lymphocytes, mononuclear phagocytes and immature myeloid dendritic cells (DC). Interaction with CD4 induces conformational changes of gp120 Env leading to the unmasking of the binding site for the second entry receptor, CCR5 (frequently) or CXCR4 (rarely). The interaction of gp120 Env with CD4 and the chemokine co-receptor results in the formation of the 6-helix bundle of gp41 Env and its insertion into the plasma membrane of the target cell, thereby permitting the fusion of virion and cell membranes.

Virion Uncoating and Reverse Transcription

Fusion between virion and cell membranes promotes the release of the virion core into the cytoplasm of the target cell. The CA disassembles releasing the viral ribonucleoprotein complex and the formation of pre-integration complex (PIC) (Coffin 1997). The reverse transcription process begins within the PIC in the cytoplasm to be completed in the nucleus. However, in the absence of the accessory virion-associated protein Vif, either Apolipoprotein B mRNA Editing enzyme, Catalytic polypeptide-like (APOBEC) 3G or 3F complex the single-stranded viral DNA and promote its degradation.

Integration

Integration of a linear form of HIV DNA into the host genome (after which it takes the name of "provirus", Fig. 3) is directed by cellular components of the PIC and occurs in host genes that are highly transcribed by RNA Pol II. This process is catalyzed by the viral integrase and involves terminal cleavage and ligation reactions (Coffin 1997). The provirus is defined by two non-coding identical sequences at the 5' and 3' ends, termed long terminal repeats (LTRs) and encoding all structural and non-structural viral genes (Fig. 3).

Proviral Transcription

Proviral transcription is influenced by host and viral factors. In particular, host factors bind to specific sequences present in the 5' HIV LTR. Although the 3'-LTR is identical to that in 5', its main function is the addition of the polypurine tract required for the translation of viral mRNAs (Coffin 1997). The HIV LTR is conventionally divided into three distinct regions: U3, R and U5. Both the enhancer and core promoter regions are located upstream of the transcription start site (+1)

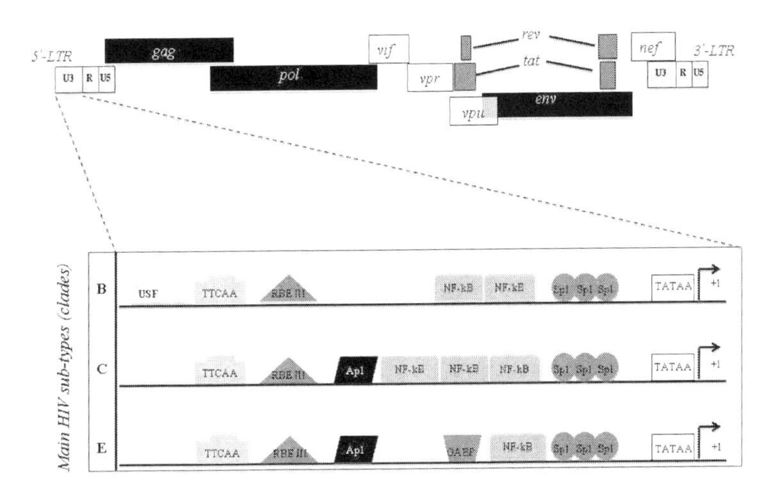

Fig. 3 Configuration of the HIV-1 provirus. Black boxes, structural genes; grey boxes, regulatory genes; open boxes, accessory genes. Composition of the U3 region of the HIV-1 LTR differs among different viral subtypes (subtype B dominant in Europe, North America and Australia; subtype C dominant in Africa; subtype E mostly in South East Asia) (Jeeninga *et al.* 2000).

in the U3 region, while R is downstream to the transcription start site (+1) and harbors the sequence encoding the RNA structure binding the regulatory protein Tat (TAR region). The core enhancer region in U3 contains Sp-1 and TATA binding sequences defining the binding site of the RNA polymerase II complex. Enhancement of HIV basal transcription is triggered by the virus-encoded Tat protein and/or by several host transcription factors, including Nuclear Factor-kB (NF-kB), NF of Activated T cells (NFAT), Upstream Stimulatory Factor-1 (USF-1) and Activated Protein-1 (AP-1) (Coffin 1997) (Fig. 3).

HIV Protein Translation, Assembly and Virion Maturation

Cytoplasmic export of genomic and mRNAs is controlled by the regulatory protein Rev (except for the fully spliced 2Kb mRNAs coding for Tat, Rev and Nef that can exit the nucleus in the absence of Rev), while Tat, like Rev, possesses a nuclear localization signal that allows its reentry in the nucleus to up-regulate HIV transcription after binding to the TAR RNA region. After synthesis of the viral proteins, HIV assembly and release in $CD4^+$ T lymphocytes occurs at the plasma membrane, as for all type C retroviruses. However, in macrophages a significant fraction of this process takes place in intracellular vacuolar compartment(s) (Gould *et al.* 2003) (Fig. 4). Tetherin represents an interferon (IFN)-inducible host restriction factor that prevents the release of new virions from the plasma membrane (Strebel *et al.* 2009).

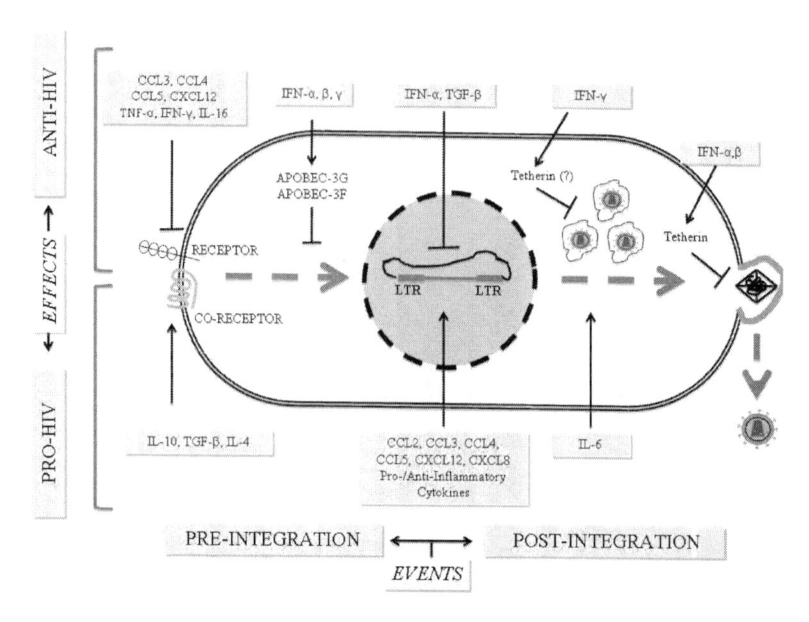

Fig. 4 Cytokines and chemokines control the HIV life cycle. In addition to the multiple effects exerted by cytokines and chemokines on virus expression, the same cytokine may exert opposite regulatory effects on early (pre-integration) or late (post-integration) events of the HIV life cycle.

Once released from the plasma membrane, virion maturation occurs as consequence of the viral protease. In addition to cell-free virion infection, a much more efficient process of infection spreading is mediated by the formation of "virological synapses" between infected and uninfected cells, in which determinants expressed at the cell surface favor the interaction between gp120 Env and the cell entry receptors and co-receptors (Haller and Fackler 2008).

CHEMOKINES AND HIV INFECTION

Chemokines and their receptors play a central role in HIV infection and disease progression, since CCR5 (and in some case CXCR4) is an obligatory co-receptor for the virus in order to gain entry into $CD4^{+}$ target cells. CCR5-binding chemokines CCL3, CCL4 and CCL5 and the CXCR4-binding molecule CXCL12 are consequently potent competitive blockers of HIV entry (Lusso 2006) (Fig. 4). In contrast, the intracellular signaling delivered upon CCR5 ligation by either chemokines or gp120 Env can up-regulate HIV expression in monocyte-derived macrophages (MDM), DC and activated peripheral blood mononuclear cells (PBMC). Similar HIV-inductive effects have been described for CCL2 and CXCL8 (Alfano and Poli 2005) (Fig. 4).

As chemokines can modulate HIV life cycle, HIV proteins can also regulate chemokine expression and production. Nef can up-regulate the secretion of CCL3 and CCL4 in MDM infected *in vitro* as well as of CXCL8 and CCL3 in DC, whereas Tat was shown to induce mostly expression of CCL2, a chemokine that up-regulates HIV replication in macrophages, astrocytes and endothelial cells. In contrast to Nef, Tat has been reported to suppress the secretion of CCL3 and CCL4 in antigen-presenting cells. Finally, gp120 Env stimulation of primary $CD4^+$ T lymphocytes and MDM induced secretion of CCL4 and CXCL12, as reviewed (Alfano and Poli 2005).

INTERFERONS (IFNS) AND HIV INFECTION

Type I IFN (IFN-α/β) has potent anti-HIV activities by inhibiting multiple steps of the virus life cycle, including reverse transcription and virus expression from integrated provirus in acutely infected primary T cells and MDM and virion release from chronically infected cell lines. These antiviral effects have been linked to the capacity of IFN-α/β of inducing cellular restriction factors, such as APOBEC3G and Tetherin (Strebel *et al.* 2009). Recent evidence indicates that HIV virions can induce the secretion of IFN-α from DC through engagement of Toll-like receptor-7 (TLR-7) by endocytosed virion RNA (Beignon *et al.* 2005).

HIV^+ individuals display high plasma levels of type I IFN during the acute phase of infection. Administration of IFN-α has been investigated as a potential therapeutic, particularly for AIDS-associated Kaposi's sarcoma, proving to reduce HIV plasma viremia in patients with relatively high $CD4^+$ T cell counts, but it has been abandoned after the advent of more specific and less toxic cART regimens.

In addition to IFN-α/β, IFN-omega has also shown potent anti-viral effect on HIV replication *in vitro* although it has not been pursued as antiviral agent.

Type II IFN (IFN-γ) is a product of activated T and natural killer cells. During HIV disease, IFN-γ expression is elevated, particularly in the germinal centers of lymph nodes that are infiltrated by activated $CD8^+$ T lymphocytes, a unique immunopathogenetic feature of HIV disease.

In vitro, IFN-γ neutralization reduced viral replication in PBMC stimulated with interleukin-2 (IL-2) (Kinter *et al.* 1995). In addition, IFN-γ induces HIV-1 expression in the chronically HIV-infected promonocytic U1 cell line; however, when U1 cells were differentiated to macrophage-like cells by phorbol myristate (PMA) IFN-γ stimulation resulted in redirection of the main site of virion budding from the plasma to intracellular vacuolar compartments, a feature reproduced with CCL2, as reviewed (Alfano and Poli 2005). In contrast, co-stimulation with IFN-γ and tumor necrosis factor-α (TNF-α) of primary MDM, leading to M1 polarization, potently inhibited R5 HIV-1 replication (Cassol *et al.* 2009).

Although Type III IFN (IFN-λ or IL-28/29) does not directly prevent virion entry, IFN-λ can induce type I IFN, CCR5-binding chemokines and host restriction factors such as APOBEC3G/3F that ultimately inhibit HIV replication (Hou *et al.* 2009).

CYTOKINES AND HIV INFECTION *IN VITRO*

To study the interaction between the regulatory viral proteins and the cytokine network, immortalized $CD4^+$ cell lines of T lymphocytic or myelo-monocytic origin play an important role. These cell lines, including U1 (derived from the acute infection of the promonocytic U937 cell line), ACH2 (from infection of the CEM-derived A3.01 $CD4^+$ T cell line), and Jurkat-derived J-LAT, are characterized by a state of relative viral latency, but inducibility of virion production by phorbol ester PMA and pro-inflammatory cytokines. One of the earliest cytokines discovered that activate HIV expression is TNF; like PMA, it up-regulates viral expression via activation of NF-kB, a host transcription factor that binds the viral LTR (Fig. 3). Monocytes and tissue macrophages are the main cell source of this cytokine after stimulation by bacterial products, such as lipopolysaccharide (LPS). Stimulation of infected cell lines or primary MDM by TNF-α up-regulates HIV transcription and replication, whereas TNF stimulation before infection causes down-regulation of CD4, therefore inhibiting viral entry and replication (Karsten *et al.* 1996). In addition, LPS stimulation of MDM leads to the secretion of CCR5-binding chemokines that inhibit R5 HIV infection (Verani *et al.* 1997). TNF-α, like IL-1β, activates HIV expression from infected cells in an autocrine/paracrine fashion in cell lines and primary PBMC activated with IL-2 (Kinter *et al.* 1995).

Other pro-inflammatory cytokines have demonstrated the capacity of enhancing HIV replication, including IL-1 (α and β), IL-6, IL-12, and IL-18 by activation of different molecular pathways. IL-1α/β, IL-6 and granulocyte-macrophage colony-stimulating factor (GM-CSF) activate the ERK-dependent Jun/Fos (AP-1) family of transcription factors that bind to specific target sequences present both in the LTR and in an intragenic enhancer in the HIV provirus. In addition, multiple cytokines, including γ-common cytokines (IL-2, IL-7 and IL-15) and β-common cytokines (IL-3 and GM-CSF), activate signal transducer and activator of transcription-5 (STAT5) that can bind to the HIV-1 LTR and trigger virus expression (Figs. 3 and 4). However, we have reported that most HIV^+ individuals carry cleaved isoforms of STAT5 retaining DNA binding, but lacking transactivating function (Crotti *et al.* 2007).

The role of anti-inflammatory cytokines in HIV replication is more complex and results in the up-regulation or inhibition of virus production as a function of the timing of cell stimulation vs. that of infection. For example, IL-4 and IL-13 up-regulate CXCR4 but down-modulate CCR5; consequently, they favor or impair the viruses using these entry co-receptors and can induce virus replication in monocytes but inhibit it in MDM (Schuitemaker *et al.* 1992). We have confirmed reduced levels of R5 HIV replication in IL-4 polarized M2a-MDM not associated with decreased levels of HIV DNA synthesis, as observed in M1-polarized MDM, vs. control cells (Cassol *et al.* 2009), suggesting interference at steps following HIV DNA integration. IL-10 can up-regulate CXCR4 and CCR5 expression and related infections of DC (Ancuta *et al.* 2001) and monocytes (Sozzani *et al.* 1998), respectively. Both up-regulation and inhibition of HIV replication in primary MDM have been associated with the capacity of IL-10 to fully suppress synthesis and release of pro-inflammatory cytokines such as TNF-α and IL-6 (Weissman *et al.* 1995). TGF-β, like retinoic acid, suppressed HIV transcription and replication in primary MDM infected with HIV, but enhanced viral replication when the cells were stimulated before infection (Poli *et al.* 1991) (Fig. 4). Of interest, increased levels of bioactive TGF-β have been reported in the sera and PBMC of AIDS patients with <50 $CD4^+$ T lymphocytes/μl (Alonso *et al.* 1997).

CYTOKINE ACTIVATION IN HIV-INFECTED INDIVIDUALS

The onset of clinical symptoms during HIV infection occurs during the first or second week after the primary infection and generally coincides with high levels of viral replication (acute retroviral syndrome in primary HIV infection, PHI). This acute syndrome usually subsides within 2-8 wk after the onset of the symptoms in association with a curtailment of viremia levels due to the immune response to the infection, and a subsequent period of clinical latency (Fig. 5). The levels of cytokines and chemokines have been shown to mirror these variations in plasma viremia. In particular, levels of IL-15, IFN-α/β and CXCL10 increase rapidly but transiently; IL-18, TNF-α, IFN-γ and IL-22 also increase rapidly but are sustained for prolonged times, whereas the increase in IL-10 is slightly delayed. This cytokine storm following PHI is described as greater than that observed in acute hepatitis B and C virus infections (Stacey *et al.* 2009). Thus, the magnitude of this systemic cytokine response is not a predictor *per se* of viral clearance that does not occur with HIV infection, unlike what is observed with viral hepatitis. The intense cytokine response in acute HIV infection may instead fuel viral replication and likely mediates immunopathology, as observed in other acute viral infections (Cameron *et al.* 2007).

HIV disease progression is associated with a decrease of Th1 cytokines, although still in the presence of increased expression of several pro-inflammatory cytokines. In parallel, higher levels of anti-inflammatory (IL-10, TGF-β), Th2 (IL-4, IL-5, IL-13) cytokines and IFNs are frequently observed in advanced chronic infection (Alfano *et al.* 2008). The expression of chemokines and cytokines in the

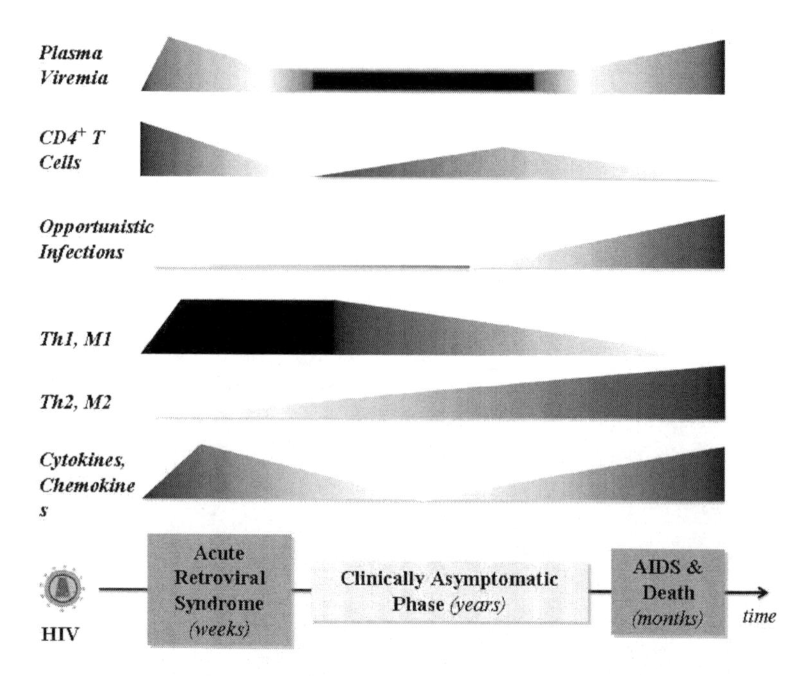

Fig. 5 Main immunological and virological events associated with HIV infection *in vivo*. The described trends are referred to the natural history of HIV infection in the absence of cART.

peripheral lymphoid tissue (pLT) and gut-associated lymphoid tissue (GALT) during symptomatic acute HIV infection has been investigated (Nilsson *et al.* 2007). Immune activation in GALT was found concomitant to or preceding activation in pLT during acute HIV infection with an increase expression of both Th1 and Th2 cytokines. The expression of IL-2 was up-regulated in GALT from acute HIV-infected patients with a trend towards a higher expression of IL-4 and IL-10 in comparison to pLT. This combined expression of Th1 and Th2 cytokines has been proposed to represent a reflection of a non-coordinated HIV antigen-driven immune response (Appay *et al.* 2002).

Although cART has dramatically improved the life quality and expectancy of HIV^+ individuals, side effects have emerged with prolonged therapy. Among these is the Immune Reconstitution Inflammatory Syndrome (IRIS), encompassing symptoms compatible with an inflammatory response to infectious agents. Its pathogenesis has been hypothesized to involve an alteration of the ratio between pro-inflammatory Th17 cells and immunosuppressive T regulatory cells (Seddiki *et al.* 2009) (Fig. 6). IRIS can be controlled by anti-inflammatory molecules (Dhasmana *et al.* 2008), but also by combinations of certain cytokines together

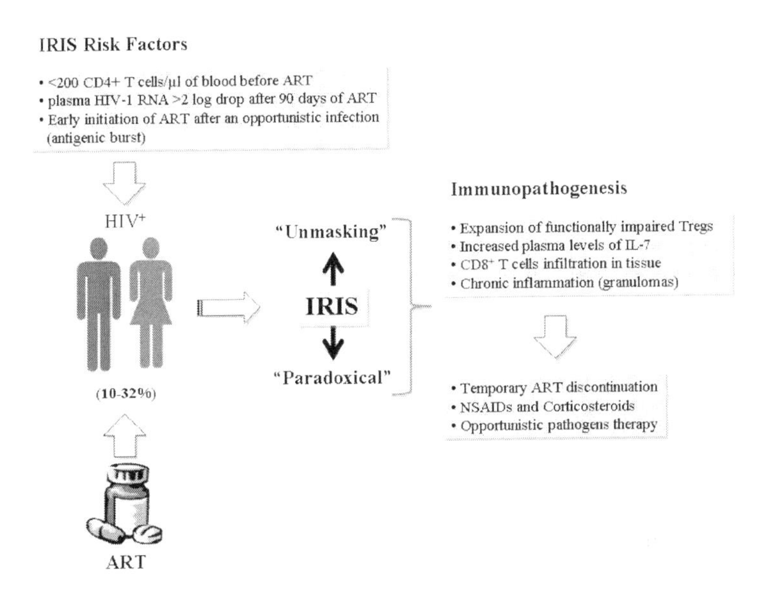

Fig. 6 Schematic summary of IRIS occurring in up to 32% of individuals assuming cART. See text for key references.

with ART. The combination of ART with IL-2 and GM-CSF has been reported to induce a better clearance of *Mycobacterium avium* causing a boost of the T cell response, likely associated with either the proliferation of immature thymocytes and/or with the rescue of anergic $CD4^+$ T cells (Pires *et al.* 2005). Moreover, since both cytokines induce maturation and activation of macrophages, this cytokine might enhance the degradation of intracellular pathogens (Pires *et al.* 2005).

CONCLUDING REMARKS

The capacity of HIV to exploit several cells of the human immune system for regulating its own state of latency or replication encompasses profoundly the network of cytokines and chemokines. The novel challenges of cART-treated HIV^+ individuals developing new syndromes (such as IRIS) and facing different causes of mortality (including solid tumors, hepatitis C complications and cardiovascular diseases) suggest that cytokines and chemokines could still be important agents either as biomarkers of disease progression or as therapeutic or preventive adjuvants.

Table I Key features of HIV infection

• HIV is a human pathogenic lentivirus.
• Pathogenesis of HIV infection is characterized by acute phase (week to months) and chronic phase (years) resulting, in patients not receiving combination anti-retroviral therapy (cART), in Acquired Immunodeficiency Disease Syndrome (AIDS).
• Death of infected individual generally occurs at the AIDS stage, characterized by profound immunodeficiency allowing occurrence of opportunistic infections and malignancies.
• As for other viral infections, boost of "anti-viral" cytokines such as interferons occurs during the acute phase, whereas the constant progression of HIV disease and immunodeficient status is associated with a progressive shift from a Th1 (pro-inflammatory) to Th0/Th2 (anti-inflammatory) immune responses.
• cART reduces viral load to undetectable levels, corrects, at least partially, the immune deficiency, and enhances the life expectancy of infected individuals.
• One of the side effects of cART is linked to the partial immune-reconstitution following the decreased viral load. The Immune Reconstitution Inflammatory Syndrome (IRIS) is associated with dysregulated equilibrium between the different subsets of Th cells boosting the production of inflammatory cytokines.
• Cytokines and chemotactic cytokines regulate all steps of the HIV life cycle.
• Control of cytokine levels (i.e., during IRIS) or reconstitution of normal levels of cytokines in HIV infected-individuals represents a challenge in the endeavor to improve immunological reconstitution and enhance quality of life in infected individuals.

Abbreviations

AIDS : Acquired Immunodeficiency Syndrome
APOBEC : Apolipoprotein B mRNA Editing enzyme, Catalytic polypeptide-like
ART : Anti-retroviral therapy
CA : capsid
DC : dendritic cells
GALT : gut-associated lymphoid tissue
HIV : Human Immunodeficiency Virus
IFN-α : Interferon-α
IFN-β : Interferon-β
IRIS : Immune Reconstitution Inflammatory Syndrome
LTR : long terminal repeat
NF-κB : Nuclear Factor-κB
PIC : pre-integration complex
pLT : peripheral lymphoid tissue
TAR : transactivation response region
TNF-α : tumor necrosis factor-α

Definition of Terms

CCR5 and CXCR4: Receptors for the chemokines CCL3, CCL4, CCL5 and CXCL12, respectively. They also represent the co-receptor required for HIV to gain entry into the target cells after binding to the primary receptor CD4. Based on the co-receptor use (CCR5 vs. CXCR4), viral strains are classified as R5 (frequent), X4 (rare) or dualtropic R5X4 (emerging in ca. 50% of infected individuals).

Macrophages: Cellular subset resident exclusively in tissues and derived from the terminal differentiation of circulating monocytes. Macrophages contribute to matrix deposition and tissue remodeling, but also represent the first line of defense against invading pathogens, as well as helping initiate specific adaptive immunity. Since monocyte-macrophages express CD4 and the chemokine co-receptors (see below), they are infectable by HIV.

Retrovirus: Virus bearing RNA genome that is transcribed into cDNA by the viral enzyme reverse transcriptase (RT). The viral DNA is then stably inserted into the host genome (provirus) and takes advantage of the transcriptional and translational machinery of the cell to replicate.

T lymphocytes: Cellular subset of the adaptive immune response involved in cell-mediated immunity as well as regulating B lymphocyte activity. Expression of the CD4 antigen on T lymphocytes identifies them as T "helper" cells recognizing peptides in the context of MHC Class II antigens and makes them susceptible to HIV infection.

Acknowledgments

This study has been partially supported by the "Europrise" and "Neat" projects of the 6th FP of the EC (to GP). L.C. performed this study as part of his PhD in Molecular Medicine, Vita-Salute San Raffaele University of Milan, Italy.

References

Alfano, M. and A. Crotti, E. Vicenzi, and G. Poli. 2008. New players in cytokine control of HIV infection. Curr. HIV/AIDS Rep. 5: 27-32.

Alfano, M. and G. Poli. 2005. Role of cytokines and chemokines in the regulation of innate immunity and HIV infection. Mol. Immunol. 42: 161-182.

Alonso, K. and P. Pontiggia, R. Medenica, and S. Rizzo. 1997. Cytokine patterns in adults with AIDS. Immunol. Invest. 26: 341-350.

Ancuta, P. and Y. Bakri, N. Chomont, H. Hocini, D. Gabuzda, and N. Haeffner-Cavaillon. 2001. Opposite effects of IL-10 on the ability of dendritic cells and macrophages to replicate primary CXCR4-dependent HIV-1 strains. J. Immunol. 166: 4244-4253.

Appay, V. and L. Papagno, C.A. Spina, P. Hansasuta, A. King, L. Jones, G.S. Ogg, S. Little, A.J. McMichael, D.D. Richman, *et al.* 2002. Dynamics of T cell responses in HIV infection. J. Immunol. 168: 3660-3666.

Arthos, J. and C. Cicala, E. Martinelli, K. Macleod, D. Van Ryk, D. Wei, Z. Xiao, T.D. Veenstra, T.P. Conrad, R.A. Lempicki, *et al.* 2008. HIV-1 envelope protein binds to and signals through integrin alpha4beta7, the gut mucosal homing receptor for peripheral T cells. Nat. Immunol. 9: 301-309.

Beignon, A.S. and K. McKenna, M. Skoberne, O. Manches, I. DaSilva, D.G. Kavanagh, M. Larsson, R.J. Gorelick, J.D. Lifson, and N. Bhardwaj. 2005. Endocytosis of HIV-1 activates plasmacytoid dendritic cells via Toll-like receptor-viral RNA interactions. J. Clin. Invest. 115: 3265-3275.

Cameron, M.J. and L. Ran, L. Xu, A. Danesh, J.F. Bermejo-Martin, C.M. Cameron, M.P. Muller, W.L. Gold, S.E. Richardson, S.M. Poutanen, *et al.* 2007. Interferon-mediated immunopathological events are associated with atypical innate and adaptive immune responses in patients with severe acute respiratory syndrome. J. Virol. 81: 8692-8706.

Cassol, E. and L. Cassetta, C. Rizzi, M. Alfano, and G. Poli. 2009. M1 and M2a polarization of human monocyte-derived macrophages inhibits HIV-1 replication by distinct mechanisms. J. Immunol. 182: 6237-6246.

Coffin, J.M. and S.H. Hughes, H.E. Varmus. 1997. Retroviruses. Cold Spring Harbor, New York, USA.

Crotti, A. and M. Lusic, R. Lupo, P.M. Lievens, E. Liboi, G. Della Chiara, M. Tinelli, A. Lazzarin, B.K. Patterson, M. Giacca, *et al.* 2007. Naturally occurring C-terminally truncated STAT5 is a negative regulator of HIV-1 expression. Blood 109: 5380-5389.

Dhasmana, D.J. and K. Dheda, P. Ravn, R.J. Wilkinson, and G. Meintjes. 2008. Immune reconstitution inflammatory syndrome in HIV-infected patients receiving antiretroviral therapy : pathogenesis, clinical manifestations and management. Drugs 68: 191-208.

Gould, S.J. and A.M. Booth, and J.E. Hildreth. 2003. The Trojan exosome hypothesis. Proc Natl. Acad. Sci. USA 100: 10592-10597.

Haller, C. and O.T. Fackler. 2008. HIV-1 at the immunological and T-lymphocytic virological synapse. Biol. Chem. 389: 1253-1260.

Hou, W. and X. Wang, L. Ye, L. Zhou, Z.Q. Yang, E. Riedel, and W.Z. Ho. 2009. Lambda interferon inhibits human immunodeficiency virus type 1 infection of macrophages. J. Virol. 83: 3834-3842.

Jeeninga, R.E. and M. Hoogenkamp, M. Armand-Ugon, M. de Baar, K. Verhoef, and B. Berkhout. 2000. Functional differences between the long terminal repeat transcriptional promoters of human immunodeficiency virus type 1 subtypes A through G. J. Virol. 74: 3740-3751.

Karsten, V. and S. Gordon, A. Kirn, and G. Herbein. 1996. HIV-1 envelope glycoprotein gp120 down-regulates CD4 expression in primary human macrophages through induction of endogenous tumour necrosis factor-alpha. Immunology 88: 55-60.

Kinter, A.L. and G. Poli, L. Fox, E. Hardy, and A.S. Fauci. 1995. HIV replication in IL-2-stimulated peripheral blood mononuclear cells is driven in an autocrine/paracrine manner by endogenous cytokines. J. Immunol. 154: 2448-2459.

Lusso, P. 2006. HIV and the chemokine system: 10 years later. Embo. J. 25: 447-456.

Nilsson, J. and S. Kinloch-de-Loes, A. Granath, A. Sonnerborg, L.E. Goh, and J. Andersson. 2007. Early immune activation in gut-associated and peripheral lymphoid tissue during acute HIV infection. AIDS 21: 565-574.

Pires, A. and M. Nelson, A.L. Pozniak, M. Fisher, B. Gazzard, F. Gotch, and N. Imami. 2005. Mycobacterial immune reconstitution inflammatory syndrome in HIV-1 infection after antiretroviral therapy is associated with deregulated specific T-cell responses: beneficial effect of IL-2 and GM-CSF immunotherapy. J. Immune Based Ther. Vaccines 3: 7.

Poli, G. and A.L. Kinter, J.S. Justement, P. Bressler, J.H. Kehrl, and A.S. Fauci. 1991. Transforming growth factor beta suppresses human immunodeficiency virus expression and replication in infected cells of the monocyte/macrophage lineage. J. Exp. Med. 173: 589-597.

Rerks-Ngarm, S. and P. Pitisuttithum, S. Nitayaphan, J. Kaewkungwal, J. Chiu, R. Paris, N. Premsri, C. Namwat, M. de Souza, E. Adams, *et al.* 2009. Vaccination with ALVAC and AIDSVAX to prevent HIV-1 infection in Thailand. N. Engl. J. Med. 361: 2209-2220.

Schuitemaker, H. and N.A. Kootstra, M.H. Koppelman, S.M. Bruisten, H.G. Huisman, M. Tersmette, and F. Miedema. 1992. Proliferation-dependent HIV-1 infection of monocytes occurs during differentiation into macrophages. J. Clin. Invest. 89: 1154-1160.

Seddiki, N. and S.C. Sasson, B. Santner-Nanan, M. Munier, D. van Bockel, S. Ip, D. Marriott, S. Pett, R. Nanan, D.A. Cooper, *et al.* 2009. Proliferation of weakly suppressive regulatory CD4+ T cells is associated with over-active CD4+ T-cell responses in HIV-positive patients with mycobacterial immune restoration disease. Eur. J. Immunol. 39: 391-403.

Sozzani, S. and S. Ghezzi, G. Iannolo, W. Luini, A. Borsatti, N. Polentarutti, A. Sica, M. Locati, C. Mackay, T.N. Wells, *et al.* 1998. Interleukin 10 increases CCR5 expression and HIV infection in human monocytes. J. Exp. Med. 187: 439-444.

Stacey, A.R. and P.J. Norris, L. Qin, E.A. Haygreen, E. Taylor, J. Heitman, M. Lebedeva, A. DeCamp, D. Li, D. Grove, *et al.* 2009. Induction of a striking systemic cytokine cascade prior to peak viremia in acute human immunodeficiency virus type 1 infection, in contrast to more modest and delayed responses in acute hepatitis B and C virus infections. J. Virol. 83: 3719-3733.

Strebel, K. and J. Luban, and K.T. Jeang. 2009. Human cellular restriction factors that target HIV-1 replication. BMC Med. 7: 48.

Verani, A. and G. Scarlatti, M. Comar, E. Tresoldi, S. Polo, M. Giacca, P. Lusso, A.G. Siccardi, and D. Vercelli. 1997. C-C chemokines released by lipopolysaccharide (LPS)-stimulated human macrophages suppress HIV-1 infection in both macrophages and T cells. J. Exp. Med. 185: 805-816.

Weissman, D. and G. Poli, and A.S. Fauci. 1995. IL-10 synergizes with multiple cytokines in enhancing HIV production in cells of monocytic lineage. J. Acquir. Immune Defic. Syndr. Hum. Retrovirol. 9: 442-449.

CHAPTER 11

NK Cells and Cytokines in Health and Disease

Rizwan Romee[1] and Jeffrey S. Miller[2,*]

[1]Department of Medicine, University of Minnesota, 420 Delaware St SE, Minneapolis, Minnesota 55455, USA

[2]University of Minnesota Cancer Center, Division of Hematology, Oncology, and Transplantation Harvard Street at East River Road, Minneapolis Minnesota 55455, USA

ABSTRACT

NK cells are important lymphocytes characterized by a CD56$^+$CD3$^-$ phenotype. Based on the expression of CD56 and CD16, NK cells are divided into CD56^{bright}CD16$^-$ and CD56^{dim}CD16$^+$ subsets. These two NK cell subsets differ in cytokine secretion and cytoxicity; CD56^{bright}CD16$^-$ cells are more proficient in cytokine secretion, while CD56^{dim}CD16$^+$ cells are better at mediating cytotoxicity. IL-15 transpresented by dendritic cells plays a critical role in the normal development, homeostasis and activation of NK cells. NK cells secrete several cytokines and chemokines, which play a key role in orchestrating an immune response early after exposure to pathogens. The importance of NK cells in clearing viral infections is underscored by the recurrent life-threatening viral infections in patients with absent or dysfunctional NK cells. This has led to studies on NK cells in viral infections such as HIV, cytomegalovirus and hepatitis. Yet, not all NK cell reactions are beneficial. There is some evidence of their overzealous role in a mouse model of diabetes mellitus and other autoimmune diseases. We have been interested in the role of NK cells in immune surveillance against malignant cells to treat cancers that have failed standard therapy. NK cells are thought to contribute to what is called a graft versus leukemia (GvL) effect in the setting of allogeneic hematopoietic cell transplantation (allo-HCT). The goal of this chapter

*Corresponding author

is to provide an overview of NK cell biology and ways in which to exploit this knowledge to treat human disease.

INTRODUCTION

NK cells are defined morphologically as large granular lymphocytes. Phenotypically they are characterized by the absence of T cell markers such as CD3 and by the presence of CD56 (NCAM). Functionally, they are unique in their ability to lyse target cells and produce cytokines without prior sensitization; hence the name "natural killer" cells. NK cells have traditionally been considered part of the innate immune system, though there is some recent evidence suggesting they may have an ability to remember previous pathogenic encounters and thus exert a memory response, a function usually associated with the adaptive immune system (Sun *et al.* 2009). NK cells comprise 10-15% of peripheral blood lymphocytes and are present in a wide variety of lymphoid and non-lymphoid body tissues. They play an important role in immune surveillance of malignant cells by their ability to preferentially kill transformed cells while sparing normal tissue. This unique ability makes them an attractive immune cell population for use in adoptive cell therapy in patients with cancer.

BASIC BIOLOGY OF NK CELLS

Human NK cells can be divided into two major subsets based on the relative expression of CD56 (NCAM) and co-expression of CD16 (FcγRIII). In the peripheral blood and spleen, around 90% of the NK cells express CD16 and have low-density expression of CD56 ($CD56^{dim}$ $CD16^{+}$). These $CD56^{dim}$ NK cells express perforin and granzyme and are cytotoxic to tumor cells even without stimulation with cytokines. Due to their expression of CD16, a low-affinity Fc receptor, $CD56^{dim}$ NK cells also participate in killing target cells coated with antibody, a process called antibody-dependent cell cytotoxicity (ADCC). $CD56^{bright}CD16^{-}$ NK cells lack perforin and granzyme expression and poorly kill tumor target cells without activation by cytokines. $CD56^{bright}$ NK cells are believed to produce more cytokines than $CD56^{dim}$ NK cells. The two NK cell subtypes also differ in their homing receptor expression: CCR7 is expressed on $CD56^{bright}CD16^{-}$ cells, favoring their homing to the lymph nodes, and CXCR1 is expressed on $CD56^{dim}CD16^{+}$ cells, favoring their homing to inflamed sites. Contact mediated killing usually involves perforin and granzyme but also can use FasL and TRAIL expression. Stimulation by cytokines such as IL-2, IL-15 or IL-12 significantly increases the cytotoxic ability of both the NK subtypes. Recent studies suggest that $CD56^{bright}$ NK cells are immediate precursors to $CD56^{dim}$ NK cells as suggested by longer telomerase length in $CD56^{bright}$ cells and their ability to differentiate into $CD56^{dim}$ cells when cultured with synovial or skin fibroblasts *in vitro* or upon transfer into NOD-SCID mice (Chan *et al.* 2007).

Human NK cell function is defined by the expression of a number of different receptors, most of which belong to one of two structural classes, the immunoglobulin (Ig) or the C-type lectin families. Functionally the receptors can be classified into inhibitory or activating. KIRs are type 1 trans-membrane proteins belonging to the Ig super-family and encoded by the KIR gene locus on chromosome 19q13.4. The KIR gene family consists of 15 genes encoding individual KIR and 2 psuedogenes that are not expressed. Based on the gene content, KIR haplotypes are classified into type A and type B. KIR haplotype A is characterized by the presence of only one activating gene, KIR2DS4, and fewer inhibitory genes. KIR haplotype B, originally defined by the presence of KIR2DL5, has one to five activating KIRs (i.e., KIR2DS1, KIR2DS2, KIR2DS3, KIR2DS5 and KIR3DS1). Individual KIR genes show an extreme degree of polymorphism and new alleles continue to be discovered. While some of the KIR bind to class 1 HLA molecules, the natural ligands for many of the activating KIR have yet to be discovered.

NKG2A is another MHC-specific inhibitory receptor expressed by NK cells and belongs to C-type lectin group of proteins. It is expressed in combination with CD94 as a heterodimer complex and binds to HLA-E on the target cells. NKG2D belongs to the NKG2A family and is expressed as a homodimer associated with an adaptor protein DAP10 through which it is able to send activating signals upon binding to appropriate ligands. NKG2D recognizes various stress proteins such as MHC class 1 polypeptide-related sequence A/B (MICA and MICB) and the class 1-like CMV-homologous ULBP proteins (ULBP1-4), which are often up-regulated on malignant and infected cells .

NKp30, NKp44 and NKp46 are potent activating NK cell receptors and are called natural cytotoxicity receptors (NCRs). NKp44 and NKp46 are able to recognize virus-specific hemagglutinin, but the endogenous ligands for these receptors are largely unknown. Recently an interesting study demonstrated the ability of NKp46 to recognize ligands on salivary and pancreatic beta cells and highlighted the importance of this interaction for the development of diabetes mellitus (Gur *et al.* 2010). Several other NK cell receptors such as 2B4 and DNAM-1 are important for killing malignant cells, the description of which is beyond the scope of this chapter. Major activating and inhibitory receptors expressed on human NK cells are depicted in Fig. 1.

The mechanism by which NK cells recognize targets and mediate their function is important for understanding how to manipulate them in health and disease. It is thought that KIR interactions with MHC class 1 ligands may be dominant clinically. Targets cells with decreased MHC class 1 expression are unable to ligate inhibitory KIR, making the cell sensitive to NK cell killing. This "missing self" phenomenon explains the recognition of various malignant and virally infected cells, which have low expression of HLA molecules (Ljunggren and Karre 1985) (Fig. 2).

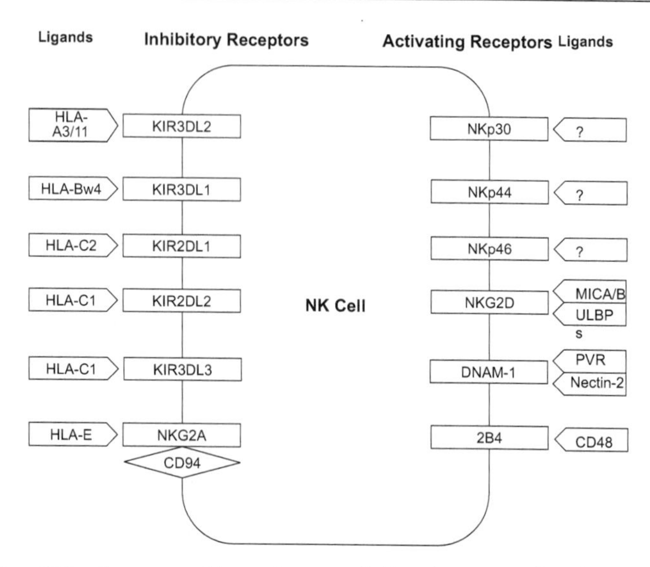

Fig. 1 Major NK cell receptors. Major inhibitory (left) and activating (right) NK cell receptors with corresponding ligands.

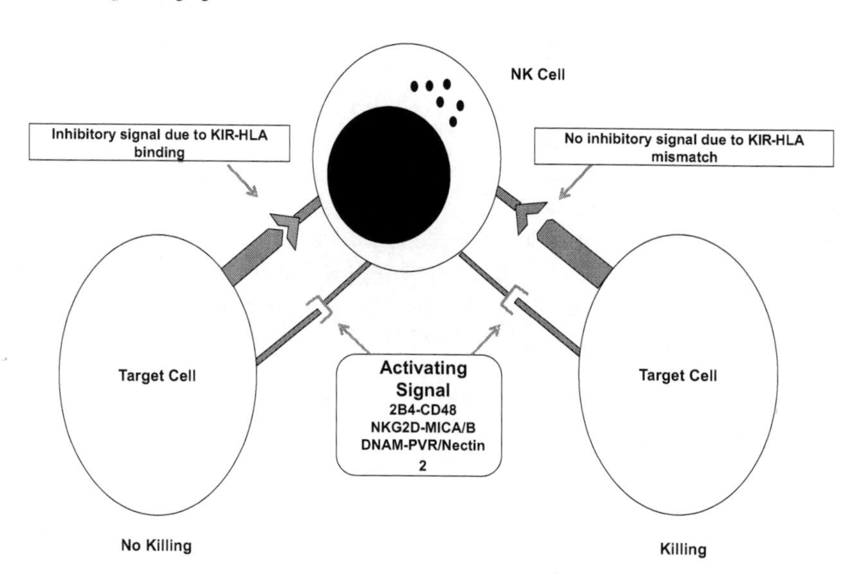

Fig. 2 Missing self hypothesis. On left, the inhibitory signal delivered by the KIR-HLA binding overcomes the activating signal from the activating receptors such as 2B4. On right, no inhibitory signal is delivered due to inability of KIR to bind the HLA; therefore, the activating signal dominates and leads to target all killing. This mechanism explains both tolerance of the NK cells towards the normal host cells and also the alloreactivity of NK cells in KIR-Ligand mismatched allo-HCT setting.

CYTOKINES IN NK CELL DEVELOPMENT AND HOMEOSTASIS

In vitro, NK cells can be derived from $CD34^+$ hematopoietic stem cells of bone marrow or cord blood origin when cultured on a mouse embryonic cell line, EL08-1D2, in presence of a cocktail of cytokines that include IL-15, IL-7, IL-3, SCF and FLT-3 Ligand (Cichocki and Miller 2010). NK cells have been successfully generated in stroma-free culture systems involving use of high concentrations of human cytokines, but these NK cells tend to have a lower expression of KIR and may not be fully developed (Yu *et al.* 1998). Analysis of knockout mice (IL-$2^{-/-}$, IL2R$\alpha^{-/-}$, IL-$7^{-/-}$, $SCT^{-/-}$ and FLT-$3L^{-/-}$) fails to have significant dysfunction in NK cell development (Schorle *et al.* 1991; He and Malek 1996; Colucci and Di Santo 2000; McKenna *et al.* 2000). In contrast, IL-$15^{-/-}$ or IL15R$\alpha^{-/-}$, IL2R$\beta^{-/-}$, and $\gamma c^{-/-}$ mice have significantly reduced numbers of mature NK cells, highlighting the critical role of these cytokines in NK cell development (Kawamura *et al.* 2003; Vosshenrich *et al.* 2005). IL-15 belongs to what is called γ-chain cytokine family, which includes IL-2, IL-4, IL-7, IL-9, and IL-21, all sharing the γ-chain of the receptor complex. These cytokines have an α-chain, which is specific to each cytokine. IL-2 and IL-15 also share their β-chain (IL-2/IL-15Rβ), which may help explain some of the overlapping actions of these two cytokines. IL-15 may be unique among the cytokines by more efficient signaling when membrane bound compared to its secreted form. IL-15 is chaperoned by IL-15Rα subunit to the cell membrane, where it remains bound to it. IL-15 binds to IL-15Rα via its sushi domain and this interaction is thought to be one of the strongest known non-covalent interactions. The membrane-bound IL-15 is then able to interact with the IL2/IL15Rβ and γc complex present on the neighboring cells. This unique mode of presentation is called transpresentation (Fig. 3).

Dendritic cells in the bone marrow and lymph nodes are thought to be the main cell type responsible for transpresentation of IL-15. This transpresentation of IL-15 is not only important for the development of NK cells from hematopoietic progenitors but is also critical for maintaining the normal homeostasis of mature NK cells. The latter is suggested by poor survival of adoptively transferred mature NK cells into IL-$15^{-/-}$ or IL15R$\alpha^{-/-}$ mice (Cooper *et al.* 2002; Prlic *et al.* 2003). In contrast to IL-15, other cytokines such as IL-2 and IL-7 are not thought to play a major role in maintaining NK cell survival as suggested by a relatively normal life span of mature NK cells adoptively transferred into IL-$2^{-/-}$ or ILR2$\alpha^{-/-}$ mice (Cooper *et al.* 2002). Although NK cell development is less clear than T or B cell development, at least some stages of the NK cell development are thought to take place in lymph nodes (Freud *et al.* 2006). Developing NK cells express surface protein characteristics that define five developmental stages (Fig. 4).

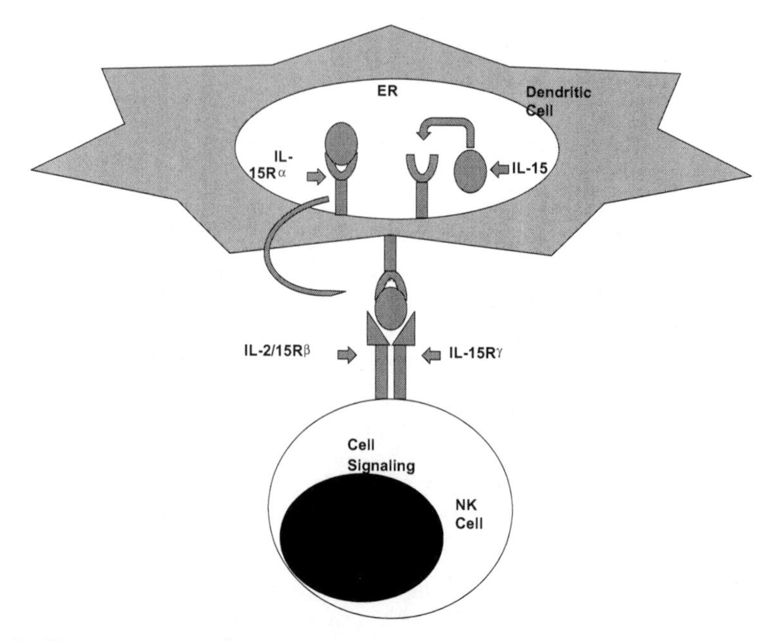

Fig. 3 Transpresentation of IL-15. In the dendritic cell, IL-15 is presumed to meet IL-15Rα in the endoplasmic reticulum (ER), which then chaperones it to the cell membrane. IL-15, while bound to IL-15Rα at the cell surface, is able to bind to the IL-2/15Rβ and IL-15Rγ complex on the neighboring NK cell.

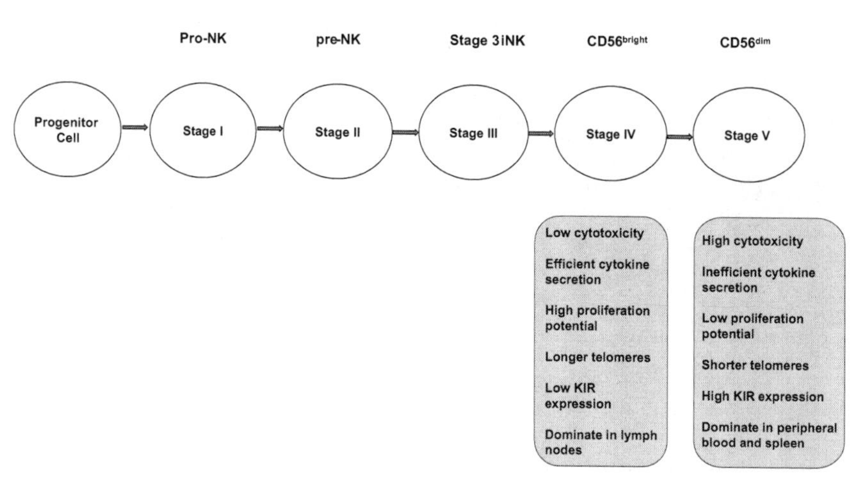

Fig. 4 Proposed five stages of NK cell development. The various stages are characterized by various surface markers. The last "maturation" step is thought to take place in the lymph nodes.

NK CELL EDUCATION

Mechanisms explaining how NK cells acquire effector function are hotly debated. Three main models of this NK cell education differ in their implied mechanisms and in whether the process is one of activation or loss of function. "Disarming" refers to the suppression of effector function in maturing NK cells that receive stimulatory signals unopposed by inhibitory signals through self MHC class I receptors, analogous to T cell anergy (Gasser and Raulet 2006). "Licencing" refers to a terminal differentiation step by maturing NK cells when they receive appropriate inhibitory signal by engaging self-MHC-1 molecules (Kim *et al.* 2005). The "Rheostat" model compares the NK cell responsiveness to a rheostat being tuned to an optimal set point that depends on the sum of stimulatory and inhibitory signals a NK cell encounters from its local environment (Joncker *et al.* 2009). What is agreed on between these models is that NK cells lacking inhibitory receptors are hypo-responsive and thus self-tolerant.

THE ROLE OF CYTOKINES IN NK CELL EFFECTOR FUNCTION

Mature NK cells respond to cytokines and are able to secrete a variety of cytokines upon activation. The first part of this section describes the cytokines acting on NK cells, while the second part describes the most prominent cytokines and chemokines secreted by mature NK cells when they become activated.

Effect of Cytokines on NK Cells

IL-2 is not responsible for NK cell stimulation under physiological conditions. Enhancement of cytokine production by NK cells requires dendritic cell–bound IL-15Rα and IL-12, whereas NK cytotoxic function is IL-15 dependent but independent of IL-12 (Koka *et al.* 2004). Experiments involving blocking the interaction of dendritic and NK cells by anti-IL-15Rα but not IL-2Rα antibodies prevent NK cell priming by the dendritic cells.

The effect of IL-15 on NK cells resembles those of IL-2 in terms of activation, proliferation and increased expression of activating receptors such as NKG2D and TRAIL (Decot *et al.* 2010). Interestingly, this cytotoxicity of NK cells is accompanied by increased expression of inhibitory KIRs (Decot *et al.* 2010). Use of IL-2 in animal models is often accompanied by increased expansion of regulatory T cells and activation-induced cell death of the T cells. These effects are seen less with IL-15; therefore, IL-15 may be optimal for *in vivo* stimulation of endogenous and adoptively transferred NK cells. IL-15 is currently under development for humans and may become available for trials using adoptive NK cell therapies in the near future. Several cytokines such as IL-12, IL-18 and IL-21 are able to stimulate NK cells *in vitro*, though the physiological significance of this stimulation is not clear.

Cytokines Mediating the Effector Functions of NK Cells

Several cytokines and chemokines are secreted by activated NK cells and these play a key role in mediating their effector functions. During infection NK cells are recruited to lymph nodes, where they encounter dendritic cells. This interaction leads to IFN-γ secretion by NK cells, which helps differentiate Th-0 helper T cells into a Th-1 phenotype. IFN-γ is essential for Th-1 immune response and regulates T cell differentiation, activation, homeostasis and survival. IFN-γ also stimulates dendritic cell expression of MHC molecules, antigen presentation and co-stimulatory molecules. In addition, IFN-γ acts on NK cells to enhance their cytotoxicity. NK cells are also thought to release TNF-α in inflamed tissues acting locally on immature dendritic cells to promote their maturation.

NK cells are able to support hematopoiesis in a GM-CSF-dependent cell line in presence of IL-2, which is thought to enhance GM-CSF secretion by NK cells (Cuturi *et al.* 1989). However, there is evidence for the hematopoiesis inhibitory effect of NK cells under certain circumstances thought to be mediated by their secretion of TNF-α and IFN-γ (Cuturi *et al.* 1989). The effects of NK cells on hematopoiesis are therefore complex but their ability to make GM-CSF suggests that NK cells might play a role supporting hematopoiesis.

NK cells have been shown to secrete IL-5 *in vitro* and this secretion is significantly augmented by IL-4 and reduced by IL-10 and IL-12 (Warren *et al.* 1995). In addition, freshly isolated NK cells are able to secrete IL-5 in presence of susceptible target cells and recombinant IL-2 (Warren *et al.* 1995). By secreting IL-5, NK cells may play a role in orchestrating an immune response against parasitic infections.

Chemokines are a group of low molecular weight cytokines important for regulating migration of leukocytes into inflamed tissues. NK cells have been shown to secrete multiple chemokines such as MIP-α, MIP-1β, IL-8 and RANTES. NK cells produce these chemokines on stimulation with cytokines or after encountering target cells. The early chemokine production by NK cells may play an important role in promoting infiltration of pro-inflammatory cells into infected tissues or tumors. Figure 5 summarizes the major cytokines and chemokines that mediate the effector functions of NK cells.

EXPLOITING THE USE OF NK CELLS IN CANCER

During recent years there has been interest in developing novel therapies relying on the prowess of NK cells to kill malignant cells. Initial human studies using

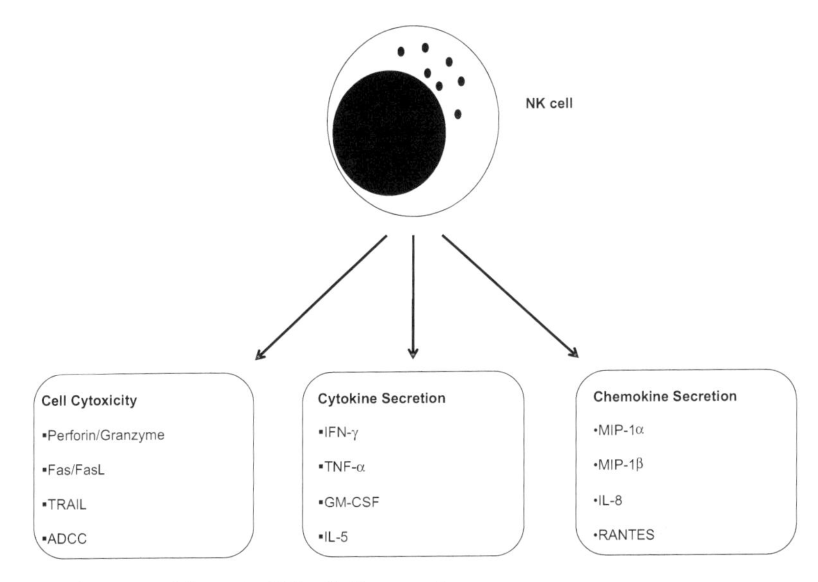

Fig. 5 Summary of the major NK cell effector pathways.

autologous NK cells were performed in the 1980s. Significant side effects were observed in the studies using high-dose IL-2 and the clinical benefit from these studies was limited. Both preclinical models and human trials using adoptive cell transfer of T cells and NK cells have highlighted the importance of prior lymphodepletion for effective *in vivo* expansion. This often involves use of chemotherapeutic agent(s) with or without total body radiation with the intention to deplete host lymphocytes to create "space" for the adoptively transferred cells. This is because the host lymphocytes compete with infused cells for access to cytokines and other growth factors, thus limiting their potential proliferation. In addition to lymphodepletion, in recent years the focus has shifted to the use of allogeneic NK cells for therapy as autologous NK cells are tolerant to host tissues due to their expression of inhibitory KIR recognizing MHC class I molecules. Safety and efficacy of the adoptive transfer approach was established in a clinical trial using *in vivo* expanded haploidentical related-donor NK cell infusions to treat patients with advanced cancer (Miller *et al.* 2005). In this trial successful expansion of the donor-derived NK cells was only seen in AML patients who had received a lymphodepleting preparatory regimen. Approximately 30% of the patients with poor prognosis AML achieved complete remission. Based on the success of this approach adoptive NK cell therapy is being studied for several other tumors including breast cancer, ovarian cancer, CLL and non-Hodgkin's lymphoma.

Allogeneic hematopoietic stem cell transplantation offers hope to thousands of cancer patients each year who otherwise have a very poor prognosis. This beneficial effect of allo-HCT is from a preparatory regimen (chemotherapy and/or radiation therapy) as well as an immune effect often called the graft versus leukemia effect (GvL) reaction from the graft-derived T and NK cells. The problem with T cells is that they also mediate GvHD, which can be life-threatening. In contrast, NK cells do not cause GvHD, they may promote engraftment, and they improve immune reconstitution. It is thought that the presence of KIR on NK cells educated in the donor and in the absence of cognate ligand in the recipient determines allo-reactivity of the donor NK cells. The Perugia group was one of the first groups to test the effect of KIR-ligand mismatch effect in the context of a potently T cell-depleted haplo-identical allo-HCT. They demonstrated improved engraftment, less AML relapse, less GvHD and superior disease-free survival (Ruggeri *et al.* 2002). In addition to the KIR-ligand mismatch, several studies have looked at the effects of donor activating KIRs on transplant outcomes. One recent study showed that the use of donors with a KIR B haplotype is associated with significant improvements in overall and relapse-free survival with more than a 30% improvement in these clinical outcomes (Cooley *et al.* 2009).

Although current protocols use IL-2 to promote *in vivo* expansion of NK cells, several approaches aimed at *ex vivo* expansion of NK cells are under development. One such method involves use of K562 cells transduced with 41BB-ligand and IL-15 (Imai *et al.* 2005). However, it is unclear at this point whether an extensive *ex vivo* expansion will have negative effect(s) on homing and cytotoxic properties of

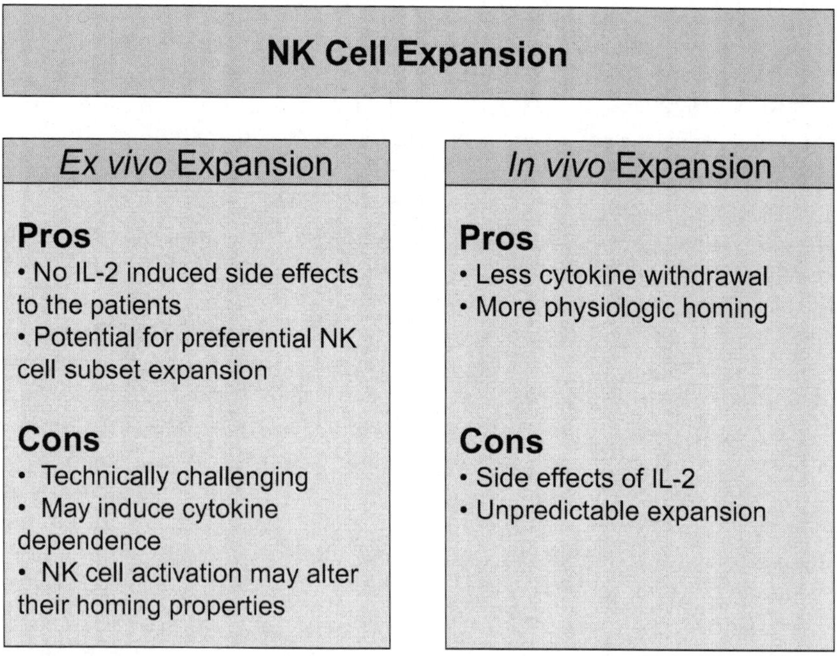

NK Cell Expansion

Ex vivo Expansion

Pros
• No IL-2 induced side effects to the patients
• Potential for preferential NK cell subset expansion

Cons
• Technically challenging
• May induce cytokine dependence
• NK cell activation may alter their homing properties

In vivo Expansion

Pros
• Less cytokine withdrawal
• More physiologic homing

Cons
• Side effects of IL-2
• Unpredictable expansion

Fig. 6 Pros and cons of *ex vivo vs. in vivo* NK cell expansion.

these cells after adoptive transfer. Figure 6 highlights pros and cons of *in vivo* and *ex vivo* NK cell expansion approaches.

Recent studies have highlighted the importance of NK cells in physiological and pathological processes varying from immune surveillance against tumors to successful placentation in pregnancy. Furthermore, our increased knowledge about NK cells puts us in a position to design studies aimed at harnessing the potential of NK cells to treat patients. The beneficial role of NK cells to prevent HIV progression to AIDS represents another opportunity worth pursuing. As we better understand the role of NK cells in autoimmune diseases such as diabetes mellitus, it might be possible to design studies aiming at reducing NK cell activity. The key to manipulating NK cells for therapeutic purposes is to better understand NK cell biology and to apply this knowledge through carefully designed translational studies.

Table 1 Key features of NK cells

- NK cells comprise 10-15% of the lymphocytes in peripheral blood.
- Phenotypically they are characterized by the absence of T cells markers such as CD3 and presence of CD56.
- Functionally NK cells are unique in their ability to lyse target cells without prior sensitization.
- NK cells originate from bone marrow–derived precursors with several stages of development taking place in lymph nodes and peripheral blood.
- IL-15 plays a crucial role in the development, homeostasis and activation of NK cells.
- NK cells produce several cytokines (IFN-γ, TNF-α, GM-CSF and IL-5) and chemokines (MIP-1α, MIP-1β, IL-8 and RANTES).
- NK cells play an important role in immune surveillance against malignant cells, clearing viral infections and placentation.
- In recent years there has been a great interest in developing novel protocols using NK cells to treat patients with malignant disorders.

Summary Points

- NK cells are important lymphocytes recognized morphologically as large granular lymphocytes and phenotypically by lack of the T cell markers (CD3) and presence of CD56.
- Functionally, they are characterized by their ability to directly lyse target cells and produce cytokines without previous sensitization to the target cells.
- In the peripheral blood, $CD56^{dim}CD16^{+}$ cells are the predominant NK cell subset while 10% of NK cells are $CD56^{bright}CD16^{-}$ and likely are immature.

- $CD56^{dim}$ and $CD56^{bright}$ NK cells differ in their capacity to proliferate, produce cytokines, kill targets and the express NK cell receptors and receptors involved in tissue homing.
- KIRs represent an important family on the NK cells and play an important role in NK cell education.
- KIR genes have a high degree of polymorphism paralleling that seen in MHC genes.
- There are two main KIR gene haplotypes, A and B, which are distributed differently among ethnic populations.
- IL-15 transpresented by dendritic cells plays a physiological role in NK cell development, homeostasis and effector function.
- NK cells secrete several cytokines and chemokines, which play an important role in shaping the early immune response.
- NK cells are important for tumor immune surveillance and in controlling viral infections and play a role in pregnancy success.
- Allogeneic NK cell therapy has the potential to treat cancer and infectious diseases.

Abbreviations

AIDS	:	Acquired Immunodeficiency Syndrome
AML	:	acute myeloid leukemia
CD	:	cluster of differentiation
FasL	:	Fas ligand
GM-CSF	:	granulocyte macrophage colony-stimulating factor
GvHD	:	graft-versus-host disease
HLA	:	human leukocyte antigen
HSCs	:	hematopoietic stem cells
KIR	:	killer-cell immunoglobulin-like receptor
MIP-1α	:	macrophage inhibitory protein 1α
MIP-1β	:	macrophage inhibitory protein 1β
MHC	:	major histocompatibility complex
NK cells	:	natural killer cells
NOD	:	non-obese diabetes
RANTES	:	Regulated on Activation Normal T Cell Expressed and Secreted
SCID	:	severe combined immunodeficiency
TRAIL	:	tumor necrosis factor-related apoptosis-inducing ligand

Definition of Terms

Antibody Dependent Cell Toxicity (ADCC): An immune mechanism whereby effector cells are able to recognize antibody-coated target cells by their Fc receptors (CD16 in case of NK cells), leading to lysis of the target cells.

Chemokines: A group of low molecular weight cytokines that regulate the movement of leukocytes in the tissues and across the endothelium.

Granzyme B: A serine protease encoded by *GZMB* gene, which like perforin is found in the granules of CD 8^{+} T cells and NK cells. Upon degranulation it plays an important role in initiating apoptosis of the target cell.

Haploidentical allogeneic HCT: A form of allo-HCT in which the donor-recipient pairs are related and mismatched at half of the HLA loci.

Major histocompatibility complex molecule (MHC) class 1 molecules: Highly polymorphic heterodimeric transmembrane proteins that display peptide fragments at the cell surface of most of the nucleated cells.

Perforin: A cytolytic protein coded by the *PRF1* gene and found in the granules of $CD8^{+}$ T cells and NK cells. Upon degranulation Perforin is able to insert into the cell membrane of the target cell leading to pore formation, which in the presence of granzyme B induces apoptosis of the cell.

References

Chan, A. and D.L. Hong, A. Atzberger, S. Kollnberger, A.D. Filer, C.D. Buckley, A. McMichael, T. Enver, and P. Bowness. 2007. CD56bright human NK cells differentiate into CD56dim cells: role of contact with peripheral fibroblasts. J. Immunol. 179: 89-94.

Cichocki, F. and J.S. Miller. 2010. In vitro development of human Killer-Immunoglobulin Receptor-positive NK cells. Methods Mol. Biol. 612: 15-26.

Colucci, F. and J.P. Di Santo. 2000. The receptor tyrosine kinase c-kit provides a critical signal for survival, expansion, and maturation of mouse natural killer cells. Blood 95: 984-991.

Cooley, S. and E. Trachtenberg, T.L. Bergemann, K. Saeteurn, J. Klein, C.T. Le, S.G. Marsh, L.A. Guethlein, P. Parham, J.S. Miller, and D.J. Weisdorf. 2009. Donors with group B KIR haplotypes improve relapse-free survival after unrelated hematopoietic cell transplantation for acute myelogenous leukemia. Blood 113: 726-732.

Cooper, M.A. and J.E. Bush, T.A. Fehniger, J.B. VanDeusen, R.E. Waite, Y. Liu, H.L. Aguila, and M.A. Caligiuri. 2002. In vivo evidence for a dependence on interleukin 15 for survival of natural killer cells. Blood 100: 3633-3638.

Cuturi, M.C. and I. Anegon, F. Sherman, R. Loudon, S.C. Clark, B. Perussia, and G. Trinchieri. 1989. Production of hematopoietic colony-stimulating factors by human natural killer cells. J. Exp. Med. 169: 569-583.

Decot, V. and L. Voillard, V. Latger-Cannard, L. Aissi-Rothe, P. Perrier, J.F. Stoltz, and D. Bensoussan. 2010. Natural-killer cell amplification for adoptive leukemia relapse immunotherapy: comparison of three cytokines, IL-2, IL-15, or IL-7 and impact on NKG2D, KIR2DL1, and KIR2DL2 expression. Exp. Hematol. 38: 351-362.

Freud, A.G. and A. Yokohama, B. Becknell, M.T. Lee, H.C. Mao, A.K. Ferketich, and M.A. Caligiuri. 2006. Evidence for discrete stages of human natural killer cell differentiation in vivo. J Exp. Med. 203: 1033-1043.

Gasser, S. and D.H. Raulet. 2006. Activation and self-tolerance of natural killer cells. Immunol. Rev. 214: 130-142.

Gur, C. and A. Porgador, M. Elboim, R. Gazit, S. Mizrahi, N. Stern-Ginossar, H. Achdout, H. Ghadially, Y. Dor, T. Nir, V. Doviner, O. Hershkovitz, M. Mendelson, Y. Naparstek, and O. Mandelboim. 2010. The activating receptor NKp46 is essential for the development of type 1 diabetes. Nat. Immunol. 11: 121-128.

He, Y. W. and T.R. Malek. 1996. Interleukin-7 receptor alpha is essential for the development of gamma delta + T cells, but not natural killer cells. J. Exp. Med. 184: 289-293.

Imai, C. and S. Iwamoto, and D. Campana. 2005. Genetic modification of primary natural killer cells overcomes inhibitory signals and induces specific killing of leukemic cells. Blood 106: 376-383.

Joncker, N.T. and N.C. Fernandez, E. Treiner, E. Vivier, and D.H. Raulet. 2009. "NK cell responsiveness is tuned commensurate with the number of inhibitory receptors for self-MHC class I: the rheostat model." J. Immunol. 182: 4572-4580.

Kawamura, T. and R. Koka, A. Ma, and V. Kumar. 2003. Differential roles for IL-15R alpha-chain in NK cell development and Ly-49 induction. J. Immunol. 171: 5085-5090.

Kim, S. and J. Poursine-Laurent, S.M. Truscott, L. Lybarger, Y.J. Song, L. Yang, A.R. French, J.B. Sunwoo, S. Lemieux, T.H. Hansen, and W.M. Yokoyama. 2005. Licensing of natural killer cells by host major histocompatibility complex class I molecules. Nature 436: 709-713.

Koka, R. and P. Burkett, M. Chien, S. Chai, D.L. Boone, and A. Ma. 2004. Cutting edge: murine dendritic cells require IL-15R alpha to prime NK cells. J. Immunol. 173: 3594-3598.

Ljunggren, H.G. and K. Karre. 1985. Host resistance directed selectively against H-2-deficient lymphoma variants. Analysis of the mechanism. J. Exp. Med. 162: 1745-1759.

McKenna, H.J. and K.L. Stocking, R.E. Miller, K. Brasel, T. De Smedt, E. Maraskovsky, C.R. Maliszewski, D.H. Lynch, J. Smith, B. Pulendran, E.R. Roux, M. Teepe, S.D. Lyman, and J.J. Peschon. 2000. Mice lacking flt3 ligand have deficient hematopoiesis affecting hematopoietic progenitor cells, dendritic cells, and natural killer cells. Blood 95: 3489-3497.

Miller, C.H. and S.G. Maher and H.A. Young. 2009. Clinical Use of Interferon-gamma. Ann NY Acad Sci 1182: 69-79.

Miller, J.S. and Y. Soignier, A. Panoskaltsis-Mortari, S.A. McNearney, G.H. Yun, S.K. Fautsch, D. McKenna, C. Le, T.E. Defor, L.J. Burns, P.J. Orchard, B.R. Blazar, J.E. Wagner, A. Slungaard, D.J. Weisdorf, I.J. Okazaki and P.B. McGlave. 2005. Successful adoptive transfer and in vivo expansion of human haploidentical NK cells in patients with cancer. Blood 105: 3051-3057.

Orr, M.T. and W.J. Murphy, and L.L. Lanier. 2010. 'Unlicensed' natural killer cells dominate the response to cytomegalovirus infection. Nat. Immunol. 11: 321-327.

Prlic, M. and B.R. Blazar, M.A. Farrar, and S.C. Jameson. 2003. In vivo survival and homeostatic proliferation of natural killer cells. J. Exp. Med. 197: 967-976.

Ruggeri, L. and M. Capanni, E. Urbani, K. Perruccio, W.D. Shlomchik, A. Tosti, S. Posati, D. Rogaia, F. Frassoni, F. Aversa, M.F. Martelli, and A. Velardi. 2002. Effectiveness of donor natural killer cell alloreactivity in mismatched hematopoietic transplants. Science 295: 2097-2100.

Schorle, H. and T. Holtschke, T. Hunig, A. Schimpl, and I. Horak. 1991. Development and function of T cells in mice rendered interleukin-2 deficient by gene targeting. Nature 352: 621-624.

Sun, J.C. and J.N. Beilke, and L.L. Lanier. 2009. Adaptive immune features of natural killer cells. Nature 457: 557-561.

Vosshenrich, C.A. and T. Ranson, S.I. Samson, E. Corcuff, F. Colucci, E.E. Rosmaraki, and J.P. Di Santo. 2005. Roles for common cytokine receptor gamma-chain-dependent cytokines in the generation, differentiation, and maturation of NK cell precursors and peripheral NK cells in vivo. J. Immunol. 174: 1213-1221.

Warren, H.S. and B.F. Kinnear, J.H. Phillips, and L.L. Lanier. 1995. Production of IL-5 by human NK cells and regulation of IL-5 secretion by IL-4, IL-10, and IL-12. J. Immunol. 154: 5144-5152.

Yu, H. and T.A. Fehniger, P. Fuchshuber, K.S. Thiel, E. Vivier, W.E. Carson, and M.A. Caligiuri. 1998. Flt3 ligand promotes the generation of a distinct CD34(+) human natural killer cell progenitor that responds to interleukin-15. Blood 92: 3647-3657.

Section IV
Cancer

CHAPTER 12

Cytokines and Lung Cancer

Masaaki Tamura*, Deryl Troyer, Dharmendra K. Maurya, Atsushi Kawabata, and Raja Shekar Rachakatla
Department of Anatomy and Physiology, Kansas State University
College of Veterinary Medicine, Manhattan, KS 66506, USA

ABSTRACT

Cytokines are small secretory proteins produced *de novo* in various cell types in response to immune stimuli and serve as molecular messengers between cells. They provide important intercellular signals in inflammation, immunity, and tumorigenesis. Cytokines and their receptors are associated with cancer development and progression. The mechanisms of involvement of cytokines in cancer are diverse. They include leukocyte infiltration into cancer tissue, regulation of cell cycles, programmed cell death of cancer cells, and immune response to cancer cells. Thus, locally and/or systemically deregulated levels of cytokines and receptors can be detected in all types of cancer patients. Many types of cytokines, including various interleukins and interferons, have been studied for their relationship with various cancers, including lung cancer. Intratumoral or serum cytokine or growth factor levels can be used as valuable markers for diagnosis, prognosis, and prediction for patient types, indicating whether a patient might positively or negatively benefit from a particular treatment; these markers also help in monitoring treatment response and disease recurrence. In addition, the potential tumoricidal effects and therapeutic efficacy of various types of cytokines have been examined in pre-clinical animal studies and clinical trials. Many pre-clinical studies revealed that local injection of various interleukins and multiple cytokine gene therapies caused regression of multiple cancers, including non-small cell lung cancer. However, their effect on established tumors is limited,

**Corresponding author*

primarily due to their cytotoxicity and short half life. Therapies with antibody-cytokine fusion proteins and tumoritropic cell-based gene therapy can overcome this limitation of cytokine cancer therapy. Recent studies with mesenchymal stem cell-based interferon-β gene therapy using either human bone marrow stem cells or human umbilical cord matrix stem cells demonstrated significant tumor regression in lung bronchioloalveolar carcinoma, a subset of adenocarcinoma, in mouse studies. Stem cell-based cytokine gene therapy is feasible in lung cancer-targeted, safe cytokine gene therapy.

INTRODUCTION

Cytokines are small secreted proteins produced *de novo* in response to immune stimuli and serve as molecular messengers between cells. They provide important intercellular signals in inflammation, immunity, tumorigenesis, and endothelial cell biology. 'Cytokines' is a general term; more specific names include monokines (cytokines made by monocytes), lymphokines (cytokines made by lymphocytes), and interleukins (cytokines made by leukocytes and acting on other leukocytes). Interleukins comprise the largest group of cytokines stimulating immune cell proliferation and differentiation. About 34 distinct interleukins have been identified in mammals. Their functions significantly overlap. Since adaptive cytotoxic T-cells mediate host cell-dependent anti-tumor activity, multiple interleukins are indirectly related to cancer growth regulation. Other classifications of cytokines include interferons (IFNs) and chemokines. IFNs are produced by virus-, bacteria- or parasite-infected cells and various cancer cells. About ten distinct IFNs have been identified in mammals; seven of these have been described for humans. They inhibit virus replication in infected cells, stimulate apoptosis by up-regulating p53 expression, stimulate expression of major histocompatibility complex molecules, MHC I and MHC II, and increase immunoproteasome activity, which is an important feature for tumor rejection by immune surveillance. Chemokines attract leukocytes to infection sites. Chemokines have conserved cysteine residues that allow them to be assigned to four groups. The groups, with representative chemokines, are C chemokines (e.g., Lymphotactin α and Lymphotactin β), CC chemokines (e.g., RANTES, monocyte chemotactic protein (MCP)-1), CXC chemokines (e.g., IL-8, SDF-1), and CXXXC chemokines (e.g., Fractalkine). These chemokines also play roles in tumor development by modulating infiltration of leukocytes, stromal fibroblasts, vascular endothelial cells, and mesenchymal stem cell (MSC) in tumor tissue.

Cytokines and their receptors are apparently associated with lung cancer development and progression by multiple mechanisms. Lung cancer is the leading cause of cancer-related morbidity and mortality in developed countries. Lung cancer-dependent deaths constituted approximately 30% of men and 26% of women of the estimated total cancer-related deaths in the United States in 2009 (Jemal *et al.* 2009). Lung cancer is categorized into the following two histologic groups that have distinct clinical behaviors: non-small cell lung cancer (NSCLC) and small cell lung cancer (SCLC). NSCLC accounts for 80 to 85% of all lung cancer cases and includes adenocarcinoma (35 to 45%), squamous cell carcinoma (25 to 35%), and large cell carcinoma (10%). The possible treatment procedures for lung cancer include surgical resection, chemotherapy, radiotherapy, and/or combination therapy and have been advanced. However, the relative 5-year survival rate of patients who had lung or bronchus cancer from 1995 to 2001 was still quite low (15%) and was not improved very much compared to the 1970s (12%). In addition to primary lung cancer, the lung is the primary site for cancer metastasis, and lung metastasis is often fatal. Therefore, it is clear that novel treatment strategies for lung cancer are urgently needed. Since expression levels of cytokines and their receptors appear to be tightly associated with lung cancer development, it is conceivable that regulation of cytokine productions might be a reasonable approach to control and treat lung cancer. Novel therapies, such as those regulating cytokine levels through recombinant cytokines, cytokine gene therapy, or immune therapy, hold great potential. In addition, local and systemic cytokine levels can be used as biomarkers for diagnosis and prognosis of lung cancer.

ROLE OF CYTOKINES IN LUNG CANCER

Cytokines play a variety of roles in lung cancer development as positive or negative factors. Multiple cytokines such as IL-4, -6, -8, and -10 are positively involved in lung cancer development through a variety of mechanisms. For instance, IL-4 increases tumor cell resistance to apoptosis (Redente *et al.* 2009), IL-6 up-regulates cyclooxygenase-2 (COX-2) expression and resultant prostaglandin production, increases resistance to apoptosis, and induces vascular endothelial cell growth factor (VEGF) expression (Dalwadi *et al.* 2005), and IL-10 suppresses T-cell and antigen presenting cell (APC) function (Sharma *et al.*, 1999) and stimulates tumor vascularization (Hatanaka *et al.*, 2001). In contrast, IL-1, -2, -3, -5, -6, -7, -12, -21, -24 and IFN-α, -β and -γ attenuate lung tumor growth. Tumor attenuation mechanisms include recruitment of lymphocytes (IL-3, -6 and -7), CD8 positive cytotoxic T cells (IL-5, -6 and -21) and NK cells (IL-5) at the tumor sites, induction of multiple programmed cell death (IFN-β, Zhang *et al.* 1999) and production of IFN-γ (IL-12). IL-6 over-expression in murine lung cancer cells reduces tumorigenicity and stimulates recruitment of CD8 positive cytotoxic T cells to the tumor site. However, IL-6 also increases cancer cells' resistance to apoptosis, thus stimulating overall cancer cell growth (Dalwadi *et al.* 2005). Indeed, higher serum levels of IL-6 in human NSCLC patients are commonly detected and are correlated with a poor prognosis (Chechlinska *et al.* 2008); it appears that IL-6 stimulates lung cancer progression. Although higher expression of other cytokines such as transforming growth factor (TGF)-β, VEGF, epidermal growth factor (EGF), human epidermal growth factor receptor 2 (HER2), fibroblast growth factor (FGF)-2, and their receptors are positively associated with lung cancer development, the complex roles of these growth factors in lung cancer and potential targeted therapies are described elsewhere (Hodkinson *et al.* 2008).

CYTOKINES AS LUNG CANCER BIOMARKERS

Cytokine and growth factor levels can be used as valuable markers for diagnosis, prognosis, histopathological type, or treatment responses that indicate whether a patient might positively or negatively respond to a particular treatment; these markers also help in monitoring treatment response and disease recurrence (Chechlinska *et al.* 2008). For example, IL-2, -4, -8, -10, -13, tumor necrosis factor alpha (TNF-α), VEGF, HER2, and EGF receptor (EGFR) are commonly deregulated in cancer tissues of NSCLC. Higher expression of IL-8 and VEGF at tumor sites are significantly associated with tumor progression. Interestingly, poor outcomes for patients with early stage NSCLC was independently associated with the lack of intratumoral IL-10 expression (Soria *et al.* 2003). In addition to local cytokine levels, serum levels of the following cytokines and receptors were found to be significantly higher in lung cancer patients than in controls: IL-1β, -2, -6, -8, -10, -12, GM-CSF, IFN-γ, TNF-α, VEGF, FGF-2, and receptors for IL-2, -6, -10, monocyte colony-stimulating factor (M-CSF), TNF, and VEGF (Chechlinska *et al.* 2008). Levels of a few cytokines, such as IL-6 and M-CSF, were reported to increase significantly with advancement of the NSCLC clinical stage. Serum levels of these two cytokines were shown to be associated with histopathological types of lung cancer; their serum levels are significantly higher in large cell carcinoma. Furthermore, lower serum levels of IL-6 are positively correlated with a better response to chemotherapy. In contrast, persistence of higher soluble IL-2 receptor levels in the postoperative period was found to be correlated with a higher early relapse rate in operable NSCLC patients. A recent case-control study of NSCLC discovered that higher serum levels of IL-6 (> 4.0 pg/ml) are associated with significantly poor survival in both African American and Caucasian populations in the United States (Enewold *et al.* 2009). Although higher serum levels of IL-10 (> 12.7 pg/ml) and IL-12 (> 5.9 pg/ml) were also associated with significantly poorer survival in African Americans, serum levels of IL-1β, -4, -5, -8, -12, GM-CSF, IFN-γ, and TNF-α did not show significant correlation with survival of NSCLC patients (Enewold *et al.* 2009). Another study analyzing a panel of 120 cytokines in the exhaled breath of lung cancer patients and control non-cancer subjects revealed that nine cytokines including eotaxin, FGFs, IL-10, and macrophage inflammatory protein (MIP)-3 were present with more than a two-fold difference between the two groups (Kullmann *et al.* 2008). These studies clearly indicate that determination of local tumor and systemic cytokine levels can provide valuable diagnostic parameters for lung cancer prognosis, recurrence, and potential response to particular treatments. Among many cytokines, however, IL-6 appears to be the only reliable independent biomarker for survival of NSCLC patients.

APPLICATION OF CYTOKINES IN TREATMENT OF CANCER

Cytokines are intercellular messengers and have diverse autocrine and paracrine functions to modulate the immune system and cell death. Since cancer cells express tumor antigens and can be distinguished from normal cells, the host immune response is an intrinsic mechanism to control cancer growth. Therefore, modulation of host immune responses by the use of recombinant cytokines or cytokine genes is a viable strategy for cancer therapy. Several cytokines have therapeutic potential when administered alone or in combination with other cytokines, chemotherapeutics, or other biological agents (Margolin 2008). Some of the promising cytokines currently under investigation for NSCLC therapies are IL-2, -4, -7 -12, -15, -24, IFN-α, -β, -γ, TNF-α and GM-CSF (Table 2, Chada *et al.* 2003, Margolin 2008). However, among these cytokines, IL-2 and IFN-α2b are the two cytokines approved by the FDA for treatment of multiple cancers including lymphoma, melanoma, renal cell carcinoma, and leukemia, but not for lung cancer (Yasumoto *et al.* 1987). IL-2 generally stimulates lymphocyte recruitment at the tumor sites and causes tumor regression (Yasumoto *et al.* 1987, Lissoni *et al.* 1993). IFN-α2b acts against the same cancers as IL-2, as well as against chronic myelogenous leukemia, Kaposi's sarcoma, and hairy cell leukemia. Co-treatment with IL-2 and IL-3 has been shown to be effective in attenuation of NSCLC in a human clinical study (Lissoni *et al.* 1993). However, since cytokines are relatively unstable *in vivo*, cancer patients undergoing cytokine-based cancer therapy have to receive a large amount of the recombinant cytokine protein to maintain the required blood concentration for biological activity. Administration of therapeutic levels of cytokines is thereby often toxic to the patients. In contrast, secretion of the cytokine from tumor or vehicle cells by gene transfer is a therapeutic avenue with which fewer side effects are expected. Indeed, the tissue-targeted delivery of therapeutic cytokine genes has become a viable therapeutic option. Although expression of sufficient amounts of cytokines in appropriate target cells remains an issue, the potential of cytokine gene therapy alone or in combination with additional therapeutic genes is being explored and appears to be very high. Kwong *et al.* have evaluated the therapeutic efficacy of gene therapy in metastatic lung cancer using adenovirus vector-mediated delivery of herpes simplex virus thymidine kinase and IL-2 plasmid DNA to treat a murine model of metastatic lung cancer in the liver (Kwong *et al.* 1997). They demonstrated that suicide and cytokine genes can be used in combination to treat metastatic lung cancer in a mouse model.

Another approach using cytokines in cancer therapy includes antibody-cytokine fusion technology, in which therapeutic cytokines are fused with a cancer-targeted antibody to improve antibody-targeted cancer immunotherapy. These molecules have the capacity to enhance the tumoricidal activity of the antibodies and/or to activate a secondary anti-tumor immune response (Ortiz-Sanchez *et al.* 2008). Although this application has not been successful in lung cancer treatment, antibody-cytokine fusion proteins can be a rational lung cancer-targeted treatment.

Chemokines are smaller chemotactic cytokines that induce migration of leukocytes, activate inflammatory responses, and are implicated in the regulation of tumor development and growth. Production of chemokines such as CXC chemokine, CXCL12, in cancer cells recruits cells expressing chemokine receptors (e.g., CXCR4) to cancer tissues, thus stimulating tumor growth. The recruited cells include leukocytes, stromal fibroblasts, vascular endothelial cells, and/or MSC. Chemokine- and VEGF-dependent stimulation of tumor-associated angiogenesis is critically important for solid tumor growth. Therefore, disrupting the association between chemokines and tumor cells provides a good opportunity to treat cancers (Chada *et al.* 2003). Indeed, anti-neovascularization therapy by VEGF inhibitors has been developed, although the effect in preclinical and clinical studies is somewhat transitory (Bergers and Hanahan 2008). EGFR abnormality is found to be associated with progression of multiple cancers including NSCLC. This discovery has stimulated development of EGFR tyrosine kinase inhibitors. Humanized anti-EGFR monoclonal antibodies and small molecule tyrosine kinase inhibitors have been developed, and several such inhibitors have been in clinical use (Sequist and Lynch 2008).

APPLICATIONS OF CELL-BASED IFN-β GENE THERAPY ON NSCLC

The cytokine IFN-β is known to have a strong ability to inhibit tumor cell growth and induce apoptosis (Zhang *et al.* 1999). However, the application of IFN-β to animal studies has been unsuccessful because of its short half life, and because the maximally tolerated dose is lower than the effective dose. Previous studies demonstrated that IFN-β gene therapy using adenoviral vectors is effective in several cancers, including lung cancer, although viral vector-based gene delivery is not cancer tissue-specific. Wilderman *et al.* recently demonstrated that tracheal administration of an adenovirus vector encoding the IFN-β gene significantly prolonged survival of mice with *K-rasG12D* mutation-induced lung adenocarcinoma (Wilderman *et al.* 2005). However, repetitive adenoviral vector-based gene delivery to tumor tissues is potentially limited. Indeed, intratumoral injection of virus vectors showed limited target protein expression in the cells adjacent to the injection site (Lang *et al.* 2003), and recipients will eventually develop neutralizing antibodies against the vectors. To overcome nonspecific gene delivery and the short half life of IFN-β, human bone marrow-derived MSC have been used as biological vehicles for IFN-β gene delivery. This MSC-based IFN-β therapy via systemic administration has been shown to be effective in attenuation of lung metastasis of breast cancer (Studeny *et al.* 2002).

Recent studies by Rachakatla *et al.* and Matsuzuka *et al.* have also shown successful expression of IFN-β in human umbilical cord matrix stem cells (hUCMSC) and have demonstrated that systemically administered IFN-β gene-transduced hUCMSC (IFN-β-hUCMSC) successfully homed to lung tumor sites (Fig. 1) and attenuated growth of lung-metastasized breast cancer (Fig. 2, Rachakatla *et al.* 2007, 2008) and human bronchioloalveolar carcinoma in mice (Fig. 3, Matsuzuka *et al.* 2010). A unique feature of these IFN-β-hUCMSC is that the efficacy of IFN-β-dependent tumor growth attenuation appears to be significantly

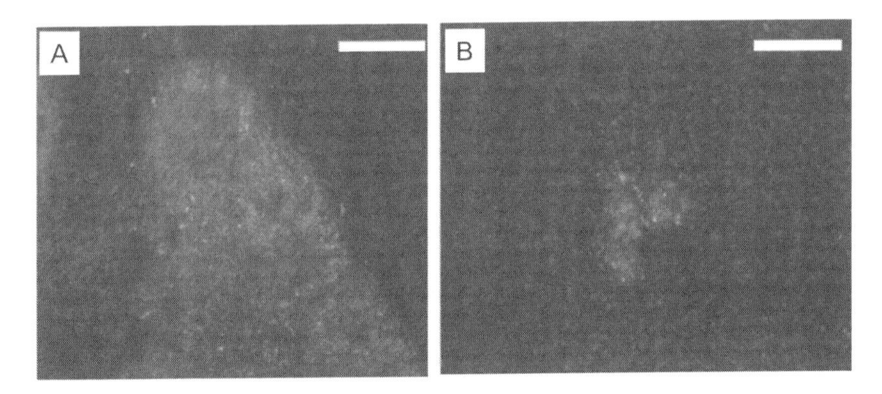

Fig. 1 Selective engraftment of IFN-β-over-expressing hUCMSC to lung metastasized MDA-231 breast carcinoma xenografts in SCID mice. MDA-231 lung metastatic breast carcinoma-bearing SCID mice were treated systemically via IV injection with IFN-β-hUCMSC (5×10^5 cells) labeled with SP-DiI red fluorescent dye, three times with one week interval. One week after the last treatment, lungs were dissected and subjected to immunohistochemical analysis. MDA-231 cells in the lung were identified by anti-human mitochondrial antibodies conjugated with a green fluorophore. Sections were counterstained with Hoechst 33342 nuclear stain (blue). These pictures clearly indicate that systemically administered IFN-β-hUCMSC selectively engrafted within a microtumor or in close proximity to the lung tumor (green). (Scale bar = 100 μm). Figure is modified from the original figure by Rachakatla *et al.* (2008).

Color image of this figure appears in the color plate section at the end of the book.

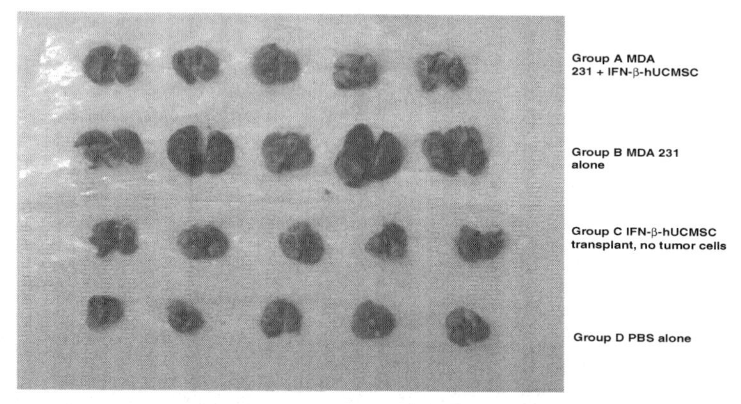

Fig. 2 Systemic administration of IFN-β-over-expressing hUCMSC significantly attenuated the growth of lung metastasized MDA-231 breast carcinoma in SCID mice. MDA-231 lung metastatic breast carcinoma-bearing SCID mice were treated systemically via IV injection with un-engineered or IFN-β-hUCMSC (5×10^5 cells) three times with one week interval. One week after the last treatment, lungs were dissected and overall macroscopic morphology was observed. Group A, MDA-231 tumor-bearing lungs treated with IFN-β-hUCMSC; Group B, MDA-231 tumor-bearing lungs without treatment; Group C, IFN-β-hUCMSC alone without tumor cell inoculation; and Group D, PBS alone without tumor cell inoculation. Figure is modified from the original figure by Rachakatla *et al.* (2007).

enhanced by the intrinsic tumoricidal ability of the hUCMSC themselves (Fig. 4, Ayuzawa *et al.* 2009, Matsuzuka *et al.* 2010). Furthermore, poor immunogenicity (Weiss *et al.* 2008) and non-tumorigenicity (Rachakatla *et al.* 2007) of hUCMSC make this cellular vehicle potentially applicable to human patients. This live cell-based cytokine gene therapy appears to be significantly safe and effective in controlling NSCLC, lung metastasis of other types of cancers, and various other types of primary cancers (Fig. 5). Cell-based cytokine gene therapy should also be applicable to various inflammatory diseases. This procedure seems to have high potential for application to human lung cancer patients.

Table 1 Key features of lung cancer

- Lung cancer is the leading cause of cancer-related death in developed countries. Lung cancer deaths constitute approximately 30% of total cancer-related deaths; this is more than the next four most common cancer-dependent deaths combined.
- The relative 5-year survival rate of lung cancer patients is approximately 15%. This rate has not improved much in the last four decades.
- Lung cancer is categorized into two histologic groups of non-small cell lung cancer (NSCLC) and small cell lung cancer (SCLC). NSCLC accounts for 80 to 85% of all lung cancer cases and includes adenocarcinoma, and squamous cell and large cell carcinomas.
- In addition to primary lung cancer, the lung is the primary site for cancer metastasis, and lung metastasis is often fatal.
- The possible treatment procedures for lung cancer include surgical resection, chemotherapy, radiotherapy, and/or combination therapy. But these are only effective when lung cancer is treated in its early stages.

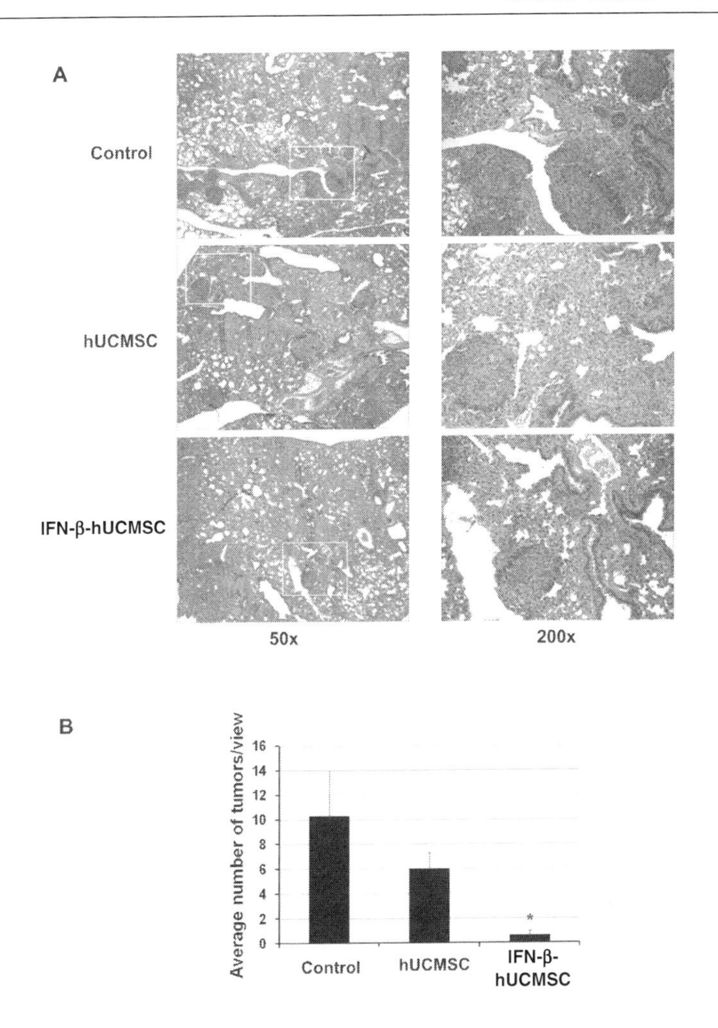

Fig. 3 Systemic administration of IFN-β-over-expressing hUCMSC significantly attenuated the growth of H358 orthotopic lung adenocarcinoma xenografts in SCID mice. H358 orthotopic lung adenocarcinoma xenografts were treated systemically via IV injection with IFN-β-hUCMSC (4×10^5 cells) four times with 5 d interval. Two weeks after the last treatment, lungs were dissected and fixed in 10% formalin. Paraffin-embedded lung sections were stained by H&E staining. Morphologies of lung tumors in three different treatments are presented in the upper panels (A). Multiple sizes of tumors, from as small as a cluster of several cancer cells to relatively large tumors, were counted under the microscope at 50x magnification. White squares in the 50x pictures indicate the area magnified in the 200x pictures. The average number of tumors in three view areas was expressed in the bar graph. *, $p < 0.05$ as compared to the level of PBS-treated control. Figure is modified from the original figure by Matsuzuka *et al.* (2010).

Color image of this figure appears in the color plate section at the end of the book.

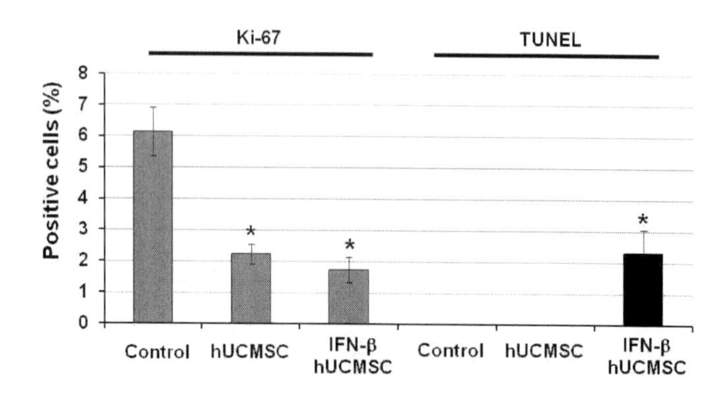

Fig. 4 Analysis of cell division and apoptosis in H358 orthotopic lung adenocarcinoma xenografts in SCID mouse lungs treated with either PBS, hUCMSC alone or IFN-β-over-expressing hUCMSC. Tumor-bearing lungs were collected as described in the caption for Fig. 3. Cell division in H358 lung adenocarcinoma xenografts was analyzed by evaluating the percentage of anti-Ki-67 positive cells. Apoptosis was analyzed by counting TUNEL positive cells. The average percentages of Ki-67 positive cells (gray bars) and TUNEL positive cells (solid bars) in 10 tumor areas in each treatment group were expressed in the bar graph. These results suggest that un-engineered hUCMSC significantly attenuate cell division but do not stimulate apoptosis, whereas IFN-β-hUCMSC attenuate cell division and stimulate apoptosis of tumor cells. *, $p < 0.05$ as compared to the level of PBS-treated control. Figure is modified from the original figure by Matsuzuka *et al.* (2010).

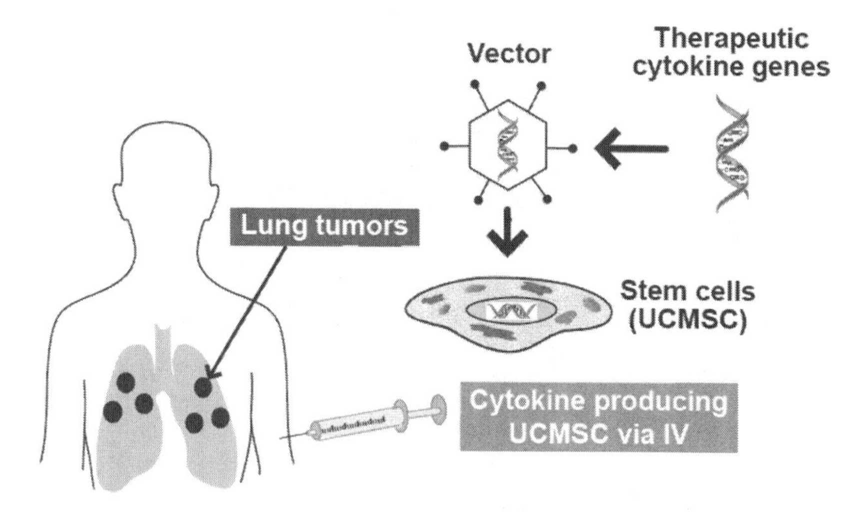

Fig. 5 Schematic illustration of cellular vehicle-based targeted gene therapy. Systemically administered stem cells transfected with therapeutic cytokine genes specifically migrate to cancer site and produce cytokine effectively. This procedure overcomes cytokine short half life and cytotoxicity to normal tissues.

Color image of this figure appears in the color plate section at the end of the book.

Table 2 Some selected cytokines potentially effective for lung cancer therapy

Cytokine	*Tumor growth*	*preclinical trial*	*clinical trial*
IL-4, -8, -10	↑	formation of significantly large tumors, apoptosis resistance, suppression of T cell and APC function (Redente *et al.* 2009, Dalwadi *et at.* 2005, Sharma *et al.* 1999)	
IL-2, -3, -7	↓	Marked reduction of tumor burden with extensive lymphocytic infiltration of the tumors, enhancement survival (Sharma *et al.* 1999, Porgador *et al.* 1992).	Attenuation of the tumor by IL-2 (9 patients with complete regression, from phase II trial).
IL-5, -21	↓	Antitumor effects through NK cells and CD8+ cells, promotion of innate and adaptive immune responses (Kasaian *et al.* 2002).	
IL-6	↕	Reduction of tumorigenicity and recruitment of CD8 positive cytotoxic T-cells to the tumor sit (Porgador *et al.* 1992). Resistance to apoptosis, stimulation of overall cancer cell growth (Dalwadi *et al.* 2005)	
IL-12	↓	Antitumor activity through the production of IFN-y (Sumimoto *et al.* 1998).	
IFN-α, -β, -γ	↓	Enhancement of immune response against parental lung tumor cells (Redente *et al.* 2009. Krejcova *et al.* 2009, Park *et al.* 2009).	

Upward (↑) and downward arrows (↓) indicate stimulation and attenuation, respectively.

Summary Points

- All kinds of cytokines are involved in lung cancer development and progression.
- Mechanisms of cytokine involvement in lung cancer include various leukocyte infiltration into cancer tissue, regulation of cell cycles, programmed cell death of cancer cells, and immune response to cancer cells.
- Local and/or systemic levels of cytokines can be used as biomarkers for lung cancer diagnosis and prognosis, patient responses to a particular treatment, and monitoring of treatment response and disease recurrence.

- Systemic treatment with cytokine alone is generally ineffective because of the short half life of cytokines; it also may cause serious side effects. However, therapies using systemic IL-2 or VEGF, VEGFR or EGFR inhibitors are effective in reducing tumor size in a subset of human lung cancer patients.
- Cancer-targeted cytokine gene therapy is potentially effective for lung cancer treatment and prevention of recurrence.
- Cell-based cytokine gene therapy using bone marrow and umbilical cord matrix MSC is a potentially effective therapy for primary and metastasized lung cancer.

Abbreviations

APC	:	antigen-presenting cells
CD8	:	cluster of differentiation 8
COX	:	cyclooxygenase
CXC Chemokine	:	wo N-terminus cysteines of CXC chemokines are separated by one random amino acid (X)
CXXXC Chemokine	:	two N-terminus cysteines of CX_3C chemokines are separated by three random amino acids (XXX)
EGF	:	epidermal growth factor
EGFR	:	epidermal growth factor receptor
FDA	:	Food and Drug Administration
FGF-2	:	fibroblast growth factor-2
GM-CSF	:	granulocyte monocyte colony-stimulating factor
Her2/neu (also known as ErbB-2, *ERBB2*)	:	human epidermal growth factor receptor 2, commonly known as the epidermal growth factor receptor family
hUCMSC	:	human umbilical cord matrix stem cells
IFN	:	interferon
IL	:	interleukin
K-rasG12D	:	Kirsten rat sarcoma viral oncogene mutant
MCP-1	:	monocyte chemotactic protein 1
M-CSF	:	monocyte colony-stimulating factor
MHC	:	major histocompatibility complex
MIP-3	:	macrophage inflammatory protein 3
MSC	:	mesenchymal stem cells
NSCLC	:	non-small cell lung cancer
RANTES	:	chemokine (C-C motif) ligand 5, CCL5

SCLC	:	small cell lung cancer
TGF	:	transforming growth factor
TNF-α	:	tumor necrosis factor alpha
TUNEL	:	termilnal deoxynucleotidyl transferase dUTP nick end labeling
VEGF	:	vascular endothelial cell growth factor

Definition of Terms

Interleukins (IL): Various proteins with low molecular weight that are produced primarily by various leukocytes and that function especially in regulation of the immune system, particularly cell-mediated immunity, including tumor immunity.

IL-6: An interleukin that acts as both pro-inflammatory and anti-inflammatory cytokine. It is secreted by T cells and macrophages to stimulate immune response. IL-6 serum levels in NSCLC are a valuable biomarker for prognosis and survival of patients.

Interferons (IFN): Various proteins, produced by virus-, bacterium-, or parasite-infected cells and cancer cells, that inhibit reproduction of the invading virus/bacteria and induce resistance to further infection or tumor growth. IFN-α, IFN-β and IFN-γ are relevant to lung carcinogenesis.

IFN-β: A cytokine produced by a variety of cells. IFN-β is strongly cytotoxic to many types of cancer cells as well as normal epithelial cells. Half life is very short in circulation ($t^{1/2}$= approximately 5 hr).

MSC: Mesoderm-originated embryonic connective tissue-derived multipotent stem cells; these are associated with tumor stromal development and can be used as gene delivery vehicles.

UCMSC: Umbilical cord matrix (Wharton's jelly)-derived stem cells. Human UCMSC have tumoritropic properties and are non-tumorigenic. Human UCMSC are poorly immunogenic and have good potential as gene carriers.

Tumor tropism: The turning or movement of a cell or a part toward or away from an external stimulus, such as cytokines, prostaglandins or growth factors.

References

Ayuzawa, A. and C. Doi, R.S. Rachakatla, M.M. Pyle, D.K. Maurya, D. Troyer, and M. Tamura. 2009. Naïve human umbilical cord matrix derived stem cells significantly attenuate growth of human breast cancer cells in vitro and in vivo. Cancer Lett. 280: 31-37.

Bergers, G. and D. Hanahan. 2008. Modes of resistance to anti-angiogenic therapy. Nat. Rev. Cancer 8: 592-603.

Chada, S. and R. Ramesh, and A.M. Mhashilkar. 2003. Cytokine- and chemokine-based gene therapy for cancer. Curr Opin Mol Ther. 5: 463-474.

Chawla-Sarkar, M. and D.W. Leaman, and E.C. Borden. 2001. Preferential induction of apoptosis by interferon (IFN)-beta compared with IFN-alpha2: correlation with TRAIL/Apo2L induction in melanoma cell lines. Clin. Cancer Res. 7(6): 1821-1831.

Chechlinska, M. and M. Kowalska, and J. Kaminska. 2008. Cytokines as potential tumour markers. Expert Opin. Med. Diagn. 2: 691-711.

Dalwadi, H. and K. Krysan, N. Heuze-Vourc'h, M. Dohadwala, D. Elashoff, S. Sharma, N. Cacalano, A. Lichtenstein, and S. Dubinett. 2005. Cyclooxygenase-2-dependent activation of signal transducer and activator of transcription 3 by interleukin-6 in non-small cell lung cancer. Clin Cancer Res. 11: 7674-7682.

Enewold, L. and L.E. Mechanic, E.D. Bowman, Y.L. Zheng, Z. Yu, G. Trivers, A.J. Alberg, and C.C. Harris. 2009. Serum concentrations of cytokines and lung cancer survival in African Americans and Caucasians. Cancer Epidemiol. Biomarkers Prev. 18: 215-222.

Hatanaka, H. and Y. Abe, M. Naruke, T. Tokunaga, Y. Oshika, T. Kawakami, H. Osada, J. Nagata, J. Kamochi, T. Tsuchida, H. Kijima, H. Yamazaki, H. Inoue, Y. Ueyama, and M. Nakamura. 2001. Significant correlation between interleukin 10 expression and vascularization through angiopoietin/TIE2 networks in non-small cell lung cancer. Clin. Cancer Res. 7: 1287-1292.

Hodkinson, P.S. and A. Mackinnon, and T. Sethi. 2008. Targeting growth factors in lung cancer. Chest. 133: 1209-1216.

Jemal, A. and R. Siegel, E. Ward, Y. Hao, J. Xu, and M.J. Thun. 2009. Cancer Statistics, 2009. CA Cancer J. Clin. 59: 225-249.

Kullmann, T. and I. Barta, E. Csiszer, B. Antus, and I. Horvath. 2008. Differential cytokine pattern in the exhaled breath of patients with lung cancer. Pathol. Oncol. Res. 14: 481-483.

Kwong, Y.L. and S.H. Chen, K. Kosai, M. Finegold, and S.L. Woo. 1997. Combination therapy with suicide and cytokine genes for hepatic metastases of lung cancer. Chest 112(5): 1332-1337.

Lang, F.F. and J.M. Bruner, G.N. Fuller, K. Aldape, M.D. Prados, S. Chang, M. Berger, M.W. McDermott, S.M. Kunwar, L.R. Junck, W. Chandler, J.A. Zwiebel, R.S. Kaplan, and W.K. Yung. 2003. Phase I trial of adenovirus-mediated p53 gene therapy for recurrent glioma: biological and clinical results. J. Clin. Oncol. 21: 2508-2518.

Lissoni, P. and S. Barni, E. Tisi, F. Rovelli, S. Pittalis, R. Rescaldani, L. Vigore, A. Biondi, A. Ardizzoia, and G. Tancini. 1993. In vivo biological results of the association between interleukin-2 and interleukin-3 in the immunotherapy of cancer. Eur. J. Cancer 29A: 1127-1132.

Matsuzuka, M. and R.S. Rachakatla, C. Doi, D.K. Maurya, N. Ohta, A. Kawabata, M. Pyle, L. Pickel, J. Reischman, F. Marini, D. Troyer, and M. Tamura. 2010. Human umbilical cord matrix–derived stem cells expressing interferon-β gene significantly attenuate bronchioloalveolar carcinoma xenografts in SCID mice. Lung Cancer. 70: 28-36.

Margolin, K. 2008. Cytokine therapy in cancer. Expert Opin. Biol. Ther. 8: 1495-1505.

Ortiz-Sanchez, E. and G. Helguera, T.R. Daniels, and M.L. Penichet. 2008. Antibody-cytokine fusion proteins: applications in cancer therapy. Expert Opin. Biol. Ther. 8: 609-632.

Rachakatla, R.S. and F. Marini, M.L. Weiss, M. Tamura, and D. Troyer. 2007. Development of human umbilical cord matrix stem cell-based gene therapy for experimental lung tumors. Cancer Gene Ther. 14: 828-835.

Rachakatla, R.S. and M.M. Pyle, R. Ayuzawa, S.M. Edwards, F.C. Marini, M.L. Weiss, M. Tamura, and D. Troyer. 2008. Combination treatment of human umbilical cord matrix stem cell-based interferon-beta gene therapy and 5-fluorouracil significantly reduces growth of metastatic human breast cancer in SCID mouse lungs. Cancer Invest. 26: 662-670.

Redente, E.F. and L.D. Dwyer-Nield, B.S. Barrett, D.W. Riches, and A.M. Malkinson. 2009. Lung tumor growth is stimulated in IFN-gamma-/- mice and inhibited in IL-4Ralpha-/- mice. Anticancer Res. 29: 5095-5101.

Sequist, L.V. and T.J. Lynch. 2008. EGFR tyrosine kinase inhibitors in lung cancer: an evolving story. Annu. Rev. Med. 59: 429-442.

Sharma, S. and M. Stolina, Y. Lin, B. Gardner, P.W. Miller, M. Kronenberg, and S.M. Dubinett. 1999. T cell-derived IL-10 promotes lung cancer growth by suppressing both T cell and APC function. J. Immunol. 163: 5020-5028.

Soria, J.C. and C. Moon, B.L. Kemp, D.D. Liu, L. Feng, X. Tang, Y.S. Chang, L. Mao, and F.R. Khuri. 2003. Lack of interleukin-10 expression could predict poor outcome in patients with stage I non-small cell lung cancer. Clin. Cancer Res. 9: 1785-1791.

Studeny, M. and F.C. Marini, R.E. Champlin, C. Zompetta, I.J. Fidler, and M. Andreeff. 2002. Bone marrow-derived mesenchymal stem cells as vehicles for interferon-beta delivery into tumors. Cancer Res. 62: 3603-3608.

Sumimoto, H. and K. Tani, Y. Nakazaki, T. Tanabe, H. Hibino, M.S. Wu, K. Izawa, H. Hamada, and S. Asano. 1998. Superiority of interleukin-12-transduced murine lung cancer cells to GM-CSF or B7-1 (CD80) transfectants for therapeutic antitumor immunity in syngeneic immunocompetent mice. Cancer Gene Ther. 5: 29-37.

Weiss, M.L. and C. Anderson, S. Medicetty, K.B. Seshareddy, R.J. Weiss, I. VanderWerff, D. Troyer, and K.R. McIntosh. 2008. Immune properties of human umbilical cord Wharton's jelly-derived cells. Stem Cells 26: 2865-2874.

Wilderman, M.J. and J. Sun, A.S. Jassar, V. Kapoor, M. Khan, A. Vachani, E. Suzuki, P.A. Kinniry, D.H. Sterman, L.R. Kaiser, and S.M. Albelda. 2005. Intrapulmonary IFN-beta gene therapy using an adenoviral vector is highly effective in a murine orthotopic model of bronchogenic adenocarcinoma of the lung. Cancer Res. 65: 8379-8387.

Wong, V.L. and D.J. Rieman, L. Aronson, B.J. Dalton, R. Greig, and M.A. Anzano. 1989. Growth-inhibitory activity of interferon-beta against human colorectal carcinoma cell lines. Int. J. Cancer 43(3): 526-530.

Yasumoto, K. and K. Mivazaki, A. Nagashima, T. Ishida, T. Kuda, T. Yano, K. Sugimachi,. and K. Nomoto. 1987. Induction of lymphokine-activated killer cells by intrapleural instillations of recombinant interleukin-2 in patients with malignant pleurisy due to lung cancer. Cancer Res. 47: 2184-2187.

Zhang, H. and P.P. Koty, J. Mayotte, and M.L. Levitt. 1999. Induction of multiple programmed cell death pathways by IFN-beta in human non-small-cell lung cancer cell lines. Exp. Cell Res. 247: 133-141.

CHAPTER 13

Cytokines and Breast Cancer

Jonna Frasor
Department of Physiology and Biophysics, University of Illinois at Chicago, Chicago, IL, USA 60612

ABSTRACT

Breast tumors develop within a specialized tumor microenvironment that consists of numerous cell types, including cancer cells, stromal cells, adipose tissue, and infiltrating immune cells. These cells release a wide range of factors that can modulate tumor development by regulating cancer cell proliferation, survival, invasion and motility, as well as local blood vessel formation, or angiogenesis. In particular, cytokines are highly expressed in the breast tumor microenvironment and can play a crucial role in many of these processes during breast tumorigenesis. Cytokines also appear to be important in breast tumor metastasis, particularly to the bone, where cytokines are abundantly expressed. Interestingly, the effects of cytokines can be pro-tumorigenic and lead to enhanced tumor development and progression to a more aggressive disease, or anti-tumorigenic with reduced tumor growth and inhibition of tumor angiogenesis. Furthermore, the effects of cytokines can result from both modulation of the immune system and direct effects of cytokines on the tumor itself, independent of the immune system. The direct effects, rather than the immuno-modulatory effects, of cytokines on breast tumor growth, invasion, and metastasis are the focus of this chapter.

INTRODUCTION

The mammary gland is composed of an extensive epithelial ductal structure that ends in clusters of terminal ductal lobular units, which become capable of producing milk following pregnancy during lactation. Myoepithelial cells form a second layer of epithelial cells that surround the luminal epithelial cells and contract to promote milk ejection from the ducts. The entire ductal system is embedded within the stroma, which contains a variety of cell types including fibroblasts and adipocytes. Mammary tissue is highly responsive to hormonal status with ovarian and pituitary hormones playing essential roles in controlling the development of the mammary gland during puberty, as well as during pregnancy, and having important functions during breastfeeding, or lactation. In particular, estrogen and progesterone regulate ductal outgrowth and branching, as well as differentiation into functional alveolar structures during pregnancy. After pregnancy, prolactin and oxytocin are important hormones that control milk production and secretion during lactation. When lactation is terminated, the mammary gland undergoes a remodeling process, known as involution, where the gland returns to the pre-pregnancy state. This process involves extensive cell death, macrophage infiltration and the creation of an inflammatory environment similar to that seen in wound healing. It has been proposed that this inflammatory environment found during involution may contribute to pregnancy-associated breast cancer (Schedin 2006).

Breast cancer is a highly complex and heterogeneous disease. It is thought that breast tumors can arise from malignant transformation of different types of progenitor cells, which can lead to formation of tumors with different histological phenotypes and pathological courses (Visvader 2009). Breast tumors can remain encapsulated within the duct (i.e., ductal carcinoma *in situ*) or can invade the surrounding stroma. As invasive breast cancers progress, they can metastasize to various organs in the body, including the liver, lung, brain and bone. Gene expression profiling of human breast tumors has revealed that there are at least five intrinsic subtypes of breast cancer based on molecular signatures (Sorlie *et al.* 2001). Two subtypes, called Luminal A and Luminal B, have signatures similar to that of luminal epithelial cells and tend to retain hormone-responsiveness but have very different patient survival, with Luminal A having a better prognosis than Luminal B tumors. Other subtypes are characterized by over-expression of the epidermal growth factor family member, Her2, or by a basal-like gene expression profile, which is based on a specific basal cell cytokeratin expression pattern. It has been proposed that each of these breast cancer subtypes arises from a different cell population within the breast and is associated with differential disease progression, clinical outcome, and response to therapeutic regimens.

Interestingly, estrogen, which controls development and function of the normal mammary gland, also plays an important role in breast cancer. Over 70% of breast tumors, primarily of the Luminal A and Luminal B subtypes, express the estrogen receptor (ER) and are dependent on estrogen for tumor growth, making ER an important therapeutic target. Similarly, the enzyme responsible for the production of estrogens, aromatase, is also highly expressed in the breast and is responsible for increased local production of estrogen within the breast tumor microenvironment, particularly in post-menopausal women. Targeting aromatase with specific inhibitors has also proven to be an effective therapeutic strategy for women with breast cancer. Several lines of evidence, discussed below, suggest that crosstalk between cytokines and estrogen may play an important role in breast cancer progression.

SOURCES OF CYTOKINES IN THE BREAST TUMOR MICROENVIRONMENT

Not only is breast cancer a heterogeneous disease, but a heterogeneous population of cells can be found within each tumor. These cells, which include the malignant cancer cells themselves and a variety of stromal cells, such as immune cells, fibroblasts, and endothelial cells, secrete factors, allow for cell-to-cell communication, provide support and structure through production of extracellular matrix proteins, and create what is called the tumor microenvironment (Fig. 1).

A wide array of pleiotropic cytokines can be found within the tumor microenvironment, with a number of different cytokines that appear to be over-expressed in breast tumors compared to normal tissue. Interestingly, a cluster of chemokine genes, including IL-8 (CXCL8), which are located on 4q21, appear to be co-expressed in breast tumors and distant metastases in a coordinated manner

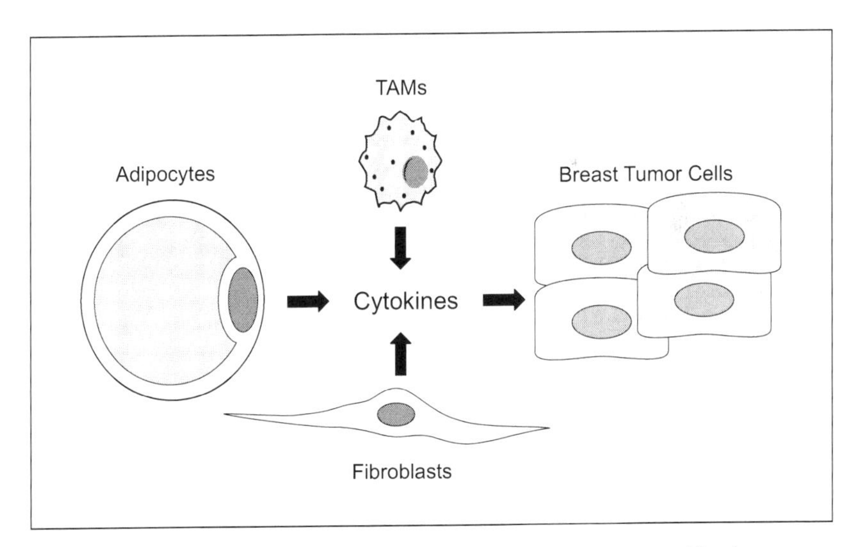

Fig. 1 Sources of cytokines in the breast tumor microenvironment. The breast tumor microenvironment consists of numerous cell types that can produce cytokines, including the malignant cancer cells, surrounding stromal cells such as fibroblasts and adipocytes, and also infiltrating tumor-associated macrophages (TAMs). In addition to modulating the immune system, cytokines can act directly on cancer cells to have either a pro-tumorigenic or anti-tumorigenic impact (unpublished figure).

(Bieche *et al.* 2007). TNFα, which is one of the most widely studied cytokines in breast cancer, is expressed in ~50% of breast tumors but at very low to undetectable levels in normal breast tissue (Miles *et al.* 1994, Pusztai *et al.* 1994). Within the tumors, TNFα appears to be localized to the stroma and more specifically to macrophage-like cells, while TNFα receptors are expressed on most cell types, including breast cancer cells (Leek *et al.* 1998, Miles *et al.* 1994, Pusztai *et al.* 1994).

One of the most important sources of cytokines, such as TNFα, within the breast tumor microenvironment is infiltrating tumor-associated macrophages (TAMs). In response to environmental stimuli or substances secreted by the tumor, TAMs can be recruited within the tumor, where they can have either pro-tumor or anti-tumor activities (Leek and Harris 2002). On balance, however, the presence of macrophages within a breast tumor is associated with more aggressive disease and a reduced relapse-free survival (Leek and Harris 2002). In particular, TAMs are associated with increased angiogenesis in human tumors and play an essential role in angiogenesis in an animal model of breast cancer (Leek and Harris 2002, Lin *et al.* 2006). It has also been proposed that infiltrating macrophages may also be involved in pregnancy-associated breast cancer, which is defined as the subset of breast cancers diagnosed within 5 yr of a pregnancy (O'Brien and Schedin 2009). As involution occurs and tissue remodeling takes place following lactation, there is an influx of macrophages into mammary tissue and expression of a number of cytokines and chemokines is increased in the mouse mammary gland (Clarkson *et al.* 2004, Stein *et al.* 2004).

An additional source of cytokines that can affect breast tumors is the surrounding adipose tissue. A proteomic approach was used to examine proteins found in the breast tumor interstitial fluid and in mammary adipose tissue surrounding the tumor (Celis *et al.* 2005). A large number of cytokines were identified and suggest that cytokines produced by local adipose tissue may act as paracrine factors to influence breast tumors. This has important implications given the increasing prevalence of obesity worldwide and the correlation between obesity and the risk of post-menopausal breast cancer.

CYTOKINES INFLUENCE ESTROGEN PRODUCTION AND ESTROGEN RECEPTOR ACTIVITY

In post-menopausal women, breast cells are capable of producing estrogen at levels up to 100-fold that seen in the circulation. The highly localized pool of estrogen can then bind to and act through its receptor (ER), which is expressed in up to 75% of breast tumors, and stimulate hormone-dependent tumor growth. This reliance of the majority of breast tumors on estrogen makes the enzymes responsible for estrogen production, as well as the ER itself, major therapeutic targets in breast cancer. Estrogens are produced from precursor androgens by the enzymes aromatase, which converts the androgen androstenedione to the weak estrogen estrone, and 17β-hydroxysteroid dehydrogenase (17β-HSD), which converts estrone to the highly active estrogen estradiol. In addition, estrone sulfatase (E1-STS) plays a role in estrogen production in the breast, where it is responsible for the formation of estrone from the inactive metabolite estrone sulfate. All three of these enzymes are expressed in the breast and intriguingly all can be up-regulated by cytokines (Purohit *et al.* 2002) through various mechanisms (Fig. 2). For example, a number of cytokines including IL-11, oncostatin-M, IL-6, and leukemia inhibitory factor acting through the Jak/Stat pathway can up-regulate aromatase expression in human adipose tissue (Zhao *et al.* 1995), whereas TNFα can stimulate aromatase expression in the same tissue via the AP-1 pathway (Zhao *et al.* 1996). Activity of 17β-HSD, on the other hand, is regulated in breast cancer cells themselves rather than stromal adipocytes by cytokines IL-6 and IL-4 (Purohit *et al.* 2002). The stimulatory effect of cytokines IL-6 and TNFα on E1-STS appears to be at the level of enzyme activity rather than enzyme expression (Purohit *et al.* 2002). Taken together, these findings indicate that a microenvironment rich in cytokines, as is found in many breast tumors and the involuting breast, may lead to high local levels of estrogens, which can act on the ER to promote tumor growth.

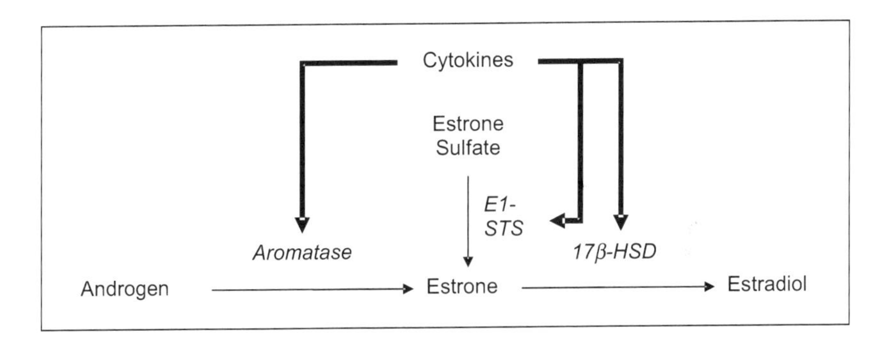

Fig. 2 Cytokines stimulate local estrogen production in the breast. Cytokines stimulate (thick arrows) expression and/or activity of three enzymes (italics) that are responsible for the local production of the potent estrogen, 17β-estradiol in the breast. Aromatase converts androgens into the weak estrogen estrone, E1-STS converts the inactive metabolite estrone sulfate in estrone, and 17β-HSD converts estrone into estradiol (unpublished figure).

Expression of the ER is a major prognostic and therapeutic marker in breast cancer. Cytokines not only increase local estrogen production, but can also affect ER-positive breast tumors through other mechanisms. For example, recent studies indicate that pro-inflammatory cytokines TNFα and IL-1β signaling through the NFκB pathway in breast cancer cells can modulate ER activity on specific target genes (Fig. 3) and produce a gene expression profile associated with the more

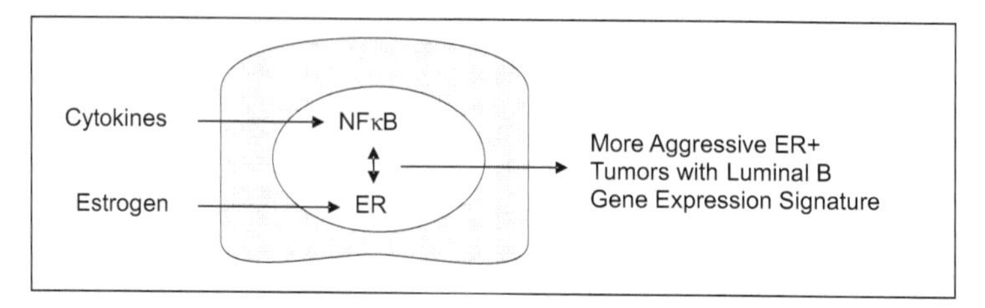

Fig. 3 Crosstalk between the estrogen receptor and the NFκB pathway in breast cancer cells. Cytokines acting through the NFκB pathway and estrogen activation of the ER lead to a gene expression profile that correlates with ER-positive Luminal B tumors, which are more aggressive and associated with a worse patient prognosis than the ER-positive Luminal A tumors (unpublished figure).

aggressive ER-positive Luminal B breast tumors but not the less aggressive ER-positive Luminal A tumors (Frasor *et al.* 2009). Other studies have shown that estrogen can regulate the expression of cytokines in breast cancer cells to influence cell proliferation. For example, estrogen can increase production of CXCL12/SDF-1, which not only acts as a potent mitogen to stimulate proliferation of breast cancer cells but also feeds back to regulate ER activity (Hall and Korach 2003, Sauve *et al.* 2009). Furthermore, the presence of pro-inflammatory cytokines may influence the ability of the ER to respond properly to the therapeutic drug tamoxifen. Tamoxifen acts as a Selective Estrogen Receptor Modulator (SERM), meaning that it can act as an ER antagonist or agonist depending on the cellular background. It has been suggested that IL-1β can alter ER response to tamoxifen by turning the normal antagonistic activity of tamoxifen into agonist activity and abrogating the growth inhibitor effects of tamoxifen (Zhu *et al.* 2006). Therefore, many close links have been established between the presence of cytokines in the breast tumor microenvironment and both estrogen production and ER activity, suggesting an important role for cytokines in hormone-dependent breast cancer progression.

CYTOKINES CAN HAVE PRO- OR ANTI-TUMOR ACTIVITIES

Cytokines also have a multitude of effects on breast tumors that are independent of not only the estrogen/ER pathway but also independent of the immune system. Interestingly, a large and complex body of literature suggests that the effects of cytokines can be positive or negative, meaning that individual cytokines can either promote or inhibit tumor growth and progression. The exact understanding of why a cytokine may have these differential effects is not clear but studies suggest that the stage of tumor development or signaling pathways activated by cytokines play important roles in determining the overall effect of a cytokine on breast tumors.

Two examples of cytokines with complex effects on breast tumors are described below, TNFα and TGFβ.

It is well established that TNFα can induce cell death in breast cancer cells and have anti-angiogenic activity in models of breast tumors. Studies have shown that cells sensitive to TNFα-induced death display a prolonged activation of the JNK pathway (Deng *et al.* 2003). Paradoxically, TNFα and other pro-inflammatory cytokines are found at high levels in the tumor microenvironment. TNFα has been found to stimulate production of additional cytokines and chemokines and can regulate breast cancer cell proliferation, survival, motility and invasiveness, as well

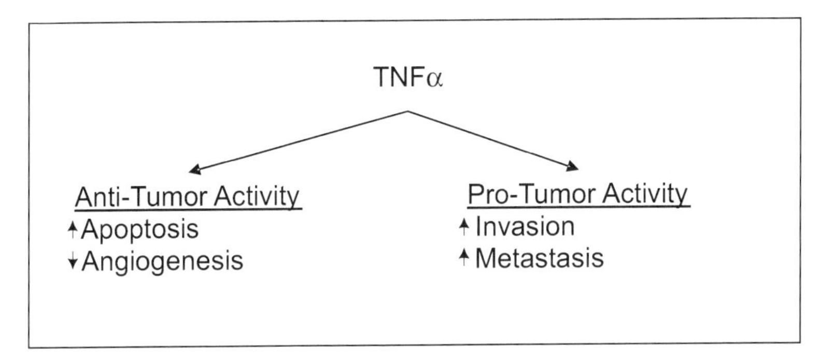

Fig. 4 TNFα has both anti- and pro-tumor activities. While TNFα has been shown to induce apoptosis in breast cancer cells and inhibit breast tumor angiogenesis, it can also stimulate breast cancer cell invasion and is associated with metastasis (unpublished figure).

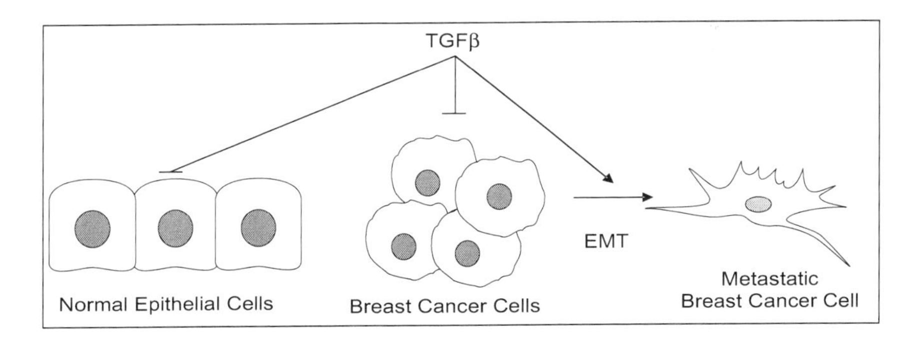

Fig. 5 TGFβ has both anti- and pro-tumorigenic activities. In both normal breast epithelial cells and breast cancer cells, TGFβ can inhibit proliferation. However, it can also drive metastasis of breast cancer cells, potentially by inducing EMT in breast cancer cells (unpublished figure).

as tumor angiogenesis and metastasis. Many of these pro-tumorigenic effects of TNFα are mediated by activation of NFκB in cancer cells. Elegant studies using macrophages in co-culture with breast cancer cells demonstrated that macrophages can stimulate cancer cell invasiveness in a TNFα-, NFκB- and JNK-dependent manner (Hagemann *et al.* 2005). The fact that TNFα can have diverse effects on breast cancer may explain the disappointing results in clinical trials targeting TNFα in breast cancer patients. For example, a phase II study with Etanercept in women with advanced disease showed no effect on disease progression but substantially lower content of pro-inflammatory markers in serum (Madhusudan *et al.* 2004). A more complete understanding of the molecular mechanisms of TNFα action in both promoting and inhibiting breast cancer may lead to more effective and targeted strategies for breast cancer.

As with TNFα, many early studies showed that TGFβ has anti-tumorigenic activity through inhibition of proliferation and induction of apoptosis in both normal and transformed mammary epithelial cells (Barcellos-Hoff and Akhurst 2009). This was demonstrated in TGFβ1+/- mice, which have accelerated mammary gland development (Tan *et al.* 2009), a finding that suggested that TGFβ plays an important role in controlling or limiting hormonally stimulated proliferation of mammary epithelial cells. Additional studies have demonstrated that TGFβ acts on mammary epithelial cells to limit estrogen-stimulated proliferation through activation of Smad2/3 specifically in cells that express ER (Barcellos-Hoff and Akhurst 2009). Additional mechanisms of crosstalk between estrogen action through the ER and the TGFβ pathway may underlie some of the effects of TGFβ in both normal mammary gland and breast cancer (Tan *et al.* 2009). Many studies have found that ER-positive, hormone-sensitive breast cancer cell lines are highly sensitive to the growth-suppressive effects of TGFβ, while ER-negative or hormone-independent cell lines are less sensitive. Furthermore, it has been suggested that TGFβ may underlie some of the growth-inhibitory effects of SERMs, such as tamoxifen. In addition to regulating proliferation, activation of TGFβ and Smad3 is associated with massive apoptosis that occurs during involution and in response to DNA damage induced by agents such as ionizing radiation (Barcellos-Hoff and Akhurst 2009).

However, inactivation of TGFβ in mouse models does not lead to spontaneous mammary tumor formation and further studies indicate additional oncogenic events are required for cancer initiation (Tan *et al.* 2009). In the case of the polyoma middle T expression in the mammary gland, tumors are accelerated by loss of TGFBR2 (Forrester *et al.* 2005). Absence or deficiency of TGFβ signaling accelerates not only mammary gland development but also tumor initiation, progression and metastasis, confirming an important role for TGFβ in limiting growth of breast cancer. However, other studies clearly establish a role for TGFβ in promoting tumor progression and metastasis in particular. Studies that demonstrated TGFβ inhibits early tumor initiation and progression also demonstrated that TGFβ over-expression can enhance invasion and metastasis. This complex role for TGFβ in breast cancer is further highlighted by the fact that both increased TGFβ activity and loss of TGFβ activity can enhance mammary tumorigenesis in mouse models (Bierie *et al.* 2008). Several theories may explain this promoting role of TGFβ in cancer. For example, TGFβ is thought to play a role in the epithelial to mesenchymal transition (EMT), an embryonic developmental process that is reactivated and used by cancer cells during invasion and metastasis. Additionally, TGFβ appears to be involved in maintaining stem cells or progenitor cell populations (Tang *et al.* 2007). As is the case with TNFα, careful dissection of the molecular pathways required for the growth-promoting versus the growth-inhibiting effects of TGFβ is necessary to more effectively take advantage of TGFβ signaling for breast cancer therapy.

ROLE OF CYTOKINES IN BREAST CANCER METASTASIS

Over 90% of cancer deaths are the result of tumor metastasis to distal organs. Metastasis is a complex process during which tumor cells leave the original microenvironment and travel through the circulation to other organs. The cell must then enter the new tissue, a process called extravasation, and adapt to grow within the new metastatic tissue microenvironment. Interestingly, cytokines have been implicated in the majority of these steps, and they are particularly important in bone metastasis. When breast cancer cells colonize in bone, crosstalk between the cancer and bone cells occurs via release of paracrine factors that can stimulate tumor growth as well as induce bone remodeling, usually resulting in net bone degradation (Bussard *et al.* 2008). For example, preclinical models of metastasis in animals with reduced TGFβ signaling show reduced metastasis to bone (Padua and Massague 2009). It may be that TGFβ present with the bone microenvironment that is important for metastasis. TGFβ is present within the extracellular matrix and can become activated during osteoclastic bone remodeling. The released TGFβ can then act to cause further bone remodeling as well as regulate target gene expression in cancer cells. In addition to TGFβ, metastatic breast cancer cells secrete a number of cytokines into the bone microenvironment, including IL-8, IL-6 and IL-11, that can act on osteoclasts to further stimulate bone degradation. Thus, cytokines play an essential role in a vicious cycle of signaling between breast cancer cells and bone that occurs during metastasis.

Table 1 Key features of cytokine action in breast cancer

- Cytokines are over-expressed in the breast tumor microenvironment compared to normal breast tissue.
- Tumor-associated macrophages infiltrate into tumors and are a rich source of cytokine production in the breast tumor microenvironment.
- Involution of the mammary gland following lactation is associated with an inflammatory environment, high cytokine production and TAM infiltration, and formation of aggressive pregnancy-associated breast cancers.
- The majority of breast tumors express the estrogen receptor and are dependent on estrogen for growth. Cytokines enhance estrogen production and influence estrogen receptor activity.
- Cytokines can act on breast cancer cells directly and have either pro- or anti-tumorigenic activities.
- On one hand, TNFα induces apoptosis in breast cancer cells and inhibits angiogenesis to reduce breast tumor growth; on the other hand, TNFα can stimulate breast cancer cell invasiveness.
- TGFβ plays an important role in limiting hormone-stimulated mammary epithelial and breast cancer cell proliferation but also acts to increase breast cancer cell metastasis, particularly to bone.
- Metastasis to bone involves a vicious cycle of crosstalk between bone cells and breast cancer cells, which is mediated by cytokines.

Table 2 Key features of breast cancer

- Breast cancer is a heterogeneous disease with at least five different intrinsic subtypes based on gene expression profiling.
- Approximately 70% of breast tumors express the estrogen receptor (ER), which is a major molecular prognostic marker and therapeutic target of breast cancer.
- Aromatase is an enzyme that is responsible for estrogen production locally within the breast and is also a therapeutic target.
- There are two subtypes of ER-positive tumors: Luminal A, which are less aggressive and have better patient outcomes, and Luminal B, which are more aggressive and have poor patient outcomes.
- Her2 is also over-expressed in a subset of more aggressive breast tumors and can be targeted with drugs.
- The most aggressive tumors, basal-like breast tumors, are most often triple negative because they lack the estrogen receptor, progesterone receptor, and Her2.

Summary Points

- Cytokines are over-expressed in the breast tumor microenvironment compared to normal breast tissue.
- Tumor-associated macrophages infiltrate into tumors and are a rich source of cytokine production in the breast tumor microenvironment.

- Involution of the mammary gland following lactation is associated with an inflammatory environment, high cytokine production and TAM infiltration, and formation of aggressive pregnancy-associated breast cancers.
- The majority of breast tumors express the estrogen receptor and are dependent on estrogen for growth. Cytokines enhance estrogen production and influence estrogen receptor activity.
- Cytokines, such as TNFα and TGFβ, can act on breast cancer cells directly and have either pro- or anti-tumorigenic activities.

Abbreviations

17β-HSD	:	17β-hydroxysteroid dehydrogenase
AP-1	:	activator protein 1
CXCL	:	chemokine ligand
E1-STS	:	estrone sulfatase
EMT	:	epithelial mesenchymal transition
ER	:	estrogen receptor
IL	:	interleukin
Jak	:	Janus kinase
JNK	:	c-Jun N-terminal kinase
NFκB	:	nuclear factor κ B
SDF-1	:	stromal cell-derived factor 1
SERM	:	selective estrogen receptor modulator
Stat	:	signal transducers and activators of transcription
TAMs	:	tumor-associated macrophages
TGFβ	:	transforming growth factor β
TNFα	:	tumor necrosis factor α

Definition of Terms

Aromatase: A key enzyme in estrogen synthesis that is up-regulated by cytokines.

Breast tumor microenvironment: A number of different cell types and substances that make up the unique environment in which the breast tumor grows.

Epithelial to mesenchymal transition: An embryonic developmental process that occurs in invasive and metastatic cancer cells.

Estrogen receptor: A protein that is expressed in the majority of breast tumors and responds to estrogen by driving proliferation and survival of estrogen-dependent cells.

Osteoclastic bone remodeling: Process of bone degradation that causes release of cytokines.

Tumor-associated macrophages: Macrophages that infiltrate the tumor and are associated with increased angiogenesis and increased metastasis.

References

Barcellos-Hoff, M.H. and R.J. Akhurst. 2009. Transforming growth factor-beta in breast cancer: too much, too late. Breast Cancer Res. 11: 202.

Bieche, I. and C. Chavey, C. Andrieu, M. Busson, S. Vacher, L. Le Corre, J.-M. Guinebretiere, S. Burlinchon, R. Lidereau and G. Lazennec. 2007. CXC chemokines located in the 4q21 region are up-regulated in breast cancer. Endocr Relat Cancer. 14: 1039-1052.

Bierie, B. and D.G. Stover, T.W. Abel, A. Chytil, A.E. Gorska, M. Aakre, E. Forrester, L. Yang, K.U. Wagner and H.L. Moses. 2008. Transforming growth factor-beta regulates mammary carcinoma cell survival and interaction with the adjacent microenvironment. Cancer Res. 68: 1809-1819.

Bussard, K.M. and C.V. Gay and A.M. Mastro. 2008. The bone microenvironment in metastasis; what is special about bone? Cancer Metastasis Rev. 27: 41-55.

Celis, J.E. and J.M.A. Moreira, T. Cabezon, P. Gromov, E. Friis, F. Rank and I. Gromova. 2005. Identification of Extracellular and Intracellular Signaling Components of the Mammary Adipose Tissue and Its Interstitial Fluid in High Risk Breast Cancer Patients: Toward Dissecting The Molecular Circuitry of Epithelial-Adipocyte Stromal Cell Interactions. Mol Cell Proteomics. 4: 492-522.

Clarkson, R.W. and M.T. Wayland, J. Lee, T. Freeman, and C.J. Watson. 2004. Gene expression profiling of mammary gland development reveals putative roles for death receptors and immune mediators in post-lactational regression. Breast Cancer Res. 6: R92-109.

Deng, Y. and X. Ren, L. Yang, Y. Lin, and X. Wu. 2003. A JNK-dependent pathway is required for TNFalpha-induced apoptosis. Cell 115: 61-70.

Forrester, E. and A. Chytil, B. Bierie, M. Aakre, A.E. Gorska, A.R. Sharif-Afshar, W.J. Muller and H.L. Moses. 2005. Effect of conditional knockout of the type II TGF-beta receptor gene in mammary epithelia on mammary gland development and polyomavirus middle T antigen induced tumor formation and metastasis. Cancer Res. 65: 2296-2302.

Frasor, J. and A. Weaver, M. Pradhan, Y. Dai, L.D. Miller, C.Y. Lin, and A. Stanculescu. 2009. Positive cross-talk between estrogen receptor and NF-kappaB in breast cancer. Cancer Res. 69: 8918-8925.

Hagemann, T. and J. Wilson, H. Kulbe, N.F. Li, D.A. Leinster, K. Charles, F. Klemm, T. Pukrop, C. Binder, and F.R. Balkwill. 2005. Macrophages induce invasiveness of epithelial cancer cells via NF-kappa B and JNK. J. Immunol. 175: 1197-1205.

Hall, J.M. and K.S. Korach. 2003. Stromal cell-derived factor 1, a novel target of estrogen receptor action, mediates the mitogenic effects of estradiol in ovarian and breast cancer cells. Mol. Endocrinol. 17: 792-803.

Leek, R.D. and A.L. Harris. 2002. Tumor-associated macrophages in breast cancer. J. Mammary Gland Biol. Neoplasia 7: 177-189.

Leek, R.D. and R. Landers, S.B. Fox, F. Ng, A.L. Harris, and C.E. Lewis. 1998. Association of tumour necrosis factor alpha and its receptors with thymidine phosphorylase expression in invasive breast carcinoma. Br. J. Cancer 77: 2246-2251.

Lin, E.Y. and J.F. Li, L. Gnatovskiy, Y. Deng, L. Zhu, D.A. Grzesik, H. Qian, X.N. Xue, and J.W. Pollard. 2006. Macrophages regulate the angiogenic switch in a mouse model of breast cancer. Cancer Res. 66: 11238-11246.

Madhusudan, S. and M. Foster, S.R. Muthuramalingam, J.P. Braybrooke, S. Wilner, K. Kaur, C. Han, S. Hoare, F. Balkwill, D.C. Talbot, T.S. Ganesan, and A.L. Harris. 2004. A phase II study of etanercept (Enbrel), a tumor necrosis factor alpha inhibitor in patients with metastatic breast cancer. Clin. Cancer Res. 10: 6528-6534.

Miles, D.W. and L.C. Happerfield, M.S. Naylor, L.G. Bobrow, R.D. Rubens, and F.R. Balkwill. 1994. Expression of tumour necrosis factor (TNF alpha) and its receptors in benign and malignant breast tissue. Int. J. Cancer 56: 777-782.

O'Brien, J. and P. Schedin. 2009. Macrophages in breast cancer: do involution macrophages account for the poor prognosis of pregnancy-associated breast cancer? J. Mammary Gland Biol. Neoplasia 14: 145-157.

Padua, D. and J. Massague. 2009. Roles of TGFbeta in metastasis. Cell Res. 19: 89-102.

Purohit, A. and S.P. Newman, and M.J. Reed. 2002. The role of cytokines in regulating estrogen synthesis: implications for the etiology of breast cancer. Breast Cancer Res. 4: 65-69.

Pusztai, L. and L.M. Clover, K. Cooper, P.M. Starkey, C.E. Lewis, and J.O. McGee. 1994. Expression of tumour necrosis factor alpha and its receptors in carcinoma of the breast. Br. J. Cancer 70: 289-292.

Sauve, K. and J. Lepage, M. Sanchez, N. Heveker, and A. Tremblay. 2009. Positive feedback activation of estrogen receptors by the CXCL12-CXCR4 pathway. Cancer Res. 69: 5793-5800.

Schedin, P. 2006. Pregnancy-associated breast cancer and metastasis. Nat. Rev. Cancer 6: 281-291.

Sorlie, T. and C.M. Perou, R. Tibshirani, T. Aas, S. Geisler, H. Johnsen, T. Hastie, M.B. Eisen, M. van de Rijn, S.S. Jeffrey, T. Thorsen, H. Quist, J.C. Matese, P.O. Brown, D. Botstein, P. Eystein Lonning, and A.L. Borresen-Dale. 2001. Gene expression patterns of breast carcinomas distinguish tumor subclasses with clinical implications. Proc. Natl. Acad. Sci. USA 98: 10869-10874.

Stein, T. and J.S. Morris, C.R. Davies, S.J. Weber-Hall, M.A. Duffy, V.J. Heath, A.K. Bell, R.K. Ferrier, G.P. Sandilands, and B.A. Gusterson. 2004. Involution of the mouse mammary gland is associated with an immune cascade and an acute-phase response, involving LBP, CD14 and STAT3. Breast Cancer Res. 6: R75-91.

Tan, A.R. and G. Alexe, and M. Reiss. 2009. Transforming growth factor-beta signaling: emerging stem cell target in metastatic breast cancer? Breast Cancer Res. Treat. 115: 453-495.

Tang, B. and N. Yoo, M. Vu, M. Mamura, J.S. Nam, A. Ooshima, Z. Du, P.Y. Desprez, M.R. Anver, A.M. Michalowska, J. Shih, W.T. Parks and L.M. Wakefield. 2007. Transforming growth factor-beta can suppress tumorigenesis through effects on the putative cancer stem or early progenitor cell and committed progeny in a breast cancer xenograft model. Cancer Res. 67: 8643-8652.

Visvader, J.E. 2009. Keeping abreast of the mammary epithelial hierarchy and breast tumorigenesis. Genes Dev. 23: 2563-2577.

Zhao, Y. and J.E. Nichols, S.E. Bulun, C.R. Mendelson, and E.R. Simpson. 1995. Aromatase P450 gene expression in human adipose tissue. Role of a Jak/STAT pathway in regulation of the adipose-specific promoter. J. Biol. Chem. 270: 16449-16457.

Zhao, Y. and J.E. Nichols, R. Valdez, C.R. Mendelson, and E.R. Simpson. 1996. Tumor necrosis factor-alpha stimulates aromatase gene expression in human adipose stromal cells through use of an activating protein-1 binding site upstream of promoter 1.4. Mol. Endocrinol. 10: 1350-1357.

Zhu, P. and S.H. Baek, E.M. Bourk, K.A. Ohgi, I. Garcia-Bassets, H. Sanjo, S. Akira, P.F. Kotol, C.K. Glass, M.G. Rosenfeld, and D.W. Rose. 2006. Macrophage/cancer cell interactions mediate hormone resistance by a nuclear receptor derepression pathway. Cell 124: 615-629.

CHAPTER 14

Cancer Immunotherapy and Cytokines

Masoud H. Manjili*, Maciej Kmieciak and Rose H. Manjili
Department of Microbiology & Immunology
Virginia Commonwealth University School of Medicine,
Massey Cancer Center, Box 980035, Richmond,
Virginia 23298, USA

ABSTRACT

Cytokines, as messengers of the immune system, are being used in cancer therapy to enhance anti-tumor immunity. IL-2 and IFN-α are two FDA-approved cytokines for the treatment of cancers. GM-CSF and IL-12 are also being used as vaccine adjuvant to enhance immune responses against cancers. However, side effects associated with the injection of these cytokines and their limited efficacy in only a fraction of cancer patients remain major issues. Recently, certain cytokines with potential activity on expanding tumor-reactive T cells have been tested *ex vivo* rather than *in vivo* in order to overcome safety issues and facilitate the differentiation of tumor-specific T cells for adoptive T cell immunotherapy (AIT). Identification of key cytokines that enhance proliferation and differentiation of tumor-reactive T cells for producing an objective response upon AIT is crucial for manipulating immune responses against cancer. Thus far, common gamma chain cytokines (IL-2, IL-7, IL-15, IL-21) have been shown to be promising candidates for the generation of T cells that can induce tumor regression upon AIT. The present review will focus on *ex vivo* use of common gamma chain cytokines either alone

*Corresponding author

or in combination by introducing different *ex vivo* protocols for the expansion of tumor-specific T cells for AIT of cancer.

INTRODUCTION

Adoptive T cell immunotherapy (AIT), or the adoptive transfer of antigen-sensitized T cells activated and/or expanded *ex vivo* in the presence of a T cell growth factor IL-2, has been pursued as a way to amplify T cell responses to tumor antigen. Early attempts of AIT using tumor-infiltrating lymphocytes (TIL) and IL-2 produced some objective responses (Rosenberg *et al.* 1994). However, generation of a lymphopenic condition is an effective preconditioning regimen that can enhance the efficacy of AIT, perhaps by minimizing cellular cytokine sinks (Johnson *et al.* 2009). It has been shown that T cell phenotypes, i.e., effector (T_E), effector/memory (T_{EM}), or central/memory (T_{CM}), exhibit differential effects against the tumor (Klebanoff *et al.* 2005) (Table 2). Therefore, development of an *ex vivo* cytokine formulation for the differentiation and expansion of a suitable T cell phenotype is another critical factor for predicting objective responses against metastatic tumors. The T cell-associated cytokines, including those signaling through T-cell membrane receptors that contain the common cytokine-receptor gamma chain, have been shown to be promising in altering differentiation of the

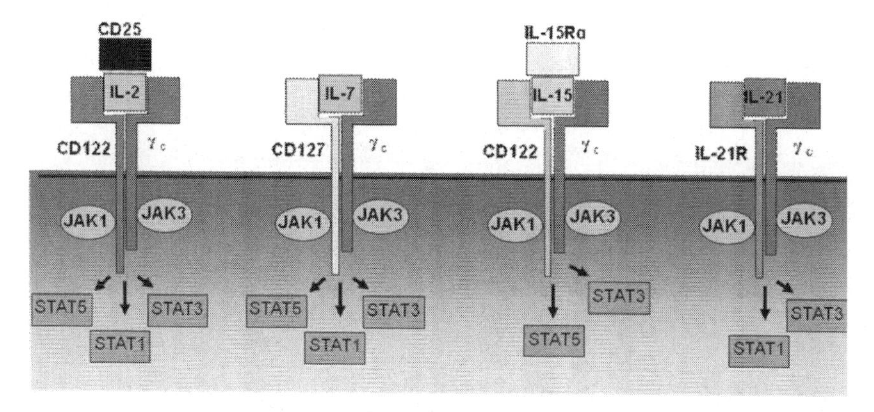

Fig. 1 Cytokine receptors on T cells. Common gamma chain cytokines (IL-2, IL-7, IL-15, IL-21) being used in adoptive T cell immunotherapy of cancer both *ex vivo* and *in vivo,* and their shared receptors, the IL-2 Rγ (γ c). Predominant STAT signaling for each cytokine is shown.

Color image of this figure appears in the color plate section at the end of the book.

Table 1 Key features of common gamma chain cytokines for AIT

- Common gamma chain cytokines can expand T cells in the absence of any antigenic stimulation.
- Control of T cell differentiation to a desired phenotype for AIT is possible by use of common gamma chain cytokines *ex vivo*.
- Common gamma chain cytokines can be used both *ex vivo* and *in vivo* for immunotherapy of cancer.
- Good Manufacturing Practice recombinant IL-2, IL-7, IL-15, and IL-21 are available for use in the clinic.

Table 2 T cell phenotypes.

T cell phenotypes	*Markers*
Naive	CD44-CD62L+
T_E	CD44 + CD62L–
T_{EM}	CD44 + CD62^{low}
T_{CM}	CD44 + CD62L^{high}

tumor-specific CD8+ T cells. Gamma chain is a critical component of the receptors for IL-2, IL-4, IL-7, IL-9, IL-15, and IL-21, which together regulate lymphocyte development and control adaptive immune responses (Fig. 1).

CURRENT PROTOCOLS FOR THE GENERATION OF TUMOR-SPECIFIC CD8+ T CELLS FOR AIT

There are three major *ex vivo* protocols that have been evaluated for selective growth of tumor-specific CD8+ T cells with optimal affinity for the tumor antigen. Source of T cells could be TIL, tumor-draining lymph nodes (TDLN), or peripheral blood lymphocytes (PBL) from cancer patients. The oldest protocol uses tumor-antigen-pulsed dendritic cells for selective stimulation of the tumor-specific T cells. Anti-CD3 antibody has also been used for non-specific proliferation of the pre-selected tumor-specific clones from TIL (Zhou *et al.* 2005). The second protocol uses bryostatin-1/ionomycin (B/I) for selective activation of tumor-specific T cells regardless of the need for identification of the tumor antigens. Bryostatin 1 is a macrocyclic lactone derived from *Bugula neritina*, a marine invertebrate. Bryostatin activates protein kinase C (PKC) and ionomycin increases intracellular calcium (Morales *et al.* 2009). Together, B/I mimic signaling through the CD3/TcR complex and lead to activation and proliferation of effector T cells. Finally, genetic engineering of PBL of cancer patients with the tumor-antigen specific T cell receptors (TcR) can enhance the affinity of T cells for the tumor antigen (Johnson *et al.* 2009). All these approaches require appropriate cytokine formulations *ex vivo* for the differentiation and expansion of T cells for AIT. This review will focus on recent development in the use of common gamma chain cytokines for *ex vivo* expansion and differentiation of tumor-reactive T cells.

Interleukin 2 (IL-2)

IL-2 is the major T cell growth factor cytokine that is produced exclusively by activated T cells and dendritic cells (DCs). The IL-2 receptor (IL-2 R) is expressed by T cells, B cells and NK cells (Table 3). The receptor comprises three polypeptide subunits: IL-2 receptor α chain (IL-2 Rα) or CD25, IL-2 Rβ chain or CD122, and IL-2 Rγ chain or CD132 (Fig. 1). IL-2R β and γ chains are critical for signal transduction, whereas IL-2 Rα chain augments binding affinity for IL-2 and does not seem to contribute to IL-2 signaling (Leonard 1999).

Table 3 Sources and targets of common gamma-chain receptor cytokines.

Cytokines	*Produced by*	*Receptors expressed by*
IL-2	T cells and dendritic cells	T cells, B cells and NK cells
IL-7	Stromal cells, epithelial cells and fibroblasts	T cells, pre-B cells and dendritic cells
IL-15	Monocytes, dendritic cells, epithelial cells	T cell NK cells
IL-21	CD4+ T cells and NK T cells	T cells, B cells, NK cells, dendritic cells

In addition to being a T cell growth factor, IL-2 supports differentiation of CD8+ T cells towards effector and effector/memory phenotypes (T_E and T_{EM}) by down-regulation of lymph node homing receptors CD62L and CCR7. Such IL-2-expanded T cells traffic to the tumor site and are excluded from the lymphoid organs. Therefore, they may induce early anti-tumor responses but may not lead to a long-term memory response. Our group reported that IL-2 induced expansion of HER-2/neu-specific CD8+ T_E and T_{EM} cells, and reduced viability of T cells after a 6 d *ex vivo* expansion prior to AIT for the treatment of breast cancer in FVBN202 transgenic mice (Morales *et al.* 2009). Such reduced viability (56% Annexin V negative cells) was perhaps due to activation-induced cell death (AICD), which is one limitation of IL-2. Similar results were reported using Pmel-1-specific CD8+ T cells (Klebanoff *et al.* 2004). In melanoma patients, concentration of IL-2 during *ex vivo* expansion of TIL determined T cell function such that T cells grown in high-dose IL-2 (600-6000 U/ml) expanded massively and produced high titer IFN-γ in response to antigenic stimulation, but displayed very weak cytotoxicity. T cells grown in low-dose IL-2 (10-120 U/ml), on the other hand, expanded weakly and secreted very low titer of antigen-specific IFN-γ, but displayed high cytotoxic activity. Taking a combined approach by using low-dose IL-2 during the first week and continuing T cell culture with high-dose IL-2 for another week resulted in a massive expansion, high *IFN-γ* production and cytotoxic activity of T cells against the tumor (Besser *et al.* 2009). Although IL-2 enhances cytolytic function of tumor-specific T cells *in vitro*, AIT by means of IL-2-expanded T cells do not show objective responses to the extent predicted by *in vitro* experiments (Hinrichs *et al.* 2008). This could be due to the use of high-dose IL-2 during T cell expansion or administration of high-dose IL-2 following AIT. Culture of the

B/I-activated human T cells from healthy donors with IL-2 *ex vivo* resulted in the *de novo* generation of CD4+CD25+FoxP3+ T cells, though without any detectable suppressive function (Kmieciak *et al.* 2009). Strengths and weaknesses of IL-2 and other common gamma chain cytokines are listed in Table 4.

Table 4 Strengths and weaknesses of common gamma-chain cytokines for AIT

Cytokine	*Strength*	*Weakness*
IL-2	-T cells grown with IL-2 will traffic to the tumor site upon AIT	-induces ADCC -Supports Tregs
	-Expands NK cells and T cells	-Fails to provide long-term recall responses
IL-7	-Supports viability and maintenance of tumor-specific T cells	-Limited effect on T-cell expansion because of reduced expression of CD127 on effector cells
	-Does not support Tregs	
IL-15	-Expands T_{cm} cells that will provide long-term recall responses after AIT	-Increases Tregs in the presence of TGF-β
	-Supports secondary memory T cells	
IL-21	-Inhibits Tregs -Induces expression of perforin	-Represses expression of CD25 that could inhibit IL-2-induced T cell growth
	-Prevents T cell exhaustion	-Suppresses secretion of IFN-γ

Interleukin 7 (IL-7)

IL-7 is a hematopoietic growth factor cytokine which is secreted by the stromal cells, epithelial cells and fibroblasts but is not produced by lymphocytes. The *IL-7 R* is expressed by T cells, pre-B cells and DCs (Table 3). The receptor comprises two polypeptides, an affinity binding receptor IL-7 Rα or CD127 and a signaling gamma chain receptor CD132 (Fig. 1). IL-7 is crucial for the survival and homeostatic expansion of naive and memory CD8+ T cells. Because of the important role of IL-7 in all stages of T cell development and maintenance, it has been introduced recently into clinical trials as an immunotherapeutic agent for cancer patients in an attempt to increase T cell replenishment. Injection of IL-7 resulted in the expansion of both CD4+ and CD8+ T cells as well as a relative reduction of CD4+ Tregs (Rosenberg *et al.* 2006). This cytokine is also being tested for the *ex vivo* expansion of tumor-reactive T cells. IL-7 alone, however, does not support significant proliferation of PBMC. Using particular culture conditions, i.e., a higher density of T cells (3-5 x 10^6 cells/ml) and higher concentration of IL-7 (150 ng/ml), proliferation of tumor-reactive CD8+ memory T cells was detected in an IL-7-dependent fashion (Kittipatarin and Khaled 2009). The TcR signaling or IL-7 signaling results in down-regulation of CD127, and it has been suggested that this allows for efficient sharing of an important limiting resource for T cells. However, blocking of the IL-2 Rα, CD25, leads to an increased IL-7-mediated

STAT-5 phosphorylation, a delayed IL-7-induced down-regulation of CD127, and an enhanced IL-7-mediated T cell proliferation (Monti *et al.* 2009). The $CD127^{low}$ population that is vulnerable to apoptosis following TcR activation has high expression of pro-apoptotic Bim and low expression of anti-apoptotic Bcl-2. This $CD127^{low}$ population is highly responsive to IL-15 in order to expand memory pool. Increased expression of CD127 in naive T cells and in T_{CM} render these T cells susceptible and highly responsive to IL-7-induced expansion *ex vivo*.

Interleukin 15 (IL-15)

IL-15 is produced by monocytes, DCs and epithelial cells. IL-15 R is expressed by T cells and NK cells, and consists of three polypeptide subunits: an IL-15 Rα chain, which determines its binding to IL-15, and shared IL-2 β (CD122) and γ (CD132) chains (Table 3 and Fig. 1).

About 20% of the human CD8+ T cell pool in peripheral blood has low expression of CD127. Although CD127 is a specific receptor for IL-7, its expression on T cells also determines responsiveness of T cells to common gamma chain cytokines other than IL-7. For example, IL-15 decreases the AICD in CD8+CD127+ T cells but not in CD8+CD127- T cells while inducing comparable proliferation of the two subsets (Crawley *et al.* 2009). Such effects were in part mediated by IL-15-induced expression of anti-apoptotic Bcl-2 as well as inhibition of the pro-apoptotic Bim in CD8+CD127+ T cells but not in CD8+CD127- T cells (Inoue *et al.* 2009). Although IL-15 can induce expression of Bcl-2 in naive and memory T cells, its best-defined role in supporting memory T cells is to promote proliferation of these T cells (Sandau *et al.* 2010). It has been shown that naive T cells and T_{CM} cells are CD8+CD127+, while T_E and T_{EM} cells are mostly CD8+CD127- (Paiardini *et al.* 2005). Culture of antigen-experienced T cells with IL-15 *ex vivo* restores their ability to respond to TcR-mediated proliferation (Paiardini *et al.* 2005).

Recent studies demonstrated that primary memory (generated after the first encounter with antigen) and secondary memory (generated after activation of primary memory or during recall response) CD8+ T cells differ in their differentiation state and functional properties. Generation of secondary memory CD8+ T cells as well as acquisition of IL-2 production after the antigen stimulation take longer than those of primary memory T cells. Unlike primary memory, secondary memory T cells show an impaired lymphopenia-driven proliferation *in vivo*. While IL-7 and IL-15 are important in supporting primary memory CD8+ T cells, only IL-15 is crucial in supporting secondary memory CD8+ T cells (Sandau *et al.* 2010). Concerning the effects of IL-15 on Tregs, it was shown that IL-15 could enhance the acquisition of regulatory functions by CD4+ T cells following activation and in the presence of TGF-β. However, IL-15 renders T_{EM} cells refractory to suppressive function of Tregs via activation of the PI3K signaling pathway but does not involve the rescue of these T cells from apoptosis (Ben Ahmed *et al.* 2009). IL-15-expanded CD8+ T cells display only moderately effector function *in vitro*; however, they are highly effective after AIT *in vivo*. This is perhaps because of IL-5-induced differentiation of T_{CM} subset responsible for generating prolonged memory responses, whereas T_E cells undergo apoptosis after several cycles of proliferation and/or cytolytic function against the tumor.

Interleukin 21 (IL-21)

IL-21 is produced by activated CD4+ T cells only. IL-21 receptor is expressed in T cells, B cells, DCs and NK cells, and is similar to IL-2 receptor β chain (IL-2Rβ), a component of the receptors for IL-2 and IL-15 (Table 3 and Fig. 1). IL-21 represses expression of CD25 (Hinrichs *et al.* 2008) and FoxP3 (Li and Yee 2008), thereby inhibiting Tregs. The TcR stimulation of CD8+ T cells increases expression of the IL-21 R, rendering them responsive to IL-21-induced proliferation. It was reported that antigen-specific CD8+ T cells from melanoma patients were substantially expanded *ex vivo* in the presence of IL-21, and this effect could not be mimicked by IL-2, IL-7 or IL-15 (Li *et al.* 2005). Early differentiation of tumor-reactive CD8+ T cells into cytolytic effector cells, *ex vivo*, has a negative impact on the ability of these cells to produce objective responses upon AIT. In fact, pre-clinical studies revealed that IL-21 did inhibit early effector differentiation of CD8+ T cells *ex vivo*, but it did not prevent the effector differentiation after the AIT, because of the enhanced ability of IL-21-expanded T cells to undergo secondary expansion *in vivo* and to mediate tumor regression (Hinrichs *et al.* 2008). On the other hand, IL-21 has been reported to enhance cytolytic function of antigen-induced CD8+ T cells that were refractory to IL-2-induced acquisition of cytolytic function (Casey and Mescher 2007). In fact, IL-21 can increase perforin-mediated cytotoxicity of CD8+ memory T cells without increased phosphorylation of STAT-1, 3, 5 (Ebert 2009). Interestingly, antigenic stimulation of T cells in the presence of IL-21 resulted in the generation of antigen-reactive CD8+ T cells that failed to produce IFN-γ upon stimulation (Casey and Mescher 2007). In addition, IL-21 could partially inhibit IL-12-induced IFN-γ response by CD8+ T cells both at the mRNA and protein levels. IL-21 preferentially expands melanoma antigen-specific high-affinity CD8+ T cells obtained from melanoma patients or healthy donors (Li *et al.* 2005).

Recently, it has been reported that IL-21 R signaling prevents T cell exhaustion and that IL-21 is required during recall T cell responses. This effect of IL-21 on long-term survival of antigen-specific memory T cells has made this cytokine a potential candidate for immunotherapy of cancer. In patients with metastatic melanoma and renal cell carcinoma, administration of IL-21 induced activation of NK cells and CD8+ T cells *in vivo*.

SYNERGISTIC EFFECTS OF COMMON GAMMA CHAIN CYTOKINES

Current protocols for AIT of cancer patients involve using common gamma chain cytokines alone or in different combinations for the expansion of tumor-reactive T cells *ex vivo*. It is crucial to develop a combined cytokine formulation that will expand T cells with a greater efficacy against cancer. It has been reported that IL-15, but not IL-2, could induce robust cytotoxic responses in antigen-experienced T cells while only a minor effect was detectable when IL-7 was used *ex vivo*. In contrast, naive CD8+ T cells from healthy donors, specific for a melanoma-associated antigen, displayed cytotoxic responses after *in vitro* expansion in the presence of IL-2 (Rosenthal *et al.* 2009). Tumor-sensitized T cells are present in the metastatic lymph nodes and at tumor sites, whereas naive self/tumour-specific naive T cells are found in the blood. Therefore, depending on the source of T cells for AIT, different cytokine combination may be used for the *ex vivo* expansion. While PBL require IL-2, TDLN and TIL would grow better in the presence of IL-7 and IL-15. Although naive T cells are mainly characterized by a low-affinity TcR, high-affinity TcRs can be cloned and transduced into the PBL of patients with cancer in order to enhance anti-tumor activity upon AIT (Johnson *et al.* 2009).

IL-15 or IL-2 induces proliferation of CD8+CD127^{low} T cells, whereas IL-7 favors the accumulation of CD8+CD127^{high} T cells (Rubinstein *et al.* 2008). The former are thought to be short-lived T_{EM}, whereas the latter are long-lived T_{CM}. While IL-2 and IL-15 induce similar levels of tumor-reactive T cell expansion, combination of these cytokines induces higher levels of T cell expansion than IL-15 alone (Liu *et al.* 2006). IL-7 synergistically acts with IL-2 to increase the expression of CD25 in tumor antigen-specific T cells (Liu *et al.* 2006). Liu *et al.*(2006) showed that combination of all three cytokines (IL-2, IL-7, IL-15) did not improve T cell expansion or CD25 expression any further than combination of the two cytokines (IL-2+IL-15 or IL-2+IL-7). Therefore, distinct cytokine combination would be better than simply adding all cytokines together for the *ex vivo* expansion of T cells. For choosing the best combination one must take into consideration the status of gamma chain-cytokine receptor expression at different stage of T cell differentiation. For example, T cells begin to down-regulate (IL-7 Rα) or CD127 upon activation. While IL-7 induces robust expansion of naive and memory T cells, reduced expression of CD127 by cytotoxic T cells precludes their response to IL-7. Presence of IL-7 by itself decreases expression of CD127 by human CD8+ T cells. IL-2 also negatively regulates expression of CD127 via a PI3K/Akt-dependent mechanism (Leonard 1999). Therefore, IL-7 alone does not seem to induce substantial expansion of tumor-sensitized T cells *ex vivo* because of down-regulation of CD127 expression in activated T cells as well as the negative regulatory effects of IL-7 on CD127 expression. It has been reported that IL-21 can significantly increase the IL-7-induced expansion of cytotoxic T cells by preventing the IL-7-induced down-regulation of CD127 on antigen-stimulated T cells. Synergistic effect of IL-21 with IL-7 also promotes cytolytic activity of T cells.

Naive CD8+ T cells express low levels of IL-21 R, and IL-21 does not by itself induce proliferation of this naive subset. However, IL-21 can act synergistically with IL-7 or IL-15 to induce proliferation of both naive and memory phenotype $CD8^+$ T cells (Zeng *et al.* 2005). IL-21 synergistically acts with IL-15 to increase the yield of CD8+ memory T cells *ex vivo*. Combination of IL-21 and IL-15 resulted in an increased yield of central-memory and effector-memory CD8+ T cells compared with using IL-5 alone. IL-21 also synergizes, albeit more weakly, with IL-7, but not with IL-2, to expand CD8+ memory T cells (Zeng *et al.* 2005). Very recently, it was reported that IL-21 directs the IL-15-induced proliferation of CD8+ T cells preferentially to the CD28+ subset, by suppressing proliferation of the CD28- subset driven by IL-7 or IL-15 as well as enhancing the STAT-5 phosphorylation that was initiated by IL-15 in CD28+ subset (Nguyen and Weng 2010). In fact, IL-21 functions as a key regulator to support T cell proliferation in favor of the CD28+ competent memory cells and against the CD28- T cells. The latter T cells are impaired in their capacity to respond to antigens during the recall-response. Importantly, IL-21 also counteracts the effect of IL-15 to down-regulate expression of CD28 in memory CD8+ T cells (Nguyen and Weng 2010).

We have reported that IL-7+IL-15 resulted in approximately 5- to 10-fold greater yields of viable T cells and more prolonged proliferation of the activated CD8+ T cells than IL-2 (Cha *et al.* 2009). IL-7+IL-15 also supported expansion of CD8+ T_{CM} subset. The CD8+ T cells grown in IL-2 had higher levels of IFN-γ secretion upon stimulation with the tumor antigen *in vitro* compared to the cells grown in IL-7+IL-15. However, AIT by means of T cells grown in IL-7+IL-15 was more effective against 4T1 mammary tumors than cells grown in IL-2 (Cha *et al.* 2009). Alternating common gamma chain cytokines protocol *ex vivo* (IL-7+IL-15 with a one-time dose of IL-2 on day two during a 7 d expansion) (Table 5) has been found to be superior to the IL-2-expanded T cells or IL-7+IL-15-expanded T cells, showing a higher overall viability, as well as a higher proportion of effector

Table 5 AIT by means of alternating "common gamma-chain cytokines" protocol

	Ex vivo expansion of T cells
• Day 0:	Isolation of T cells from tumor-bearing and *ex vivo* activation of T cells with B/I + IL-7 for 16 hr.
• Day 1:	*Ex vivo* of culture with T cells with IL-y + IL-15 after washing of cells with complete medium at room temperature.
• Day 2:	Pulse T cell culture with IL-2.
• Day 3:	Wash and split cells and culture with IL-7 + IL-15 for another 3 d.
• Day 5:	Cyclophosphamide chemotherapy or total body radiation of the recipient.
• Day 6:	Evaluation of the tumor-specificity and viability of T cells, begin AIT.

T cell phenotypes, which could readily exhibit anti-tumor activity (Morales *et al.* 2009). In fact, different combination of common gamma chain cytokines leads to different phenotypic expansion of T cells with distinct anti-tumor function *in vitro* and *in vivo.*

Summary Points

- While IL-2 is the best cytokine for proliferation of naive T cells from PBL, IL-15 and IL-21 are more effective on tumor-sensitized T cells obtained from TDLN or TIL.
- Alternating gamma chain cytokines during *ex vivo* expansion of T cells result in the production of T cells with the improved objective responses after AIT, compared to other cytokine formulations.
- While IL-2 and to some extent IL-15 may induce Tregs, IL-7 and IL-21 repress Tregs.
- T_{CM} cells are more effective than T_E or T_{EM} cells in producing objective responses in cancer patients. While IL-2 supports T_E and T_{EM} cells, IL-15 and IL-21 support mainly T_{CM} cells.
- Combination of cytokine-expanded T cells *ex vivo* with lymphodepletion prior to AIT and administration of IL-21 after AIT can improve therapeutic efficacy of AIT.

Abbreviations

AICD	:	activation-induced cell death
AIT	:	adoptive T cell immunotherapy
B/I	:	Bryostatin 1/Ionomycin
DCs	:	dendritic cells
FDA	:	Food and Drug Administration
GM-CSF	:	*granolucyte* monocyte colony stimulating factor
IFN-γ	:	Interferon gamma
IL-2	:	Interleukin 2
MHC	:	major histocompatibility complex
PBL	:	peripheral blood lymphocytes
PBMC	:	peripheral blood mononuclear cells
PKC	:	protein kinase C
T_{CM}	:	CD8+ central/memory T cells
T_E	:	CD8+ effector T cells

T_{EM}	:	CD8+ effector/memory T cells
TcR	:	T cell receptor
TDLN	:	tumor draining lymph node
TIL	:	tumor infiltrating lymphocytes
TRAIL	:	tumor necrosis factor-related apoptosis inducing ligand
Tregs	:	CD4+ regulatory T cells

Definition of Terms

Adoptive T cell immunotherapy: A biological therapeutic protocol that isolates T cells from cancer patients, expands and differentiates them *ex vivo* and then infuses them back into cancer patients.

Common gamma chain cytokines: A total of 6 cytokines (IL-2, IL-4, IL-7, IL-9, IL-15, IL-21) that all share CD132 receptor for their signal transduction.

Differentiation of lymphocytes: A process through which naive T cells become effector, effector/memory or central/memory T cells.

Tumor-draining lymph nodes: Lymph nodes that lie immediately downstream of tumors. Activation of T cells by tumors and antigen-presenting cells occurs in these lymph nodes. Therefore, tumor-draining lymph nodes are one of the best sources for isolating tumor-reactive T cells for AIT.

Tumor-infiltrating lymphocytes: Lymphocytes that infiltrate into the tumor lesions and can be used as one of the sources for AIT of cancer.

Acknowledgements

This work was supported by NIH R01 CA104757 grant (M.H. Manjili). We gratefully acknowledge the support of VCU Massey Cancer Centre and the Commonwealth Foundation for Cancer Research.

References

Ben Ahmed, M. and N. Belhadj Hmida, N. Moes, S. Buyse, M. Abdeladhim, H. Louzir, and N. Cerf-Bensussan. 2009. IL-15 renders conventional lymphocytes resistant to supperssive function of regulatory T cells through activation of the phosphatidylinositol 3-kinase pathway. J. Immunol. 182: 6763-6770.

Besser, M.J. and E. Schallmach, K. Oved, A.J. Treves, G. Markel, Y. Reiter, and J. Schachter. 2009. Modifying interleukin-2 concentrations during culture improves function of T cells for adoptive immunotherapy. Cytotherapy 11: 206-217.

Casey, K.A. and M.F. Mescher. 2007. IL-21 promotes differentiation of naive CD8 T cells to a unique effector phenotype. J. Immunol. 178: 7640-7648.

Cha, E. and L. Graham, M.H. Manjili, and H.D. Bear. 2009. IL-7 + IL-15 are superior to IL-2 for the *ex vivo* expansion of 4T1 mammary carcinoma-specific T cells with greater efficacy against tumors *in vivo*. Breast Cancer Res. Treat. [Epub ahead of print].

Crawley, A.M. and T. Katz, K. Parato, and J.B. Angel. 2009. IL-2 receptor gamma chain cytokines differentially regulate human CD8+CD127+ and CD8+CD127- T cell division and susceptibility to apoptosis. Int. Immunol. 21: 29-42.

Ebert, E.C. 2009. Interleukin 21 up-regulates perforin-mediated cytotoxic activity of human intra-epithelial lymphocytes. Immunology 127: 206-215.

Hinrichs, C.S. and R. Spolski, C.M. Paulos, L. Gattinoni, K.W. Kerstann, D.C. Palmer, C.A. Klebanoff, S.A. Rosenberg, W.J. Leonard, and N.P. Restifo. 2008. IL-2 and IL-21 confer opposing differentiation programs to CD8+ T cells for adoptive immunotherapy. Blood 111: 5326-5333.

Inoue, S. and J. Unsinger, C.G. Davis, J.T. Muenzer, T.A. Ferguson, K. Chang, D.F. Osborne, A.T. Clark, C.M. Coopersmith, J.E. McDunn, and R.S. Hotchkiss. 2009. IL-15 prevents apoptosis, reverses innate and adaptive immune dysfunction, and improves survival in sepsis. J. Immunol. [Epub ahead of print].

Johnson, L.A. and R.A. Morgan, M.E. Dudley, L. Cassard, J.C. Yang, M.S. Hughes, U.S. Kammula, R.E. Royal, R.M. Sherry, J.R. Wunderlich, C.C. Lee, N.P. Restifo, S.L. Schwarz, A.P. Cogdill, R.J. Bishop, H. Kim, C.C. Brewer, S.F. Rudy, C. VanWaes, J.L. Davis, A. Mathur, R.T. Ripley, D.A. Nathan, C.M. Laurencot, and S.A. Rosenberg. 2009. Gene therapy with human and mouse T-cell receptors mediates cancer regression and targets normal tissues expressing cognate antigen. Blood 114: 535-546.

Kittipatarin, C. and A.R. Khaled. 2009. *Ex vivo* expansion of memory CD8 T cells from lymph nodes or spleen through *in vitro* culture with interleukin-7. J. Immunol. Methods 344: 45-57.

Klebanoff, C.A. and S.E. Finkelstein, D.R. Surman, M.K. Lichtman, L. Gattinoni, M.R. Theoret, N. Grewal, P.J. Spiess, P.A. Antony, D.C. Palmer, Y. Tagaya, S.A. Rosenberg, T.A. Waldmann, and N.P. Restifo. 2004. IL-15 enhances the *in vivo* antitumor activity of tumor-reactive CD8+ T cells. Proc. Natl. Acad. Sci. USA 101: 1969-1974.

Klebanoff, C.A. and L. Gattinoni, P. Torabi-Parizi, K. Kerstann, A.R. Cardones, S.E. Finkelstein, D.C. Palmer, P.A. Antony, S.T. Hwang, S.A. Rosenberg, T.A. Waldmann, and N.P. Restifo. 2005. Central memory self/tumor-reactive CD8+ T cells confer superior antitumor immunity compared with effector memory T cells. Proc. Natl. Acad. Sci. USA 102: 9571-9576.

Kmieciak, M. and M. Gowda, L. Graham, K. Godder, H.D. Bear, F.M. Marincola, and M.H. Manjili. 2009. Human T cells express CD25 and Foxp3 upon activation and exhibit effector/memory phenotypes without any regulatory/suppressor function. J. Transl. Med. 7: 89.

Leonard W.J. Type I cytokines and interferons and their receptors. pp. 701-749. In: W.E. Paul [ed] 2003. Fundamental Immunology. Lippincott–Raven, Philadelphia.

Li, Y. and M. Bleakley, and C. Yee. 2005. IL-21 influences the frequency, phenotype, and affinity of the antigen-specific CD8 T cell response. J. Immunol. 175: 2261-2269.

Liu, S. and J. Riley, S. Rosenberg, and M. Parkhurst. 2006. Comparison of common gamma-chain cytokines, interleukin-2, interleukin-7, and interleukin-15 for the *in vitro* generation of human tumor-reactive T lymphocytes for adoptive cell transfer therapy. J. Immunother. 29: 284-293.

Monti, P. and C. Brigatti, A.K. Heninger, M. Scirpoli, and E. Bonifacio. 2009. Disengaging the IL-2 receptor with daclizumab enhances IL-7-mediated proliferation of CD4(+) and CD8(+) T cells. Am. J. Transplant. 9: 2727-2735.

Morales, J.K. and M. Kmieciak, L. Graham, M. Feldmesser, H.D. Bear, and M.H. Manjili. 2009. Adoptive transfer of HER2/neu-specific T cells expanded with alternating gamma chain cytokines mediate tumor regression when combined with the depletion of myeloid-derived suppressor cells. Cancer Immunol. Immunother. 58: 941-953.

Nguyen, H. and N-P. Weng. 2010. IL-21 preferentially enhances IL-15 mediated homeostatic proliferation of human CD28+CD8 memory T cells throughout the adult age span. J. Leukoc. Biol. 87: 43-49.

Paiardini, M. and B. Cervasi, H. Albrecht, A. Muthukumar, R. Dunham, S. Gordon, H. Radziewicz, G. Piedimonte, M. Magnani, M. Montroni, S.M. Kaech, A. Weintrob, J.D. Altman, D.L. Sodora, M.B. Feinberg, and G. Silvestri. 2005. Loss of CD127 expression defines an expansion of effector CD8+ T cells in HIV-infected individuals. J. Immunol. 174: 2900-2909.

Rosenberg, S.A. and C. Sportès, M. Ahmadzadeh, T.J. Fry, L.T. Ngo, S.L. Schwarz, M. Stetler-Stevenson, K.E. Morton, S.A. Mavroukakis, M. Morre, R. Buffet, C.L. Mackall, and R.E. Gress. 2006. IL-7 administration to humans leads to expansion of CD8+ and CD4+ cells but a relative decrease of CD4+ T-regulatory cells. J. Immunother. 29: 313-319.

Rosenberg, S.A. and J.R. Yannelli, J.C. Yang, S.L. Topalian, D.J. Schwartzentruber, J.S. Weber, D.R. Parkinson, C.A. Seipp, J.H. Einhorn, and D.E. White. 1994. Treatment of patients with metastatic melanoma with autologous tumor-infiltrating lymphocytes and interleukin 2. J. Natl. Cancer Inst. 86: 1159-1166.

Rosenthal, R. and C. Groeper, L. Bracci, M. Adamina, C. Feder-Mengus, P. Zajac, G. Iezzi, M. Bolli, W.P. Weber, D.M. Frey, U. von Holzen, D. Oertli, M. Heberer, and G.C. Spagnoli. 2009. Differential responsiveness to IL-2, IL-7, and IL-15 common receptor gamma chain cytokines by antigen-specific peripheral blood naive or memory cytotoxic CD8+ T cells from healthy donors and melanoma patients. J. Immunother. 32: 252-261.

Rubinstein, M.P. and N.A. Lind, J.F. Purton, P. Filippou, J.A. Best, P.A. McGhee, C.D. Surh, and A.W. Goldrath. 2008. IL-7 and IL-15 differentially regulate CD8+ T-cell subsets during contraction of the immune response. Blood 112: 3704-3712.

Sandau, M.M. and J.E. Kohlmeier, D.L. Woodland, and S.C. Jameson. 2010. IL-15 regulates both quantitative and qualitative features of the memory CD8 T cell pool. J. Immunol. 184: 35-44.

Zeng, R. and R. Spolski, S.E. Finkelstein, S. Oh, P.E. Kovanen, C.S. Hinrichs, C.A. Pise-Masison, M.F. Radonovich, J.N. Brady, N.P. Restifo, J.A. Berzofsky, and W.J. Leonard. 2005. Synergy of IL-21 and IL-15 in regulating CD8+ T cell expansion and function. J. Exp. Med. 201: 139-148.

Zhou, J. and M.E. Dudley, S.A. Rosenberg, and P.F. Robbins. 2005. Persistence of multiple tumor-specific T-cell clones is associated with complete tumor regression in a melanoma patient receiving adoptive cell transfer therapy. J. Immunother. 28: 53-62.

CHAPTER 15

Chemotherapeutic Drugs and Cytokines

Elieser Gorelik, Anna Lokshin and Vera Levina*

University of Pittsburgh Cancer Institute, Hillman Cancer Center
5117 Centre Ave. Room 1.19d, Pittsburgh, PA 15213 USA

ABSTRACT

This review summarizes our understanding of the role that the cancer cytokine network plays in tumor survival and growth. The tumor cytokine network comprises tumor and stroma-produced factors and receptors. Cytokines produced by tumor cells and stroma cells are critical for tumor cell proliferation and for the formation of neovasculature that provides the oxygen and nutrients necessary for progressive tumor growth. Increasing amounts of experimental evidence indicate that chemotherapeutic drug treatment could stimulate the production of multiple cytokines. This drug-induced cytokine production enrichment is the adaptive response by which tumor cells attempt to protect themselves from the genotoxic stress induced by these drugs. Numerous growth factors and chemokines share angiogenic and growth-stimulating properties; thus, reducing a single factor is insufficient for complete blockage of tumor growth. Instead, a broad disruption of the tumor cytokine network should improve the efficacy of current anti-cancer strategy. Drug-resistant and self-renewing cancer stem cells (CSCs) are thought to be responsible for the failure of current cancer chemotherapy. CSCs, in comparison to bulk tumor cells, manifest higher levels of growth and angiogenic factor production and over-express certain receptors. Because these characteristics are unique to CSCs, they may be potent targets for cancer therapy. Combining standard chemotherapy with the targeting of specific axes of cytokine network in bulk tumor cells and CSCs could increase the efficacy of cancer therapy.

*Corresponding author

INTRODUCTION

Cytokines are cell-produced soluble factors that are involved in intercellular communications and stimulate cell proliferation, differentiation, migration and survival. They contribute to a chemical signaling language that regulates tissue repair, hematopoiesis, inflammation, and immune response. Cytokine signaling results in the transcription of target genes that are involved in cell invasion, motility, interactions with the extracellular matrix (ECM) and survival (review in Balkwill 2004, Mantovani *et al.* 2008). Cytokines and their receptors represent a complex network in which one cytokine can influence the production and response of many other cytokines.

"Cytokine" is a heterogeneous group of cytokines, chemokines, growth and angiogenic factors. Many growth factors were initially identified based on their cell origin or cell targeting (i.e., FGF, PDGF, HGF, EGF). It was quickly recognized that growth factors could regulate the growth of various histologically distinct cells that express receptors specific to these factors. Thus, cytokine division for growth factors and angiogenic factors is very conditional because many growth factors, such as FGF, HGF, and EGF, stimulate proliferation of endothelial cells and play an important role in angiogenesis. On the contrary, angiogenic factors such as VEGF and PDGF could stimulate proliferation of different type of cells and thus also could be considered as growth factors.

Chemokines are a large group of factors that stimulate the mobility and migration of different types of cells. Increasing amounts of evidence indicate that chemokines are able to stimulate the proliferation of normal and malignant cells. Because chemokines are able to stimulate endothelial cell proliferation, they have potent angiogenic activity (Balkwill 2004, Mantovani *et al.* 2008).

CANCER CYTOKINE NETWORK

A complex network of cytokines and their receptors influences the development of primary tumors and metastases. The ability of tumor cells to produce various chemokines, angiogenic and growth factors is crucial for tumor cell proliferation and for the formation of the stroma and neovasculature that provide the oxygen and nutrients needed to support progressive tumor growth. Tumor cells are highly efficient producers of various chemokines, growth and angiogenic factors that have autocrine and paracrine effects on tumor and stromal cells and create the tumor microenvironment that allows for progressive tumor growth (Robinson and Coussens 2005). Stroma cells also produce various cytokines that support tumor growth. Thus, cancer cells are engaged in a complex cross-talk with stroma cells (mesenchymal fibroblasts, inflammatory cells, endothelial cells of tumor vasculature and others), and extracellular matrix (ECM) components, collectively creating the tumor "microenvironment" (Mantovani *et al.* 2008, Orimo and Weinberg 2006).

Malignant transformation is often associated with the over-expression of various growth factors, angiogenic factors and chemokines that stimulate tumor and stromal cell proliferation and manifest potent angiogenic effects (Robinson and Coussens 2005). Angiogenesis and tumor blood vessel network formation are governed by numerous pro- and anti-angiogenic factors, including VEGF, bFGF, EGF, PDGF, TGF-β, angiostatin, endostatin and thrombospondin. Tumor-produced VEGF is a major angiogenic factor that stimulates the migration and proliferation of endothelial cell and blood vessel formation that are essential for oxygen and nutrient supply, tumor growth and metastatic spread (Ferrara 1999). PDGF is also over-expressed in various human malignancies. PDGF stimulates the migration and proliferation of pericytes/smooth muscle cells. Pericytes form an association with endothelial cells and help stabilize newly formed endothelial cell tubes. Inhibiting this association might block blood vessel formation and tumor growth (Jain and Booth 2003). The FGF family is involved in diverse processes including embryonic development, wound healing, tissue regeneration and angiogenesis, as well as in autonomous tumor growth and tumor vascularization. FGFb is one of the most potent angiogenic factors and is also mitogenic for a wide range of mesoderm- and ectoderm-derived normal or malignant cells (Bikfalvi *et al.* 1998). G-CSF and HGF stimulate tumor cell proliferation, as well as possess potent pro-angiogenic abilities. TGF-β is an important tumor-produced factor that regulates tumor cell proliferation, invasiveness and angiogenesis. TGF-β signaling pathways play a pivotal role in diverse cellular processes and can serve as an example of the cytokine network's contextual response within each tumor (Gorelik and Flavell 2002). It has been shown that TGF-β changes its role from a tumor suppressor in normal or dysplastic cells to a tumor promoter in advanced cancers (review in Massague 2008). It is widely believed that the Smad-dependent pathway is involved in TGF-β tumor-suppressive functions, whereas activating Smad-independent pathways, coupled with TGF-β's loss of tumor-suppressor functions, is important for its pro-oncogenic functions. TGF-β, as well as IL-10, has potent immunosuppressive activity and helps tumors escape immune-mediated destruction (Massague 2008).

IL-6 is a pleiotropic pro-inflammatory cytokine that affects B and T cell differentiation, induces acute phase reactant production, and stimulates hematopoiesis. It also can directly stimulate tumor growth and has potent angiogenic effects (Ben-Baruch 2008). Tumor cells also secrete various chemokines, e.g., IL-8, MCP-1, and RANTES, that are able to stimulate cell proliferation, migration and invasion (Ben-Baruch 2008). They also manifest potent angiogenic activity and could work in concert with other angiogenic factors to stimulate tumor angiogenesis (Ben-Baruch 2008). IL-8 is a chemokine with a wide range of pro-inflammatory effects. It stimulates neutrophils, monocytes and lymphocyte migration, as well as promotes tumor cell proliferation and metastasis (Bar-Eli 1999). In addition, IL-8 has strong angiogenic activity via transactivation of VEGFR2 (Petreaca *et al.* 2007).

MCP-1 and RANTES are able to stimulate the migration of normal and malignant cells as well as promote tumor angiogenesis (Balkwill 2004, Mantovani *et al.* 2008). Stroma cells such as endothelial cells, fibroblasts, lymphocytes and macrophages can also produce various cytokines, chemokines, growth and angiogenic factors that may affect the survival and proliferation of tumor and stroma cells. Tumor- and stroma-produced soluble factors constitute a tumor cytokine network that plays an important role in tumor growth and protection from endogenous (hypoxia, oxygen free radicals) and exogenous (drugs, x-irradiation) damage.

Using Luminex multiplex technology (Luminex, Austin), high content screening (HCS) and a high content analysis (HCA) platform from Cellomics (ThermoFisher Scientific, Pittsburgh), we performed a comprehensive analysis of the cytokine network of eight human tumor cell lines (erythroleukemia, non small lung cancer, melanoma, breast cancer, and ovarian cancer). An analysis of 30 different cytokines in their conditioned media revealed that human tumor cells of different histological origin are able to produce numerous similar cytokines (IL-6, IL-8, IL-10, G-CSF, HGF, VEGF, bFGF, MIP-1a, MIP-1b, IFN-a) (Levina *et al.* 2008a). However, tumor cells of different histological origin vary in their ability to produce the different molecules essential for tumor cell proliferation, tumor angiogenesis and migration, and inflammatory cells. For example, OVCAR-3, ovarian cancer cells, produced the highest levels of VEGF and the lowest levels of bFGF, in comparison to the other cell lines investigated. However, tumor cell lines of the same histological origin also showed different capacities in factor production. For example, NSCLC H460 cells produced 60x the amount of IL-6 and 30x the amount of IL-8 produced by NSCLC A549 cells, whereas CCL2 was produced in A549 cells at a level 40-fold that in H460 NSCLC cells (Levina *et al.* 2008a).

The cytokine profile differences in tumor cells growing *in vivo* indicate the complexity of the cytokine network and its contextual response during tumor development (Levina *et al.* 2008b).

Cytokines may accumulate in the blood when growing tumors overproduce them. Increased blood levels of some cytokines such as VEGF, bFGF, HGF, IL-8, and IL-6 are associated with therapeutic resistance and poor prognosis (Hartsell *et al.* 2007). It is possible that local, intratumoral concentrations of these factors are higher than in circulation, and thus, tumor-produced cytokines could play an important role in tumor growth and resistance to drug therapy.

ANTI-TUMOR CHEMOTHERAPY AND THE CANCER CYTOKINE NETWORK

Anti-tumor chemotherapy is based on the ability to kill or inhibit the proliferation of cancer cells. Chemotherapeutic drugs can be divided into several groups based on the mechanisms of their actions and their chemical structure (Table 1). DNA-damaging drugs are commonly used for treating different types of cancer. They include alkylating agents, platinum drugs (cisplatin), anthracyclines (doxorubicin); antitumor antibiotics (actinomycin-D, bleomycin, mitomycin-C), and topoisomerase inhibitors (etoposide). However, one of the major unresolved problems of cancer therapy is tumor drug resistance. Multiple mechanisms of drug resistance in tumor cells have been identified, including decreased cellular uptake or increased cellular efflux of chemotherapeutic drugs. In addition, alterations in the signaling pathways controlling the cell cycle or induction of cell death by apoptosis were identified (Airley 2009). Although cytokines are abundantly expressed in tumors, the effect of drugs on the tumor cell's ability to produce cytokines growing *in vitro* did not receive the attention it deserved and actually remains poorly investigated. We hypothesize that cytokines play a role in tumor survival by providing proliferative signals that could counterbalance the apoptotic effects of chemotherapeutic drugs (Levina *et al.* 2008a).

Table 1 Anti-cancer chemotherapeutic drugs

Chemotherapeutic drug classification	*Mechanism of action*	*Examples of drugs*
Alkylating agents	Chemically modify cell's DNA.	Cisplatin, carbiplatin, cyclophosphamide,
Anti-metabolites	Interfere with DNA production and therefore cell division	Mercaptopurine, azathioprine, pyridine
Vinca alkaloids and Taxanes	Bind to specific sites on tubulin, inhibiting the assembly of tubulin into microtubules	Vincristine,vinblastine, vinorelbine, vindesine; paclitaxel, docetaxel
Anti-tumour antibiotics	Chemically modify cell's DNA.	Doxorubicin, epirubicin, bleomycin
Topoisomerase inhibitors	Type I topoisomerase inhibitors Type II topoisomerase inhibitors	Irinotecan, topotecan Etoposide, amsacrine

Drugs may reduce cytokine production by killing tumor cells and inhibiting DNA/RNA and protein synthesis. It is also possible that drugs initiate an adaptive cellular response, resulting in an augmented production of pro-survival growth factors.

Studies confirming the up-regulation of cytokines in different cancer cells upon drug treatment are summarized in Table 2. It was shown that treating melanoma cells with dacarbazine leads to IL-8 and VEGF over-expression (Lev *et al.* 2003).

Table 2 Drug-induced augmentation of cytokine production by cancer cells

Human cancer cells	*Chemotherapeutic drug*	*Cytokine up-regulation*	*References*
Non small cell lung cancer	Doxorubicin, Cisplatin, Bleomycin, Etoposide	IL-6, CXCL8, MCP-1, G-CSF, RANTES, bFGF, VEGF	Levina *et al.* 2008a
Small cell lung cancer	Doxorubicin	IL-8, MCP-1	Shibakura *et al.* 2003
Breast cancer	Doxorubicin	IL-6, CXCL8, MCP-1, RANTES, bFGF, G-CSF, VEGF	Levina *et al.* 2008a
Melanoma	Dacarbazine	IL-8 and VEGF	Lev *et al.* 2003
Colorectal cancer	Oxaliplatin	VEGF	Fan *et al.* 2008
Epidermal carcinoma	Etoposide, Mitomycin C	IL-8, TNF-α	Darst *et al.* 2004

IL-8 gene transfection created cell resistance to the cytotoxic effect of the drug, whereas the IL-8-neutralizing antibody lowered cell sensitivity to dacarbazine (Lev *et al.* 2003). Acute exposure of human colorectal cancer cells to oxaliplatin led to a marked induction of VEGF mRNA, protein and also increased VEFGR1 expression (Fan *et al.* 2008). Augmentation of IL-8 and TNF-α production was observed in human epidermal carcinoma cell lines treated with etoposide and mitomycin C (Darst *et al.* 2004). Doxorubicin increased IL-8 and MCP-1 production in human small cell lung cancer cell lines (Shibakura *et al.* 2003). Such studies were limited by the conventional ELISA that allows testing of only one cytokine at a time.

We tested the effects of chemotherapeutic drugs on the tumor cell cytokine network using multiplexed (xMAP^{TM}) technology (Luminex Inc) for a simultaneous testing of 30 cytokines (Levina *et al.* 2008a). First, we demonstrated that the level of cytokine production is a function of tumor cell amount; reducing tumor cell amount *in vitro* resulted in a proportional reduction in cytokine production (Levina *et al.* 2008a). Therefore, to normalize cytokine production according to the cell amount, we calculated the cytokine production per $1x10^6$ cells/ml/day. Treating H460 NSCLC cells with chemotheraupetic drugs (cisplatin, doxorubicin, etoposide and bleomycin) resulted in a different response in cytokine production. Cisplatin and doxorubicin significantly increased the production of multiple cytokines, whereas etoposide and bleomycin induced a smaller effect (Levina *et al.* 2008a). Doxorubicin and cisplatin treatment substantially increased the production of IL-6, CXCL8, CCL2, CCL5, bFGF, G-CSF and VEGF by H460 cells as well as by tumor cells of different histological origin (Levina *et al.* 2008a).

Further proof of the link between chemotherapy treatment and cytokine production came from our study of intracellular cytokine accumulation in H460 cells using novel HCS and HCA platform from Cellomics (Fig. 1-3, Levina *et al.* 2008a). Doxorubicin increased the production of IL-8, CCL2, G-CSF, CCL5 and bFGF in H460 cells (Fig. 1). However, the cells are very heterogeneous in their cytokine production. For instance, H460 cells consist of low and high producers

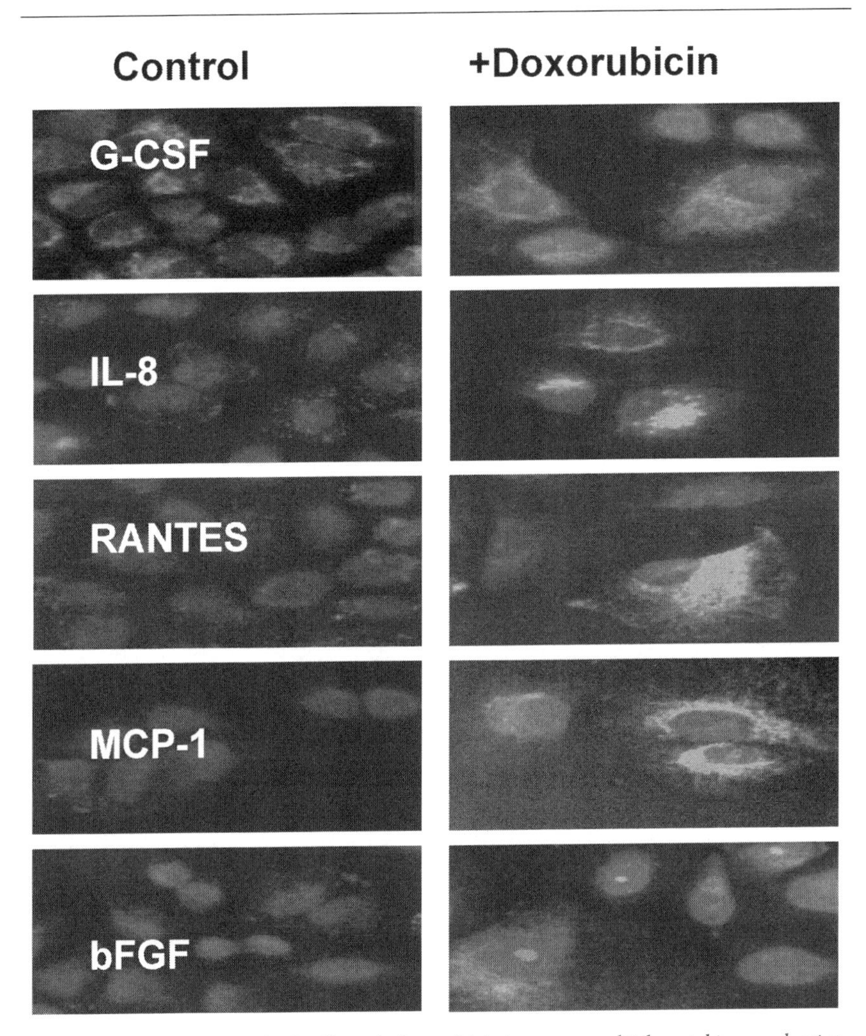

Fig. 1 Treating H460 NSCLC cells with doxorubicin increases multiple cytokine production. Intracellular accumulation of cytokines in doxorubicin-treated and untreated cells was analyzed. H460 cells were treated with doxorubicin (0.125 μg/ml) for 16 hr and were then incubated with monensin (2 μM) for 3 hr. Cells were incubated with primary Abs against CXCL8, CCL2, CCL5, G-CSF, bFGF or VEGF and then with secondary Alexa Fluor 488- or Alexa 680-conjugated Abs. Cell nuclei were stained with Hoechst 33342 and cell images were acquired using the Cellomics ArrayScan HCS Reader and analyzed using the Target Activation BioApplication Software Module. Images of immunofluorescently stained cells for CXCL8, CCL2, CCL5, G-CSF or bFGF (Alexa488) are presented.

Color image of this figure appears in the color plate section at the end of the book.

of VEGF (Fig. 2). Doxorubicin treatment increased VEGF production in both remaining distinct subpopulations (Fig. 2). A cell's double fluorescent labeling, applied to test whether the same cells produce different cytokines, revealed that a majority of cells produced both CXCL8 and CCL5 cytokines, but did so in a preferential manner. After doxorubicin treatment, production of both CCL5 and CXCL8 substantially increased. Furthermore, cells that preferentially produced CXCL8 started to produce CCL5 (Fig. 3). Similarly, double staining showed that a vast majority of cells produce both CXCL8 and bFGF or CCL2, and doxorubicin treatment increased the intracellular production of these factors (Fig. 3). These findings produced additional confirmation that drug treatment stimulates tumor cells to produce cytokine and demonstrated that more detailed studies are required to understand the interaction between different cytokine-receptor axes and the complexity of the tumor cytokine network.

The increase in cytokine production was associated with a drug-induced rapid activation of multiple transcription factors as well as expression of the corresponding cytokine genes (Levina *et al.* 2008a).

Thus, drug-induced augmentations of multiple cytokines in tumor cells could reflect common features of cellular stress response in treated cells.

What are the consequences of drug-induced augmentation of the cytokine network? The drug-induced cytokines may act in an autocrine/paracrine manner to protect surviving tumor cells and induce their proliferation. Indeed, culture media conditioned by tumor cells protected tumor cells from drug-induced apoptosis and stimulated their proliferation. Antibody-neutralizing IL-6, IL-8, MCP-1 and RANTES blocked this stimulation (Levina *et al.* 2008a).

Some studies indicate that IL-4 and IL-10 are involved in epithelial tumor cell drug resistance (Conticello *et al.* 2004). A medium conditioned by thyroid cancer cells protected cells from anticancer drug–induced apoptosis (doxorubicin, cisplatin) or FasL. This protection was abrogated by anti-IL-4 and IL-10 antibodies (Conticello *et al.* 2004). It was found that primary epithelial cancer cells from colon, breast and lung carcinomas express high levels of IL-4. Treating tumor cells with IL-4 increased their resistance to doxorubicin, etoposide and TRAIL-induced apoptosis associated with increased levels of the antiapoptotic proteins PED, cFLIP, Bcl-xL and Bcl-2 (Todaro *et al.* 2009). IL-4 blockage resulted in a significant decrease in epithelial cancer cell growth rate and sensitized them, both *in vitro* and *in vivo*, to apoptosis induction by TRAIL and chemotherapy via decreasing the antiapoptotic factors PED, cFLIP, Bcl-xL and Bcl-2 (Todaro *et al.* 2009).

The correlation between the chemoresistance of neuroblastoma cells and their more pro-angiogenic genotype was recently demonstrated (Michaelis *et al.* 2009). A comparative human gene expression analysis of 14 chemosensitive and chemoresistant neuroblastoma cell lines identified a consistent shift to a more pro-angiogenic phenotype of chemoresistant neuroblastoma cells. In addition,

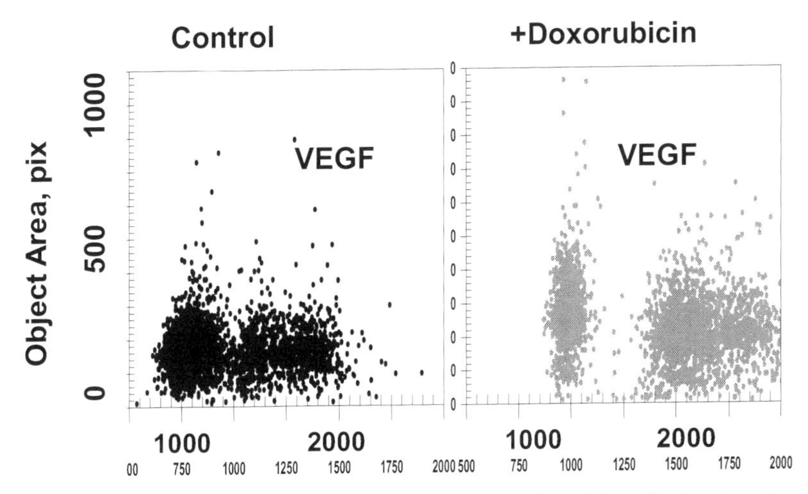

Fig. 2 Tumor cell population is heterogeneous: analysis of VEGF production in naïve and doxorubicin-treated cells. Cells were prepared as described for Fig. 1. Fluorescence intensity of VEGF (Alexa488) is plotted against object area, pix. Each point represents a single cell.

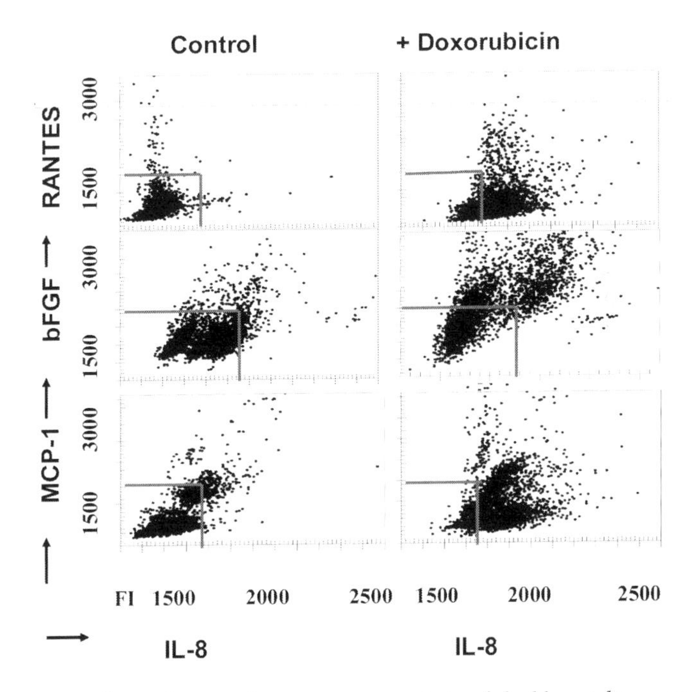

Fig. 3 Tumor cell population is heterogeneous: analysis of double cytokine accumulation in untreated and doxorubicin-treated cells. Cells were prepared as described for Fig. 1 and double-stained for CCL5, bFGF or CCL2 (Alexa488) and CXCL8 (Alexa680). Fluorescence intensities of CCL5, bFGF or CCL2 were plotted against fluorescence intensity of CXCL8. Each point represents a single cell. The red lines show the boundaries of the fluorescence intensity of untreated cells.

media conditioned by chemoresistant cells proved more proficient in HUVEC endothelial cell growth stimulation and endothelial cell tube formation than media conditioned by chemosensitive cells (Michaelis *et al.* 2009). However, the molecular mechanisms underlying the increased pro-angiogenic activity of neuroblastoma cells are individual and differ between the investigated resistant cell lines (Michaelis *et al.* 2009).

IL-6 and IL-8 cytokines were found to be involved in tumor cell radioresistance (Efimova *et al.* 2009). Radioresistance of Stat1 over-expressing tumor cells was associated with a suppressed apoptotic response to drugs and increased IL6-IL8 signaling. A combined IL6 and IL8 signaling inhibition via neutralizing antibodies led to tumor cell sensitization to x-irradiation. These data suggest IL6 and IL8 as potential targets for tumor radio sensitization (Efimova *et al.* 2009).

Cellular stress responses could be induced not only by drugs, but also by various apoptosis-induced molecules, resulting in cytokine production augmentation. FasL-Fas signaling stimulated the production of IL-6 in tumor cells (Choi *et al.* 2002). We found that TRAIL, by binding to cognate death receptors, stimulates the production of IL-8, MCP-1, RANTES and bFGF in breast, NSCLC and ovarian tumor cells. This stimulation was associated with an activation of caspases 1 and 8 and various transcription factors. Media conditioned by tumor cells protected tumor cells from the apoptotic effect of TRAIL and this effect was blocked by antibody-neutralizing IL-8, CCL5 and bFGF (Levina *et al.* 2008c).

TGF-β was found to be a major factor in media conditioned by prostate cancer cells that protected cells from TNF-induced apoptosis. TGF-β signaling has been considered a useful therapeutic target (Massague 2008).

In general, these findings demonstrate that different genotoxic stresses induced by chemotheraupetic drugs, death receptor signaling, or ionizing radiation may lead to an augmentation of major pro-inflammatory cytokines, such as IL-6, IL-8, IL-10, TGF-β CCL2 and CCL5, that bind to their receptors and activate downstream signaling. This signaling could change the balance between pro-apoptotic and anti-apoptotic molecules, leading to an increase in tumor cell drug resistance and survival.

A clear understanding that cytokines are involved in the growth and progression of tumors has stimulated the development of new therapeutic modalities aimed at neutralizing these factors or inhibiting their receptor signal transduction. Two classes of angiogenic and growth factor targeting compounds are currently used in clinical practice: cytokine or cytokine receptor neutralizing monoclonal antibodies and receptor tyrosine kinase (RTK) inhibitors. The era of such targeted therapy began with the approval of Trastuzumab, a monoclonal, anti-HER2 antibody designed to treat metastatic NSCLC cancer. Targeting VEGF with monoclonal antibodies (Bevacizumab, Avastin) has been recently approved for ovarian and advanced NSCLC treatment (Airley 2009). Moreover, successful Imatinib (RTK inhibitor) treatment for chronic myeloid leukemia (Airley 2009) has stimulated the development of various RTK inhibitors for treating diverse malignancies. However, the beneficial effects of these strategies have been limited and only slightly increase patients' overall survival. It is possible that chemotherapeutic drugs stimulated the production of growth and angiogenic factors and may have counterbalanced the effects of the therapy. In addition, numerous growth factors and chemokines exhibit shared angiogenic and growth-stimulating properties, and thus inhibiting a single factor is enough to completely block tumor growth. Chemotherapy efficacy could be improved by modalities capable of broad inhibition or inactivation of the multiple factors that are involved in autocrine/paracrine regulation of tumor growth.

CYTOKINE NETWORK OF CSCS

Within the last decade, resistance to chemotherapy and high recurrence rates have been ascribed to the presence of cancer stem cells (CSCs), a rare cell subpopulation that maintain their tumorigenic potential after cytotoxic therapy (Clarke and Fuller 2006, Visvader and Lindeman 2008). CSCs share a variety of biological properties with normal somatic stem cells in terms of self-renewal, propagation of differentiated progenies, expression of specific cell markers and stem cell genes, use of common signaling pathways, and residence within the specific niche. CSCs have very high tumorigenic activity and thus have been termed "cancer-initiating" cells. CSCs have been identified in various human malignancies including breast, brain, prostate, pancreatic, colon, and lung cancers (review in Clarke and Fuller 2006, Visvader and Lindeman 2008). CSCs manifest higher levels of drug resistance and are thought to restore tumor growth after chemotherapy (Visvader and Lindeman 2008). Our experiments showed that *in vitro* treatment of NSCLC cells with chemotheraupetic drugs such as cisplatin, doxorubicin, and etoposide resulted in the elimination of the vast majority of tumor cells. However, a small fraction of tumor cells survived and developed fast-growing colonies of tumor cells with all the characteristics of CSCs (Levina *et al.* 2008b). Drug-surviving NSCLC cells expressed stem cell markers (CD133, CD117, SSEA-3, TRA1-81, Oct-4) and were capable of self-renewal, pluripotency, and differentiation *in vitro*. Lung CSCs demonstrated high tumorigenic and metastatic potential following SCID mice inoculation, supporting their classification as CSCs. These lung CSCs were able to differentiate and became drug sensitive. However, in the presence of drugs, they did not differentiate and maintained drug resistance (Levina *et al.* 2008c).

We found that the highly tumorigenic and metastatic properties of CSCs are correlated with their high ability to produce angiogenic and growth factors VEGF, bFGF, IL-6, IL-8, HGF, PDGF, G-CSF, and SCGF-β and to over-express VEGFR2, FGFR2, and CXCR1,2, and four receptors in lung CSC–derived tumors (Levina *et al.* 2008c). A further investigation of the CSC cytokine network helped us identify an important role for SCF and its receptor c-kit that were preferentially expressed by

CSCs cells. The SCF-c-kit signaling axis plays an important role in self-renewal and proliferation of lung CSCs (Levina *et al.* 2010). Blocking the SCF–c-kit autocrine loop with a neutralizing SCF antibody or with the c-kit RTK inhibitor, Imatinib (Gleevec), resulted in the complete elimination of CSCs (Levina *et al.* 2010). If conventional chemotherapy is able to eliminate a majority of tumor cells and leave CSCs, a combination of chemotherapy and specific tumor cytokine network axis targeting CSCs might be very beneficial for improving cancer therapy. Indeed, treating bulk NSCLC cells with cisplatin killed a vast majority of tumor cells but surviving CSCs quickly restored the tumor cell population. Imatinib treatment alone did not lead to significant reduction of tumor cells, probably because CSCs are a very small population of cells. However, a combined treatment of cisplatin and imatinib resulted in the elimination of CSCs and bulk NSCLC cells (Levina *et al.* 2010).

Recently, it was found that the IL-8–CXCR1 axis plays an important role in the survival of human breast CSCs. CXCR1 blockage via an anti-CXCR1 antibody or repertaxin, a small-molecule CXCR1 inhibitor, selectively eliminated the CSC population in two human breast cancer cell lines *in vitro* (Ginestier *et al.* 2010).

This data suggests that targeting specific axes of cytokine network in CSCs may provide a novel approach in targeting and eliminating CSCs, the most aggressive tumor population.

Summary Points

- The cancer cytokine network is a complex of tumor- and stroma-produced soluble factors and receptors and is involved in almost every aspect of tumorigenesis.
- Treating tumor cells with chemotheraupetic drugs up-regulated multiple cytokines production in surviving cells.
- Up-regulation of cytokine production in drug-treated cells represents an adaptive stress response through which tumor cells protect themselves from drug-mediated toxicity.
- Targeting the cancer cytokine network might improve the efficacy of chemotherapy.
- Treating tumor cells with chemotherapeutic drugs could select for drug-resistant CSCs that are capable of restoring tumor cell population.
- The high tumorigenic properties of CSCs are associated with their increased ability to produce angiogenic and growth factors, aiding in tumor establishment in different anatomical locations.
- The efficacy of chemotherapy could be improved by targeting specific axes of CSC cytokine network.

Abbreviations

CSCs	:	cancer stem cells
FGF	:	fibroblast growth factor
G-CSF	:	granulocyte colony-stimulating factor
HGF	:	hepatocyte growth factor
IL	:	interleukin
MCP-1	:	monocyte chemotactic protein 1
MIP	:	macrophage inflammatory protein
PDGF	:	platelet-derived growth factor
RANTES	:	regulated upon activation normal T cell expressed and presumably secreted
SCF	:	stem cell factor
SCGF-β	:	stem cell growth factor beta
SDF-1	:	stromal derived factor
TGF-β	:	transforming growth factor beta
RTK	:	receptor tyrosine kinase
VEGF	:	vascular endothelial growth factor

Definition of Terms

Cancer stem cells: A rare subpopulation of undifferentiated cells responsible for tumor initiation, maintenance, and spread. They are drug-resistant and display the ability to self-renew and generate a progeny of differentiated cells that constitute the large majority of cells in the tumor.

Cytokine network: A complex network of chemokines, cytokines, growth factors, angiogenic factors and their receptor that regulate different aspects of tumorigenesis.

References

Airley, R. 2009. Cancer Chemotherapy. Willey Blackwell, West Sussex, UK.

Balkwill, F. 2004. Cancer and the chemokine network. Nat Rev Cancer 4: 540-550.

Bar-Eli, M. 1999. Role of interleukin-8 in tumor growth and metastasis of human melanoma. Pathobiology 67: 12-18.

Ben-Baruch, A. 2006. The multifaceted roles of chemokines in malignancy. Cancer Metastasis Rev. 25: 357-371.

Bikfalvi, A. and C. Savona, C. Perollet, and S. Javerzat. 1998. New insights in the biology of fibroblast growth factor-2. Angiogenesis 1: 155-173.

Choi, C. and G.Y. Gillespie, N.J. Van Wagoner, and E.N. Benveniste. 2002. Fas engagement increases expression of interleukin-6 in human glioma cells. J. Neurooncol. 56: 13-19.

Clarke, M.F. and M. Fuller. 2006. Stem cells and cancer: two faces of eve. Cell 124: 1111-1115.

Conticello, C. and F. Pedini, A. Zeuner, M. Patti, M. Zerilli, G. Stassi, A. Messina, C. Peschle, and R. De Maria. 2004. IL-4 protects tumor cells from anti-CD95 and chemotherapeutic agents via up-regulation of antiapoptotic proteins. J. Immunol. 172: 5467-5477.

Darst, M. and M. Al-Hassani, T. Li, Q. Yi, J.M. Travers, D.A. Lewis, and J.B Travers. 2004. Augmentation of chemotherapy-induced cytokine production by expression of the platelet-activating factor receptor in a human epithelial carcinoma cell line. J. Immunol. 172: 6330-6335.

Efimova, E.V. and H. Liang, S.P. Pitroda, E. Labay, T.E. Darga, V. Levina, A. Lokshin, B. Roizman, R.R. Weichselbaum, and N.N. Khodarev. 2009. Radioresistance of Stat1 over-expressing tumour cells is associated with suppressed apoptotic response to cytotoxic agents and increased IL6-IL8 signaling. Int. J. Radiat. Biol. 85: 421-431.

Fan, F. and M.J. Gray, N.A. Dallas, A.D. Yang, G. Van Buren 2nd, E.R. Camp, and L.M. Ellis. 2008. Effect of chemotherapeutic stress on induction of vascular endothelial growth factor family members and receptors in human colorectal cancer cells. Mol. Cancer Ther. 7: 3064-3070.

Ferrara, N. 1999. Role of vascular endothelial growth factor in the regulation of angiogenesis. Kidney Int. 56: 794-814.

Ginestier, C. and S. Liu, M.E. Diebel, H. Korkaya, M. Luo, M. Brown, J. Wicinski, O. Cabaud, E. Charafe-Jauffret, D. Birnbaum, J.L. Guan, G. Dontu, and M.S. Wicha. 2010. CXCR1 blockade selectively targets human breast cancer stem cells in vitro and in xenografts. J. Clin. Invest. 20: 485-497.

Gorelik, L. and R.A. Flavell. 2002. Transforming growth factor-beta in T-cell biology. Nat. Rev. Immunol. 2: 46-53.

Hartsell, W.F. and C.B. Scott, G.S. Dundas, M. Mohiuddin, R.F. Meredith, P. Rubin, and I.J. Weigensberg. 2007. Can serum markers be used to predict acute and late toxicity in patients with lung cancer? Analysis of RTOG 91-03. Am. J. Clin. Oncol. 30: 368-376.

Jain, R.K. and M.F. Booth. 2003. What brings pericytes to tumor vessels? J. Clin. Invest. 112: 1134-1136.

Lev, D.C. and M. Ruiz, L. Mills, E.C. McGary, J.E. Price, and M. Bar-Eli. 2003. Dacarbazine causes transcriptional up-regulation of interleukin 8 and vascular endothelial growth factor in melanoma cells: a possible escape mechanism from chemotherapy. Mol. Cancer Ther. 2: 753-763.

Levina, V. and Y. Su, B. Nolen, X. Liu, Y. Gordin, M. Lee, A. Lokshin, and E. Gorelik. 2008a. Chemotherapeutic drugs and human tumor cells cytokine network. Int. J. Cancer 123: 2031-2040.

Levina, V. and A.M. Marrangoni, R. DeMarco, E. Gorelik, and A.E. Lokshin. 2008b. Drug-selected human lung cancer stem cells: cytokine network, tumorigenic and metastatic properties. PLoS ONE. 3: e3077.

Levina, V. and A.M. Marrangoni, R. DeMarco, E. Gorelik, and A.E. Lokshin. 2008c. Multiple effects of TRAIL in human carcinoma cells: Induction of apoptosis, senescence, proliferation, and cytokine production. Exp. Cell Res. 314: 1605-1616.

Levina, V. and A. Marrangoni, T. Wang, S. Parikh, Y. Su, R. Herberman, A. Lokshin, and E. Gorelik. 2010. Elimination of human lung cancer stem cells through targeting of the stem cell factor-c-kit autocrine signaling loop. Cancer Res. 70: 338-346.

Mantovani, A. and P. Allavena, A Sica, and F. Balkwill. 2008. Cancer-related inflammation. Nature 454: 436-444.

Massague, J. 2008. TGFbeta in Cancer. Cell 134: 215-230.

Michaelis, M. and D. Klassert, S. Barth, T. Suhan, R. Breitling, B. Mayer, N. Hinsch, H.W. Doerr, J. Cinatl, and J. Cinatl. 2009. Chemoresistance acquisition induces a global shift of expression of aniogenesis-associated genes and increased pro-angogenic activity in neuroblastoma cells. Mol. Cancer 8: 80-97.

Orimo, A. and R.A. Weinberg. 2006. Stromal fibroblasts in cancer: a novel tumor-promoting cell type. Cell Cycle 5: 1597-1601.

Petreaca, M.L., and M. Yao, Y. Liu, K. Defea, and M. Martins-Green. 2007. Transactivation of vascular endothelial growth factor receptor-2 by interleukin-8 (IL-8/CXCL8) is required for IL-8/CXCL8-induced endothelial permeability. Mol. Biol. Cell 18: 5014-5023.

Robinson, S.C. and L.M. Coussens. 2005. Soluble mediators of inflammation during tumor development. Adv. Cancer Res. 93: 159-187.

Shibakura, M. and K. Niiya, T. Kiguchi, I. Kitajima, M. Niiya, N. Asaumi, N.H. Huh, Y. Nakata, M. Harada, and M. Tanimoto. 2003. Induction of IL-8 and monoclyte chemoattractant protein-1 by doxorubicin in human small cell lung carcinoma cells. Int. J. Cancer 103: 380-386.

Todaro, M. and Y. Lombardo, M.G. Francipane, M.P. Alea, P. Cammareri, F. Iovino, A.B. Di Stefano, C. Di Bernardo, A. Agrusa, G. Condorelli, H. Walczak, and G. Stassi. 2008. Apoptosis resistance in epithelial tumors is mediated by tumor-cell-derived interleukin-4. Cell Death Differ. 15: 762-772.

Visvader, J.E. and G.J. Lindeman. 2008. Cancer stem cells in solid tumours: accumulating evidence and unresolved questions. Nat. Rev. Cancer 8: 755-768.

Section V
Cardiovascular and Metabolic Disease

CHAPTER 16

Inflammatory Cytokines and Cardiovascular Risk

Gordon Lowe*, Paul Welsh, Ann Rumley and Naveed Sattar
Division of Cardiovascular and Medical Sciences, Faculty of Medicine
BHF Glasgow Cardiovascular Research Centre, University of Glasgow
126 University Place, Glasgow G12 8TA, UK

ABSTRACT

There is increasing interest in the roles of inflammation and of pro-inflammatory cytokines in cardiovascular disease (CVD), including coronary heart disease (CHD), stroke, peripheral arterial disease (PAD) and venous thromboembolism. Worldwide, inter-individual differences in CVD risk are largely explained by well-established cardiovascular risk factors. These may be non-modifiable (age, male sex) or modifiable (tobacco smoke exposure, arterial blood pressure, blood cholesterol, obesity, diabetes, psychosocial factors).

We have hypothesized that the effects of these risk factors on CVD may be partly mediated by pro-inflammatory cytokines and their effects on atherosclerosis, thrombosis and blood rheology. In this chapter, we review large epidemiological studies of the associations of their circulating levels (and their functional genotypes) with risk factors, and with risk of CVD. Most published information is available for circulating interleukin-6 (IL-6), levels of which are associated with most risk factors, risk of CHD, outcome of stroke, progression of PAD, and blood viscosity. These associations of IL-6 may partly explain the associations of circulating "downstream" inflammatory markers—C-reactive protein (CRP), fibrinogen, white cell count, erythrocyte sedimentation rate (ESR), blood viscosity—with risk of CVD: a hypothesis that can be tested by Mendelian randomization studies, or by randomized controlled trials of cytokine-antagonist drugs. Less epidemiological information is currently available for other

*Corresponding author

pro-inflammatory cytokines such as interleukin-18 (IL-18), tumour necrosis factor alpha (TNFα), matrix metalloproteinase-9 (MMP-9) and leptin, and anti-inflammatory cytokines such as adiponectin and interleukin-10 (IL-10). Ongoing studies should clarify their associations with risk factors and with risk of CVD and, where appropriate, assess their causality.

INTRODUCTION

Cardiovascular disease (CVD), including coronary heart disease (CHD), stroke, and peripheral arterial disease (PAD) of the lower limbs, is a major cause of death and disability (http://www.who.int/cardiovascular_diseases/). Classical risk factors include increasing age and male sex and three major modifiable risk factors: tobacco smoking, arterial blood pressure, and blood cholesterol. High density lipoprotein (HDL) cholesterol is protective, in contrast to low density lipoprotein (LDL) cholesterol. Clinical risk scores, such as the Framingham scores including these factors, have been widely used to predict CVD risk and to advise on lifestyle and drug treatments to reduce risk.

Emerging risk factors for CVD include obesity, measured by body mass index (BMI), markers of socioeconomic position, circulating markers of inflammation (such as white blood cell count, viscosity, fibrinogen, CRP and inflammatory cytokines) and circulating markers of haemostasis and thrombosis (fibrinogen, von Willebrand factor, tissue plasminogen activator and its major inhibitor, plasminogen activator inhibitor type 1, and fibrin D-dimer). International collaborative meta-analyses of individual participant data, performed initially by the Fibrinogen Studies Collaboration (http://www.phpc.cam.ac.uk/MEU/FSC/), are being performed by the Emerging Risk Factors Collaboration (http://www.phpc.cam.ac.uk/MEU/ERFC/).

Risk factors may promote atherosclerosis; arterial plaque rupture, which precipitates arterial thrombosis and major clinical CHD and PAD events and stroke; platelet-fibrin thrombosis not only at the site of arterial plaque rupture, but also in other parts of the circulation including the venous circulation; and ischaemic organ damage distal to an atherothrombotic occlusion (Fig. 1) (Hansson 2005).

Inflammation and pro-inflammatory cytokines play key roles in all of these pathological processes. The pioneering pathological and experimental studies of Ross, Libby and Hansson (Hansson 2005) have established an important role for inflammation in atherogenesis and in arterial plaque rupture. C-reactive protein (CRP) may promote ischaemic necrosis, in acute myocardial infarction (Pepys *et al.* 2006).

Circulating levels of inflammatory markers in clinical use (ESR, plasma viscosity, white blood cell count, fibrinogen or CRP) all show associations with risks of CHD, stroke and PAD (Fibrinogen Studies Collaboration 2005; Lowe 2005; Emerging Risk Factors Collaboration 2010) (Table 1). The causality of fibrinogen or CRP in

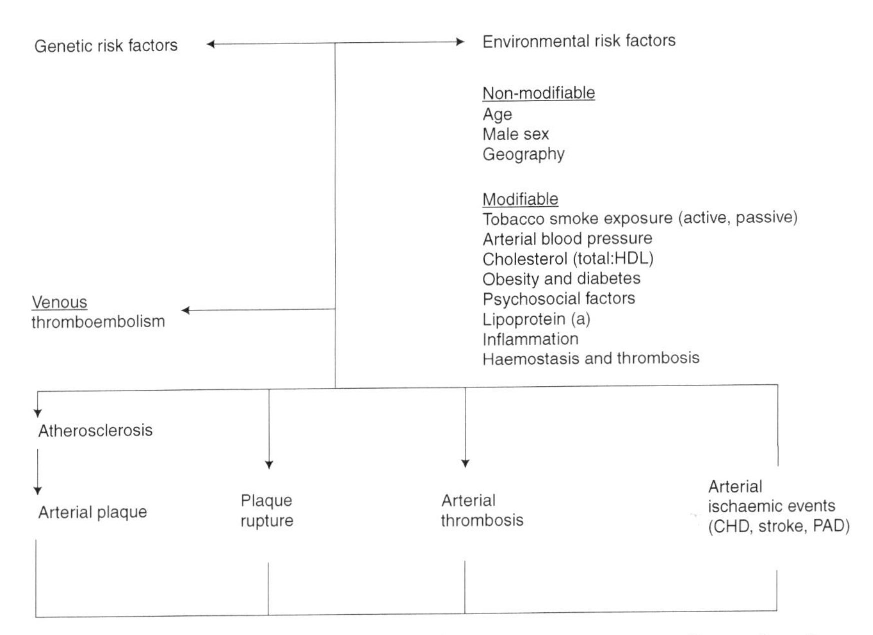

Fig. 1 Mechanisms through which risk factors may promote cardiovascular disease (unpublished).

Table I Indirect comparisons of associations of inflammatory markers with risk of major CHD events in available meta-analyses of prospective studies (unpublished).

	Risk association comparing top vs bottom third of population distributions (95% CI)	*Reference*
Routine clinical inflammatory markers		
Fibrinogen	HR 1.75 (1.59-1.92)	(Danesh *et al.* 2005)
White cell count	RR 1.49 (1.38-1.61)	(Danesh *et al.* 1998)
Plasma viscosity	RR 1.57 (1.34-1.85)	(Danesh *et al.* 2000a)
ESR	RR 1.33 (1.15-1.54)	(Danesh *et al.* 2000a)
CRP	HR 1.58 (1.48-1.68)	(Emerging Risk Factors Collaboration 2010)
Albumin	RR 1.47 (1.27-1.70)	(Danesh *et al.* 1998)
Inflammatory cytokines		
IL-6	HR 1.67 (1.35-2.05)	(Danesh *et al.* 2008)
IL-18	HR 1.30 (0.99-1.69)	Jefferies *et al.* in preparation
Leptin	HR 1.44 (0.95-2.16)	(Sattar *et al.* 2009b)
Adiponectin	HR 0.84 (0.70-1.01)	(Sattar *et al.* 2006)

All models are adjusted for "classical risk factors" as defined in each meta-analysis. HR = hazard ratio, RR = relative risk.

promoting CVD has not been established, by either Mendelian randomized trials of genotypes associated with increased levels of CRP, or clinical trials of drugs that selectively lower their levels or antagonize their effects. They are associated with non-vascular as well as vascular mortality (Danesh *et al.* 2005; Sattar *et al.* 2009a; Emerging Risk Factors Collaboration 2010). However, the magnitude and duration of the inflammatory response may influence clinical outcomes (Lowe 2005).

INFLAMMATORY CYTOKINES AND CVD

As well as "downstream" inflammatory markers, inflammatory cytokines have been proposed to play a key role in CVD. Several are present in arteries and up-regulated in atherosclerotic plaques, with roles in plaque rupture, thrombogenesis, and arterial remodelling (Hansson 2005) (Table 2). Several are expressed in adipose tissue (adipokines), including tumour necrosis factor alpha (TNFα), IL-6, leptin and adiponectin, and may play a role in obesity-related CVD and type 2 diabetes (Yudkin *et al.* 1999). IL-6 from a variety of body organs may play a key role in mediating the associations of "downstream" inflammatory markers (e.g., white cell count, plasma viscosity, fibrinogen, CRP) with risk of CVD (Woodward *et al.* 1999; Welsh *et al.* 2008a) (Fig. 2).

Table 2 Key inflammatory cytokines in CVD

IL-6	Key "messenger" cytokine, mediating hepatic synthesis of inflammatory markers and release of leucocytes from bone marrow
IL-18	Expressed in arteries, potential role in plaque rupture
TNFα	Expressed in arteries, potential role in plaque rupture Adipokine
MMP-9	Expressed in arteries, potential role in plaque rupture
Leptin	Adipokine with pro-inflammatory effects
Adiponectin	Adipokine with anti-inflammatory effects
IL-10	Anti-inflammatory cytokine
IL-1	Expressed in arteries, potential role in plaque rupture
Interferon γ	Expressed in arteries, potential role in plaque rupture

Most published studies investigating CVD risk in humans have measured circulating cytokine concentrations. Limitations to these studies include high variability including diurnal, seasonal and acute reactive variations. Serial measurement in individuals, and adjustment for regression dilution bias, can minimize such bias in studies of associations with CVD risk (Danesh *et al.* 2008). Existing immuno-assays do not distinguish between free cytokine levels and levels influenced by cytokine binding to circulating receptors (Rafiq *et al.* 2007). Sample processing (e.g., serum versus plasma, type of anticoagulant, time to sample spinning) may greatly influence levels. Some are detectable only at the extreme

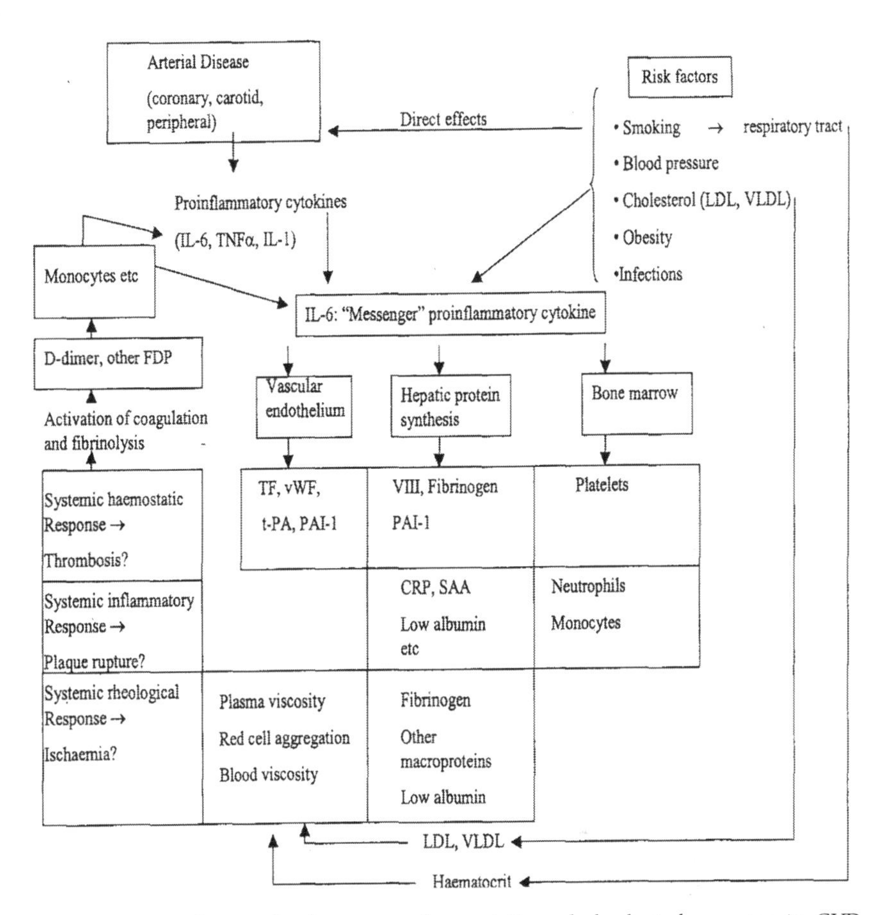

Fig. 2 IL-6 regulation of inflammatory, haemostatic and rheological response in CVD (unpublished).

lower limit of circulating blood analytes in healthy subjects (e.g., IL-6 and TNFα at c. 1 pmol/L). Finally, levels sampled from peripheral veins do not necessarily reflect local expression in tissues relevant to CVD. Indeed, some cytokines are highly reactive (e.g., TNFα) and exert most of their effects in local tissues.

INTERLEUKIN-6 AND CVD

IL-6 is a key pro-inflammatory cytokine, expressed in several body tissues including arteries, lungs and adipose tissue. It is a major regulator of circulating levels of "downstream" inflammatory markers (Fig. 2), including hepatic synthesis of CRP and fibrinogen, and bone marrow release of white blood cells. It has been hypothesized that IL-6 might play a key role in mediating the pro-inflammatory effects of risk factors on cardiovascular risk (Woodward *et al.* 1999).

In a systematic review of 17 prospective studies of circulating IL-6 levels and CHD risk (Danesh *et al.* 2008), the odds ratio for CHD per two SD increase in baseline IL-6 levels was 1.61 (95% CI 1.42, 1.83). The year-to-year variability of IL-6 values within individuals was high: using this data to correct for regression dilution bias, the odds ratio increased to 3.34 (2.45, 4.45). Figure 3 shows that the strength of this adjusted association of IL-6 levels with CHD risk appeared stronger than that of some downstream inflammatory markers (CRP and ESR), and comparable with that of cholesterol and systolic blood pressure (Danesh *et al.* 2008). In a more recent report IL-6 level showed a stronger association with risk of CVD events than a wide range of downstream inflammatory markers (Patterson *et al.* 2009). IL-6 levels are associated with age and most CVD risk factors (Woodward *et al.* 1999; Wannamethee *et al.* 2007b; Danesh *et al.* 2008; Welsh *et al.* 2008c; Jefferis *et al.* 2010).

In the Edinburgh Artery Study, progression of occlusive atherosclerosis in the lower limb was monitored over 15 yr. Several circulating inflammatory and haemostatic biomarkers were associated with progression of atherosclerosis, but IL-6 showed the strongest association (Tzoulaki *et al.* 2006). IL-6 showed stronger associations with incident MI or CHD death compared to incident stable angina (Wannamethee *et al.* 2009) and with fatal compared to non-fatal CVD events (Sattar *et al.* 2009a).

The high intra-individual variability of IL-6 levels makes it unlikely that single measures of IL-6 concentrations could add clinical utility to currently used CVD risk prediction scores (Woodward *et al.* 2009). Hence the main interest in the association of IL-6 levels with CVD risk is in targeting IL-6 for prevention or treatment of CVD. Mendelian randomization studies are limited by the lack of functional genotypes, apart from the IL-6 receptor gene. Ongoing randomized controlled trials of IL-6 receptor antagonists (Maini *et al.* 2006) may be more informative for the potential causality of IL-6 in CVD.

INTERLEUKIN-18 AND CVD

Second to IL-6, interleukin-18 (IL-18) (Gracie *et al.* 2003) has been the interleukin most studied for its association with CVD risk. It has been identified in human atherosclerotic plaques and may play a role in plaque instability. Cross-sectional epidemiological studies have shown associations of IL-18 level with CHD and several risk factors (Welsh *et al.* 2008c, 2009b).

Several studies of circulating IL-18 levels and CVD risk in generally healthy persons have been published (for references, see Woodward *et al.* 2009) and a systematic review is currently in progress. This shows a significant association after adjustment for age, CVD risk factors and other inflammatory markers, which appears about half as strong as that of IL-6 and CVD risk (Jefferies *et al.*, personal communication, 2010).

It is unlikely that IL-18 level could add clinical utility to currently used CVD risk prediction scores (Woodward *et al.* 2009). Mendelian randomization studies are progressing, although currently neither genetic variants of IL-18 nor its receptor are associated with increased CVD risk (Grisoni *et al.* 2009).

TUMOUR NECROSIS FACTOR ALPHA AND CVD

Tumour necrosis factor alpha (TNFα) is expressed in arterial plaques and may play a role in atherosclerosis; it is also an adipokine (Hansson 2005). Cross-sectional epidemiological studies have shown associations of circulating TNFα concentrations with CHD and with risk factors (Welsh *et al.* 2008c, 2009b).

To date, prospective associations of circulating TNFα concentration with CVD risk have been inconsistent (for references, see Woodward *et al.* 2009). It is unlikely that TNFα level would add clinical utility to CVD risk prediction scores (Woodward *et al.* 2009).

For causality, MR studies have not yet shown convincing evidence that elevated circulating TNFα increases CVD risk. However, TNFα antagonists in treatment of rheumatoid arthritis have been reported to reduce the two-fold increase in risk of CVD events in such patients (Dixon *et al.* 2007).

MATRIX METALLOPROTEINASE-9 AND CVD

Matrix metalloproteinase-9 (MMP-9) is one of a family of zinc-containing zymogen endoproteinases, which have functions in the normal and injury-induced turnover of the extracellular matrix, and may play a role in atherosclerotic plaque rupture. We have reported moderate associations of serum MMP-9 level with risk of CHD (Welsh *et al.* 2008b, Jefferis *et al.* 2010b), as well as associations with tobacco-smoking (a major cause of plaque rupture). Plasma MMP-9 concentrations and genetic variations have also been associated with CVD risk among people with existing CVD (Blankenberg *et al.* 2003).

LEPTIN AND CVD

Leptin is a pleiotropic adipokine, whose circulating levels correlate positively with body fat mass. Leptin has been suggested as a biochemical mediator of the link between adiposity and CVD, since recombinant leptin promotes atherosclerosis and thrombosis in a mouse model. However, in a recent prospective study and meta-analysis of seven previous studies (comprising 1,335 CHD cases), circulating

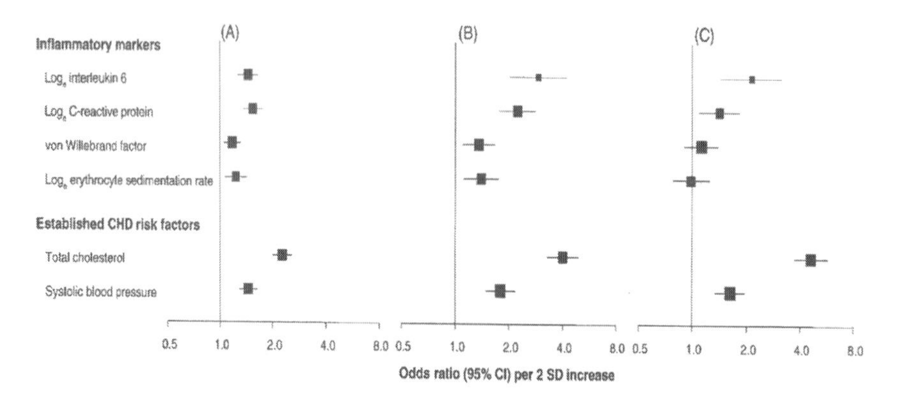

Fig. 3 Associations of circulating levels of IL-6, CRP, VWF, ESR and total cholesterol, and of systolic blood pressure, with risk of CHD in the Reykjavik Study and the British Regional Heart Study. From Danesh *et al.*, PLoS Med 2008;5:e78 (open access publication). (A) Baseline exposure adjusted for baseline confounders. (B) Usual exposure adjusted for baseline confounders. (C) Usual exposure adjusted for usual confounders.

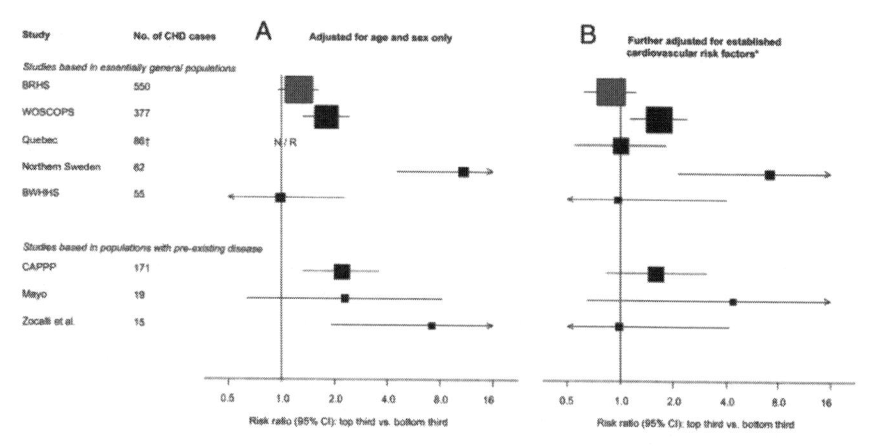

Fig. 4 Systematic review of associations of leptin with CVD risk. From Sattar *et al.* 2009b (open access publication). Risk ratios for CHD in a comparison of extreme thirds of leptin levels for each study. Horizontal lines represent 95% confidence intervals. BRHS, British Regional Heart Study; BWHHS, British Women's Heart Health Study; CAPPP, Captopril Prevention Project; CHD, coronary heart disease; N/R, not reported; WOSCOPS, West of Scotland Coronary Outcomes Prevention Study.

leptin level had only weak borderline significant associations with CHD risk (Sattar *et al.* 2009b) (Fig. 4). Circulating leptin is a far stronger risk factor for diabetes than for CVD (Welsh *et al.* 2009a).

ADIPONECTIN AND CVD

Adiponectin is an unusual adipokine in that circulating levels are inversely associated with fat mass in healthy individuals. As such, adiponectin has been proposed to play anti-inflammatory and anti-atherogenic roles. However, in a recent meta-analysis of seven studies (comprising 1,318 CHD cases) the OR comparing extreme thirds of adiponectin was 0.84 (95% CI 0.70-1.01) (Sattar *et al.* 2006) (Fig. 5). Like leptin, the association of adiponectin with incident diabetes is much stronger than the association with CVD (Wannamethee *et al.* 2007a).

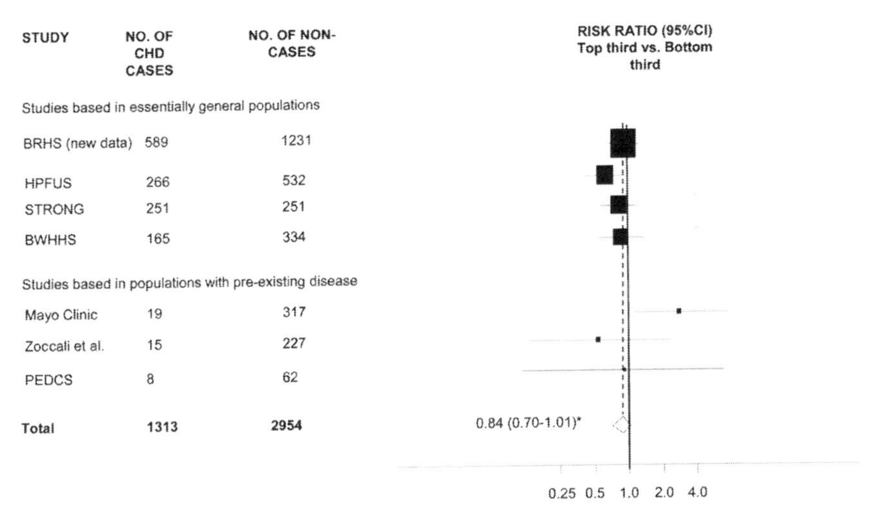

Fig. 5 Systematic review of associations of adiponectin with CVD risk. From Sattar *et al.* 2006 (open access publication). BWHHS, British Women's Heart Health Study; HPFUS, Health Professionals Follow-up Study; PEDCS, Pittsburgh Epidemiology of Diabetes Complications Study. Horizontal lines represent 95% CIs of the risk estimates from each study. *Calculated using a fixed effects model.

Genetic studies are needed to help define causal pathways: early evidence suggests that adiponectin expression may influence both CHD and diabetes risk (Richards *et al.* 2009).

INTERLEUKIN-10 AND CVD

Interleukin-10 (IL-10) is well known for its anti-inflammatory role in immune regulation. In keeping with this, over-expression of IL-10 reduces atherosclerosis development in mouse models, and IL-10-deficient atherosclerosis-prone mice more rapidly develop plaques with a phenotype conducive to rupture, compared to control mice.

Epidemiological data relating IL-10 to CVD risk is limited. In a recent prospective study in middle-aged women with prevalent atherosclerosis, a *positive* association of circulating levels of IL-10 with risk of CHD was reported with a hazard ratio of 1.34 (95% CI 1.06-1.68) per SD increment, although there were only 71 CHD events (Lakoski *et al.* 2008).

CONCLUSION

Circulating inflammatory cytokines, particularly IL-6 and IL-18, are generally moderately associated with an increased risk of CVD, independently of the classical CVD risk factors. They may also mediate part of the effects of classical risk factors (e.g., tobacco smoking) on CVD risk. Current evidence suggests any incremental clinical utility for risk prediction is likely to be slight, and such measurements are probably not a cost effective way of screening for CVD risk. Intervention studies are underway to assess the ability of anti-inflammatory biologics (such as TNFα or IL-6 blockers) to reduce cardiovascular risk in patients with rheumatoid arthritis. However, use of such agents in generally healthy people is problematic because of the complications of systemic immunosuppression. Future treatments may attempt to dampen inflammatory responses, such as blocking inflammatory cytokine signalling and cellular recruitment within arterial vessels in a targeted manner. The issue of targeting inflammatory pathways to prevent CVD in the general population remains open.

Table 3 Key facts about inflammatory cytokines and cardiovascular risk

- Several inflammatory cytokines are expressed in arteries and may play potential roles in rupture of atherosclerotic plaques and superadded thrombosis, the commonest cause of coronary heart disease and strokes.
- In addition, interleukin-6 (IL-6) is a key "messenger" cytokine, which mediates hepatic synthesis of inflammatory markers and release of leucocytes into the circulation.
- Prospective studies of circulating levels of IL-6 have established that they show similar associations with risk of coronary heart disease to circulating levels of "downstream" inflammatory markers (fibrinogen, white cell count, plasma viscosity, erythrocyte sedimentation rate, C-reactive protein).
- Ongoing prospective studies are currently evaluating the associations of other cytokines (e.g., interleukin-18, leptin, adiponectin) with cardiovascular risk.
- While these associations are potentially relevant to pathogenesis and treatment, at present they do not appear sufficiently strong to add to predicting an individual's cardiovascular risk.

Summary Points

- Inflammation is increasingly recognized to play a key role in CVD, via atherogenesis, plaque rupture, thrombosis and ischaemia.

- Circulating levels of pro-inflammatory cytokines, in particular IL-6 and IL-18, are associated with increased risk of CVD events in generally healthy populations, as well as in cohorts of persons with acute or chronic CVD.
- There is also some evidence that circulating levels of anti-inflammatory cytokines, in particular adiponectin, are associated with decreased risk of CVD events.
- These associations are not sufficiently strong to indicate that addition of assays of cytokines will improve currently used risk scores for CVD in a cost-effective manner.
- Their causality in CVD is currently being assessed in Mendelian randomization studies, and in trials of inhibitors of pro-inflammatory cytokines.

Abbreviations

BMI : body mass index
CHD : coronary heart disease
CRP : C-reactive protein
CVD : cardiovascular disease
ESR : erythrocyte sedimentation rate
HDL : high-density lipoprotein
HR : hazard ratio
IL-1 : interleukin-1
IL-6 : interleukin-6
IL-10 : interleukin-10
IL-18 : interleukin-18
LDL : low-density lipoprotein
Lp (a) : lipoprotein (a)
MMP-9 : matrix metalloproteinase-9
PAD : peripheral arterial disease
PAI-1 : plasminogen activator inhibitor type 1
RR : relative risk
TNFα : tumour necrosis factor alpha
t-PA : tissue plasminogen activator
VWF : von Willebrand factor

Definition of Terms

Atherosclerosis: A progressive, focal disease affecting the coronary, cranial and lower limb arteries.

Cardiovascular risk factors: A measurable biological phenotype, for which levels are associated with future risk of developing cardiovascular disease (coronary heart disease or stroke).

*Coronary heart disease:*Atherosclerosis in the coronary arteries may result in intermittent chest pain on exercise (stable angina), chest pain that is irregular (unstable angina), or myocardial infarction.

Epidemiology: The study of factors that cause health and disease within populations.

Ischaemia: A reduced blood supply causing cell death. Often caused by a thrombus occluding an afferent blood vessel, resulting in myocardial infarction in the heart, or stroke within brain tissue.

Mendelian randomization: Causality of a biomarker in multifactorial diseases, such as CVD, cannot be inferred from conventional epidemiological studies. Mendelian randomization relates genetic polymorphisms, which cause well-defined phenotypic traits, to risk of disease outcomes. If the genotypes of interest are not confounded by other factors (due to random allocation of genetic material at conception), and the genotypes are not pleiotropic, any relationship between genotype and outcome may be evidence of a causal association.

Peripheral arterial disease: Narrowing of lower limb arteries may result in intermittent leg pain on exercise (claudication).

Regression dilution bias: Variability in biomarkers over time means that one measure of a biomarker in an individual is likely to be an incorrect estimation of their "usual" level. This will result in an underestimation in the association of a biomarker with risk of disease: the more variable the marker, the greater the underestimate. By correcting for variation over time (using serial measures), the association with disease risk can be more accurately estimated, reducing the regression dilution bias.

Stroke: In the arteries supplying the brain, atherothrombosis is a common cause of acute focal brain damage (stroke), although stroke may also be caused by blood clots arising from the heart (atrial fibrillation, heart valve disease or replacement heart valves) or by small vessel disease of the brain (lacunar stroke).

Thrombosis: The main clinical consequences of atherosclerosis that cause death and disability result from acute rupture of an atherosclerotic plaque, and arterial occlusion by platelet-fibrin clot.

Acknowledgements

We acknowledge funding of our studies from the British Heart Foundation and the Chief Scientist Office, Scottish Government. We thank all our collaborators in epidemiological studies, and Helen Mosson for secretarial assistance in preparing this manuscript.

References

Blankenberg, S. and H.J. Rupprecht, O. Poirier, C. Bickel, M. Smieja, G. Hafner, *et al.* 2003b. Plasma concentrations and genetic variation of matrix metalloproteinase 9 and prognosis of patients with cardiovascular disease. Circulation 107: 1579-1585.

Danesh, J. and S. Kaptoge, A.G. Mann, N. Sarwar, A. Wood, S.B. Angleman, *et al.* 2008. Long-term interleukin-6 levels and subsequent risk of coronary heart disease: two new prospective studies and a systematic review. PLoS Med. 5: e78.

Dixon, W.G. and K.D. Watson, M. Lunt, K.L. Hyrich, A.J. Silman, and D.P. Symmons. 2007. Reduction in the incidence of myocardial infarction in patients with rheumatoid arthritis who respond to anti-tumor necrosis factor alpha therapy: results from the British Society for Rheumatology Biologics Register. Arthritis Rheum. 56: 2905-2912.

Emerging Risk Factors Collaboration. 2010. C-reactive protein concentration and risk of coronary heart disease, stroke, and mortality: an individual participant meta-analysis. Lancet 375: 132-140.

Fibrinogen Studies Collaboration. 2005. Plasma fibrinogen level and the risk of major cardiovascular diseases and nonvascular mortality: an individual participant meta-analysis. JAMA 294: 1799-1809.

Gracie, J.A. and S.E. Robertson, and I.B. McInnes. 2003. Interleukin-18. J. Leukoc. Biol. 73: 213-224.

Grisoni, M.L. and C. Proust, M. Alanne, M. Desuremain, V. Salomaa, K. Kuulasmaa, *et al.* 2009. Lack of association between polymorphisms of the IL18R1 and IL18RAP genes and cardiovascular risk: the MORGAM Project. BMC Med. Genet. 10: 44.

Hansson, G.K. 2005. Inflammation, atherosclerosis, and coronary artery disease. N. Engl. J. Med. 352: 1685-1695.

Jefferis, B.J. and P. Whincup, P. Welsh, G. Wannamethee, A. Rumley, L. Lennon, *et al.* 2010. Prospective study of matrix metalloproteinase-9 and risk of myocardial infarction and stroke in older men and women. Atherosclerosis 208: 557-563.

Lakoski, S.G. and Y. Liu, K.B. Brosnihan, and D.M. Herrington. 2008. Interleukin-10 concentration and coronary heart disease (CHD) event risk in the estrogen replacement and atherosclerosis (ERA) study. Atherosclerosis 197: 443-447.

Lowe, G.D. 2005. Circulating inflammatory markers and risks of cardiovascular and non-cardiovascular disease. J. Thromb. Haemost. 3: 1618-1627.

Maini, R.N. and P.C. Taylor, J. Szechinski, K. Pavelka, J. Broll, G. Balint, *et al.* 2006. Double-blind randomized controlled clinical trial of the interleukin-6 receptor antagonist, tocilizumab, in European patients with rheumatoid arthritis who had an incomplete response to methotrexate. Arthritis Rheum. 54: 2817-2829.

Patterson, C.C. and A.E. Smith, J.W. Yarnell, A. Rumley, Y. Ben-Shlomo, and G.D. Lowe. 2010. The associations of interleukin-6 (IL-6) and downstream inflammatory markers with risk of cardiovascular disease: The Caerphilly Study. Atherosclerosis 209: 551-557.

Pepys, M.B. and G.M. Hirschfield, G.A. Tennent, J.R. Gallimore, M.C. Kahan, V. Bellotti, *et al.* 2006. Targeting C-reactive protein for the treatment of cardiovascular disease. Nature 440: 1217-1221.

Rafiq, S. and T.M. Frayling, A. Murray, A. Hurst, K. Stevens, M.N. Weedon, *et al.* 2007. A common variant of the interleukin 6 receptor (IL-6r) gene increases IL-6r and IL-6 levels, without other inflammatory effects. Genes Immun. 8: 552-559.

Richards, J.B. and D. Waterworth, S. O'Rahilly, M.F. Hivert, R.J. Loos, J.R. Perry, *et al.* 2009. A genome-wide association study reveals variants in ARL15 that influence adiponectin levels. PLoS Genet. 5: e1000768.

Sattar, N. and H.M. Murray, P. Welsh, G.J. Blauw, B.M. Buckley, S. Cobbe, *et al.* 2009a. Are markers of inflammation more strongly associated with risk for fatal than for nonfatal vascular events? PLoS Med. 6: e1000099.

Sattar, N. and G. Wannamethee, N. Sarwar, J. Chernova, D.A. Lawlor, A. Kelly, *et al.* 2009b. Leptin and coronary heart disease: prospective study and systematic review. J. Am. Coll. Cardiol. 53: 167-175.

Sattar, N. and G. Wannamethee, N. Sarwar, J. Tchernova, L. Cherry, A.M. Wallace, *et al.* 2006. Adiponectin and coronary heart disease: a prospective study and meta-analysis. Circulation 114: 623-629.

Tzoulaki, I. and G.D. Murray, J.F. Price, F.B. Smith, A.J. Lee, A. Rumley, *et al.* 2006. Hemostatic factors, inflammatory markers, and progressive peripheral atherosclerosis: the Edinburgh Artery Study. Am. J. Epidemiol. 163: 334-341.

Wannamethee, S.G. and G.D. Lowe, A. Rumley, L. Cherry, P.H. Whincup, and N. Sattar. 2007a. Adipokines and risk of type 2 diabetes in older men. Diabetes Care 30: 1200-1205.

Wannamethee, S.G. and P.H. Whincup, A. Rumley, and G.D. Lowe. 2007b. Inter-relationships of interleukin-6, cardiovascular risk factors and the metabolic syndrome among older men. J. Thromb. Haemost. 5: 1637-1643.

Wannamethee, S.G. and P.H. Whincup, A.G. Shaper, A. Rumley, L. Lennon, and G.D. Lowe. 2009. Circulating inflammatory and hemostatic biomarkers are associated with risk of myocardial infarction and coronary death, but not angina pectoris, in older men. J. Thromb. Haemost. 7: 1605-1611.

Welsh, P. and G.D. Lowe, J. Chalmers, D.J. Campbell, A. Rumley, B.C. Neal, *et al.* 2008a. Associations of proinflammatory cytokines with the risk of recurrent stroke. Stroke 39: 2226-2230.

Welsh, P. and H.M. Murray, B.M. Buckley, A.J. de Craen, I. Ford, J.W. Jukema, *et al.* 2009a. Leptin predicts diabetes but not cardiovascular disease: results from a large prospective study in an elderly population. Diabetes Care 32: 308-310.

Welsh, P. and P.H. Whincup, O. Papacosta, S.G. Wannamethee, L. Lennon, A. Thomson, *et al.* 2008b. Serum matrix metalloproteinase-9 and coronary heart disease: a prospective study in middle-aged men. QJM 101: 785-791.

Welsh, P. and M. Woodward, A. Rumley, and G. Lowe. 2008c. Associations of plasma pro-inflammatory cytokines, fibrinogen, viscosity and C-reactive protein with cardiovascular risk factors and social deprivation: the fourth Glasgow MONICA study. Br. J. Haematol. 141: 852-861.

Welsh, P. and M. Woodward, A. Rumley, and G. Lowe. 2009b. Associations of circulating TNFalpha and IL-18 with myocardial infarction and cardiovascular risk markers: the Glasgow Myocardial Infarction Study. Cytokine 47: 143-147.

Woodward, M. and A. Rumley, H. Tunstall-Pedoe, and G.D. Lowe. 1999. Associations of blood rheology and interleukin-6 with cardiovascular risk factors and prevalent cardiovascular disease. Br J Haematol 104: 246-257.

Woodward, M. and P. Welsh, A. Rumley, H. Tunstall-Pedoe, and G.D. Lowe. 2009. Do inflammatory biomarkers add to the discrimination of cardiovascular disease after allowing for social deprivation? Results from a 10 year cohort study in Glasgow, Scotland. Eur Heart J doi: 10.1093/eurheartj/ehp115 First published online: April 10, 2009.

CHAPTER 17

Angiogenic Cytokine Therapy for Ischemic Heart Disease

William Hiesinger and Y. Joseph Woo*

Division of Cardiovascular Surgery, Department of Surgery
University of Pennsylvania School of Medicine, 3400 Spruce Street
6 Silverstein Pavilion, Philadelphia, PA 19104, USA

ABSTRACT

Ischemic heart disease comprises an escalating national and global health challenge. Current therapy consists of pharmacologic optimization and limited revascularization, reconstructive or replacement options. These modalities are highly effective for only a fraction of patients and do not address the significant microvascular deficiencies that persist even when a diseased artery is stented or bypassed. Endogenous machinery to repair injured and ischemic myocardium is inadequate and tremendous resources have been devoted to developing molecular therapies that enhance both the microvascular perfusion and the function of ischemic or infarcted myocardium. A significant portion of this work has focused on angiogenic cytokines, primarily fibroblast growth factor (FGF), vascular endothelial growth factor (VEGF), hepatocyte growth factor (HGF), placental growth factor (PGF), and stromal cell-derived factor-1α (SDF). Angiogenic cytokine based therapies act, at least in part, through endothelial progenitor stem cell (EPC) activity, and those cells have the intrinsic ability to mature into the endothelial cells and support network vital for vasculogenesis and functional myocardial vasculature. Phase I and II clinical trials, while not conclusive, have achieved encouraging results using both recombinant protein cytokines and single-agent gene therapies. The data, accumulated over more than a decade, has demonstrated both clinical feasibility and an acceptable safety profile for angiogenic cytokine therapy. In this chapter, we examine both the preclinical and

*Corresponding author

clinical development of angiogenic cytokine therapy for ischemic heart disease and suggest areas for future expansion of this burgeoning field.

INTRODUCTION

More people die annually from cardiovascular disease than from any other cause. The World Health Organization estimates that more than 17 million people per year die from cardiovascular disease, and by 2030, the number will increase to almost 24 million. Of these deaths, an estimated 7.2 million were due to ischemic heart disease and low- and middle-income countries are disproportionally affected. An estimated 80,000,000 American adults (approximately 1 in 3) have at least one type of cardiovascular disease with more than 22,000,000 suffering from either coronary heart disease or heart failure (Lloyd-Jones *et al.* 2009). As the related epidemics of obesity and physical inactivity continue to spread, and as global life expectancy lengthens, the incidence of morbidity and mortality related to cardiovascular disease is expected to increase dramatically.

Current therapeutic options for ischemic heart disease include pharmacologic optimization using β-adrenergic receptor antagonists, diuretics, HMG-CoA reductase inhibitors, and ACE inhibitors, percutaneous and surgical revascularization, mechanical restraints, and surgical resections that attempt to improve myocardial efficiency and function. Unfortunately, most therapies are instituted relatively late in the overall course of the disease process and complete revascularization with either percutaneous coronary intervention (PCI) or coronary artery bypass grafting (CABG) is possible in only 63-80% of patients with ischemic heart disease (Boodhwani *et al.* 2006). Even when successful, immediate post-infarction treatments often fail to prevent disease progression, subsequent adverse ventricular remodeling, and heart failure because these treatments do not restore microvascular perfusion and do not address the underlying ventricular cellular pathophysiology (Araszkiewicz *et al.* 2006). Microvascular dysfunction has been shown to be an important independent predictor of ventricular remodeling as well as reinfarction, heart failure, and death (Bolognese *et al.* 2004). This has spurred the search for an innovative microrevascularization strategy that can serve as a primary therapy and/or as a supplemental, adjunctive therapy to traditional coronary revascularization methods. Angiogenic cytokine therapy represents one such avenue for the potential microvascular augmentation of ischemic myocardium.

New blood vessels are formed by two mechanisms: angiogenesis and vasculogenesis. Angiogenesis refers to new vessel growth from mature, differentiated endothelial cells (ECs) that are able to migrate and proliferate, forming sprouts from existing vessels. Vasculogenesis involves the homing of bone-marrow-derived endothelial progenitor cells (EPCs) to areas of vascular expansion, where they subsequently differentiate into mature ECs and capillaries *in situ*. Angiogenic cytokines, either expressed *in vivo* in response to a variety of stimuli or administered exogenously as a therapeutic modality, form vessels via vasculogenesis. They up-regulate and mobilize EPCs from the bone marrow and promote their targeting and incorporation into nascent vessels (Yoon *et al.* 2004).

Angiogenic cytokine therapy has primarily involved either the delivery of a recombinant protein or gene therapy. Protein therapy is typically delivered via direct intramyocardial or intracoronary injection. Established safety, predictable pharmacokinetics, a lack of long-term unexpected side effects, and ease of delivery are some of recombinant protein therapy's advantages over gene therapy. However, the short protein half-life and high cost are not optimal for a widely translatable therapy. Angiogenic cytokine gene therapy uses either plasmid DNA coding a desired gene or viral vectors targeting specific tissues. The advantages of gene transfer include prolonged and sustained expression of the protein in the target tissues, ability to express transcriptional factors, potential for regulated expression, and ability to express multiple genes simultaneously. However, the short- and long-term toxicities of various vectors are incompletely understood and gene-based therapy can theoretically cause detrimental sustained expression leading to pathologic angiogenesis or inflammatory reactions (Boodhwani *et al.* 2006). The adenovirus vector has been the most extensively investigated for cardiovascular application (Atluri and Woo 2008).

OVERVIEW OF ANGIOGENIC CYTOKINES

Over the years, angiogenic cytokine therapy has demonstrated the ability to increase collateral vessel formation, augment perfusion, limit infarct progression, and enhance cardiac function in a broad spectrum of experimental models. Initial investigations into the therapeutic potential of angiogenic cytokines focused on FGF and VEGF.

FGF-2, a highly potent cytokine, is known to modulate numerous cardiac cellular functions, including cell proliferation, differentiation, survival, adhesion, migration, motility, apoptosis, vasculogenesis, angiogenesis and blood vessel remodeling. It is expressed in the myocardium at all developmental stages by a wide variety of cell types, most notably, cardiac myocytes and vascular cells. FGF-2, highly conserved among species, is one of 23 structurally related polypeptide growth factors (FGF-1 to FGF-23) and is also considered to be a member of the larger heparin binding growth factor family, which includes VEGF. The effects of FGF-2 in the heart are mediated primarily by specific cell surface receptors of the tyrosine kinase family, predominantly FGF receptor-1 (FGFR-1). The mechanisms involved in FGF-2-mediated angiogenesis have not been fully elucidated; however, it is known that FGF-2 induces VEGF expression in vascular endothelial cells via both paracrine and autocrine pathways and that the roles of FGF-2 and VEGF are inextricably linked. This relationship is believed to relate to the observed synergistic response when the two growth factors are used in combination (Detillieux *et al.* 2003).

VEGF is a potent angiogenic cytokine in its own right and is considered to be one of the more critical factors involved in angiogenesis and vasculogenesis during embryonic development and adult life. The VEGF family of cytokines includes five members: VEGF-A, B, C, D, E with VEGF165 being the best-characterized and most abundant of the VEGF species major splice variants. The major effector of the angiogenic actions of VEGF is the VEGFR-2 (KDR/Flk-1), which is expressed primarily on endothelial cells. The downstream production of nitric oxide (NO) through the VEGF/VEGFR-2 pathway plays a critical angiogenic role in a variety of different vascular beds including fetal myocardium and the coronary collateral system. The generation of NO initiates and modulates the angiogenic effects of various growth factors and is a promoter of VEGF expression, indicating a feed forward mechanism between NO and VEGF in regulating angiogenesis. VEGF is involved in a myriad of vasculogenic processes including mobilization and induction of endothelial progenitor cells, enhancement of endothelial cell survival, vasodilation, promotion of vascular permeability and leakage, induction of integrin expression, modulation of endothelial fibrinolysis- and coagulation-related agents, and secretion of matrix metalloproteinases (Toyota *et al.* 2004).

The encouraging results yielded from investigation into FGF and VEGF sparked interest in other proangiogenic cytokines including placental growth factor (PGF), hepatocyte growth factor (HGF), and stromal cell-derived factor-1α (SDF). PGF, like VEGF and its homologs, belongs to the so-called cystein-knot superfamily of growth factors, whose members are all characterized by a common motif of eight spatially conserved cysteines, which allow the formation of dimers. The three-dimensional structure of human PGF is strikingly similar to that of VEGF despite sharing only 42% of its amino acid sequence. As previously described, VEGF regulates angiogenesis primarily by interacting with two tyrosine kinase receptors, VEGF receptor-1 (VEGFR-1 or Flt-1) and VEGF receptor-2 (VEGFR-2 or Flk-1). However, PGF specifically binds VEGFR-1, but not VEGFR-2. PGF seems to be particularly important for angiogenesis under pathological conditions (e.g., myocardial infarction or ischemia) and contributes by affecting endothelial cells, amplifying the angiogenic activity of VEGF, stimulating vessel maturation and stabilization, recruiting inflammatory cells, and mobilizing vascular and hematopoietic stem cells from the bone marrow. In addition, PGF can form heterodimers with VEGF which induce angiogenesis almost to the same extent as VEGF/VEGF homodimers (Autiero *et al.* 2003).

Hepatocyte growth factor, a multifunctional heparin-binding glycoprotein, is a large disulfide-linked heterodimetric protein composed of a 69-kDa a-subunit and a 34-kDa b-subunit. The biological activity of HGF is mediated by the stimulation of the HGF-specific tyrosine kinase receptor (HGFR or c-MET) encoded by the c-met proto-oncogene (Jin *et al.* 2004). HGF is widely expressed and was originally identified and characterized as a potent mitogen for hepatocytes. This mesenchyme-derived cytokine regulates cell growth, cell motility, and the morphogenesis of a broad variety of cell types and also stimulates endothelial cell growth without inducing the replication of vascular smooth muscle cells. These properties, along with some of its other physiological activities such as antiapoptosis and antifibrosis, make HGF an interesting candidate for therapeutic angiogenesis (Funatsu *et al.* 2002).

Stromal cell-derived factor 1α, a key regulator of physiological cell motility during embryogenesis and after birth, is constitutively expressed in a wide variety of cells including endothelial cells, dendritic cells, and stromal cells. SDF is a 68 amino acid peptide belonging to the CXC family of chemokines and is the sole ligand for the G protein–coupled receptor, CXCR4, through which its biologic effects are mediated. It is also remarkably conserved among species; a single amino acid substitution is all that differentiates the human and murine sequences (De La Luz Sierra *et al.* 2004). This powerful chemoattractant is considered to be one of the key regulators of hematopoietic stem cell trafficking between the peripheral circulation and bone marrow and has been shown to effect EPC proliferation and mobilization to induce vasculogenesis (Yamaguchi *et al.* 2003). In addition, SDF has been shown to be significantly up-regulated in response to both myocardial ischemia and infarction (Pillarisetti and Gupta 2001).

EXPERIMENTAL ANGIOGENIC CYTOKINE THERAPY

Cytokine driven vasculogenic treatments act by enticing bone marrow derived stem/progenitor cell populations to areas of myocardial ischemia where they exert their therapeutic effect by the direct cellular differentiation of stem cells into neovasculature as well as from their capacity to release more angiogenic factors. The body of evidence that endogenous mobilization of EPCs could be induced to grow new blood vessels in ischemic myocardium by local or systemic delivery of angiogenic cytokines (e.g., FGF, VEGF, PGF, HGF, and SDF) continues to expand (Renault and Losordo 2007). To this effect, it has been demonstrated that intrapericardial injections of FGF-2 in a porcine model of chronic myocardial ischemia resulted in a significant increase in angiographic collaterals and blood flow in an experimentally occluded coronary artery, improved myocardial perfusion and function in the ischemic territory, and histologic evidence of increased myocardial vascularity (Laham *et al.* 2000b). Transgenic myocardial overexpression of FGF, perivascular, intrapericardial, and intramyocardial injections of recombinant FGF protein, and adenoviral FGF-5 expression have all been safely attempted in a broad range of preclinical animal models. The FGF cytokine therapy resulted in significant increases in myocardial coronary vascular density, branching, and perfusion, as well as improvements in myocardial function and left ventricular preservation (Battler *et al.* 1993, Fernandez *et al.* 2000, Giordano *et al.* 1996, Harada *et al.* 1994, Hughes *et al.* 2004, Liu *et al.* 2006, Sakakibara *et al.* 2003, Sato *et al.* 2000, Yanagisawa-Miwa *et al.* 1992). However, despite these impressive results, it has been found that hypercholesteremia-induced endothelial dysfunction (a hallmark of coronary artery disease) leads to a marked impairment of the myocardial angiogenic response in chronically ischemic swine treated with perivascular FGF-2 suggesting that FGF therapy may be less efficacious in a clinical setting without interventions to improve coronary endothelial function (Ruel *et al.* 2003).

VEGF has been repeatedly demonstrated to mobilize and recruit EPCs resulting in subsequent neovascularization (Asahara *et al.* 1999, Hattori *et al.* 2001, Kalka *et al.* 2000a, b). In adult sheep, human VEGF165 gene transfer reduced infarct size after acute coronary artery occlusion via neovascular formation, reduced fibrosis and increased cardiomyocyte regeneration (Vera Janavel *et al.* 2006). Like FGF, the vasculogenic properties of VEGF may not be optimized in patients with ischemic vascular disease. The decreased bioavailability of nitric oxide and the endothelial dysfunction present in the clinical population with ischemic heart disease theoretically limit VEGF-mediated activity.

The vasculogenic potential of the VEGF homolog PGF has also been extensively studied. Direct intramyocardial injection of PGF following a large myocardial infarction in a murine model was associated with significant angiogenesis, preserved ventricular geometry, and improved myocardial function (Kolakowski *et al.* 2006) (Fig. 1). It has also been found that prolonged systemic elevation of circulating PGF after myocardial infarction in a mouse stimulates angiogenesis in the infarct border zone and improves LV contractility, without causing undesired side effects (hypotension, edema, etc.) (Roncal *et al.* 2008). Despite the fact that embryonic angiogenesis in mice is not affected by a congenital deficiency of PGF, its loss severely impaired angiogenesis, plasma extravasation and collateral vessel growth during ischemia and inflammation. Transplantation of wild-type bone marrow rescued the impaired angiogenesis and collateral growth in the PGF-deficient mice, arguing strongly that PGF contributes to vessel growth in the adult by mobilizing bone marrow–derived EPCs. In addition, there is a specific synergism between PGF and VEGF and, by up-regulating PGF, endothelial cells are able to amplify their responsiveness to VEGF during vasculogenesis in many pathological disorders (Carmeliet *et al.* 2001). Because of the synergistic effect, it might prove beneficial to co-administer VEGF and PGF for maximal vasculogenic potential.

Hepatocyte growth factor has also demonstrated the potential to mobilize bone marrow–derived progenitor cells to areas of ischemic myocardium. Potential cardiac and vasculogenic precursors are mobilized into the peripheral blood after MI and are capable of migrating towards an HGF gradient (expressed in infarcted myocardium) based on their high expression of the c-MET receptor (Kucia *et al.*

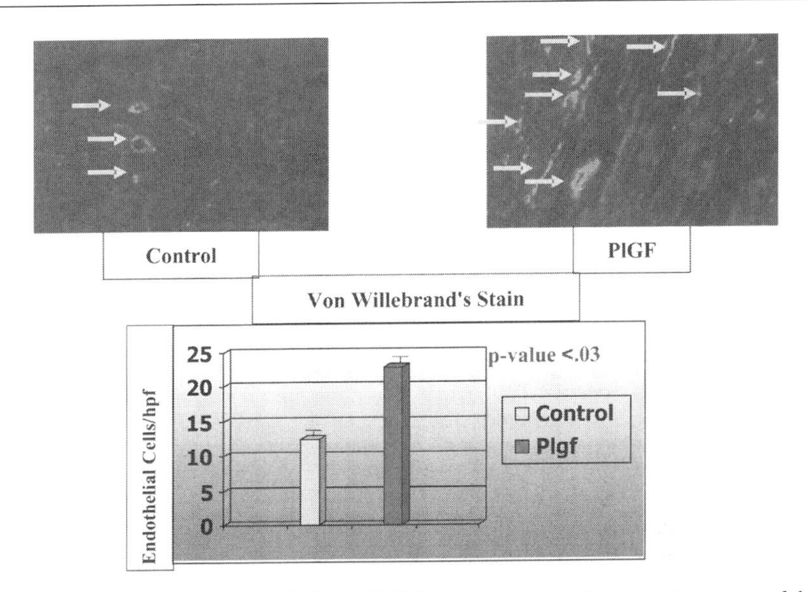

Fig. 1 Impact of placental growth factor (PGF) treatment on angiogenesis in a rat model of ischemic cardiomyopathy. Microscopic analysis of immunofluorescent expression for von Willebrand's factor along border zone regions revealed higher endothelial cell density in PGF-treated animals than in control animals. The border zone myocardium in the PGF group averaged 22.8 ± 3.5 endothelial cells per high-powered field compared to 12.4 ± 3.2 endothelial cells per high-powered field in the control group ($p < 0.03$) (reproduced from Kolakowski *et al.*).

2004). This potent chemokine has also proven to effectively attenuate post-infarction heart failure. HGF gene transfer following a large acute myocardial infarction in a rat model resulted in significantly increased angiogenesis and reduced apoptosis yielding preserved myocardial function and geometry (Jayasankar *et al.* 2003) (Fig. 2). HGF was also effective in a murine model of chronic heart failure. Over-expression of HGF 3 wk post-MI enhanced angiogenesis, reduced apoptosis, and preserved ventricular geometry and cardiac contractile function (Jayasankar *et al.* 2005). As further proof of concept, it was shown in a large animal model that adenovirus-mediated human HGF promotes angiogenesis, reduces cardiocyte apoptosis and ameliorates post-infarction heart function in a dose-dependent manner (Yang *et al.* 2009).

SDF is among the most potent and specific angiogenic cytokines discussed so far. Its sole target is the CXCR4 cell surface antigen expressed in significant levels on CD34+ EPCs and expression of this receptor is related to efficient SDF-induced transendothelial migration (Mohle *et al.* 1998). In addition, serum SDF serves as a strong and independent predictor of increased circulating EPCs (Chang *et al.* 2009). However, in an experimental femoral artery ligation hindlimb ischemia model, intramuscular SDF administration was unable to prevent limb autoamputation

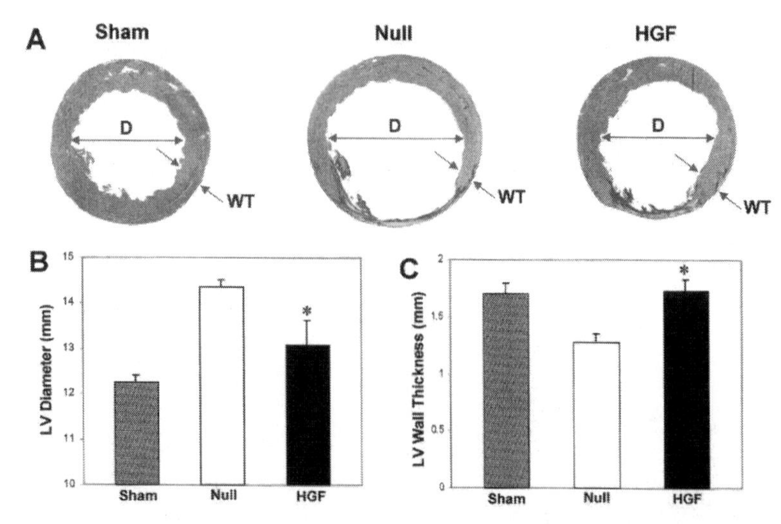

Fig. 2 Effects of hepatocyte growth factor (HGF) adenoviral gene transfer on myocardial preservation in a rat model of ischemic cardiomyopathy. (a) Representative cross-sections of a single rat heart from each of three experimental groups (sham [unligated]), adeno-null post-LAD ligation, and adenoHGF-treated post-LAD ligation) 6 wk after initial surgery. The sections are hematoxylin- and eosin-stained, and the arrows depict the diameter (D) and wall thickness (WT) measurement areas. (b) Graph of average left ventricular (LV) chamber diameter in each experimental group at the study endpoint. (c) Graph of average LV free wall thickness in each experimental group at the conclusion of the study (sham n = 10, null n = 9, HGF n = 10) (reproduced from Jayasankar *et al.*). * p = 0.01 vs null control.

(Yamaguchi *et al.* 2003). This may have been due to insufficient numbers of circulating EPCs. Clinical trials aimed at boosting bone marrow production of EPCs with granulocyte colony stimulating factor after acute myocardial infarction have had mixed results and randomized double-blind studies were unable to demonstrate efficacy (Ince *et al.* 2005, Kang *et al.* 2006, Kuethe *et al.* 2005, Ripa *et al.* 2006a, Zohlnhofer *et al.* 2006). This may have been due to the lack of specific targeting of the expanded EPCs to ischemic myocardium. One specific strategy to combat these issues involves expanding the native EPC pool with GMCSF-mediated bone marrow stimulation and delivering intramyocardial SDF as a highly specific and localized chemotactic signal for the expanded EPCs (Fig. 3). This approach, studied in a chronic rat model of ischemic cardiomyopathy, up-regulated circulating EPCs, yielded significantly enhanced myocardial endothelial progenitor cell density, increased vasculogenesis, and augmented myocardial function by enhancing perfusion, reversing cellular ischemia, increasing cardiomyocyte viability, and preserving ventricular geometry (Atluri *et al.* 2006, Woo *et al.* 2005). In addition, a minimized peptide analog of SDF containing both the N- and C-termini was shown to induce EPC migration, improve ventricular function after acute MI, and may eventually provide translational advantages over recombinant human SDF-1α (Hiesinger *et al.* 2010).

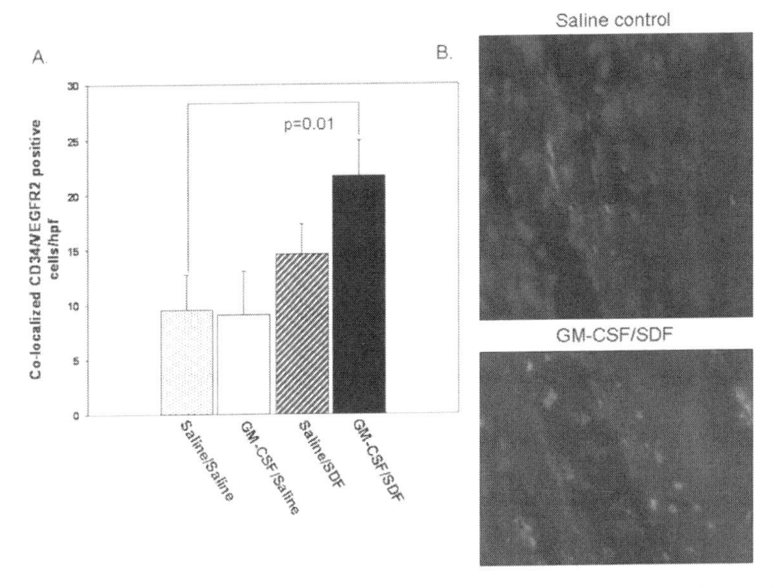

Fig. 3 Effect of stromal cell-derived factor 1α (SDF) and granulocyte-macrophage colony-stimulating factor (GM-CSF) on EPC migration in a rat model of ischemic cardiomyopathy. (a) Graph of the mean number of EPC markers CD34+/VEGFR2+ coexpressing cells per high-power field in hearts from the four groups. (b) Representative immunostains of saline control and SDF/GM-CSF-treated hearts for CD34 (green), VEGFR2 (red; merged = orange), and DAPI nuclei (blue) (reproduced from Woo *et al.*).

Color image of this figure appears in the color plate section at the end of the book.

As the encouraging pre-clinical data mounts, the strategy of EPC mobilization and targeting by angiogenic cytokines has emerged as a viable therapeutic option for ischemic cardiovascular diseases. This therapeutic strategy might be ideal for those patients with chronic myocardial ischemia who have no option for revascularization, patients who cannot tolerate maximally invasive treatment, or as an adjunct to coronary bypass surgery.

CLINICAL TRIALS OF ANGIOGENIC CYTOKINE THERAPY FOR ISCHEMIC HEART DISEASE

Bolstered by the pre-clinical data, a number of Phase I/II clinical trials have been performed to assess the safety, feasibility, and efficacy of angiogenic cytokine therapy for ischemic heart disease (Table 1). These studies have used both recombinant protein therapy and gene therapy as well as numerous different delivery methods including intracoronary, direct intramyocardial injections, and catheter-based endocardial injections targeted with the NOGA electromechanical cardiac mapping system. The vast majority of the clinical trials have investigated various

Table I Clinical trials of angiogenic cytokine therapy for ischemic heart disease

Trial	*Therapy*	*Indication*	*Delivery route*	*Phase*	*Result*
Henry *et al.* 2001	rVEGF	CMI-N/O	IC	I	Well tolerated, improvements in collateral density score.
VIVA Trial (Henry *et al.* 2003)	rVEGF	CMI-N/O	IC + IV	II	No improvement beyond placebo by day 60. By day 120, highdose rVEGF resulted in significant improvement in angina.
Vale *et al.* 2001	$phVEGF_2$	CMI	IM-NOGA	I	Reduction in angina, nitroglycerin consumption, ischemia by electro-mechanical mapping, and improved myocardial perfusion.
Losordo *et al.* 2002	$phVEGF_2$	CMI-N/O	IM-NOGA	I/II	Significant improvement in CCS angina class and strong trends favoring improvements in exercise duration and function.
Losordo *et al.* 1998	$phVEGF_{165}$	CMI-N/O	IM-Thor	I	Well tolerated; significant reduction in angina, reduced ischemia, improved collaterals.
Symes *et al.* 1999	$phVEGF_{165}$	CMI-N/O	IM-Thor	I	Improved collateral and significant symptomatic improvement with phVEGF.
Vale *et al.* 2000	$phVEGF_{165}$	CMI-N/O	IM-Thor	I	Augmentation of perfusion of ischemic myocardium; reduction in ischemic defect size.
Euroinject One (Kastrup *et al.* 2005)	$phVEGF_{165}$	Angina-N/O	IM-NOGA	II	No improvement of stressinduced myocardial perfusion; significantly improved regional wall motion.
Tio *et al.* 2004	$phVEGF_{165}$	CAD/Angina	IM-NOGA	I/II	VEGF therapy reduced myocardial ischemia compared to TMR and medical therapy.
Ripa *et al.* 2006	$phVEGF_{165}$ + G-CSF	CMI	IM-NOGA	II	Increased CD34+ stem cells, no change in perfusion.

Contd...

Contd...

Trial	*Therapy*	*Indication*	*Delivery route*	*Phase*	*Result*
NORTHERN Trial (Stewart *et al.* 2009)	phVEGF$_{165}$	CMI-N/O or single vessel occlusive disease	IM-NOGA	II	No difference between groups.
Rosengart *et al.* 1999	AdenoVEGF$_{121}$	CMI±N/O	IM during CABG or via Thor	I	Improvement of wall motion in the area of vector administration and improvement in angina class.
REVASC Trial (Stewart *et al.* 2006)	AdenoVEGF$_{121}$	Angina-N/O	IM-Thor	II	Significant improvements in exercise time to 1 mm ST-segment depression, total exercise duration, time to moderate angina, and in angina symptoms.
Fuchs *et al.* 2006	AdenoVEGF$_{121}$	CMI-N/O	IM-NOGA	I	Well tolerated.
KAT Trial (Hedman *et al.* 2003)	phVEGF$_{165}$/	AdenoVEGF$_{165}$	CAD for PC	I	IC II No differences in clinical restenosis rate; a significant increase in myocardial perfusion w/ AdVEGF.
Sellke *et al.* 1998	rFGF-2	CAD	IM during CABG	I	Well tolerated.
Ruel *et al.* 2002	rFGF-2	CAD	IM during CABG	I	Significantly less angina, decreased residual myocardial ischemia, and trend toward higher late LVEF with FGF treatment.
Schumacher *et al.* 1998	rFGF	CAD	IM during CABG	I	Increased capillaries by angiography.
Laham *et al.* 1999	rFGF-2	CAD	IM during CABG	I	Well tolerated; Improved myocardial perfusion in high dose group.
Udelson *et al.* 2000	rFGF-2	CMI-N/O	IC or IV	I	Attenuation of stressinduced ischemia and improvement in myocardial perfusion.

Contd...

Contd...

Trial	*Therapy*	*Indication*	*Delivery route*	*Phase*	*Result*
Laham *et al.* 2000	rFGF-2	CMI-N/O	IC	I	Well tolerated; improvement in QOL, exercise tolerance, and regional wall thickening.
Unger *et al.* 2000	rFGF-2	Stable angina	IC	I	Well tolerated.
FIRST Trial (Simons *et al.* 2002)	rFGF-2	CMI-N/O	IC	II	No improvement in exercise tolerance or myocardial perfusion.
AGENT (Grines *et al.* 2002)	AdenoFGF-4	Chronic Angina	IC	I	Well tolerated.
Grines *et al.* 2003	AdenoFGF-4	Chronic Angina	IC	II	Trend towards improved myocardial perfusion with AdFGF-4 treatment.
AGENT 3/4 (Henry *et al.* 2007)	AdenoFGF-4	Chronic Angina	IC	II	Pooled data analysis demonstrated a significant, gender-specific beneficial effect of AdFGF-4 on total ETT time, time to 1 mm ST-segment depression, time to angina in women.
Yuan *et al.* 2009	AdenoHGF	CAD	IM during CABG	I	Well tolerated, possible improved myocardial perfusion with HGF.

permutations of VEGF therapy. While Phase I data has convincingly determined both the safety and viability of VEGF cytokine therapy (as a recombinant protein and adenovirus or plasmid-encoded VEGF) (Fuchs *et al.* 2006, Henry and Abraham, 2000, Henry *et al.* 2001, Rosengart *et al.* 1999, Symes *et al.* 1999), the Phase II results are far less certain and often contradictory.

Losordo *et al.* (2002) completed a Phase I/II study investigating the percutaneous catheter-based gene transfer of $phVEGF_2$ to left ventricular myocardium in a prospective, randomized, double-blind, placebo-controlled, dose-escalating study of inoperable patients with class III or IV angina. Patients were randomized to receive injections of placebo or $phVEGF_2$ in doses of 200 μg, 800 μg, or 2000 μg via NOGA electroanatomical guidance catheter. Compared to placebo, patients receiving $phVEGF_2$ demonstrated a statistically significant improvement in Canadian Cardiovascular Society (CCS) angina class in phVEGF2-treated versus placebo-treated patients and showed strong trends favoring efficacy in other clinical endpoints including exercise duration, functional improvement, and Seattle Angina Questionnaire data. Stewart *et al.* (2006) found similarly encouraging results in the REVASC (Randomized Evaluation of VEGF for Angiogenesis in Severe Coronary disease) trial. This Phase II randomized, controlled study found that patients receiving $AdVEGF_{121}$ (administered by direct intramuscular injection via a minithoracotomy) demonstrated objective improvement in exercise-induced ischemia compared to patients managed with maximum medical therapy. However, subsequent studies by some of the same investigators are far less promising.

The NOGA angiogenesis Revascularization Therapy: assessment by RadioNuclide imaging (NORTHERN) trial (Stewart *et al.* 2009) was a multicenter double-blind, placebo-controlled study of intramyocardial $phVEGF_{165}$ gene therapy versus placebo specifically designed to overcome the major limitations of previous proangiogenic gene therapy trials. A total of 93 patients with chronic myocardial ischemia and no option for revascularization or single vessel occlusion/in stent restenosis based on the coronary anatomy and suitability for standard revascularization procedures were randomized to receive 2,000 μg of VEGF plasmid DNA or placebo (buffered saline) delivered via endocardial injections using a NOGA electroanatomical guidance catheter. The study demonstrated no difference between the VEGF-treated and the placebo groups in myocardial perfusion (assessed by single photon emission tomography (SPECT) imaging), exercise treadmill time, and anginal symptoms at 3 or 6 mon.

The results of clinical trials for FGF angiogenic cytokine therapy are also ambiguous. The FIRST trial (Simons *et al.* 2002) (FGF Initiating RevaScularization Trial) was a Phase II, multicenter, randomized, double-blind, placebo-controlled trial of a single intracoronary infusion of $rFGF_2$ at multiple doses that enrolled 337 patients with chronic myocardial ischemia and no option for revascularization. Despite the fact that $rFGF_2$ reduced angina symptoms after 90 d compared to placebo, no significant differences remained at 180 d due to continued improvement in the placebo group. It should also be noted that hypotension occurred with higher frequency in the highest dose rFGF$_2$ group.

In a much smaller study, Ruel *et al.* (2002) reported long-term benefits in a randomized, double-blind study, in which FGF$_2$ was delivered to the ungraftable myocardial territory of patients concomitantly undergoing coronary artery bypass grafting. Patient follow-up averaged nearly 3 yr and, while mean CCS angina class improved at late follow-up in all groups, patients treated with FGF$_2$ had significantly more freedom from angina recurrence than those treated with placebo. In addition, late SPECT scan revealed a persistent reversible or a new, fixed perfusion defect in the ungraftable territory of 80% of patients who received placebo versus only 11% of patients treated with FGF$_2$. Left ventricular stress perfusion defect scores were lower and a trend toward a higher late left ventricular ejection fraction was noted in FGF$_2$-treated patients.

FGF gene therapy was also extensively studied in the AGENT (Angiogenic GENe Therapy) 3 and 4 trials (Henry *et al.* 2007). These studies enrolled a combined 532 patients suffering from chronic angina in a randomized, double-blind, placebo-controlled fashion to determine whether intracoronary administration of low and high dose Ad5FGF4 could induce angiogenesis and provide a new clinical approach to the treatment of chronic angina pectoris. Two preliminary clinical trials provided evidence that it could improve exercise treadmill test (ETT) time and myocardial perfusion. Enrollment was stopped for both trials when a planned interim analysis of the AGENT-3 trial revealed the study was unlikely to yield a statistically significant result on the primary endpoint of change from baseline in ETT. A pooled data analysis from the two nearly identical trials was performed and demonstrated a significant, gender-specific beneficial effect of Ad5FGF4 on total ETT time, time to 1 mm ST-segment depression, time to angina, and CCS class in women.

There have been very few clinical trials investigating HGF as a potential therapy for ischemic heart disease. Yuan *et al.* (2008) recently conducted a Phase I, open-label, safety and tolerance trial in 18 patients suffering from CAD who received direct multipoint myocardial injections of AdHGF during CABG. The intramyocardial administration of AdHGF was well tolerated and no evidence of systemic or cardiac adverse events was reported.

While therapeutic angiogenesis clearly appears safe and feasible, there has been no definitive answer about its efficacy. The body of clinical data for the use of angiogenic cytokine therapy for ischemic heart disease is large and lacks clear consensus. There is little uniformity between the clinical trials, and many of them suffer from a lack of blinding or randomization. There are many questions to answer before angiogenic cytokines can become a widely used and efficacious treatment. An ideal agent needs to be identified, followed by the ideal form (e.g., protein, naked DNA or viral encoded gene) of said angiogenic cytokine. In addition, the optimal dose of each angiogenic cytokine needs to be determined (Renault and Losordo, 2007).

ADJUNCTIVE ANGIOGENIC STRATEGIES

Angiogenic cytokines are very attractive agents for amplifying native microvasculature in ischemic myocardium. Despite justified optimism, adjunctive therapies that work in concert with or augment these cytokines may be necessary to optimize their angiogenic potential. Possible adjunctive strategies include transmyocardial laser revascularization (TMR), direct progenitor cell transplantation, and bioengineered delivery systems.

There have been a number of proposed mechanisms of action for the functional cardiac benefits and anginal relief seen with TMR therapy including the creation of blood delivering transmural sinusoids, inflammation, and sympathetic denervation (Horvath 2008). However, it seems as if the most likely underlying mechanism for the efficacy of TMR is the stimulation of angiogenic cytokines and subsequent vasculogenesis. The up-regulation of myocardial VEGF, FGF, and transforming growth factor β (and other angiogenic cytokines) following TMR has been repeatedly demonstrated, further bolstering the case for an angiogenic mechanism of action (Atluri *et al.* 2008, Horvath *et al.* 1999, Pelletier *et al.* 1998). Atluri *et al.* used a large animal model of chronic myocardial ischemia to demonstrate that TMR significantly up-regulates potent proangiogenic cytokines nuclear factor κB and angiopoietin-1 and that these angiogenic cytokine elevations were associated with increases in vasculogenesis, myocardial perfusion, viability, and both regional myocardial and global hemodynamic function (Fig. 4). These encouraging findings evoke the possibility of a synergistic approach combining the angiogenic benefits of TMR with concurrent delivery of angiogenic cytokines or essential vasculogenic components (e.g., EPCs).

The isolation, *ex vivo* propagation, and transplantation of EPCs (and other bone marrow–based progenitor cells) by either direct intramyocardial implantation or percutaneous transcatheter delivery has sparked great interest and controversy. Clinical trials to date have often demonstrated statistically significant, yet clinically insignificant functional improvements and will not be discussed in this chapter. However, the potential of these cells to act as a paracrine cytokine producer may prove to be a more efficient means of angiogenic cytokine therapy than the delivery of either recombinant protein or gene transfection. EPCs have been demonstrated to secrete a multitude of angiogenic growth factors including VEGF, HGF, granulocyte colony–stimulating factor (G-CSF), and granulocyte-macrophage colony–stimulating factor (GM-CSF) (Rehman *et al.* 2003). Cell-based cytokine therapy could potentially provide prolonged expression of angiogenic proteins in a non-immunogenic carrier (i.e., through the use of the patient's own cells) and circumvent some of the shortcomings of traditional protein and gene therapies including short protein half-life, the toxicities of various gene vectors, and inflammatory reactions.

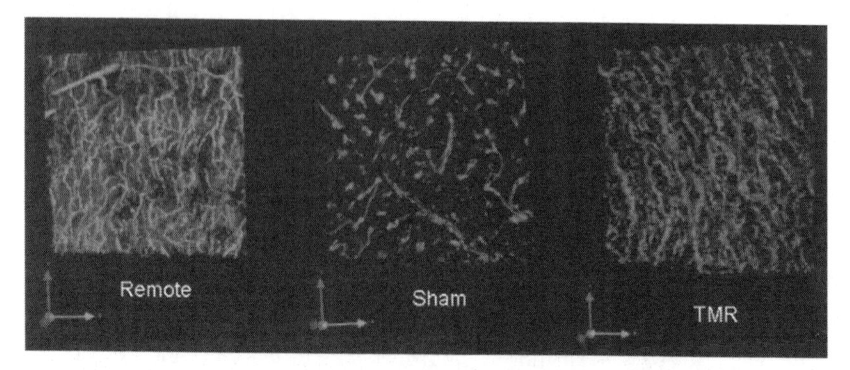

Fig. 4 Effect of transmyocardial laser revascularization (TMR) in a porcine model of ischemic heart failure. Three-dimensional microvascular lectin angiogram demonstrates enhanced myocardial perfusion after therapy with TMR. Representative angiograms from remote myocardium and ischemic myocardium from animals in sham and transmyocardial laser revascularization groups are presented (z-series, 25× oil magnification, Zeiss LSM-510 Meta Confocal Microscope; Carl Zeiss, Inc, Thornwood, NY). There was no difference in remote myocardial perfusion between sham and TMR groups. Perfusion of the ischemic myocardium was significantly increased after TMR. Orientation is presented in lower left corners. Red indicates y-axis; green indicates x-axis; blue indicates z-axis (reproduced from Atluri *et al.*).

Color image of this figure appears in the color plate section at the end of the book.

There have been many efforts to optimize the delivery of angiogenic agents using bioengineered constructs that can maintain cells or proteins in an organized manner for prolonged periods of time and prevent random dispersal within the myocardium. One novel construct used a three-dimensional dermal fibroblast mesh (Anginera™, Theregen Inc., San Francisco, CA, USA)* in an effort to convey high concentrations of angiogenic cytokines in a highly localized, highly specific manner. The mesh is constructed by seeding human dermal fibroblasts from neonatal foreskins onto a knitted vicryl mesh cultured in a medium containing essential growth factor to allow cell proliferation. Pre-clinical experiments demonstrated that scaffold-based three-dimensional human fibroblast cultures were capable of the synthesis and secretion of a number of angiogenic growth factors, including VEGF, FGF, and HGF and improved angiogenesis and myocardial preservation *in vivo* using small animal models of myocardial ischemia (Kellar *et al.* 2001, 2005, Mansbridge *et al.* 1999). Currently, Phase I clinical trials using this dermal fibroblast matrix on non-revascularizable myocardium at the time of CABG are underway. In addition, Frederick *et al.* have taken this strategy a step further by pre-treating a bioengineered endothelial progenitor cell matrix with SDF to "supercharge" its angiogenic effects. This matrix induced borderzone neovasculogenesis, attenuated adverse ventricular remodeling, and preserved ventricular function after MI in a rat model (Frederick *et al.* 2009) (Fig. 5).

The use of trade names is for product identification purposes only and does not imply endorsement.

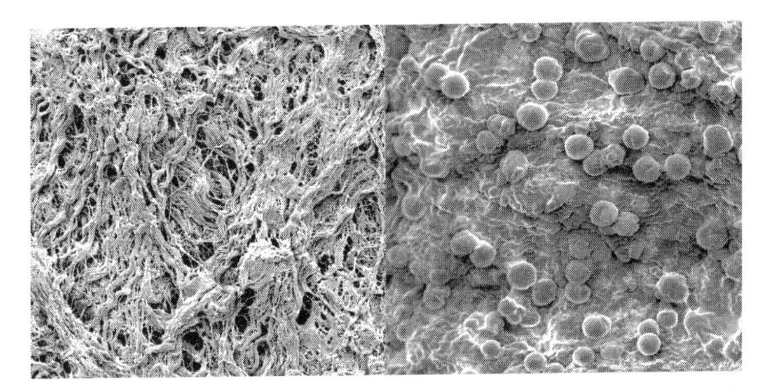

Fig. 5 Scanning electron micrographs of the vitronectin extracellular matrix before (**A**) and after (**B**) seeding with endothelial progenitor cells (EPCs) show densely populated, adherent cells.

CONCLUSION

Angiogenic cytokine therapy for ischemic heart disease has proven to have vast potential in numerous pre-clinical and clinical trials. The ability to replenish myocardial microvasculature could save the lives of the thousands of patients with myocardial ischemia and no option for surgical revascularization or who cannot tolerate a surgical procedure. The encouraging yet often contradictory Phase I and II clinical data for VEGF, FGF, and HGF illustrate the pressing need to optimize the choice, form, and dosing of agent before angiogenic cytokines can become a widely used and efficacious treatment. Continued research efforts will undoubtedly be dedicated to the discovery of novel cytokines, further elucidation of angiogenic mechanisms and upstream regulators, and adjunctive, combination, and multimodality chemokine-mediated angiogenic therapies.

Table 2 Key features of angiogenic cytokine therapy for ischemic heart disease

- Current therapeutic options for ischemic heart disease often fail to prevent disease progression, subsequent adverse ventricular remodeling, and heart failure because they do not restore microvascular perfusion.
- Microvascular dysfunction has been shown to be an important independent predictor of ventricular remodeling as well as reinfarction, heart failure, and death.
- Angiogenic cytokine therapy with FGF, VEGF, SDF, and HGF has demonstrated the ability to increase collateral vessel formation, augment perfusion, limit infarct progression, and enhance cardiac function in a broad spectrum of experimental models.
- Clinical trials have demonstrated that therapeutic angiogenesis is safe and feasible, although there has been no definitive answer about its efficacy.
- The encouraging yet often contradictory Phase I and II clinical data for VEGF, FGF, and HGF illustrate the pressing need to optimize the choice, form, and dosing of agent before angiogenic cytokines can become a widely used and efficacious treatment.
- Adjunctive therapies that work in concert with or augment angiogenic cytokines may be necessary to optimize their angiogenic potential. Potential adjunctive strategies include transmyocardial laser revascularization (TMR), direct progenitor cell transplantation, and bioengineered delivery systems.

Summary Points

- The World Health Organization estimates that more than 17 million people per year die from cardiovascular disease.
- Angiogenic cytokines form vessels via vasculogenesis. They up-regulate and mobilize EPCs from the bone marrow and promote their targeting and incorporation into nascent vessels.
- FGF-2 modulates numerous cardiac cellular functions, including cell proliferation, differentiation, survival, adhesion, migration, motility, apoptosis, vasculogenesis, angiogenesis and blood vessel remodeling. It is expressed in cardiac myocytes and vascular cells.
- VEGF is a potent angiogenic cytokine and is one of the more critical factors involved in angiogenesis and vasculogenesis during embryonic development and adult life. The downstream production of nitric oxide plays a critical angiogenic role.
- PGF is important for angiogenesis under pathological conditions (e.g., myocardial infarction or ischemia) and contributes by affecting endothelial cells, amplifying the angiogenic activity of VEGF, stimulating vessel maturation and stabilization, recruiting inflammatory cells, and mobilizing vascular and hematopoietic stem cells from the bone marrow.
- HGF regulates cell growth, cell motility, and the morphogenesis of a broad variety of cell types and also stimulates endothelial cell growth without inducing the replication of vascular smooth muscle cells.
- SDF is one of the key regulators of hematopoietic stem cell trafficking between the peripheral circulation and bone marrow and has been shown to effect EPC proliferation and mobilization to induce vasculogenesis.
- Expanding the native EPC pool with GMCSF-mediated bone marrow stimulation and delivering intramyocardial SDF as a highly specific and localized chemotactic signal yields significantly enhanced myocardial endothelial progenitor cell density, increased vasculogenesis, and augmented myocardial function by enhancing perfusion, reversing cellular ischemia, increasing cardiomyocyte viability, and preserving ventricular geometry.
- The majority of clinical trials have investigated various permutations of VEGF and FGF therapy. While Phase I data has convincingly determined both the safety and viability of angiogenic cytokine therapy, the Phase II results are far less certain and often contradictory.
- The most likely underlying mechanism for the efficacy of TMR is the stimulation of angiogenic cytokines and subsequent vasculogenesis.
- Adjunctive therapies that work in concert with or augment angiogenic cytokines may be necessary to optimize their potential. Potential adjunctive and include transmyocardial laser revascularization, direct progenitor cell transplantation, and bioengineered delivery systems.

Abbreviations

Adeno	:	adenoviral vector delivered
AGENT	:	Angiogenic Gene Therapy
CABG	:	coronary artery bypass graft
CAD	:	coronary artery disease
CCS	:	Canadian Cardiovascular Society angina class
CMI	:	chronic myocardial ischemia
EC	:	endothelial cell
EPC	:	endothelial progenitor stem cell
ETT	:	exercise treadmill test
FGF	:	fibroblast growth factor
FIRST	:	FGF Initiating RevaScularization Trial
G-CSF	:	granulocyte colony-stimulating factor
GM-CSF	:	granulocyte-macrophage colony-stimulating factor
HGF	:	hepatocyte growth factor
IC	:	intracoronary
IM	:	intramuscular
IV	:	intravenous
KAT	:	Kuopio Angiogenesis Trial
MI	:	myocardial infarction
NO	:	nitric oxide
N/O	:	no option for revascularization
NOGA™	:	electromechanical cardiac mapping system
NORTHERN	:	the NOGA angiogenesis Revascularization Therapy: assessment by RadioNuclide imaging trial
PCI	:	percutaneous coronary intervention
PGF	:	placental growth factor
ph	:	plasmid DNA encoding human gene
QOL	:	quality of life
REVASC	:	Randomized Evaluation of VEGF for Angiogenesis in Severe Coronary disease
SDF	:	stromal cell-derived factor-1α
Thor	:	thoracotomy
TMR	:	transmyocardial laser revascularization
VEGF	:	vascular endothelial growth factor
VIVA	:	vascular endothelial growth factor in ischemia for vascular angiogenesis trial

Definition of Terms

Angiogenesis: New vessel growth from mature, differentiated endothelial cells that are able to migrate and proliferate, forming sprouts from existing vessels.

Clinical trial: Biomedical or health-related research studies in human beings that follow a pre-defined protocol and test an experimental drug or treatment to evaluate its safety, determine a safe dosage range, identify side effects, and confirm its effectiveness.

Endothelial progenitor stem cell: A population of circulating cells with the ability to differentiate into endothelial cells, the cells that make up the lining of blood vessels.

Gene therapy: An experimental technique that uses genes to treat or prevent disease by replacing a mutated gene with a healthy copy of the gene, inactivating a mutated gene, or introducing a new gene into the body to help fight a disease.

Heart failure: A chronic, progressive condition in which the heart is unable to pump enough blood to meet the body's needs for blood and oxygen.

Ischemia: An absolute or relative shortage of the blood supply to an organ resulting in a shortage of oxygen, glucose and other blood-borne fuels.

Microvasculature: The portion of the circulatory system composed of the smallest vessels, such as the capillaries, arterioles, and venules.

Recombinant protein: A protein derived from recombinant DNA.

Vasculogenesis: The homing of bone-marrow–derived endothelial progenitor cells to areas of vascular expansion, where they subsequently differentiate into mature endothelial cells and capillaries *in situ*.

Ventricular remodeling: Changes in the size, shape, and function of the heart to compensate for increased cardiac strain and decreased cardiac function.

References

Araszkiewicz, A. and S. Grajek, M. Lesiak, M. Prech, M. Pyda, M. Janus, and A. Cieslinski. 2006. Effect of impaired myocardial reperfusion on left ventricular remodeling in patients with anterior wall acute myocardial infarction treated with primary coronary intervention. Am. J. Cardiol. 98: 725-728.

Asahara, T. and T. Takahashi, H. Masuda, C. Kalka, D. Chen, H. Iwaguro, Y. Inai, M. Silver, and J.M. Isner. 1999. VEGF contributes to postnatal neovascularization by mobilizing bone marrow-derived endothelial progenitor cells. EMBO J. 18: 3964-3972.

Atluri, P. and G.P. Liao, C.M. Panlilio, V.M. Hsu, M.J. Leskowitz, K.J. Morine, J.E. Cohen, M.F. Berry, E.E. Suarez, D.A. Murphy, W.M.F. Lee, T.J. Gardner, H.L. Sweeney, and Y.J. Woo. 2006. Neovasculogenic therapy to augment perfusion and preserve viability in ischemic cardiomyopathy. Ann. Thorac. Surg. 81: 1728-1736.

Atluri, P. and C.M. Panlilio, G.P. Liao, E.E. Suarez, R.C. McCormick, W. Hiesinger, J.E. Cohen, M.J. Smith, A.B. Patel, W. Feng, and Y.J. Woo. 2008. Transmyocardial revascularization to enhance myocardial vasculogenesis and hemodynamic function. J. Thorac. Cardiovasc. Surg. 135: 283-291.

Atluri, P. and Y.J. Woo. 2008. Pro-angiogenic cytokines as cardiovascular therapeutics: assessing the potential. BioDrugs 22: 209.

Autiero, M. and A. Luttun, M. Tjwa, and P. Carmeliet. 2003. Placental growth factor and its receptor, vascular endothelial growth factor receptor-1: novel targets for stimulation of ischemic tissue revascularization and inhibition of angiogenic and inflammatory disorders. J. Thromb. Haemost. 1: 1356-1370.

Battler, A. and M. Scheinowitz, A. Bor, D. Hasdai, Z. Vered, E. Di Segni, N. Varda-Bloom, D. Nass, S. Engelberg, and M. Eldar. 1993. Intracoronary injection of basic fibroblast growth factor enhances angiogenesis in infarcted swine myocardium. J Am. Coll. Cardiol. 22: 2001-2006.

Bolognese, L. and N. Carrabba, G. Parodi, G.M. Santoro, P. Buonamici, G. Cerisano, and D. Antoniucci. 2004. Impact of microvascular dysfunction on left ventricular remodeling and long-term clinical outcome after primary coronary angioplasty for acute myocardial infarction. Circulation 109: 1121-1126.

Boodhwani, M. and N.R. Sodha, R.J. Laham, and F.W. Sellke. 2006. The future of therapeutic myocardial angiogenesis. Shock 26: 332-341.

Carmeliet, P. and L. Moons, A. Luttun, V. Vincenti, V. Compernolle, M. De Mol, Y. Wu, F. Bono, L. Devy, H. Beck, D. Scholz, T. Acker, T. DiPalma, M. Dewerchin, A. Noel, I. Stalmans, A. Barra, S. Blacher, T. Vandendriessche, A. Ponten, U. Eriksson, K.H. Plate, J.M. Foidart, W. Schaper, D.S. Charnock-Jones, D.J. Hicklin, J.M. Herbert, D. Collen, and M.G. Persico. 2001. Synergism between vascular endothelial growth factor and placental growth factor contributes to angiogenesis and plasma extravasation in pathological conditions. Nat. Med. 7: 575-583.

Chang, L.T. and C.M. Yuen, C.K. Sun, C.J. Wu, J.J. Sheu, S. Chua, K.H. Yeh, C.H. Yang, A.A. Youssef, and H.K. Yip. 2009. Role of stromal cell-derived factor-1alpha, level and value of circulating interleukin-10 and endothelial progenitor cells in patients with acute myocardial infarction undergoing primary coronary angioplasty. Circ. J. 73: 1097-1104.

De La Luz Sierra, M. and F. Yang, M. Narazaki, O. Salvucci, D. Davis, R. Yarchoan, H.H. Zhang, H. Fales, and G. Tosato. 2004. Differential processing of stromal-derived factor-1alpha and stromal-derived factor-1beta explains functional diversity. Blood 103: 2452-2459.

Detillieux, K.A. and F. Sheikh, E. Kardami, and P.A. Cattini. 2003. Biological activities of fibroblast growth factor-2 in the adult myocardium. Cardiovasc. Res. 57: 8-19.

Fernandez, B. and A. Buehler, S. Wolfram, S. Kostin, G. Espanion, W.M. Franz, H. Niemann, P.A. Doevendans, W. Schaper, and R. Zimmermann. 2000. Transgenic myocardial overexpression of fibroblast growth factor-1 increases coronary artery density and branching. Circ. Res. 87: 207-213.

Frederick, J.R. and J.R. Fitzpatrick III, R.C. McCormick, D.A. Harris, A.Y. Kim, J.R. Muenzer, A. Murthy, P.F. Hsiao, A.D. Prystowsky, M.J. Smith, J. Cohen, P. Atluri, and Y.J. Woo 2009. Abstract 4221: Stromal cell-derived factor-1 {alpha} activation of tissue engineered endothelial progenitor stem cell matrix enhances ventricular function after myocardial infarction by inducing neovasculogenesis. Circulation 120: S928b-.

Frederick, J.R. and J.R. Fitzpatrick, 3rd, R.C. McCormick, D.A. Harris, A.Y. Kim, J.R. Muenzer, N. Marotta, M.J. Smith, J.E. Cohen, W. Hiesinger, P. Atluri, and Y.J. Woo. 2010. Stromal cell-derived factor-l alpha activation of tissue-engineered endothelial progenitor cell matrix enhances ventricular function after myocardial infarction by inducing neovasculogenesis. Circulation. Sep 14; 122(11 Supplement): S107-S117.

Fuchs, S. and N. Dib, B.M. Cohen, P. Okubagzi, E.B. Diethrich, A. Campbell, J. Macko, P.D. Kessler, H.S. Rasmussen, S.E. Epstein, and R. Kornowski. 2006. A randomized, double-blind, placebo-controlled, multicenter, pilot study of the safety and feasibility of catheter-based intramyocardial injection of AdVEGF121 in patients with refractory advanced coronary artery disease. Catheter Cardiovasc. Interv. 68: 372-378.

Funatsu, T. and Y. Sawa, S. Ohtake, T. Takahashi, G. Matsumiya, N. Matsuura, T. Nakamura, and H. Matsuda. 2002. Therapeutic angiogenesis in the ischemic canine heart induced by myocardial injection of naked complementary DNA plasmid encoding hepatocyte growth factor. J. Thorac. Cardiovasc. Surg. 124: 1099-1105.

Giordano, F.J. and P. Ping, M.D. McKirnan, S. Nozaki, A.N. DeMaria, W.H. Dillmann, O. Mathieu-Costello, and H.K. Hammond. 1996. Intracoronary gene transfer of fibroblast growth factor-5 increases blood flow and contractile function in an ischemic region of the heart. Nat. Med. 2: 534-539.

Grines, C.L. and M.W. Watkins, G. Helmer, W. Penny, J. Brinker, J.D. Marmur, A. West, J.J. Rade, P. Marrott, H.K. Hammond, and R.L. Engler. 2002. Angiogenic Gene Therapy (AGENT) trial in patients with stable angina pectoris. Circulation 105: 1291-1297.

Grines, C.L. and M.W. Watkins, .J.J. Mahmarian, A.E. Iskandrian, J.J. Rade, P. Marrott, C. Pratt, and N. Kleiman. 2003. A randomized, double-blind, placebo-controlled trial of Ad5FGF-4 gene therapy and its effect on myocardial perfusion in patients with stable angina. J. Am. Coll. Cardiol. 42: 1339-1347.

Harada, K. and W. Grossman, M. Friedman, E.R. Edelman, P.V. Prasad, C.S. Keighley, W.J. Manning, F.W. Sellke, and M. Simons. 1994. Basic fibroblast growth factor improves myocardial function in chronically ischemic porcine hearts. J. Clin. Invest. 94: 623-630.

Hattori, K. and S. Dias, B. Heissig, N.R. Hackett, D. Lyden, M. Tateno, D.J. Hicklin, Z. Zhu, L. Witte, R.G. Crystal, M.A. Moore, and S. Rafii. 2001. Vascular endothelial growth factor and angiopoietin-1 stimulate postnatal hematopoiesis by recruitment of vasculogenic and hematopoietic stem cells. J. Exp. Med. 193: 1005-1014.

Hedman, M. and J. Hartikainen, M. Syvanne, J. Stjernvall, A. Hedman, A. Kivela, E. Vanninen, H. Mussalo, E. Kauppila, S. Simula, O. Narvanen, A. Rantala, K. Peuhkurinen, M.S. Nieminen, M. Laakso, and S. Yla-Herttuala. 2003. Safety and feasibility of catheter-based local intracoronary vascular endothelial growth factor gene transfer in the prevention of postangioplasty and in-stent restenosis and in the treatment of chronic myocardial ischemia: phase II results of the Kuopio Angiogenesis Trial (KAT). Circulation 107: 2677-2683.

Henry, T.D. and J.A. Abraham. 2000. Review of Preclinical and Clinical Results with Vascular Endothelial Growth Factors for Therapeutic Angiogenesis. Curr. Interv. Cardiol. Rep. 2: 228-241.

Henry, T.D. and B.H. Annex, G.R. McKendall, M.A. Azrin, J.J. Lopez, F.J. Giordano, P.K. Shah, J.T. Willerson, R.L. Benza, D.S. Berman, C.M. Gibson, A. Bajamonde, A.C. Rundle, J. Fine, and E.R. McCluskey. 2003. The VIVA trial: Vascular endothelial growth factor in Ischemia for vascular angiogenesis. Circulation 107: 1359-1365.

Henry, T.D. and C.L. Grines, M.W. Watkins, N. Dib, G. Barbeau, R. Moreadith, T. Andrasfay, and R.L. Engler. 2007. Effects of Ad5FGF-4 in patients with angina: an analysis of pooled data from the AGENT-3 and AGENT-4 trials. J. Am. Coll. Cardiol. 50: 1038-1046.

Henry, T.D. and K. Rocha-Singh, J.M. Isner, D.J. Kereiakes, F.J. Giordano, M. Simons, D.W. Losordo, R.C. Hendel, R.O. Bonow, S.M. Eppler, T.F. Zioncheck, E.B. Holmgren, and

E.R. McCluskey. 2001. Intracoronary administration of recombinant human vascular endothelial growth factor to patients with coronary artery disease. Am. Heart J. 142: 872-880.

Hiesinger, W. and J.R. Frederick, P. Atluri, R.C. McCormick, N. Marotta, J.R. Muenzer, and Y.J. Woo. 2010. Spliced stromal cell-derived factor-1α analog stimulates endothelial progenitor cell migration and improves cardiac function in a dose-dependent manner after myocardial infarction. J Thorac Cardiovasc Surg. 140:1174-1180.

Horvath, K.A. 2008. Transmyocardial laser revascularization. J. Card. Surg. 23: 266-276.

Horvath, K.A. and E. Chiu, D.C. Maun, J.W. Lomasney, R. Greene, W.H. Pearce, and D.A. Fullerton. 1999. Up-regulation of vascular endothelial growth factor mRNA and angiogenesis after transmyocardial laser revascularization. Ann. Thorac. Surg. 68: 825-829.

Hughes, G.C. and S.S. Biswas, B. Yin, R.E. Coleman, T.R. DeGrado, C.K. Landolfo, J.E. Lowe, B.H. Annex, and K.P. Landolfo. 2004. Therapeutic angiogenesis in chronically ischemic porcine myocardium: comparative effects of bFGF and VEGF. Ann. Thorac. Surg. 77: 812-818.

Ince, H. and M. Petzsch, H.D. Kleine, H. Eckard, T. Rehders, D. Burska, S. Kische, M. Freund, and C.A. Nienaber. 2005. Prevention of left ventricular remodeling with granulocyte colony-stimulating factor after acute myocardial infarction: final 1-year results of the Front-Integrated Revascularization and Stem Cell Liberation in Evolving Acute Myocardial Infarction by Granulocyte Colony-Stimulating Factor (FIRSTLINE-AMI) Trial. Circulation 112: I73-80.

Jayasankar, V. and Y.J. Woo, L.T. Bish, T.J. Pirolli, S. Chatterjee, M.F. Berry, J. Burdick, T.J. Gardner, and H.L. Sweeney. 2003. Gene transfer of hepatocyte growth factor attenuates postinfarction heart failure. Circulation 108 Suppl 1, II230-236.

Jayasankar, V. and Y.J. Woo, T.J. Pirolli, L.T. Bish, M.F. Berry, J. Burdick, T.J. Gardner, and H.L. Sweeney. 2005. Induction of angiogenesis and inhibition of apoptosis by hepatocyte growth factor effectively treats postischemic heart failure. J. Card. Surg. 20: 93-9101.

Jin, H. and J.M. Wyss, R. Yang, and R. Schwall. 2004. The therapeutic potential of hepatocyte growth factor for myocardial infarction and heart failure. Curr. Pharm. Des. 10: 2525-2533.

Kalka, C. and H. Masuda, T. Takahashi, R. Gordon, O. Tepper, E. Gravereaux, A. Pieczek, H. Iwaguro, S.I. Hayashi, J.M. Isner, and T. Asahara. 2000a. Vascular endothelial growth factor(165) gene transfer augments circulating endothelial progenitor cells in human subjects. Circ. Res. 86: 1198-1202.

Kalka, C. and H. Tehrani, B. Laudenberg, P.R. Vale, J.M. Isner, T. Asahara, and J.F. Symes. 2000b. VEGF gene transfer mobilizes endothelial progenitor cells in patients with inoperable coronary disease. Ann. Thorac. Surg. 70: 829-834.

Kang, H.J. and H.Y. Lee, S.H. Na, S.A. Chang, K.W. Park, H.K. Kim, S.Y. Kim, H.J. Chang, W. Lee, W.J. Kang, B.K. Koo, Y.J. Kim, D.S. Lee, D.W. Sohn, K.S. Han, B.H. Oh, Y.B. Park, and H.S. Kim. 2006. Differential effect of intracoronary infusion of mobilized peripheral blood stem cells by granulocyte colony-stimulating factor on left ventricular function and remodeling in patients with acute myocardial infarction versus old myocardial infarction: the MAGIC Cell-3-DES randomized, controlled trial. Circulation 114: I145-151.

Kastrup, J. and E. Jorgensen, A. Ruck, K. Tagil, D. Glogar, W. Ruzyllo, H.E. Botker, D. Dudek, V. Drvota, B. Hesse, L. Thuesen, P. Blomberg, M. Gyongyosi, and C. Sylven.

2005. Direct intramyocardial plasmid vascular endothelial growth factor-A165 gene therapy in patients with stable severe angina pectoris A randomized double-blind placebo-controlled study: the Euroinject One trial. J. Am. Coll. Cardiol. 45: 982-988.

Kellar, R.S. and L.K. Landeen, B.R. Shepherd, G.K. Naughton, A. Ratcliffe, and S.K. Williams. 2001. Scaffold-based three-dimensional human fibroblast culture provides a structural matrix that supports angiogenesis in infarcted heart tissue. Circulation 104: 2063-2068.

Kellar, R.S. and B.R. Shepherd, D.F. Larson, G.K. Naughton, and S.K. Williams. 2005. Cardiac patch constructed from human fibroblasts attenuates reduction in cardiac function after acute infarct. Tissue Eng. 11: 1678-1687.

Kolakowski, S. and M.F. Berry, P. Atluri, T. Grand, O. Fisher, M.A. Moise, J. Cohen, V. Hsu, and Y.J. Woo. 2006. Placental growth factor provides a novel local angiogenic therapy for ischemic cardiomyopathy. J. Card. Surg. 21: 559-564.

Kucia, M. and B. Dawn, G. Hunt, Y. Guo, M. Wysoczynski, M. Majka, J. Ratajczak, F. Rezzoug, S.T. Ildstad, R. Bolli, and M.Z. Ratajczak. 2004. Cells expressing early cardiac markers reside in the bone marrow and are mobilized into the peripheral blood after myocardial infarction. Circ. Res. 95: 1191-1199.

Kuethe, F. and H.R. Figulla, M. Herzau, M. Voth, M. Fritzenwanger, T. Opfermann, K. Pachmann, A. Krack, H.G. Sayer, D. Gottschild, and G.S. Werner. 2005. Treatment with granulocyte colony-stimulating factor for mobilization of bone marrow cells in patients with acute myocardial infarction. Am. Heart J. 150: 115.

Laham, R.J. and N.A. Chronos, M. Pike, M.E. Leimbach, J.E. Udelson, J.D. Pearlman, R.I. Pettigrew, M.J. Whitehouse, C. Yoshizawa, and M. Simons. 2000a. Intracoronary basic fibroblast growth factor (FGF-2) in patients with severe ischemic heart disease: results of a phase I open-label dose escalation study. J. Am. Coll. Cardiol. 36: 2132-2139.

Laham, R.J. and M. Rezaee, M. Post, D. Novicki, F.W. Sellke, J.D. Pearlman, M. Simons, and D. Hung. 2000b. Intrapericardial delivery of fibroblast growth factor-2 induces neovascularization in a porcine model of chronic myocardial ischemia. J Pharmacol. Exp. Ther. 292: 795-802.

Laham, R.J. and F.W. Sellke, E.R. Edelman, J.D. Pearlman, J.A. Ware, D.L. Brown, J.P. Gold, and M. Simons. 1999. Local perivascular delivery of basic fibroblast growth factor in patients undergoing coronary bypass surgery: results of a phase I randomized, double-blind, placebo-controlled trial. Circulation 100: 1865-1871.

Liu, Y. and L. Sun, Y. Huan, H. Zhao, and J. Deng. 2006. Effects of basic fibroblast growth factor microspheres on angiogenesis in ischemic myocardium and cardiac function: analysis with dobutamine cardiovascular magnetic resonance tagging. Eur. J. Cardiothorac. Surg. 30: 103-107.

Lloyd-Jones, D. and R. Adams, M. Carnethon, G. De Simone, T.B. Ferguson, K. Flegal, E. Ford, K. Furie, A. Go, K. Greenlund, N. Haase, S. Hailpern, M. Ho, V. Howard, B. Kissela, S. Kittner, D. Lackland, L. Lisabeth, A. Marelli, M. McDermott, J. Meigs, D. Mozaffarian, G. Nichol, C. O'Donnell, V. Roger, W. Rosamond, R. Sacco, P. Sorlie, R. Stafford, J. Steinberger, T. Thom, S. Wasserthiel-Smoller, N. Wong, J. Wylie-Rosett, and Y. Hong. 2009. Heart disease and stroke statistics--2009 update: a report from the American Heart Association Statistics Committee and Stroke Statistics Subcommittee. Circulation 119: e21-181.

Losordo, D.W. and P.R. Vale, R.C. Hendel, C.E. Milliken, F.D. Fortuin, N. Cummings, R.A. Schatz, T. Asahara, J.M. Isner, and R.E. Kuntz. 2002. Phase 1/2 placebo-controlled, double-blind, dose-escalating trial of myocardial vascular endothelial growth factor

2 gene transfer by catheter delivery in patients with chronic myocardial ischemia. Circulation 105: 2012-2018.

Losordo, D.W. and P.R. Vale, J.F. Symes, C.H. Dunnington, D.D. Esakof, M. Maysky, A.B. Ashare, K. Lathi, and J.M. Isner. 1998. Gene therapy for myocardial angiogenesis: initial clinical results with direct myocardial injection of phVEGF165 as sole therapy for myocardial ischemia. Circulation 98: 2800-2804.

Mansbridge, J.N. and K. Liu, R.E. Pinney, R. Patch, A. Ratcliffe, and G.K. Naughton. 1999. Growth factors secreted by fibroblasts: role in healing diabetic foot ulcers. Diabetes Obes. Metab. 1: 265-279.

Mohle, R. and F. Bautz, S. Rafii, M.A. Moore, W. Brugger, and L. Kanz. 1998. The chemokine receptor CXCR-4 is expressed on CD34+ hematopoietic progenitors and leukemic cells and mediates transendothelial migration induced by stromal cell-derived factor-1. Blood 91: 4523-4530.

Pelletier, M.P. and A. Giaid, S. Sivaraman, J. Dorfman, C.M. Li, A. Philip, and R.C. Chiu. 1998. Angiogenesis and growth factor expression in a model of transmyocardial revascularization. Ann. Thorac. Surg. 66: 12-18.

Pillarisetti, K. and S.K. Gupta. 2001. Cloning and relative expression analysis of rat stromal cell derived factor-1 (SDF-1)1: SDF-1 alpha mRNA is selectively induced in rat model of myocardial infarction. Inflammation 25: 293-300.

Rehman, J. and J. Li, C.M. Orschell, and K.L. March. 2003. Peripheral blood "endothelial progenitor cells" are derived from monocyte/macrophages and secrete angiogenic growth factors. Circulation 107: 1164-1169.

Renault, M.A. and D.W. Losordo. 2007. Therapeutic myocardial angiogenesis. Microvasc. Res. 74: 159-171.

Ripa, R.S. and E. Jorgensen, Y. Wang, J.J. Thune, J.C. Nilsson, L. Sondergaard, H.E. Johnsen, L. Kober, P. Grande, and J. Kastrup. 2006a. Stem cell mobilization induced by subcutaneous granulocyte-colony stimulating factor to improve cardiac regeneration after acute ST-elevation myocardial infarction: result of the double-blind, randomized, placebo-controlled stem cells in myocardial infarction (STEMMI) trial. Circulation 113: 1983-1992.

Ripa, R.S. and Y. Wang, E. Jorgensen, H.E. Johnsen, B Hesse, and J. Kastrup. 2006b. Intramyocardial injection of vascular endothelial growth factor-A165 plasmid followed by granulocyte-colony stimulating factor to induce angiogenesis in patients with severe chronic ischaemic heart disease. Eur. Heart J. 27: 1785-1792.

Roncal, C. and I. Buysschaert, E. Chorianopoulos, M. Georgiadou, O. Meilhac, M. Demol, J.B. Michel, S. Vinckier, L. Moons, and P. Carmeliet. 2008. Beneficial effects of prolonged systemic administration of PlGF on late outcome of post-ischaemic myocardial performance. J. Pathol. 216: 236-244.

Rosengart, T.K. and L.Y. Lee, S.R. Patel, T.A. Sanborn, M. Parikh, G.W. Bergman, R. Hachamovitch, M. Szulc, P.D. Kligfield, P.M. Okin, R.T. Hahn, R.B. Devereux, M.R. Post, N.R. Hackett, T. Foster, T.M. Grasso, M.L. Lesser, O.W. Isom, and R.G. Crystal. 1999. Angiogenesis gene therapy: phase I assessment of direct intramyocardial administration of an adenovirus vector expressing VEGF121 cDNA to individuals with clinically significant severe coronary artery disease. Circulation 100: 468-474.

Ruel, M. and R.J. Laham, J.A. Parker, M.J. Post, J.A. Ware, M. Simons, and F.W. Sellke. 2002. Long-term effects of surgical angiogenic therapy with fibroblast growth factor 2 protein. J. Thorac. Cardiovasc. Surg. 124: 28-34.

Ruel, M. and G.F. Wu, T.A. Khan, P. Voisine, C. Bianchi, J. Li, J. Li, R.J. Laham, and F.W. Sellke. 2003. Inhibition of the cardiac angiogenic response to surgical FGF-2 therapy in a Swine endothelial dysfunction model. Circulation 108 (Suppl 1): 335-340.

Sakakibara, Y. and K. Tambara, G. Sakaguchi, F. Lu, M. Yamamoto, K. Nishimura, Y. Tabata, and M. Komeda. 2003. Toward surgical angiogenesis using slow-released basic fibroblast growth factor. Eur. J. Cardiothorac. Surg. 24: 105-111.

Sato, K. and R.J. Laham, J.D. Pearlman, D. Novicki, F.W. Sellke, M. Simons, and M.J. Post. 2000. Efficacy of intracoronary versus intravenous FGF-2 in a pig model of chronic myocardial ischemia. Ann. Thorac. Surg. 70: 2113-2118.

Schumacher, B. and T. Stegmann, and P. Pecher. 1998. The stimulation of neoangiogenesis in the ischemic human heart by the growth factor FGF: first clinical results. J. Cardiovasc. Surg. (Torino) 39: 783-789.

Simons, M. and B.H. Annex, R.J. Laham, N. Kleiman, T. Henry, H. Dauerman, J.E. Udelson, E.V. Gervino, M. Pike, M.J. Whitehouse, T. Moon, and N.A. Chronos. 2002. Pharmacological treatment of coronary artery disease with recombinant fibroblast growth factor-2: double-blind, randomized, controlled clinical trial. Circulation 105: 788-793.

Stewart, D.J. and J.D. Hilton, J.M.O. Arnold, J. Gregoire, A. Rivard, S.L. Archer, F. Charbonneau, E. Cohen, M. Curtis, C.E. Buller, F.O. Mendelsohn, N. Dib, P. Page, J. Ducas, S. Plante, J. Sullivan, J. Macko, C. Rasmussen, P.D. Kessler, and H.S. Rasmussen. 2006. Angiogenic gene therapy in patients with nonrevascularizable ischemic heart disease: a phase 2 randomized, controlled trial of AdVEGF(121) (AdVEGF121) versus maximum medical treatment. Gene Ther. 13: 1503-1511.

Stewart, D.J. and M.J. Kutryk, D. Fitchett, M. Freeman, N. Camack, Y. Su, A. Della Siega, L. Bilodeau, J.R. Burton, G. Proulx, and S. Radhakrishnan. 2009. VEGF gene therapy fails to improve perfusion of ischemic myocardium in patients with advanced coronary disease: results of the NORTHERN trial. Mol. Ther. 17: 1109-1115.

Symes, J.F. and D.W. Losordo, P.R. Vale, K.G. Lathi, D.D. Esakof, M. Mayskiy, and J.M. Isner. 1999. Gene therapy with vascular endothelial growth factor for inoperable coronary artery disease. Ann. Thorac. Surg. 68: 830-836.

Tio, R.A. and E.S. Tan, G.A. Jessurun, N. Veeger, P.L. Jager, R.H. Slart, R.M. de Jong, J. Pruim, G.A. Hospers, A.T. Willemsen, M.J. de Jongste, A.J. van Boven, D.J. van Veldhuisen, and F. Zijlstra. 2004. PET for evaluation of differential myocardial perfusion dynamics after VEGF gene therapy and laser therapy in end-stage coronary artery disease. J. Nucl. Med. 45: 1437-1443.

Toyota, E. and T. Matsunaga, and W.M. Chilian. 2004. Myocardial angiogenesis. Mol. Cell. Biochem. 264: 35-44.

Udelson, J.E. and V. Dilsizian, R.J. Laham, N. Chronos, J. Vansant, M. Blais, J.R. Galt, M. Pike, C. Yoshizawa, and M. Simons. 2000. Therapeutic angiogenesis with recombinant fibroblast growth factor-2 improves stress and rest myocardial perfusion abnormalities in patients with severe symptomatic chronic coronary artery disease. Circulation 102: 1605-1610.

Unger, E.F. and L. Goncalves, S.E. Epstein, E.Y. Chew, C.B. Trapnell, R.O. Cannon, 3rd, and A.A. Quyyumi. 2000. Effects of a single intracoronary injection of basic fibroblast growth factor in stable angina pectoris. Am. J. Cardiol. 85: 1414-1419.

Vale, P.R. and D.W. Losordo, C.E. Milliken, M. Maysky, D.D. Esakof, J.F. Symes and J.M. Isner. 2000. Left ventricular electromechanical mapping to assess efficacy of phVEGF(165) gene transfer for therapeutic angiogenesis in chronic myocardial ischemia. Circulation 102: 965-974.

Vale, P.R. and D.W. Losordo, C.E. Milliken, M.C. McDonald, L.M. Gravelin, C.M. Curry, D.D. Esakof, M. Maysky, J.F. Symes and J.M. Isner. 2001. Randomized, single-blind, placebo-controlled pilot study of catheter-based myocardial gene transfer for therapeutic angiogenesis using left ventricular electromechanical mapping in patients with chronic myocardial ischemia. Circulation 103: 2138-2143.

Vera Janavel, G. and A. Crottogini, P. Cabeza Meckert, L. Cuniberti, A. Mele, M. Papouchado, N. Fernandez, A. Bercovich, M. Criscuolo, C. Melo and R. Laguens. 2006. Plasmid-mediated VEGF gene transfer induces cardiomyogenesis and reduces myocardial infarct size in sheep. Gene Ther. 13: 1133-1142.

Woo, Y.J. and T.J. Grand, M.F. Berry, P. Atluri, M.A. Moise, V.M. Hsu, J. Cohen, O. Fisher, J. Burdick, M. Taylor, S. Zentko, G. Liao, M. Smith, S. Kolakowski, V. Jayasankar, T.J. Gardner and H.L. Sweeney. 2005. Stromal cell-derived factor and granulocyte-monocyte colony-stimulating factor form a combined neovasculogenic therapy for ischemic cardiomyopathy. J. Thorac. Cardiovasc. Surg. 130: 321-329.

Yamaguchi, J. and K.F. Kusano, O. Masuo, A. Kawamoto, M. Silver, S. Murasawa, M. Bosch-Marce, H. Masuda, D.W. Losordo, J.M. Isner and T. Asahara. 2003. Stromal cell-derived factor-1 effects on ex vivo expanded endothelial progenitor cell recruitment for ischemic neovascularization. Circulation 107: 1322-1328.

Yanagisawa-Miwa, A. and Y. Uchida, F. Nakamura, T. Tomaru, H. Kido, T. Kamijo, T. Sugimoto, K. Kaji, M. Utsuyama and C. Kurashima. 1992. Salvage of infarcted myocardium by angiogenic action of basic fibroblast growth factor. Science 257: 1401-1403.

Yang, Z.J. and B. Chen, Z. Sheng, D.G. Zhang, E.Z. Jia, W. Wang, D.C. Ma, T.B. Zhu, L.S. Wang, C.J. Li, H. Wang, K.J. Cao and W.Z. Ma. 2009. Improvement of heart function in postinfarct heart failure swine models after hepatocyte growth factor gene transfer: comparison of low-, medium- and high-dose groups. Mol. Biol. Rep. 2010 Apr; 37(4): 2075-2081.

Yoon, Y.S. and I.A. Johnson, J.S. Park, L. Diaz and D.W. Losordo. 2004. Therapeutic myocardial angiogenesis with vascular endothelial growth factors. Mol. Cell Biochem. 264: 63-74.

Yuan, B. and Z. Zhao, Y.R. Zhang, C.T. Wu, W.G. Jin, S. Zhao, W. Wang, Y.Y. Zhang, X.L. Zhu, L.S. Wang and J. Huang. 2008. Short-term safety and curative effect of recombinant adenovirus carrying hepatocyte growth factor gene on ischemic cardiac disease. In Vivo 22: 629-632.

Zohlnhofer, D. and I. Ott, J. Mehilli, K. Schomig, F. Michalk, T. Ibrahim, G. Meisetschlager, J. von Wedel, H. Bollwein, M. Seyfarth, J. Dirschinger, C. Schmitt, K. Schwaiger, A. Kastrati and A. Schomig. 2006. Stem cell mobilization by granulocyte colony-stimulating factor in patients with acute myocardial infarction: a randomized controlled trial. JAMA 295: 1003-1010.

CHAPTER 18

Cytokines and Diabetes

Kate L. Graham and Helen E. Thomas*
St Vincent's Institute, 41 Victoria Parade, Fitzroy, VIC, 3065, Australia

ABSTRACT

Type 1 diabetes (T1D) is autoimmune-mediated destruction of pancreatic beta cells resulting in insulin deficiency. Cytokines are vital in diabetes development. They drive the development of autoreactive T cells and are important for activation and maintenance of the autoimmune response. Genetic and cell intrinsic defects in patients with T1D and in animal models of the disease lead to dysregulation of cytokine production and responses, which contributes to loss of immune tolerance and destruction of beta cells. Pro-inflammatory cytokines are produced locally in the islet where they induce expression of thousands of genes, many of which promote islet inflammation. IL-1, TNF and IFNγ are toxic to beta cells *in vitro* and have been proposed to contribute to beta cell destruction by mechanisms involving free radical production and endoplasmic reticulum stress. However, *in vivo*, pro-inflammatory cytokines are likely to play an immunomodulatory role, increasing beta-cell recognition by T cells and promoting infiltration of macrophages and lymphocytes into the islets. Studying the role of cytokines in T1D has led to identification of important pathways in disease pathogenesis that could be specifically blocked to prevent diabetes.

INTRODUCTION

Type 1 diabetes (T1D) is an autoimmune disease diagnosed mainly in children (Table 1). T1D is caused by autoimmune destruction of insulin-producing pancreatic beta cells, resulting in the requirement for lifelong insulin therapy, increased risk of complications including heart failure, kidney failure and vascular disease, and reduced life expectancy.

*Corresponding author

Table 1 Key features of type 1 diabetes

- T1D is a common paediatric disease with increasing incidence.
- T1D is an autoimmune disease resulting from T cell-mediated beta-cell destruction, which develops in genetically susceptible individuals.
- T1D reflects a loss of immune tolerance caused either by a normal immune response to altered self or a loss of the normal mechanisms that prevent responses to self.
- T1D is characterized by a mononuclear cell infiltration in the pancreatic islets called insulitis, consisting of macrophages, DCs, T and B cells.
- The presence of autoantibodies against islet cell antigens is the first indication of the onset of autoimmunity.
- The strongest genetic association for T1D is within the HLA region. Polymorphisms in the insulin and PTPN22 phosphatases genes are also important and contribute about 10% genetic risk.

T1D is characterized by development of immune cell infiltration in the islets, called insulitis, consisting of monocytes, B and T lymphocytes. Insulitis is observed in mouse models of T1D and in pancreas samples from patients. In humans, autoimmunity is first indicated by the presence of autoantibodies against islet cell antigens (Pietropaolo *et al.* 2008).

T1D is a polygenic disease, with strongest association within the HLA region. In addition to HLA genes, many other genetic loci have been implicated in diabetes risk, including polymorphisms of the insulin gene and the lymphocyte specific phosphatase PTPN22, which together contribute around 10% of genetic risk.

The increased frequency of T1D over the last 50 years suggests the influence of environmental factors. Specific environmental factors remain largely unknown but may include infant diet, vitamin D deficiency, viruses and hygiene. How these factors contribute to disease remains unclear; however, it is thought that they may trigger initiation of autoimmunity, which is then accompanied by aberrant immune responses.

Much of what we know about T1D comes from study of rodent models including the non-obese diabetic (NOD) mouse, the bio-breeding (BB) rat, and transgenic or chemically induced diabetes models (Table 2). In the NOD mouse there are several

Table 2 Some of the most common mouse models of type 1 diabetes

Mouse model	*Antigen*	*Diabetes onset*	*Diabetes incidence*
NOD	Multiple	90-120 d	80% in females
CD8-dependent:			
NOD8.3	IGRP	40-50 d	80-100%
LCMV-GP	LCMV glycoprotein	14 d after LCMV infection	100%
CD4-dependent:			
BDC2.5	Chromogranin	N/A	0% in NOD
		21 d in NOD*scid*	100% in NOD*scid*
NOD4.1	Unknown	30-50 d	80-100%

N/A: not applicable.

immune aberrations that are thought to contribute to autoimmunity, including defects in central and peripheral tolerance, defective macrophage maturation and function, reduced NK cell activity, defects in NKT cells and regulatory T (Treg) cells. Many of these have also been demonstrated in patients with T1D.

T1D is a loss of immune tolerance, which results in infiltration of self-reactive T cells into the pancreatic islets and their ultimate destruction. Cytokines play a major part in development of diabetes (Fig. 1). Pro-inflammatory cytokines including interferons, interleukin-1 (IL-1) and tumor necrosis factor (TNF) are found in the islet lesion and may directly contribute to beta-cell death. Cytokine production in NOD mice may alter the balance between Th1, Th2 and Th17 responses. In addition, T1D results from a failure of immune regulation. Cytokines affect the development of Treg cell populations, and intrinsic differences between NOD mice and other strains may contribute to the failure of Treg cells.

ACTIVATION AND REGULATION OF THE IMMUNE SYSTEM IN TYPE 1 DIABETES

T1D pathogenesis involves many different cell types of the immune system. Dendritic cells (DCs) present beta-cell antigens to autoreactive $CD4^+$ and $CD8^+$ T cells. Activated $CD8^+$ T cells, known as cytotoxic T lymphocytes (CTL), directly kill beta cells in a well-characterized manner. $CD4^+$ T cells have varied roles in disease development that reflect the distinct lineages, Treg, Th1, Th2, and Th17, that develop in response to both antigen and cytokine stimulation (Fig. 2). Extensive research has linked the progression of diabetes to an imbalance between immunoregulation, Treg and Th2 subsets and effector function, Th1 and Th17 subsets.

Interleukin-12

The differentiation of naïve CD4+ T cells to Treg or effector T cells requires antigen presentation and the appropriate cytokine milieu. In the case of Th1 cells IL-12 is required to stimulate differentiation. DCs present antigen to naïve $CD4^+$ T cells providing co-stimulatory signals and the appropriate cytokines to promote differentiation. Defects in DCs have been identified in both the NOD mouse and in human diabetic patients (Tisch and Wang 2008). Several studies have shown that DCs from NOD mice have a “hyperactive” phenotype, which promotes inflammation and a bias to Th1 cell development. Bone marrow–derived DCs from NOD mice have enhanced activation of the transcription factor NFκB, which directly results in elevated levels of IL-12 and TNF secretion compared to non-autoimmune–prone strains. In addition, NOD DCs are more efficient than Balb/C DCs at stimulating naïve antigen-specific $CD8^+$ T cells. NOD

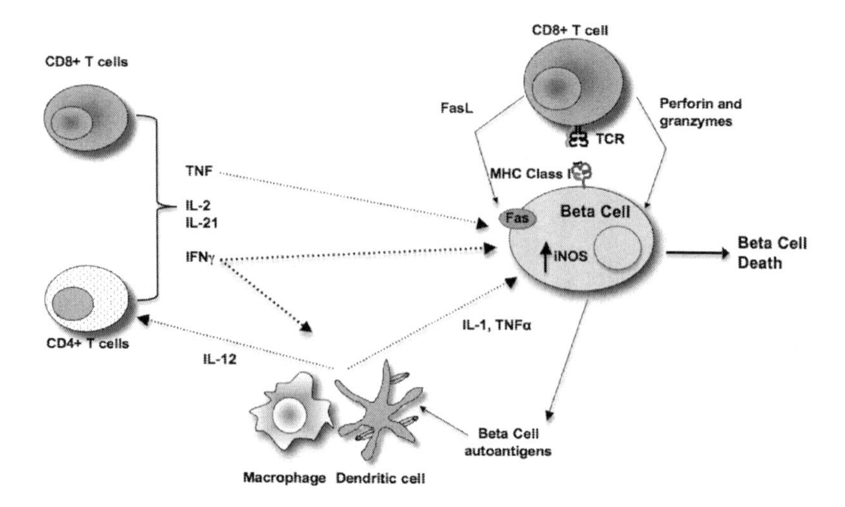

Fig. 1 Cytokine action in type 1 diabetes. Cytokines produced by T cells and antigen-presenting cells act on different cell types to promote type 1 diabetes.

mouse macrophages also secrete increased levels of IL-12 (Tisch and Wang 2008). Overall, hypersecretion of IL-12 stimulates expansion of the pathogenic Th1 subset supporting diabetes development.

Interleukin-2 and Interleukin-21

Activated Th1 cells secrete cytokines that enhance inflammation and promote $CD4^+$ and $CD8^+$ T cell survival. IL-2 acts as a T cell growth factor and an activator of CTLs and is also essential for Treg homoeostasis. In both NOD mice and humans there are multiple chromosome regions known as insulin-dependent diabetes (idd) loci, which confer susceptibility and/or protection from disease. Idd3 contains the *Il-2* and *Il-21* genes. Using genetic mapping studies, Yamanouchi *et al.* demonstrated that the NOD haplotype of *Il2* gene results in a two-fold reduction of IL-2 production by T cells compared to a congenic NOD containing the Idd3 loci from autoimmune-resistant strain C57/Bl6. This defect prevents the feedback mechanism that promotes Treg control of the immune response. Tang *et al.* showed reduced IL-2 in the islets of NOD mice, which resulted in increased apoptosis of intra-islet Treg, and an alteration in the Treg:Teff ratio facilitating disease. Correction of the IL-2 defect by administration of low-dose IL-2 increased Treg survival and protected from diabetes. These studies suggest a defect in IL-2 production by $CD4^+$ T effector cells results in declining Treg function and the promotion of tissue destruction (reviewed in Tisch and Wang 2008).

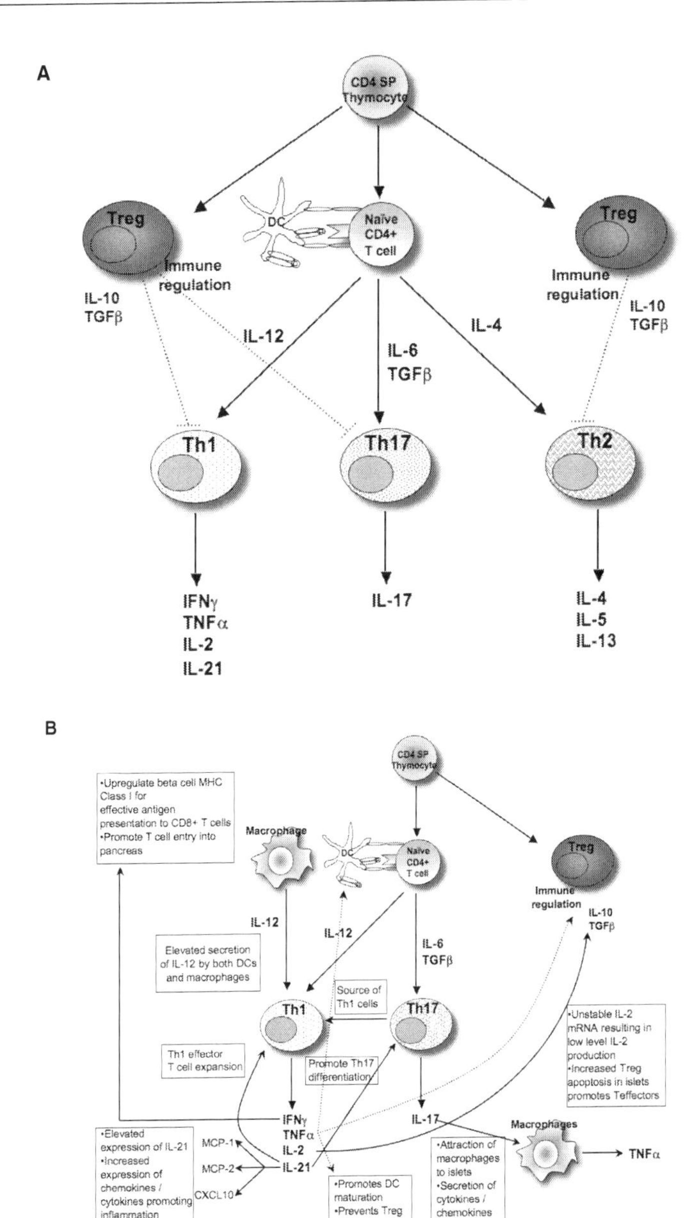

Fig. 2 T helper subsets. (A) CD4+ T cells develop into Treg, Th1, Th2 and Th17 subsets in response to antigen presented by DCs, which each produce a unique set of cytokines. Treg cells provide immunoregulation for each of these subsets. (B) Immune defects in patients with T1D and in NOD mice alter the balance of Th subsets.

The linkage of *Il21* and *Il2* within Idd3 has led to some confusion as to which cytokine is involved in diabetes. IL-21 is produced by activated $CD4^+$ T cells and NKT cells and has roles in innate and adaptive immunity. The expression of IL-21 and its receptor (IL-21R) is increased in NOD mice and a deficiency in IL-21R prevents insulitis and diabetes. The deficiency in IL-21 signalling also resulted in defective polarization to Th17 immune response, suggesting the mechanism of protection in IL-21R$^{-/-}$ mice involves prevention of Th17-driven autoimmunity (Spolski *et al.* 2008; Sutherland *et al.* 2009). In addition, transgenic expression of IL-21 in the pancreas of the non-autoimmune strain, C57BL/6, resulted in pancreatic immune infiltration, diabetes and an increase in the expression of chemokines that recruit macrophages and T cells, including MCP-1, MCP-2 and CXCL10, suggesting IL-21 promotes pancreatic inflammation leading to beta-cell destruction and diabetes (Sutherland *et al.* 2009). McGuire *et al.* found both *Il2* and *Il21* are highly expressed in NOD mice. However, IL-2 mRNA was unstable resulting in low-level expression of IL-2. In contrast, IL-21 expression was increased due to increased promoter activity. While this indicates IL-21 is key to diabetes pathogenesis, the authors suggest that the two effects may work synergistically with the defect in Treg induced by the low-level IL-2 allowing an IL-21-driven expansion of effector T cells, resulting in disease (McGuire *et al.* 2009b).

Interleukin-17

There are several members of the IL-17 family all produced by $CD4^+$ T cells, with IL-17A produced by Th17 cells. In the mouse, differentiation of Th17 cells is driven by TGFβ and IL-6 and occurs in the absence of IFNγ and IL-4. In humans, IL-6 combined with IL-1 promotes Th17 development. Th17 cells contribute to several autoimmune diseases (Cooke 2006).

Increased levels of IL-17 mRNA have been detected in the NOD pancreas prior to disease onset and treatment of NOD mice with anti-IL-17 from 10 wk of age prevented diabetes (Emamaullee *et al.* 2009). Adoptive transfer of BDC2.5 TCR transgenic $CD4^+$ T cells to NOD*scid* mice demonstrated that Th17 cells promote inflammation within islets, possibly by inducing secretion of other cytokines and chemokines (Martin-Orozco *et al.* 2009). Th17 cells can also act as a source of $CD4^+$ T cells that convert to pathogenic Th1 cells in the presence of IFNγ produced in the inflammatory cytokine milieu (Bending *et al.* 2009). In human patients, a subset of monocytes produce high levels of IL-1 and IL-6 that promote the formation of Th17 cells *in vitro*, suggesting the inflammatory environment in patients with T1D could contribute to the Th17 formation and disease development (Bradshaw *et al.* 2009).

CYTOKINE EFFECTS ON THE BETA CELL

Pro-inflammatory cytokines have been implicated in the pathogenesis of T1D. They are found in infiltrated islets from NOD mice and biopsies of patients. *In vitro*, pro-inflammatory cytokines inhibit glucose-stimulated insulin secretion and are toxic to beta cells. However, *in vivo*, they are more likely to be involved in regulation of the anti-islet immune response, and inducing gene expression in beta cells to increase their susceptibility to immune attack.

Pro-inflammatory cytokines induce activation of the transcription factors Stat1, NFκB and the MAPK pathway (Fig. 3), leading to expression of over 15,000 gene transcripts in beta cells (Ortis *et al.* 2009). Incubation of beta cells with IL-1+IFNγ or TNF+IFNγ induces different subsets of genes, which include inducible nitric oxide synthase (iNOS), the death receptor Fas, class I MHC, adhesion molecules, chemoattractants and pro- and anti-apoptotic molecules (Fig. 4).

Cytokine-Induced Beta-Cell Death

Inducible Nitric Oxide Synthase

Up-regulation of iNOS by a combination of IL-1 and IFNγ leads to production of nitric oxide (NO) and islet cell death. TNF plus IFNγ also causes beta cell death *in vitro*, mediated by intra-islet IL-1 and NO production (Mandrup-Poulsen 1996). TNF is also able to induce caspase-dependent apoptosis in beta cells (McKenzie *et al.* 2008).

Endoplasmic Reticulum Stress

Cytokine-induced NO in beta cells leads to depletion of endoplasmic reticulum (ER) Ca^{2+}, which impairs the quality of protein folding and assembly. ER stress proteins, such as CHOP, are induced by this pathway, leading to beta cell apoptosis (Eizirik *et al.* 2008). While there is no doubt that cytokines trigger activation of ER stress genes, there is some debate as to whether ER stress causes cytokine-induced apoptosis.

Mitochondrial Cell Death Pathway

Pro-inflammatory cytokines can directly activate the mitochondrial pathway of apoptosis by increasing intracellular calcium levels leading to dephosphorylation of the pro-apoptotic Bcl-2 family member Bad (Grunnet *et al.* 2009). Absence of Bax or Bak, the downstream effectors of the mitochondrial apoptosis pathway, is partially protective against apoptosis induced by TNF+IFNγ (Fig. 3) (McKenzie *et al.* 2008).

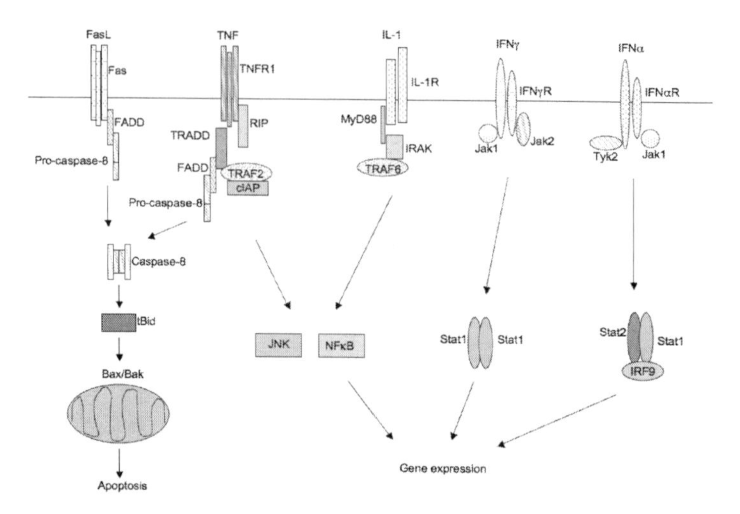

Fig. 3 Signalling pathways of pro-inflammatory cytokines. The pro-inflammatory cytokines TNF, IL-1, IFNγ and IFNα bind to cell surface receptors and activate intracellular signalling pathways. The transcription factors NFκB and Stat1 translocate to the nucleus, where they induce gene expression.

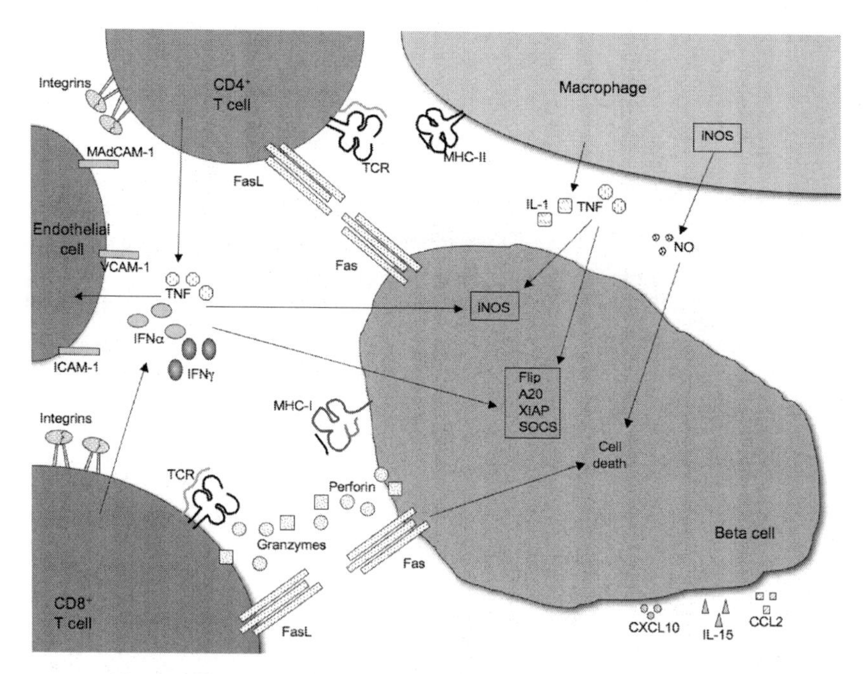

Fig. 4 Cellular and molecular mechanisms of beta cell destruction in type 1 diabetes. CD4+ and CD8+ T cells and macrophages are found in the inflammatory infiltrate and are the effectors of beta cell destruction. These different cell types produce cytokines, chemokines and effector molecules that lead to inflammation and beta cell death.

Fas

Expression of the death receptor Fas is induced by incubation of islets with combinations of IL-1 or TNF and IFNγ, making beta cells susceptible to caspase-dependent apoptosis induced by Fas ligand or anti-Fas antibodies. In beta cells, Fas signalling induces activation of the pro-apoptotic Bcl-2 family member, Bid, and subsequent downstream activation of the mitochondrial pathway of apoptosis (Fig. 3). Deficiency of Bid in mouse islets prevents Fas-mediated beta-cell death (McKenzie *et al.* 2008).

In vivo, it is becoming clear that the Fas pathway is less important as a direct mechanism of beta-cell death in mouse models of diabetes. Fas-deficiency specifically on beta cells does not prevent beta-cell death *in vivo*. Even in the absence of perforin, by far the most dominant mechanism of beta-cell death, beta-cell Fas-deficiency did not prevent diabetes. In contrast, absence of Fas signalling in immune cells is protective against diabetes (Thomas *et al.* 2009).

Cytokine-Induced NFκB in Beta Cells

NFκB can induce proliferation, differentiation and cell death. NFκB activates gene expression in beta cells and the immune system in the development of T1D.

In beta cells, NFκB is a critical regulator of the response to cytokines. The NFκB RelA and p50 subunits are activated by cytokines in beta cells, and these bind to NFκB-responsive elements in the promoter regions of genes involved in beta-cell survival, death and inflammation (Cardozo *et al.* 2001).

Inhibition of NFκB protects islets from NO-dependent death induced by IL-1+IFNγ, but accelerates TNF+IFNγ-induced death of beta cell lines and islets. Cells with intact NFκB signalling up-regulate survival proteins including the Bcl-2 family pro-survival molecule A1, the TNF signalling molecule TRAF1, the caspase-8-induced inhibitor of Fas signalling Flip, A20, BIRC3 and XIAP in response to TNF, and these prevent apoptosis (Sarkar *et al.* 2009). When NFκB is inhibited, these proteins no longer act to protect the cell.

Cytokine Effects on Immune Recognition

Class I MHC

Increased class I MHC has been observed on beta cells from humans with T1D, as well as from NOD mice and BB rats. Class I MHC is up-regulated by cytokines including IFNγ, type 1 IFN and TNF. In NOD mice, IFNγ produced by the inflammatory infiltrate is required to induce beta-cell class I MHC. Up-regulation of class I MHC expression on beta cells leads to their increased susceptibility to autoreactive T cell recognition and killing. In the absence of class I MHC, insulitis and diabetes do not occur. Studies using beta cells or T cells deficient in class I

MHC have demonstrated that a beta cell-CD8$^+$ T cell interaction is required for development of diabetes (reviewed in Tsai *et al.* 2008).

Adhesion Molecules and Chemokines

Adhesion molecules play an important role in progression of T1D (Hanninen *et al.* 1993). In mouse models, VCAM-1, E selectin and MAdCAM-1 expression is induced by IFNγ and TNF on islet endothelium in inflamed islets. Inhibition of these molecules or their corresponding integrins reduces insulitis and diabetes in NOD mice. Over-expression of TNF in beta cells increases expression of ICAM-1 on islet endothelium promoting immune infiltration, and soluble TNFR1 over-expression decreased expression of MAdCAM-1 and ICAM-1.

Chemokines direct migration and activation of immune cells during inflammation. IL-1 and IFNγ induce beta-cell expression of CCL2, CXCL10 and IL-15 among others. Over-expression of chemokines in islets induces insulitis and diabetes, and likewise, deficiency of chemokines prevents insulitis and diabetes. CXCL10, CCL2 and IL-15 have been detected in islets of NOD mice early in insulitis development, and facilitate attraction of macrophages and other immune cells to the islets (Eizirik *et al.* 2009).

Suppressors of Cytokine Signalling

The suppressors of cytokine signalling (SOCS) are intracellular molecules that potently inhibit the signalling pathway induced by cytokines. Their expression is induced by the cytokines they inhibit, and thus they act in a classical negative feedback mechanism. Over-expression of SOCS1 blocks signalling by several cytokines including interferons, IL-4, IL-15, members of the IL-6 family, TNF and toll-like receptors (Davey *et al.* 2006). SOCS3 appears to block effects of IL-1 and TNF more effectively than interferons.

SOCS gene expression in NOD mouse islets correlates with infiltration and cytokine production within infiltrated islets (Thomas *et al.* 2009a). SOCS1 is expressed in NOD mouse beta cells, but the levels are insufficient to terminate IFNγ signalling suggesting IFNγ responses in the beta cell may not normally be regulated by SOCS1. Constitutively low levels of SOCS mRNA have also been detected in human islets, and these are up-regulated by cytokines *in vitro*. SOCS1 expression has been observed in sections of pancreas from donors with T1D.

PRO-INFLAMMATORY CYTOKINES *IN VIVO*

Interferons

Type 1 interferons and IFNγ have effects on both the beta cell and the immune system during the initiation and progression of diabetes. The role of interferons has been extensively reviewed elsewhere so the remainder of this chapter will focus on other pro-inflammatory cytokines (reviewed in Thomas *et al.* 2009a).

Interleukin-1

IL-1 is produced by activated macrophages and DCs. IL-1 exerts its signal through the IL-1 receptor (IL-1R). The IL-1R antagonist protein (IL-1Ra) is a highly effective inhibitor of IL-1 signalling.

Given that it has been known for some 20 years that IL-1 is toxic to beta cells, it was somewhat surprising that IL-1R deficiency in NOD mice only slightly delayed the onset of diabetes without significantly reducing the overall incidence, indicating that IL-1 is not essential for progression to diabetes (Thomas *et al.* 2004). Using a graft model, we showed that IL-1 does not directly affect beta cells *in vivo*, but alters the immune response. Inhibition of IL-1 with soluble IL-1R or anti-IL-1 also only modestly reduced diabetes in several accelerated models (Mandrup-Poulsen 1996).

IL-1 has effects on both DCs and $CD4^+$ T effector cells. It is essential for optimal DC activation providing autocrine and paracrine stimulation to secrete IL-12p70, TNF and IL-6. IL-1 also enhances the proliferation of $CD4^+$ T effector cells, altering the proportion of Treg, allowing $CD4^+$ T effectors to escape suppression and expand.

IL-1 has attracted attention as an important inflammatory cytokine in both type 1 and type 2 diabetes. A clinical trial of IL-1Ra treatment in type 2 diabetes resulted in improved glycaemia and beta cell function after 13 wk of treatment. Because of the modest effects of IL-1 blockade on T1D, and the safety of IL-1Ra therapy, clinical trials are underway to test the effects of IL-1 blockade in T1D (Pickersgill and Mandrup-Poulsen 2009).

Tumor Necrosis Factor

TNF is secreted by activated macrophages and $CD4^+$ T lymphocytes. It signals through two distinct receptors, TNFR1 and TNFR2. TNFR1 can induce gene expression through NFκB and can also cause activation of caspase-dependent apoptosis. Because of its restricted expression on immune cells, TNFR2 is involved in T cell proliferation and anti-apoptotic pathways. TNF causes activation and differentiation of macrophages, proliferation of T cells, expression of adhesion molecules, up-regulation of MHC molecules and production of chemoattractants.

The effects of TNF on development of T1D have been studied by manipulating the amount of TNF or its receptors in mice (reviewed in Green and Flavell 1999). Neutralizing bioactive TNF, or deficiency of TNFR1 reduces insulitis and protects

NOD mice from diabetes. Further experiments suggest that the effects of TNF in diabetes development are not on T cells or direct toxicity to beta cells, but an islet response to TNF is required for T cell activation.

In neonatal NOD mice, systemic TNF or its expression in islets results in accelerated diabetes, and administration of anti-TNF Ab prevents diabetes. In contrast, expressing TNF in adult NOD mice leads to insulitis that does not progress to diabetes, and anti-TNF Ab accelerates T1D in adult NOD mice. These data suggest that the time point at which TNF exerts its effects is important.

Neonatal treatment with TNF induces activation and expression of maturation markers on DCs, resulting in increased activation of islet-specific T cells (Lee *et al.* 2005). TNF can also decrease the number and function of Treg cells in NOD mice (Wu *et al.* 2002, Green *et al.* 2002).

CONCLUSION

Cytokines are crucial in the progression of T1D. Though complex, the study of cytokines enables us to better understand the pathogenesis of T1D and ultimately discover therapies for the prevention or treatment of this disease.

Table 3 Spontaneous diabetes in NOD mice genetically modified for pro-inflammatory cytokines and their receptors

Strain	*Diabetes onset*	*Diabetes incidence*
NOD	10-20 wk	70-100%
IFNγ-/-	25 wk	35% (wt 50%)[1]
IFNγR1-/-	19 wk	75%
IFNγR2-/-	15 wk	85%
Stat1-/-	N/A[2]	0%
IRF1-/-	N/A	0%
RIP-ΔγR	17 wk	30% (wt 40%)
RIP-SOCS1	17 wk	11-45%
IL-1R-/-	18 wk (wt 14 wk)	75%
TNFR1-/-	N/A	0%

Mice deficient in cytokines or their receptors, or those expressing transgenes in beta cells to make beta cells unresponsive to cytokines, have been backcrossed onto a NOD genetic background. Results from diabetes incidence studies are presented. [1]wild type (wt) incidence was lower than the normal range for diabetes onset or incidence; [2]not applicable (N/A).

Summary Points

Cytokines play a vital role in development of type 1 diabetes, being important for activation and maintenance of the autoimmune response.

Cytokines produced by Th1 cells including TNF, IFNγ, IL-2 and IL-21 are important for diabetes development.

Cytokines lead to production of free radicals and ER stress, resulting in beta-cell apoptosis *in vitro*.

NFκB is activated by cytokines in beta cells and is an important regulator of gene transcription.

Cytokines lead to up-regulation of genes in beta cells that attract immune cell infiltration and increase recognition of beta cells by immune cells.

Deficiency of cytokine receptors on beta cells does not prevent diabetes in NOD mice, suggesting that cytokines do not directly kill beta cells *in vivo*

Definition of Terms

Antigen presentation: processing of antigens by antigen-presenting cells, which display them in class I or II MHC for recognition by CD8+ or CD4+ T cells respectively.

Apoptosis: programmed cell death involving a defined set of biochemical events and morphological features including membrane blebbing, chromatin condensation and DNA fragmentation.

Autoimmunity: the failure to recognize self resulting in an immune response to the body's own cells.

Insulitis: inflammatory infiltration around and within pancreatic islets consisting of T and B cells and antigen-presenting cells.

NOD mouse: a mouse model of spontaneous type 1 diabetes that develops disease with many of the key features of human diabetes.

Acknowledgments

This work was supported by grants and fellowships from the National Health and Medical Research Council of Australia, the Juvenile Diabetes Research Foundation and the Australian Diabetes Society.

References

Bending, D. and H. De La Pena, M. Veldhoen, J.M. Phillips, C. Uyttenhove, B. Stockinger, and A. Cooke. 2009. Highly purified Th17 cells from BDC2.5NOD mice convert into Th1-like cells in NOD/SCID recipient mice. J. Clin. Invest. 119: 565-572.

Bradshaw, E.M. and K. Raddassi, W. Elyaman, T. Orban, P.A. Gottlieb, S.C. Kent, and D.A. Hafler. 2009. Monocytes from patients with type 1 diabetes spontaneously secrete proinflammatory cytokines inducing Th17 cells. J. Immunol. 183: 4432-4439.

Cardozo, A.K. and H. Heimberg, Y. Heremans, R. Leeman, B. Kutlu, M. Kruhoffer, T. Orntoft, and D.L. Eizirik. 2001. A comprehensive analysis of cytokine-induced and nuclear factor-kappa B- dependent genes in primary rat pancreatic beta-cells. J. Biol. Chem. 276: 48879-48886.

Cooke, A. 2006. Th17 cells in inflammatory conditions. Rev. Diabet. Stud. 3: 72-75.

Davey, G.M. and W.R. Heath, and R. Starr. 2006. SOCS1: a potent and multifaceted regulator of cytokines and cell-mediated inflammation. Tissue Antigens 67: 1-9.

Eizirik, D.L. and A.K. Cardozo, and M. Cnop. 2008. The role for endoplasmic reticulum stress in diabetes mellitus. Endocr. Rev. 29: 42-61.

Eizirik, D.L. and M.L. Colli, and F. Ortis. 2009. The role of inflammation in insulitis and beta-cell loss in type 1 diabetes. Nat. Rev. Endocrinol. 5: 219-226.

Emamaullee, J.A. and J. Davis, S. Merani, C. Toso, J.F. Elliott, A. Thiesen, and A.M. Shapiro. 2009. Inhibition of Th17 cells regulates autoimmune diabetes in NOD mice. Diabetes 58: 1302-1311.

Green, E.A. and R.A. Flavell. 1999. Tumor necrosis factor alpha and the progression of diabetes in non-obese diabetic mice. Immunol. Rev. 169: 11-22.

Green, E.A. and Y. Choi, and R.A. Flavell. 2002. Pancreatic lymph node-derived CD4(+) CD25(+) Treg cells: highly potent regulators of diabetes that require TRANCE-RANK signals. Immunity 16: 183-191.

Grunnet, L.G. and R. Aikin, M.F. Tonnesen, S. Paraskevas, L. Blaabjerg, J. Storling, L. Rosenberg, N. Billestrup, D. Maysinger, and T. Mandrup-Poulsen. 2009. Proinflammatory cytokines activate the intrinsic apoptotic pathway in beta-cells. Diabetes 58: 1807-1815.

Hanninen, A. and C. Taylor, P.R. Streeter, L.S. Stark, J.M. Sarte, J.A. Shizuru, O. Simell, and S.A. Michie. 1993. Vascular addressins are induced on islet vessels during insulitis in nonobese diabetic mice and are involved in lymphoid cell binding to islet endothelium. J. Clin. Invest. 92: 2509-2515.

Lee, L.F. and B. Xu, S.A. Michie, G.F. Beilhack, T. Warganich, S. Turley, and H.O. McDevitt. 2005. The role of TNF-alpha in the pathogenesis of type 1 diabetes in the nonobese diabetic mouse: analysis of dendritic cell maturation. Proc. Natl. Acad. Sci. USA 102: 15995-16000.

Mandrup-Poulsen, T. 1996. The role of interleukin-1 in the pathogenesis of IDDM. Diabetologia 39: 1005-1029.

Martin-Orozco, N. and Y. Chung, S.H. Chang, Y.H. Wang, and C. Dong. 2009. Th17 cells promote pancreatic inflammation but only induce diabetes efficiently in lymphopenic hosts after conversion into Th1 cells. Eur. J. Immunol. 39: 216-224.

McGuire, H.M. and A. Vogelzang, N. Hill, M. Flodstrom-Tullberg, J. Sprent, and C. King. 2009. Loss of parity between IL-2 and IL-21 in the NOD Idd3 locus. Proc. Natl. Acad. Sci. USA 106: 19438-19443.

McKenzie, M.D. and E.M. Carrington, T. Kaufmann, A. Strasser, D.C. Huang, T.W. Kay, J. Allison, and H.E. Thomas. 2008. Proapoptotic BH3-only protein Bid is essential for death receptor-induced apoptosis of pancreatic beta-cells. Diabetes 57: 1284-1292.

Ortis, F. and N. Naamane, D. Flamez, L. Ladriere, F. Moore, D.A. Cunha, M.L. Colli, T. Thykjaer, K. Thorsen, T.F. Orntoft, and D.L. Eizirik. 2009. The cytokines IL-1{beta} and TNF-{alpha} regulate different transcriptional and alternative splicing networks in primary beta cells. Diabetes 59: 358-374.

Pickersgill, L.M. and T.R. Mandrup-Poulsen. 2009. The anti-interleukin-1 in type 1 diabetes action trial--background and rationale. Diabetes Metab. Res. Rev. 25: 321-324.

Pietropaolo, M. and J.M. Surhigh, P.W. Nelson, and G.S. Eisenbarth. 2008. Primer: immunity and autoimmunity. Diabetes 57: 2872-2882.

Sarkar, S.A. and B. Kutlu, K. Velmurugan, S. Kizaka-Kondoh, C.E. Lee, R. Wong, A. Valentine, H.W. Davidson, J.C. Hutton, and S. Pugazhenthi. 2009. Cytokine-mediated induction of anti-apoptotic genes that are linked to nuclear factor kappa-B (NF-kappaB) signalling in human islets and in a mouse beta cell line. Diabetologia 52: 1092-1101.

Spolski, R. and M. Kashyap, C. Robinson, Z. Yu, and W.J. Leonard. 2008. IL-21 signaling is critical for the development of type I diabetes in the NOD mouse. Proc. Natl. Acad. Sci. USA 105: 14028-14033.

Sutherland, A.P. and T. Van Belle, A.L. Wurster, A. Suto, M. Michaud, D. Zhang, M.J. Grusby, and M. von Herrath. 2009. Interleukin-21 is required for the development of type 1 diabetes in NOD mice. Diabetes 58: 1144-1155.

Thomas, H.E. and W. Irawaty, R. Darwiche, T.C. Brodnicki, P. Santamaria, J. Allison, and T.W. Kay. 2004. IL-1 receptor deficiency slows progression to diabetes in the NOD mouse. Diabetes 53: 113-121.

Thomas, H.E. and K.L. Graham, E. Angstetra, M.D. McKenzie, N.L. Dudek, and T.W. Kay. 2009a. Interferon signalling in pancreatic beta cells. Front Biosci. 14: 644-656.

Thomas, H.E. and M.D. McKenzie, E. Angstetra, P.D. Campbell, and T.W. Kay. 2009b. Beta cell apoptosis in diabetes. Apoptosis 14: 1389-1404.

Tisch, R. and B. Wang. 2008. Dysrulation of T cell peripheral tolerance in type 1 diabetes. Adv. Immunol 100: 125-149.

Tsai, S. and A. Shameli, and P. Santamaria. 2008. CD8+ T cells in type 1 diabetes. Adv. Immunol. 100: 79-124.

Wu, A.J. and H. Hua, S.H. Munson, and H.O. McDevitt. 2002. Tumor necrosis factor-alpha regulation of CD4+CD25+ T cell levels in NOD mice. Proc. Natl. Acad. Sci. USA 99: 12287-12292.

CHAPTER 19

Pro-Inflammatory Cytokines in Response to Glycemic Excursions of Hypo- and Hyperglycemia

A.E. Kitabchi[1,*], L.N. Razavi[1], B.A. Larijani[2] and E. Taheri[2]

[1]Division of Endocrinology, Department of Medicine
The University of Tennessee Health Science Center, 956 Court Ave
Suite A202, Memphis TN 38163, USA
[2]Endocrinology and Metabolism Research Center, Tehran, Iran
5th Floor, Doctor Shariati Hospital, North Kargar Avenue, Tehran 14114, Iran

ABSTRACT

Cytokines are mediators of inter- and intracellular communications. These peptides contribute to a chemical signaling language that regulates homeostasis, tissue repair and immune responses.

We have shown that glycemic excursions are associated with elevation of pro-inflammatory cytokines. We noted the increased levels of pro-inflammatory cytokines (including IL-8, IL-6, IL-1β and TNF-α) and markers of oxidative stress and cardiovascular risks concomitant with increased levels of the counter-regulatory hormones and leukocytes in hyperglycemic crises. Similarly, the acute stress of hypoglycemia by insulin-induced tolerance test in non-diabetic subjects resulted in the stimulation of pro-inflammatory cytokines, leukocytosis and markers of oxidative stress and lipid peroxidation. We conclude that the elevation of pro-inflammatory cytokines and markers of oxidative stress in glycemic excursions are the result of adaptive responses of hypothalamus-pituitary-adrenal and sympathoadrenal systems to non-inflammatory stressors such as hypo- or hyperglycemia.

**Corresponding author*

INTRODUCTION

Cytokines are mediators of inter- and intracellular communications. These peptides contribute to a chemical signaling language that regulates homeostasis, tissue repair and immune responses (Oppenheim 2001; Haddad 2002). It has been shown that the pathophysiology of inflammatory responses in autoimmune, malignant and infectious diseases can be explained by the production of cytokines and related peptides. Cytokines and their antagonists have been shown to have therapeutic potentials in treating various chronic and acute diseases (Oppenheim 2001; Ko *et al.* 2009; Lee *et al.* 2009). Cytokines and their receptors are endogenous to the brain, endocrine and immune systems. It has been suggested that the mediators of neuro-endocrine-immune interface are cytokines such as TNF-α, IL-1, IL-6 and IL-8; even the term "immune-transmitter" was suggested for these cytokines that can transmit specific signals to neurons and other cell types (Holloway *et al.* 2002).

Several cytokines are known to affect the release of anterior pituitary hormones by an action on the hypothalamus-pituitary-adrenal (HPA) axis (Turnbull *et al.* 1999; Woiciechowsky *et al.* 1999; Bumiller *et al.* 1999). The HPA axis is the key player in body responses to stressors and cytokines are the chemical mediators that stimulate the HPA axis in response to acute and chronic stressors (Heremus *et al.* 1990; Palmer *et al.* 2002). However, the exact sites of actions of the cytokines in these systems have not been fully established.

The elaboration of cytokines appears not to be restricted to injurious, inflammatory and infectious insults. Recent studies have indicated that synthesis and/or secretion of cytokines including IL-1 and IL-6 are altered during acute physical or psychological stresses (Glaser *et al.* 2005; Elenkov and Chrousos 2002; Bethin *et al.* 2000). In recent *in vitro* studies, IL-6 levels increased in rats after exposure to a novel environment, electroshock, and restraint and IL-6 mRNA synthesis increased in the midbrain of rats after exposure to these stressors (Zho *et al.* 2003; Shizuya *et al.* 1997; Kitamura *et al.* 1997). In humans, plasma IL-6 levels increase rapidly after treadmill exercise with the peak increases at 15 and 45 min (Papanicolaou *et al.* 1996). To investigate if other acute stressors can induce pro-inflammatory cytokines, we evaluated the hypothalamic-pituitary-adrenal axis, sympathetoadrenal system and cytokine production during and after the acute stress of hypoglycemia (Razavi *et al.* 2009) and hyperglycemia (Stentz *et al.* 2004).

PRO-INFLAMMATORY CYTOKINES AND INSULIN-INDUCED HYPOGLYCEMIA

The association of cytokines and hypoglycemia had not been investigated previously. Therefore, we designed a study to induce hypoglycemia in non-diabetic subjects by standard insulin tolerance test (Razavi *et al.* 2009). In this study, healthy male subjects with no history of current infection, cardiovascular risk factors, metabolic syndrome or abnormal glucose metabolism were evaluated. Table 1 shows the baseline characteristics of the participants. During and after insulin tolerance test, the levels of counter-regulatory hormones [ACTH, cortisol, growth hormone (GH), epinephrine (Epi), and norepinephrine (NE)] as well as white blood cell count (WBC) and pro-inflammatory cytokines [TNF-α, IL-6, IL-8, and IL-1β] and markers of oxidative stress and lipid peroxidation were measured at different time intervals after insulin injection (Razavi *et al.* 2009). Levels of pro-inflammatory cytokines and counter-regulatory hormones were measured in the plasma using a solid-phase, two-site sequential chemiluminescent immunometric assay. Plasma norepinephrine and epinephrine were determined by high-pressure liquid chromatography (Razavi *et al.* 2009).

Table 1 Baseline characteristics of participants of insulin-induced hypoglycemia study on admission

Subjects (n)	*13*
Age (yr)	30±4.8*
Blood pressure (mmHg)	120±0.4/76±0.3 *
BMI (kg/m^2)	23±1.3*
Temperature (F)	98.1±1.2*
HbA1c	5.6±1.06*
White blood cells x 10^3 (/μl)	6.4±1.1*

*Data are mean ± SD. Reproduced, with permission, from Razavi *et al.* (2009).

The expected elevation of counter-regulatory hormones after insulin injection in our study is demonstrated in Table 2 and Fig. 1. Pro-inflammatory cytokines, markers of lipid peroxidation and oxidative stress and WBC also increased after insulin injection concomitant with counter-regulatory hormones (Table 2). In our study, epinephrine changes significantly correlated with TNF-α and IL-8 changes (Fig. 2) and ACTH and IL-6 changes correlated with borderline significance (Fig. 3). Cortisol changes could predict changes in IL-1 significantly (Fig. 4). Of note, the changes in cortisol and norepinephrine were associated with the changes in leukocytosis (WBC) (Fig. 4).

Therefore, for the first time, we have documented the elevation of pro-inflammatory cytokines and leukocytosis in hypoglycemia along with well-known elevation of counter-regulatory hormones in non-diabetic subjects (Razavi *et al.* 2009).

Table 2 Baseline values of measured counter-regulatory hormones and cytokines, values at peak or nadir and values at 240 min after insulin injection in insulin-induced hypoglycemia. All measured parameters significantly increased after insulin injection comparing to the baseline values. Reproduced, with permission, from Razavi *et al.* (2009).

Parameters	*Baseline values ±SD*	*Values at peak or nadir ±SD, Time (min), P value*	*Values at 240'± SD*
Glucose (mg/dl)	102.2±10.2	38.23 ±19.9 (30) (P <0.001)	99.1± 13.79
Cortisol (μg/dl)	10.12 ± 4.9	15.23±5.8 (60) (p=0.004)	7.93±4.17
GH (ng/ml)	0.1±0.03	27.28±17.6 (60) (P <0.001)	2.9± 6.3
ACTH (pg/ml)	7.9±7.7	47.48±41.7 (45) (p<0.006)	5.1±3.05
NE (μg/ml)	488±174.7	807.9±282.89 (240) (p=0.005)	807.9±282.89
Epinephrine (μg/ml)	70.1±28.3	1198.92±1030.09 (45) (p<0.001)	90.4±31.1
WBC x10^3/μl	6.4±1.10	12.97±2.9 (120) (p<0.001)	12.5±2.8
TNF-α (μg/ml)	6.1±2.8	8.1 ± 2.8 (45) (p<0.001)	6.9±2.9
IL-6 (μg/ml)	2.18±1.2	3.94±1.2 (120) (p=0.0009)	3.6±1.7
IL-8 (μg/ml)	8.43±6.3	12.6±7.7 (60) (p=0.04)	8.9±5.3
IL-1β (μg/ml)	0.70±1.04	1.12 ±0.91 (240) (p=0.9)	1.12±0.91
FFA (mmol/l)	0.5±0.15	1.06±0.27 (240) (p<0.001)	1.06±0.27
TBA (lipid peroxidation) (uM)	0.6±0.1	2.7±0.2 (45) (p<0.001)**	1.1±0.1
ROS by DCF method (uM)	1.9±0.2	3.9±0.2 (45) (p<0.001)	2.2±0.1

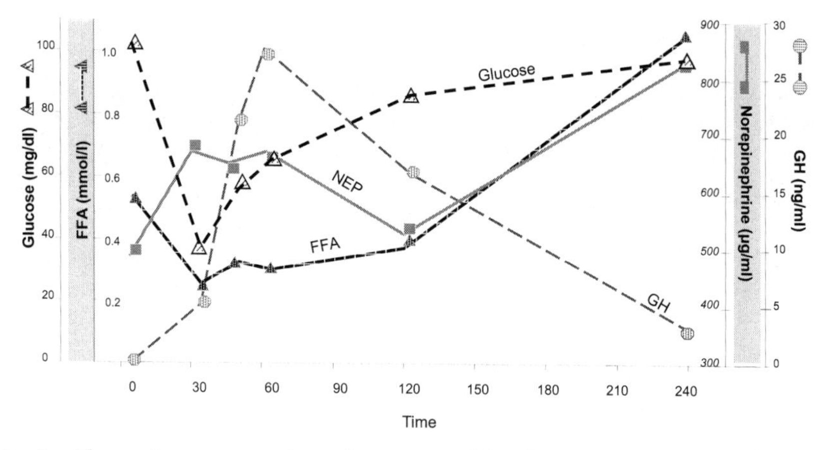

Fig. 1 Changes in counter-regulatory hormones and free fatty acids during and after insulin-induced hypoglycemia. Reproduced with permission from Razavi *et al.* (2009).

Excursions of FFA are due to: initial insulin injection → ↑ antilipolysis → ↓ FFA. Later increase of NEP leads to increased lipolysis → ↑ FFA.

* Scale for glucose (mg/dl); ** Scale for FFA (mmol/l); † Scale for GH (ng/ml); ‡ Scale for norepinephrine (μg/ml).

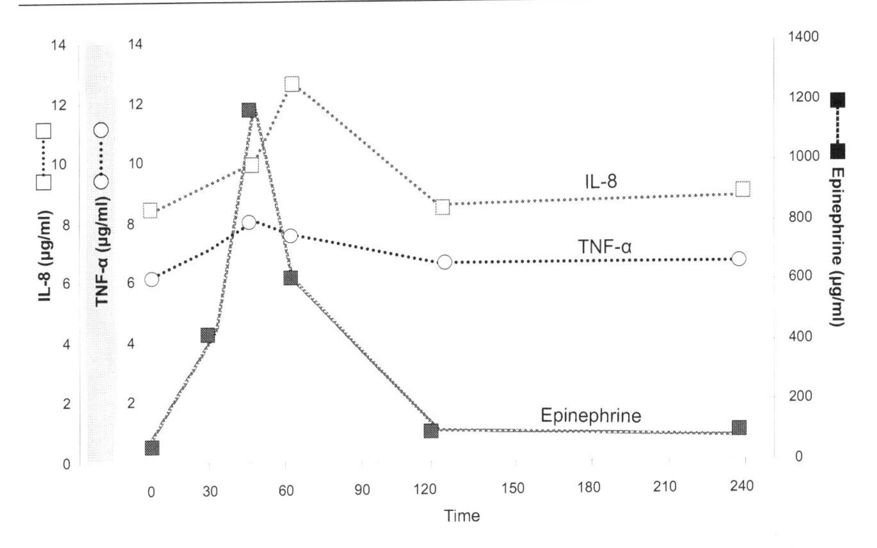

Fig. 2 Changes in epinephrine, TNF-α and IL-8 during and after insulin-induced hypoglycemia. Reproduced, with permission, from Razavi *et al.* (2009).

The area under the curve (AUC) demonstrates that EP is a predictor of TNFα and IL8. * Scale for IL-8 (µg/ml) and TNF-α (µg/ml); ** Scale for epinephrine (µg/ml).

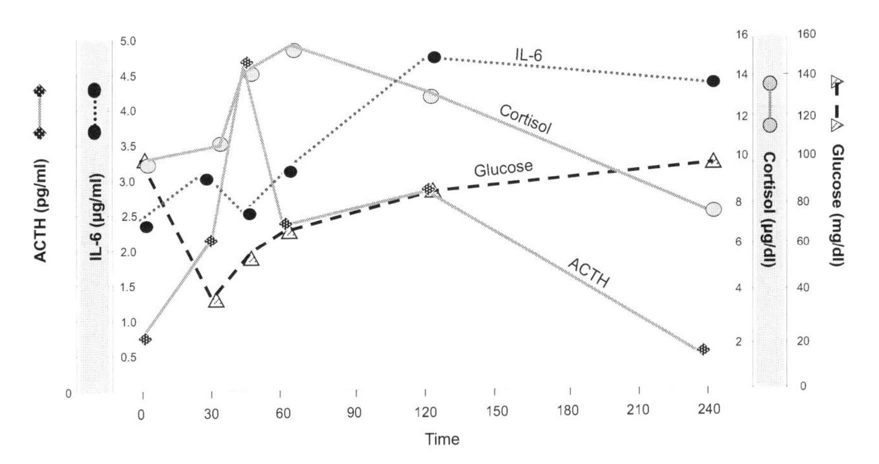

Fig. 3 Changes in ACTH, cortisol and IL-6 during and after insulin-induced hypoglycemia. Reproduced, with permission, from Razavi *et al.* (2009).

AUC for ACTH predicts IL6. Hypoglycemia is associated with elevated levels of ACTH and cortisol.

* Scale for ACTH (pg/ml); ** Scale for IL-6(µg/ml); † Scale for glucose; ‡ Scale for cortisol (µg/dl)

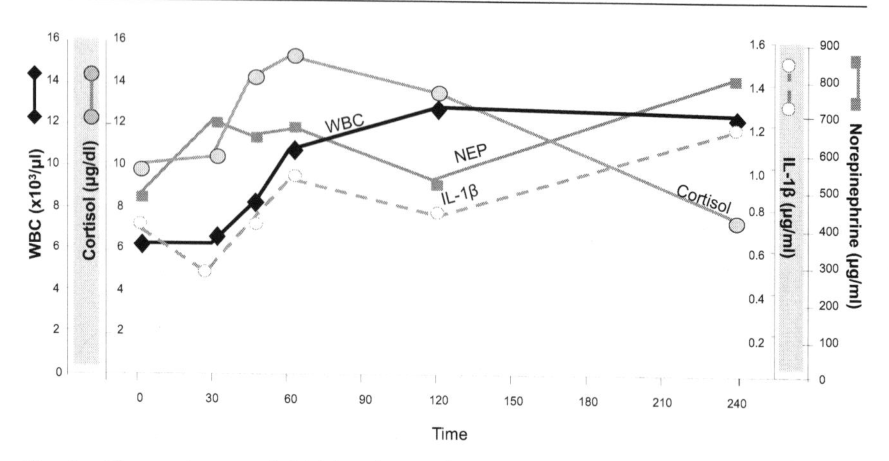

Fig. 4 Changes in cortisol, WBC and IL-1 during and after insulin-induced hypoglycemia. Reproduced, with permission, from Razavi *et al.* (2009).

The AUC for cortisol predicts IL1ß. Leukocytosis is associated with elevated levels of cortisol and norepinephrine, independent of infection.

* Scale for WBC ($x10^3$ /µl) and cortisol (µg/dl); ** Scale for IL-1B (µg/ml); † Scale for NEP (µg/ml)

PRO-INFLAMMATORY CYTOKINES AND ACUTE HYPERGLYCEMIA

We have demonstrated that hyperglycemic crises of diabetic ketoacidosis (DKA) and non-ketotic hyperglycemia (NKH) provoked the elevation of pro-inflammatory cytokines (TNF-α, IL-6, IL-8, and IL-1β), cardiovascular risk factors [homocysteine, CRP, free fatty acid (FFA), and PAI-1], markers of oxidative stress and lipid peroxidation and leukocytosis as well as the elevation of counter-regulatory hormones such as GH and cortisol (Table 3) (Stentz *et al.* 2004). In this study, we measured the above-mentioned parameters in lean and obese patients with DKA and NKH during and after the resolution of hyperglycemic crises.

All the measured parameters including pro-inflammatory cytokines were increased significantly in response to the acute hyperglycemia in both lean and obese patients with DKA and NKH compared to the control group. Interestingly, all the measured parameters returned to normal values after the resolution of DKA or NKH (Stentz *et al.* 2004, Kitabchi *et al.* 2008).

CONCLUSIONS

We have demonstrated the response of the body to the acute stress of hypoglycemia and hyperglycemia, which includes the well-known elevation of counter-regulatory

Table 3 Pro-inflammatory cytokines, cardiovascular risk factors, counter-regulatory hormones, lipid peroxidation (TBA), and DCF values on admission and resolution of hyperglycemic crises in lean and obese DKA and obese hyperglycemic patients, compared with lean and obese non-diabetic controls. All measured parameters were increased in DKA compared to non-diabetic control group. Reproduced, with permission, from Stentz *et al.* (2004).

	Lean DKA		*Obese DKA*		*Obese hyperglycemia*			
	Adm	*Resol*	*Adm*	*Resol*	*Adm*	*Resol*	*Lean Control*	*Obese Control*
TNF (pg/ml)	22.7 ± 3.6	4.6 ± 0.9*^	28.3 ± 2.8	5.9 ± 0.7*^	24 ± 3.1	5.1 ± 1.3*^	1.7 ± 0.2*^	3.9 ± 0.6*^
IL-1β (pg/ml)	9.8 ± 2.3	1 ± 0.2*^	13.7 ± 2.1	2.4 ± 0.3*^	11 ± 0.8	3.1 ± 0.8*^	1.3 ± 0.2*^	1.9 ± 0.3*^
IL-6 (pg/ml)	14.9 ± 2.6	3.9 ± 1.1*^	12.6 ± 2.1	4.3 ± 0.6*^	10 ± 1.7	3.3 ± 0.7*^	1.8 ± 0.2*^	2.1 ± 0.3*^
IL-8 (pg/ml)	29.3 ± 3.4	10.6 ± 2.3*^	27.4 ± 3.8	12 ± 2.8*^	26 ± 3.4	9.3 ± 2.8*^	4.9 ± 1.4*^	5.5 ± 1.7*^
CRP (mg/l)	51 ± 3	28 ± 1*^	59 ± 13	34 ± 9*^	28 ± 6	13 ± 3*^	1 ± 0.2*^	2 ± 0.4*^
Homocysteine (μM)	4.7 ± 0.2	3.7 ± 0.2*^	5.9 ± 0.9	5.4 ± 0.7	3.7 ± 0.4	3.1 ± 0.3*	1.8 ± 0.1*^	2.2 ± 0.3*^
FFA (mM)	1.6 ± 0.1	0.6 ± 0.1*^	1.4 ± 0.1	0.7 ± 0.1*^	1.2 ± 0.2	0.8 ± 0.1*^	0.5 ± 0.06*^	0.7 ± 0.1*^
PAI-1 (ng/ml)	42.1 ± 12.2	4.2 ± 2.1*^	40.4 ± 12.4	13.0 ± 3.4*^	35.4 ± 9.3	7.3 ± 2.4*^	1.4 ± 0.2*^	2.5 ± 0.4*^
DCF (μM)	8.6 ± 0.8	3.7 ± 0.5*^	8.9 ± 1.2	4.1 ± 0.7*^	7.8 ± 0.6	3.8 ± 0.5*^	2.3 ± 0.4*^	3.1 ± 0.6*^
TBA (μM)	3.8 ± 0.7	1.3 ± 0.4*^	4.0 ± 0.6	1.6 ± 0.2*^	3.3 ± 0.5	1.5 ± 0.4*^	0.84 ± 0.1*^	0.9 ± 0.1*^
GH (ng/ml)	12.3 ± 2.2	3.2 ± 1.0*^	10.0 ± 3.1	4.0 ± 1.2*^	1.6 ± 0.3*	0.9 ± 0.2*^	0.8 ± 0.2*^	0.8 ± 0.2*^
Cortisol	46.2 ± 2.3	21.7 ± 1.1*^	55.4 ± 5.8	24.6 ± 3.6*^	23 ± 0.9*	17.2 ± 1.4*^	14 ± 1.2*^	13 ± 1.1*^
(μg/dl)	2.3	1.1*^	5.8	3.6*^	0.9*^	1.4*^	1.2*^	1.1*^

Data are mean ± SE. Resol, resolution; PAI-1, plasminogen activator inhibitor-1; FFA, free fatty acid; CRP, C-reactive protein.

*$P < 0.01$ *vs.* lean DKA on admission (Adm). ^ $P < 0.05$ *vs.* admission value of each group.

hormones and the stimulation of pro-inflammatory cytokines, markers of lipid peroxidation and oxidative stress as well as leukocytosis.

However, hyperglycemia in DKA or NKH results in higher levels of measured hormones and cytokines compared to hypoglycemia (Tables 2, 3). This finding may be partly explained by the fact that hyperglycemic stress in DKA and NKH are more prolonged than insulin-induced hypoglycemic stress and also the baseline characteristics of participants were significantly different in patients with DKA and NKH compared to the non-diabetic participants of the insulin tolerance test (Tables 1 and 4).

Table 4 Baseline characteristics of participants in the study of pro-inflammatory cytokines and hyperglycemic crises. Patients in DKA and hyperglycemic groups had higher baseline A1c. Reproduced, with permission, from Stentz *et al.* (2004).

Clinical characteristics of hyperglycemic patients on admission

Parameter	*Lean DKA*	*Obese DKA*	*Obese Hyperglycemia*	*Lean control*	*Obese control*
Patients (n)	20	28	10	12	12
Age (years)	39 ± 27	38 ± 20	50 ± 37	357 ± 19	365 ± 31
Sex (M/F)	13/7	20/8	5/5	7/5	4/8
BMI(kg/m^2)	22 ± 0.6	33 ± 0.9*	30 ± 0.9*	22 ± 0.7	34 ± 16*
Temperature (°F)	97.3 ± 0.5	97.2 ± 0.3	98.3 ± 0.2	98.4 ± 0.1	98.5 ± 0.2
HbA_{1c}	12.5 ± 0.4	11.6 ± 0.4	10.8 ± 0.6	5.4 ± 0.5%	5.7 ± 0.7*
White blood cell × 10^8	14.4 ± 0.7	14.2 ± 1.3	11.8 ± 2.0	6.5 ± 0.7*	6.8 ± 0.6*

Data are means ± SE *P < 0.01 vs. lean DKA on admission.

At the cellular level, it has been shown that hyperglycemia in DKA can induce activation of T lymphocytes and stimulation of cytokines and markers of oxidative stress and lipid peroxidation (Kitabchi 2004 Stentz 2005). Similarly, incubation of T lymphocytes in high glucose or palmitate media could result in production of cytokines (Stentz *et al.* 2006). These findings have not been investigated in hypoglycemia yet.

However, a recent study also confirmed some of our findings and showed that insulin-induced hypoglycemia was associated with the increased levels of IL-6 and ACTH in healthy subjects (Dotson *et al.* 2008). Furthermore, it has been recently established that elaboration of pro-inflammatory cytokines may occur in the fat tissue or liver (Fain *et al.* 2004, 2006) as well as HPA axis and sympathoadrenal system (Gonzalez-Hernandez *et al.* 1996, Judd 2000, Mastorakos *et al.* 2006). Based on these findings, we proposed that the stimulation of HPA axis and sympathoadrenal system may be responsible for the increase in pro-inflammatory cytokines in response to hypoglycemic or hyperglycemic excursions. This may be an adaptive response for maintaining glucose homeostasis in glycemic excursions. However, the mechanism responsible for the elevation of pro-inflammatory cytokines in response to the metabolic stressors is still not fully elucidated.

Summary Points

- Acute hypoglycemic stress can provoke: the elevation of (1) counter-regulatory hormones, (2) pro-inflammatory cytokines, (3) markers of oxidative stress and lipid peroxidation and (4) leukocytosis.
- Acute hyperglycemia in DKA and NKH can also stimulate the elevation of (1) counter-regulatory hormones, (2) pro-inflammatory cytokines, (3) markers of cardiovascular risk, (4) markers of lipid peroxidation and oxidative stress and (5) leukocytosis.
- The elevation of pro-inflammatory cytokines and other changes have been shown in glycemic excursions of hypo- and hyperglycemia. However, hyperglycemia in DKA or NKH results in higher levels of measured hormones and cytokines compared to insulin-induced hypoglycemia.
- Prolonged glycemic derangements in hyperglycemic crises of DKA and NKH can be partially explained by the differences between hyperglycemia and hypoglycemia in elevation of pro-inflammatory cytokines. Additional investigations are required to elucidate the differences between these glycemic excursions.
- Elevation of pro-inflammatory cytokines in hypo- and hyperglycemic excursions may be an adaptive response to the acute glycemic derangements for maintaining glucose hemostasis.
- Additional studies are required to define the exact functions of cytokines in the hypothalamus-pituitary-adrenal axis and the sympathoadrenal system in response to metabolic stressors.

Abbreviations

A1c	:	hemoglobin A1c
DCF	:	dichlorofluorescein assay
DKA	:	diabetes ketoacidosis
Epi	:	epinephrine
FFA	:	free fatty acid
GH	:	growth hormone
HHS	:	hyperosmolar hypoglycemic state
IL-1β	:	interleukin-1β

IL-6	:	interleukin-6
IL-8	:	interleukin-8
NE	:	norepinephrine
ROS	:	reactive oxygen species
TBA	:	thiobarbituric acid assay
TNF-α	:	tumor necrosis factor-α

Acknowledgements

We are grateful to Ms. Brenda Scott for her secretarial assistance and Mr. John Crisler for his assistance in laboratory assays.

References

Bethin, K.E. and S.K. Vogt, and L.J. Muglia. 2000. Interleukin 6 is an essential, corticotropin-releasing hormone-independent stimulator of the adrenal axis during immune system activation. Proc. Natl. Acad. Sci. 97: 9317-9322.

Bumiller, A. and F. Gotz, W. Rhode, and G. Dorner. 1999. Effects of repeated injections of interleukin 1B or lipopolysaccharide on HPA axis in the newborn rats. Cytokine 11: 225-230.

Dotson, S. and R. Freeman, H.J. Failing, and G.K. Adler. 2008. Hypoglycemia increases serum interleukin-6 levels in healthy men and women. Diabetes Care 31: 1222-1223.

Elenkov, I.J. and G.P. Chrousos. 2002. Stress hormones, pro-inflammatory and anti-inflammatory cytokines, and auto-immunity. Ann. NY Acad. Sci. 966: 290-303.

Fain, J.N. and S.W. Bahouth, and A.K. Madan. 2004. TNF-α release by the nonfat cells of human adipose tissue. Int. J. Obes. Relat. Metab. Disord. 28: 616-622.

Fain, J.N. and D.S. Tichansky, and A.K. Madan. 2006. Most of the interleukin 1 receptor antagonist, cathepsin S, macrophage migration inhibitory factor, nerve growth factor, and interleukin 18 release by explants of human adipose tissue is by nonfat cells, not by the adipocytes. Metabolism 5: 1113-1121.

Glaser, R. and J.K. Kiecolt-Glaser. 2005. Stress-induced immune dysfunction: implications for health, Nature Reviews. Immunology 5: 243-251.

Gonzalez-Hernandez, J.A. and M. Ehrhart-Bornstein, E. Spath- Schwalbe, W.A. Scherbaum, and S. Bornstein. 1996. Human adrenal cells express tumor necrosis factor-alpha messenger ribonucleic acid: evidence for paracrine control of adrenal function. J. Clin. Endocrinol. Metab. 81: 807-813.

Haddad, J.J. and N.E. Saade, and B. Safieh-Garabedian. 2002. Cytokines and neuro-immune-endocrine interactions: a role for the hypothalamic-pituitary-adrenal revolving axis. J. Neuroimmunol. 133: 1-19.

Heremus, A.R. and C.G. Sweep. 1990. Cytokines and the hypothalamic-pituitary-adrenal axis. J. Steriods Biochem. Mol. Biol. 37: 867-871.

Holloway, A.F. and S. Rao, and M.F. Shannon. 2002. Regulation of cytokines gene transcription in the immune system. Mol. Immunol. 38: 567-580.

Judd, A.M. and G.B. Call, M. Barney, C.J. McIlmoil, A.G. Balls, A. Adams, and G.K. Oliveira. 2000. Possible function of IL-6 and TNF as intra adrenal factors in the regulation of adrenal steroid secretion. Ann. NY Acad. Sci. 917: 628-637.

Kitabchi, A.E. and F.B. Stentz, and G.E. Umpierrez. 2004. Diabetic ketoacidosis induces in vivo activation of human T- lymphocytes. Biochem. Biophys. Res. Commun. 315: 404-407.

Kitabchi, A.E. and G.E. Umpierrez, J.N. Fisher, M.B. Murphy, and F.B. Stentz. 2008. Thirty years of personal experience in hyperglycemic crises: diabetic ketoacidosis and hyperglycemic hyperosmolar state. J. Clin. Endocrinol. Metab. 93: 1541-1552.

Kitamura, H. and A. Konno, M. Morimatsu, B. Jung, K. Kimura, and M. Saito. 1997. Mobilization stress increases hepatic IL-6 expression in mice. Biochem. Biophys. Res. Commun. 238: 707-711.

Ko, H.J. and Z. Zhang, D.Y. Jung, J.Y. Jun, Z. Ma, K.E. Jones, S.Y. Chan, and J.K. Kim. 2009. Nutrient stress activates inflammation and reduces glucose metabolism by suppressing AMP-activated protein kinase in the heart. Diabetes 58: 2536-2546.

Lee, S.J. and J. Chinen, and A. Kavanaugh. 2009. Immunomodulator therapy: Monoclonal antibodies, fusion proteins, cytokines, and immunoglobulins. J. Allergy Clin. Immunol. Dec 24. [Epub ahead of print].

Mastorakos, G. and I. Ilias. 2006. Interleukin-6: a cytokine and/or a major modulator of the response to somatic stress. Ann. NY Acad. Sci. 1088: 373-381.

Oppenheim, J.J. 2001.Cytokines: past, present and future. Int. J. Hematol. 74: 3-8.

Palmer, K.L. and W.J. Kovac. 2002. Addison's disease. pp. 837-844. In: Oxford Textbook of Endocrinology and Diabetes. J.A. Williams and S.M. Sholet (eds). 2nd ed. Oxford University Press, London.

Papanicolaou, D.A. and J.S. Petrides, C. Tsigos, S. Bina, K.T. Kalogras, R. Wilder, P.W. Gold, P.A. Deutser, and G.P. Chrousos. 1996. Exercise stimulates interleukin-6 secretion: inhibition by glucocorticoids and correlation with catecholamines. Am. J. Physiol. 271: E601-E625.

Razavi Nematollahi, L. and A.E. Kitabchi, F.B. Stentz, J.Y. Wan, B.A. Larijani, M.M. Tehrani, M.H. Gozashti, K. Omidfar, and E. Taheri. 2009. Proinflammatory cytokines in response to insulin-induced hypoglycaemic stress in healthy subjects. Metabolism 58: 443-448.

Shizuya, K. and K.T. Komori, R. Fujiwara, S. Miyahara, M. Ohmori, and J. Nombra. 1997. The influence of restraint stress on the exposure of mRNAs for IL-6 and the IL-6 receptor in the hypothalamus and midbrain of the rat. Life Sci. 61: 135-140.

Stentz, F. and G.E. Umpierrez, R. Cuervo, and A.E. Kitabchi. 2004. Proinflammatory cytokines, markers of cardiovascular risks, oxidative stress, and lipid peroxidation in patients with hyperglycemic crises. Diabetes 53: 2079-2086.

Stentz, F.B. and A.E. Kitabchi. 2005. Hyperglycemia-induced activation of human T-lymphocytes with de novo emergence of insulin receptors and generation of reactive oxygen species. Biochem. Biophys. Res. Commun. 335: 491-495.

Stentz, F.B. and A.E. Kitabchi. 2006. Palmitic acid-induced activation of human T-lymphocytes and aortic endothelial cells with production of insulin receptors, reactive oxygen species, cytokines, and lipid peroxidation. Biochem. Biophys. Res. Commun. 346: 721-726.

Turnbull, A.V. and C.L. Rivier. 1999. Regulation of hypothalamic-pituitary-adrenal axis by cytokines:actions and mechanisms of action. Physiol. Rev. 79: 1-71.

Woiciechowsky, C. and B. Schoning, N. Daberkow, K. Asche, G. Stoltenburg, W.R. Lanksch, and H.D. Volk. 1999. Brain–IL-1B induces local inflammation but systemic anti-inflammatory response through stimulation of both hypothalamic-pituitary-adrenal axis and sympathetic nervous system. Brain Res. 816: 563-571.

Zho, D. and A.W. Kusnecov, M.R. Shurin, M. Depoali, and B. Rabin. 1993. Exposure to physical and psychosocial stressors elevates plasma interleukin-6: relationship to the activation of hypothalamus-pituitary-adrenal axis. Endocrinology 1133: 2523-2530.

CHAPTER 20

Cytokines in Metabolic Syndrome—Adipose Tissue as a Key Contributor

Marjukka Kolehmainen* and Pirkka Kirjavainen
[1]University of Eastern Finland, Department of Public Health and Clinical Nutrition, Clinical Nutrition, Food and Health Research Centre
PO Box 1627, FIN-70211 Kuopio, Finland

ABSTRACT

Obesity, particularly abdominal obesity, is a well-established risk factor for MetS, T2DM and CVD. The primary goal for obesity treatment is to prevent and treat these chronic diseases. MetS is a cluster of cardiovascular risk factors including abdominal obesity, elevated plasma glucose, dyslipidemia, hypertension and prothrombotic/pro-inflammatory state. Inflammation is now recognized as a central mediator in CVD and T2DM. In the state of obesity, pro-inflammatory and pro-thrombotic factors are produced in the adipose tissue. Via producing chemokines, macrophage migration to the adipose tissue is strongly amplified. Thus, it is evident that adipose tissue participates actively in the development of insulin resistance and inflammation.

Unhealthy diet, as well as sedentary lifestyle, also plays a key role in the development of inflammation. It has been hypothesized that postprandial dysmetabolism, characterized as repeated high and long-term increases of glucose and lipid concentrations after meals, may result in inflammation, endothelial dysfunction and development of atherosclerotic damage. There is solid evidence that many chronic health problems could be overcome by effective lifestyle modification.

Weight loss markedly decreases macrophage accumulation and production of inflammatory markers in adipose tissue. It has been indicated that decreasing

**Corresponding author*

energy intake and increasing physical activity may be effective in reducing overall inflammation. It is the aim of the present chapter to introduce some of the key cytokines/adipokines and their function during the development of inflammation in MetS, with a focus on adipose tissue function in the state of obesity.

INTRODUCTION

The prevalence of obesity, insulin resistance (IR), and type 2 diabetes mellitus (T2DM) is increasing worldwide because of sedentary lifestyle. MetS, T2DM and CVD share the same risk factors and seem to develop along with each other. Currently, 60% of patients with coronary artery disease have abnormal glucose homeostasis, i.e., impaired fasting glycemia, impaired glucose tolerance or T2DM. The discovery that adipose tissue is an active endocrine organ has opened a new avenue to resolve the basic mechanisms involved in the development of insulin resistance and its further development to chronic diseases. Obesity is a well-established risk factor for MetS, T2DM and CVD. The primary goal for obesity treatment is to prevent and treat these chronic diseases.

Inflammation is now recognized as a central mediator in CVD and T2DM (Herder *et al.* 2009, O'Keefe *et al.* 2008). In obesity, pro-inflammatory and prothrombotic factors are produced in the adipose tissue, but also insulin-sensitizing factors such as adiponectin. Via producing chemokines, such as monocyte chemoattractant protein 1 (MCP-1) or osteopontin (OPN), macrophage migration to the adipose tissue is strongly amplified. Thus, it is evident that adipose tissue participates actively in the development of insulin resistance and inflammation.

In the past few years, this area of research has been extensively reviewed from various perspectives (among others the elegant reviews by: Maury and Brichard 2010, Trayhurn *et al.* 2008, Vasquez-Vela *et al.* 2008, Cancello and Clement 2006, Hotamisligil 2006, Tilg and Moschen 2006, La Cava and Matarese 2004). Thus, the objective is to give an overview of these increasingly numerous reports and reviews, many of which will not, unfortunately, be mentioned due to limited space.

METABOLIC SYNDROME OR NOT?

MetS is a constellation of cardiovascular risk factors including abdominal obesity, elevated plasma glucose, dyslipidemia, hypertension and prothrombotic/pro-inflammatory state (Grundy 2008). Its clinical importance lies in identifying individuals at high risk of T2DM and CVD. Whether it forms an independent health problem has long been controversial. Various criteria or statements of MetS have been given by different bodies, such as the World Health Organisation, the National Cholesterol Education Program–Adult Treatment Panel III, The European Group for study of Insulin Resistance, American Diabetes Association,

European Association on Study of Diabetes and International Diabetes Federation as well as recent a consensus statement (Alberti *et al.* 2009).

Although obesity is a risk factor for metabolic disorders, it seems that most of the risk lies in the abdominally distributed adipose tissue. Genetic factors have a decisive role, with an estimated 35-50% of the individual's fat distribution proportions being defined by genotype. Main adipose tissue depots in humans are visceral—situated in the abdominal cavity—and subcutaneous abdominal and gluteal depots, all three differing markedly in their functional activity. In obese individuals (BMI $\geq$ 30), adipose tissue amounts to 45% or more of the total body composition (Trayhurn *et al.* 2008). Thus, white adipose tissue is then the largest endocrine organ (Trayhurn *et al.* 2008). Visceral adipose tissue has long been regarded as the strongest risk factor for metabolic disturbances. However, visceral adipose tissue comprises only some 10% of the total adipose tissue mass of obese subjects (6% in women and 18% in men), and thus, abdominal subcutaneous depot has also great impact on fat-induced metabolic problems. Lately, it has been noted that epicardial fat would bear a strong risk for defective metabolism (Arner 2007), since its local inflammatory effects could affect the vascular function of the arterial wall.

ADIPOSE TISSUE IS A KEY PLAYER IN DEVELOPMENT OF INFLAMMATION IN METABOLIC SYNDROME

Adipose tissue interacts with various metabolic pathways and physiological routes by producing adipokines and other cytokines that participate in regulation of glucose, lipid, energy metabolism as well as immunity. In fact, some 60 biologically active secretory factors have been recognized (Trayhurn *et al.* 2008).

Adipose tissue is composed of many different cell types, such as mature adipocytes, macrophages, preadipocytes, mesenchymal stem cells, and fibrinocytes, all of which may contribute to a variable extent to adipose tissue secretory function. Of note is that visceral and subcutaneous adipose tissues have quite distinct cytokine profiles (Cancello and Clement 2006). It has been suggested that the limited capacity of the subcutaneous adipose tissue to store excess energy results in overflow of fatty acids to intra-abdominal, i.e., visceral fat and "ectopic" sites such as liver, muscle, and pancreatic islets (Maury and Brichard 2010), which then causes local inflammation in the tissues and enhances the local production of cytokines. Leptin is preferentially secreted by subcutaneous adipose tissue, while the expression of adiponectin, plasminogen activator inhibitor 1 (PAI1), interleukin (IL) 8, and IL1β is more pronounced in visceral adipose tissue than in subcutaneous tissue (Maury and Brichard 2010). However, the preferential site for IL6 production still remains unclear. It seems that both visceral and subcutaneous adipose tissue of obese as well as non-obese subjects release this cytokine (Cancello and Clement 2006). It is, thus, evident that the relative amount of the different depots may contribute to the relative risks for the occurrence of metabolic and cardiovascular complications.

In the chronic state of obesity, adipocyte hypertrophy is usually seen. Lipolysis has been detected to be enhanced in the large compared to the smaller adipocytes. There are also other factors affecting this phenomenon, but ineffective response to insulin in large adipocytes, i.e., insulin resistance, leading to increased production of hormone-sensitive lipase and adipose tissue glyceride lipase, is the most important factor. This leads to increased release of free fatty acids. Especially in the visceral fat this is harmful, since the fatty acids released have direct access to hepatic circulation from visceral adipose tissue depot. This, in turn, impairs hepatic metabolism and leads eventually to increased hepatic glucose and VLDL-particle production.

It is evident that inflammation develops in the adipose tissue of obese individuals and contributes to systemic inflammation, but what initiates this inflammatory cascade?

Only limited attention has been focused on aiming to answer why growing adipocyte size and adipose mass in obesity should lead to a state of inflammation within the adipose tissue (Trayhurn *et al.* 2008). It has been suggested that the inflammation is fundamentally a response to: (1) oxidative stress, (2) endoplasmic reticulum stress, or (3) local hypoxia (Trayhurn *et al.* 2008). These proposed mechanisms complement each other and might be part of a vicious circle: hypoxia can lead to endoplasmic reticulum stress and the generation of reactive oxygen species which, in turn, leads to more pronounced hypoxia (Trayhurn *et al.* 2008).

Hypoxia has been presented as a key contributing factor for stimulation of cytokines production and secretion in adipose tissue (Trayhurn *et al.* 2008). This is linked to the capacity of adipocytes to increase their size over ten-fold during times of energy surplus, reaching 150-200 μM in diameter. Thus, the size of adipocytes can become larger than the normal diffusion range of O_2 in tissues. Moreover, it is well recognized that blood flow is not increased post-prandially in the adipose tissue of obese individuals similarly as in lean individuals, and that obesity-induced increase in cardiac output does not seem to reach adipose tissue (Trayhurn *et al.* 2008). Both of these factors enhance the hypoxia even more in the obese adipose tissue. Hypoxia induces then the initiation of an inflammatory response and angiogenesis (Trayhurn *et al.* 2008). This response has been found to be transmitted by hypoxia inducible factor 1α (HIF-1α). Other hypoxia-sensitive transcription factors are NFκB and CREBP. These transcription factors induce mRNA expression of several inflammation-related factors, such as IL1β, IL6, leptin, macrophage migration inhibitory factor (MIF), matrix metalloproteinases (MMP), PAI1, visfatin and tumor necrosis factors α (TNFα); in promoting the inflammation, they decrease adiponectin mRNA level.

Endoplasmic reticulum (ER) stress is a response to the disruption of protein folding by the ER, and following accumulation of unfolded proteins in the ER (Tilg and Moschen 2006). The ER is a network of membranes in which the secretory and membrane proteins gain their proper conformation. Folding, maturation, storage and transport of these proteins take place in this organelle (Hotamisligil 2006). Unfolded or misfolded proteins are detected, removed from the ER and degraded. Accumulation of unfolded proteins, energy and nutrient fluctuations, hypoxia, toxins, viral infections induce a high demand on the machinery that may give rise to stress (Hotamisligil 2006). ER stress leads to suppression of signaling through the insulin receptor through activation of Jun N-terminal kinases (JNK) and subsequent serine phosphorylation of the insulin receptor substrate (Tilg and Moschen 2006). Insulin receptor substrate is one of the main mediators of insulin signaling controlling the sensitivity to insulin. Insulin resistance in adipose tissue in turn leads to the consequences discussed above.

On the other hand, in obesity, the turnover rate of adipocytes is higher than in the adipose tissue of lean subjects, thus requiring greater infiltration of macrophages to clear the cellular debris. Moreover, the larger adipocytes may simply have increased capacity of basal production of certain cytokines, which may then promote the phenotypic switch of the macrophage population from non-/anti-inflammatory to pro-inflammatory and their cytokines in turn further regulate the cytokine production by adipocytes. Also, the pro-inflammatory effects of gut barrier permeating lipopolysaccharide (LPS) may contribute to obesity-associated inflammation.

ADIPOSE TISSUE IS AN AMPLE SOURCE OF CYTOKINES

In human obesity, adipose tissue is characterized by adipocyte hypertrophy (increased size) and hyperplasia (increased number), macrophage infiltration, endothelial cell activation and fibrosis (Trayhurn *et al.* 2008, Arner 2007, Cancello and Clement 2006). Adipocytes and the infiltrating macrophages are the primary cytokine producers in AT. Adipocytes produce an extensive range of immunologically active factors including adipokines, which are cytokines that are primarily (but not exclusively) produced in the adipose tissue. Adipokines include among others adiponectin, leptin, resistin and visfatin. Of these, adiponectin and leptin are produced in the highest concentrations. Other factors produced by adipocytes include the cytokines TNFα, IL1β, IL6, IL10, TGFβ and osteopontin; chemokines IL8, MCP1, MIF, and chemerin; as well as nerve growth factor (NGF); and pro-angiogenic molecules vascular endothelial growth factor (VEGF) and MMPs. Hypertrophy would appear to be associated with increased production of pro-inflammatory cytokines.

In humans, macrophage infiltration is correlated with both adipocyte size and BMI and is reduced after surgery-induced weight loss in morbidly obese subjects (Cancello and Clement 2006). Macrophages infiltrate more into the visceral adipose tissue than to the subcutaneous depots (Maury and Brichard

2010). Interestingly, studies on mice indicate that non-obese state adipose tissue macrophages exhibit an anti-inflammatory, so-called wound-healing phenotype (one of the alternatively activated (M2) macrophage phenotypes) characterized by IL10 and araginase production. In obesity, adipose tissue macrophage population is transformed to classically activated (M1), pro-inflammatory phenotype (Maury and Brichard 2010). These classically activated adipose tissue macrophages are probably the primary source of the central insulin signaling interfering cytokine, TNFα, in adipose tissue and, thus, central in the etiology of T2DM. They are also an ample source of MCP1 (CCL2), attracting more macrophages to adipose tissue, and T helper cell type 1 (Th1) differentiation promoting IL12.

Adiponectin

The circulating levels of adiponectin are the highest of all adipokines (5-10 mg/mL). In human, adiponectin is primarily expressed by mature adipocytes, but to some extent also by other cell types such as skeletal muscle cells (Tilg and Moschen 2006). Adiponectin is considered one of the most important beneficial cytokines. It exhibits insulin-sensitizing, fat-burning, cardioprotective, anti-inflammatory and anti-oxidant properties, thereby counteracting several MetS-associated pathologies (Maury and Brichard 2010). Systemic adiponectin levels are negatively correlated with BMI, and decreased in viscerally obese subjects, in patients with T2DM or CVD. Why are adiponectin levels decreased in obesity and its complications? It has been postulated that pro-inflammatory state and concomitant oxidative stress and hormonal disturbances might be the key factors (Maury and Brichard 2010). Of the obesity-associated pro-inflammatory cytokines at least TNFα would appear to suppress adiponectin expression in adipocytes.

AdipoR1 and AdipoR2 serve as major receptors for adiponectin. Interaction of adiponectin with its receptors stimulates the activation of AMP-activated protein kinase (AMPK), peroxisome-proliferated receptor (PPAR) α and p38mitogen-activated protein kinase (Maury and Brichard 2010). AdipoR1 is abundantly expressed in muscle, whereas AdipoR2 is also expressed in liver (Tilg and Moschen 2006). AdipoR1 is more tightly linked to the activation of AMPK pathways that regulate the inhibition of gluconeogenesis together with increased fatty acid oxidation, while AdipoR2 is more involved in the activation of the PPAR-α pathway, which stimulates fatty acid oxidation and inhibits oxidative stress and inflammation.

The beneficial effects of adiponectin are probably due to its anti-inflammatory/anti-oxidant properties via regulation of endothelial cell and monocyte/macrophage functions (Maury and Brichard 2010). Interaction between adiponectin and its receptors (AdipoR1 and AdipoR2) on monocytes and macrophages suppresses the NFκB-mediated synthesis of TNFα and interferon (IFN)-γ and stimulates the production of the central anti-inflammatory cytokines IL10 and IL1 receptor antagonist (IL1Ra) (Tilg and Moschen 2006). There is also indication that adiponectin may suppress the phagocytic activity of macrophages and their responsiveness to toll-like receptor 4 (TLR4)-mediated activation by LPS. Notably, TLR4-mediated signaling pathways can interfere with insulin signaling and also be activated by free fatty acids, which are commonly elevated in circulation in obesity. The regulation of LPS responsiveness may also in itself be relevant regarding MetS, as LPS translocation from the gut microbiota may be increased in particular by Western diets containing high fat (Ghoshal *et al.* 2009). Interestingly, rosiglitazone treatment of patients with T2DM was associated concomitantly with reduced serum endotoxin concentrations and increased adiponectin (Al-Attas *et al.* 2009).

On endothelium, adiponectin reduces adhesion molecule expression in response to TNFα-induced activation of NFκB in endothelial cells (Tilg and Moschen, 2006). Adiponectin also suppresses superoxide radical production in endothelial cells treated with oxidized LDL or high dose of glucose, while oxidative stress itself decreases adipose tissue adiponectin production (Tilg and Moschen 2006).

In the circulation, adiponectin exists in many forms: in low-molecular-weight form (LMW, full-length trimer), globular form (proteolytic cleavage fragment), middle-molecular-weight form (MMW, hexamer), and high-molecular-weight form (HMW, hexamer olimerised into polymer) (Tilg and Moschen 2006). It may be that it is the ratio of HMW/MMW adiponectin to LMW, rather than total adiponectin levels, that may be most relevant to development of insulin sensitivity. Weight loss would seem to result in a shift from LMW adiponectin to MMW and HMW adiponectin isoforms. These changes are also accompanied by improvements in various anthropometric and metabolic parameters. However, the immunological role of different adiponectin isomers requires further investigations. *In vitro*, HMW adiponectin has been shown to augment the production of IL6 and IL8, a neutrophil chemoattractant, in monocytes as a response to LPS, whereas LMW adiponectin exhibited anti-inflammatory effect via reduced IL6 and enhanced IL10 production (Tilg and Mochen 2006).

Vaspin (visceral adipose-tissue-derived serine protease inhibitor) is a more recently identified adipokine with comparable properties to adiponectin, including improved insulin sensitivity and potential anti-inflammatory effects seen as suppression of TNF, leptin and resistin production (Tilg and Moschen 2006).

Leptin

Leptin is a primarily adipocyte-produced, multipotential hormone/cytokine with pro-inflammatory characteristics. Circulating leptin concentrations are increased in common obesity, indicative of obesity-associated leptin resistance.

The circulating leptin levels parallel adipose tissue mass, are higher in females than in males and are reduced by testosterone, fasting and malnutrition (La

Cava and Matarese 2004). Leptin secretion is increased by insulin (and *vice versa*), endotoxins, IL1 and TNF-α. Leptin regulates energy metabolism, and is considered as a marker of energy sufficiency that controls adipose tissue growth by interacting with the central nervous system, namely the hypothalamo-pituitary-adrenal (HPA)-axis (La Cava and Matarese 2004, Vasquez-Vela *et al.* 2008). Leptin is secreted into the bloodstream with a circadian rhythm that is opposite to that of glucocorticoid secretion. In the hypothalamus, leptin regulates appetite, autonomic nervous system outflow, bone mass and the secretion of the HPA hormones (Vasquez-Vela *et al.* 2008). In the periphery, leptin increases basal metabolism, influences reproductive function, regulates pancreatic β-cell function and insulin secretion, is pro-angiogenic for endothelial cells and regulates bone-marrow hematopoiesis (Matarese *et al.* 2007).

As a cytokine, leptin would appear to have broad effects on both innate and adaptive immunity characterized by stimulation of cell-mediated, pro-inflammatory responses, thus reversely mirroring the effects of adiponectin. Some of the stimulatory effects of leptin include those on the production of pro-inflammatory cytokines such as TNF-α, IL-6 and IL-12 by monocytes, macrophages and dendritic cells, chemotaxis and activation of neutrophils and activation of natural killer cells (Tilg and Moschen 2006). Leptin also appears to have pronounced regulatory effects on T cell responsiveness at many levels. These effects include stimulation of the survival and proliferation of thymocytes and naïve T helper cells, dendritic cells (DC) to direct T cell polarization toward Th1 and the production of the Th1 cytokines IFN-γ and Il-2, but suppressing action on the production of Th2 cytokine IL-4 (La Cava and Matarese 2004). To some extent, leptin can also be produced by T cells themselves. Interestingly, while starvation-induced immune suppression seems reversible by leptin administration, in animal model of influenza virus infection, the obesity-associated high basal leptin levels are associated with defective infection-induced leptin secretion and subsequently reduced protective memory T cell responses (Karlsson *et al.* 2010). However, while the data on the immunological effects of leptin are convincing and abundant, until today, they are so far largely based on mouse and *in vitro* studies and thus the effects of leptin on human immune responses are unclear.

Receptors for leptin, expressed in four isoforms, are located in nearly all tissues in humans. Long form of the receptor, ObRb, is able to transduce complete signaling associated with energy metabolism by activation of the hypothalamic–sympathetic nervous system axis (Vasquez-Vela *et al.* 2008, Maury and Brichard 2010). It activates JAK-STAT3, leading to suppression of appetite-stimulating (orexigenic) peptides (e.g., neuropeptide Y and agouti-related protein), and increase in appetite-suppressing (anorexigenic) peptides (e.g., pro-opiomelanocortin and corticotrophin-releasing hormone) (Maury and Brichard 2010). Leptin stimulates fatty acid oxidation and glucose uptake (Vasquez-Vela *et al.* 2008, Maury and Brichard 2010) and thus prevents lipid accumulation in adipose and other tissues.

However, in common obesity, leptin resistance may enhance, in part, adipose tissue enlargement, especially in ectopic depots. Leptin receptors are also located in adipose tissue, suggesting autocrine/paracrine action for leptin.

Visfatin and Resistin

Visfatin seems to be a pro-inflammatory adipokine, but its significance in metabolic disorders remains to be confirmed. Visfatin (originally named pre-B colony enhancing factor, PBEF) was once described as a protein that exerts insulin-like functions by binding and activating insulin receptor without competing with insulin. However, these original findings were irreproducible and brought into question by others (Garten *et al.* 2009). The correlation between the circulating visfatin concentrations and MetS-related factors has been inconsistent (Maury and Brichard 2010). Studies with improved methodology indicate that visfatin levels are increased in type 2 diabetes but there is no correlation to anthropometric or metabolic parameters (Garten *et al.* 2009). Subsequent to the introduction of visfatin, it was shown that it has nicotinamide adenine dinucleotide (NAD) biosynthetic activity, which is essential for pancreatic β-cell function, and thus, as alternative name, nicotinamide phosphoribosyltransferase (Nampt) was introduced (Garten *et al.* 2009). The immunomodulatory effects of Nampt/visfatin include promotion of neutrophil and macrophage survival and production of pro-inflammatory cytokines IL6 and IL8 (Garten *et al.* 2009). Accordingly, elevated Nampt/Visfatin levels have been implicated in several acute or chronic inflammatory conditions, such as atherosclerosis and CVD (Maury and Brichard 2006).

Resistin is another pro-inflammatory adipokine that would appear to promote the formation of insulin resistance at least in mice. In humans peripheral blood mononuclear cells rather than adipose tissue cells would appear to be the primary source of circulating resistin. Resistin production by PBMCs is up-regulated by IL1 IL6, TNFα and LPS but not by IFNγ or leptin. Resistin itself would seem to promote the synthesis of IL1, IL6, TNF and IL12 via NFκB pathway in several different cell types (Tilg and Moschen 2006).

Monocyte Chemoattractant Protein I and Osteopontin

Macrophage infiltration to adipose tissue is one of the most fundamental initial events in the development of the obesity-associated inflammatory state and the consequent formation of insulin resistance (Maury and Brichard 2010). These cells are attracted to the adipose tissue by chemokines, such as MCP1 and osteopontin (OPN), the secretion of which is strongly increased in adipose tissue in obesity (Maury and Brichard 2010, Arner 2007).

MCP1, also known as chemokine (C-C motif) ligand 2 (CCL2), is probably the most relevant macrophage chemoattractant in adipose tissue. It is produced more by stromal cells than by adipocytes and more in the visceral than in subcutaneous adipose tissue, as is described for TNF-α and IL6. MCP1 is also involved in the recruitment of monocytes/macrophages into the arterial vessel wall, a major event leading to atherosclerotic lesions (Maury and Brichard 2010).

OPN is involved in cell adhesion, chemoattraction, and immunomodulation. In particular, OPN is highly secreted by macrophages at inflammation sites where it promotes monocyte adhesion, migration, and differentiation as well as phagocytosis (Maury and Brichard 2010, Arner 2007).

Tumor Necrosis Factor α

TNFα is a pro-inflammatory cytokine (Hotamisligil 2006). It acts like autocrine and/or paracrine in human adipose tissue; thus, adipose tissue does not play a major role in producing circulating pool of TNFα. TNFα is over-expressed in adipose tissue of obese individuals (Hotamisligil 2006, Arner 2007) showing increased expression especially in visceral adipose tissue. This is in accordance with the greater numbers of macrophages in visceral adipose tissue, considering that macrophages are the primary source of TNFα and considering the key role of TNFα in promoting monocyte/macrophage exit from blood vessels into the tissues. Most effects of TNFα on adipose tissue are mediated by the TNFα receptor 1 subtype (TNFR1) and subsequent activation of various transduction pathways. Overproduction of TNFα in obesity could be considered an ultimate attempt to control the harmful effects of excess adipose tissue as it results in enhanced lipolysis, insulin resistance and adipocyte differentiation, a seeming attempt to prevent further weight gain. In chronic state, however, these effects will not be beneficial (Hotamisligil 2006).

TNFα inhibits insulin signaling and induces insulin resistance via various molecular mechanisms (Maury and Brichard 2010, Hotamisligil 2006). These mechanisms involve a JNK-mediated serine phosphorylation of insulin receptor substrate (IRS) 1, which inhibits normal tyrosine phosphorylation of IRS1 and downstream insulin signaling. TNFα may also decrease the amount of glucose transporting protein 4 (GLUT4) (Hotamisligil 2006). In humans, however, increased phosphorylation and decreased expression of perilipin are the dominant factors behind the lipolytic effect of TNFα. In addition, TNFα induces insulin resistance indirectly, e.g., by stimulating adipocyte lipolysis so that the circulating fatty acid level increases. Fatty acids inhibit glucose metabolism in skeletal muscle, increase hepatic glucose production and inhibit insulin receptor signaling, as mentioned above.

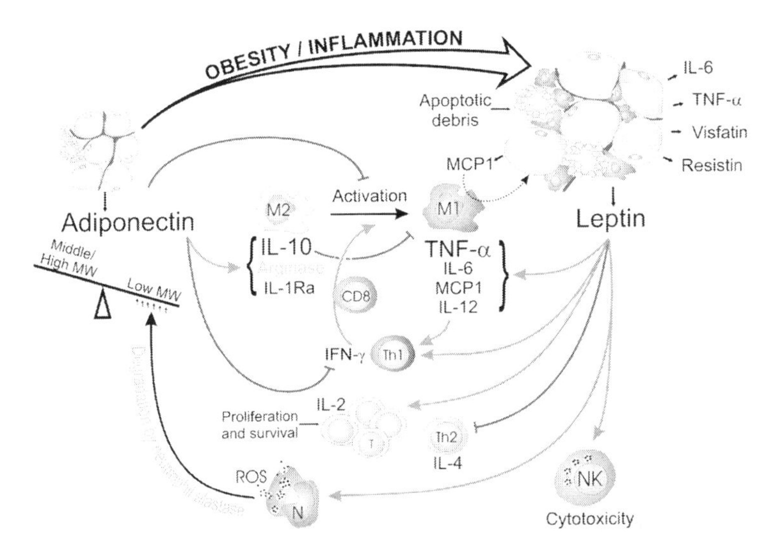

Fig. 1 Interaction of key pro- and anti-inflammatory cytokines, with particular emphasis on the effects of leptin and adiponectin. In the lean state, adipose tissue is in non-inflammatory state characterized by abundant concentrations of adiponectin, which has several anti-inflammatory effects. Also alternatively activated (M2) macrophages of wound-healing phenotype prevail and contribute to the immunological homeostasis by production of anti-inflammatory cytokines such as IL-10. With increasing obesity, changes in adipose tissue cause a shift towards pro-inflammatory secretions such as leptin and in time the adipose tissue macrophage population is increased and transformed to classically activated, pro-inflammatory phenotype (M1) characterized by pro-inflammatory cytokine production. Leptin promotes NK cytotoxicity, neutrophil chemotaxis and activation, T-cell differentiation to Th1, proliferation and survival of naïve T cells and production of pro-inflammatory cytokines such as TNF-α by monocytes/macrophages. IFN-γ producing Th1 cells promote the transition from M2 to M1 cells and may activate cytotoxic T cells (CD8+), which may also contribute to M1 activation and recruitment. Neutrophil action (production of neutrophil elastase) may result in degradation of middle and high molecular weight adiponectins, which may result in reduced anti-inflammatory action by this adipokine. Green arrows indicate promoting and red lines with blunt end inhibitory action. Th1/Th2, T-helper cell type 1/type 2; NK, natural killer cell; N, neutrophil; ROS, reactive oxygen species; M1, classically activated, pro-inflammatory, macrophage; M2, alternatively activated, wound healing/anti-inflammatory macrophage; Low/Middle/High MW, Low/Middle/High molecular weight adiponectin; CD8, CD8+ cytotoxic T-cells. (Figure produced by Pirkka Kirjavainen.)

Color image of this figure appears in the color plate section at the end of the book.

TNFα may also have an atherogenic role. It up-regulates the expressions of intracellular cell adhesion molecule 1 (ICAM1), vascular cell adhesion molecule 1 (VCAM1) and MCP1 in the vascular wall, and by inducing scavenger receptor class A expression and oxidized LDL uptake in macrophages (Hotamisligil 2006). This effect might be mediated via tight cross-link with epicardial adipose tissue, as mentioned above.

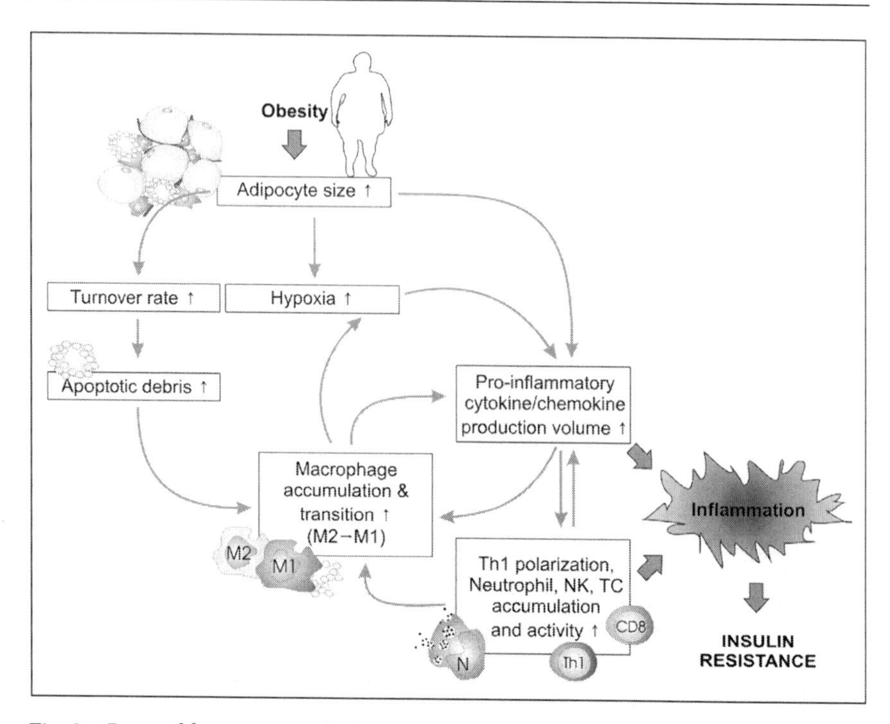

Fig. 2 Potential key events in the formation of a vicious cycle maintaining and exacerbating chronic inflammation in obesity, which interferes with insulin signaling resulting in reduced insulin sensitivity. In obesity, the size of adipocytes is increased. This causes (1) increased turnover and thus apoptosis rate and need for clearance by macrophages, (2) hypoxia, and (3) increased production volume of potentially pro-inflammatory cytokines such as leptin. The cytokine milieu together with other signals then skew the adipose tissue macrophage population to exhibit pro-inflammatory phenotype (M1), result in Th1 polarization and recruitment and activation of cytotoxic T-cells (CD8+), natural killer cells and neutrophils, which further exacerbate the inflammation. These events also result in further infiltration of monocytes/macrophages to the adipose tissue, causing further hypoxia and increase in pro-inflammatory cytokine/chemokine production; thus, a vicious cycle may be formed. Green arrows indicate promoting action. Th1, T-helper cell type 1; NK, natural killer cell; N, neutrophil; M1, classically activated, pro-inflammatory, macrophage; M2, alternatively activated, wound healing/anti-inflammatory macrophage; CD8, CD8+ cytotoxic T-cells. (Figure produced by Pirkka Kirjavainen.)

Color image of this figure appears in the color plate section at the end of the book.

What makes TNFα so powerful an pro-inflammatory molecule is its ability to stimulate production of other pro-inflammatory factors, such as IL6 (Maury and Brichard 2010, Vasquez-Vela *et al.* 2008), and its inhibition of production of anti-inflammatory factors, such as adiponectin (Arner 2007).

Table 1 Key adipokines, other cytokines and chemokines associated with inflammation in obesity

CYTOKINE/ CHEMOKINE	*Abbreviation*	*Pro- /Anti-inflammatory*	*Presumable primary source(s)*	*Preumable primary relevance in obesity*
ADIPOKINES				
Adiponectin	..	Anti	V;AC	**Insulin sensitivity ↑**
Leptin	..	Pro/Anti	S;AC	**Insulin sensitivity ↓**
Visfatin/ Nampt	..	Pro	V	**IL6, IL8 ↑ (?)**
Resistin	..	Pro	PBMC/AC	**Insulin sensitivity ↓**
Vaspin			V	**Insulin sensitivity ↑**
OTHER CYTOKINES				
Tumor necrosis factor α	TNFα	Pro	M1/AC	**Insulin sensitivity ↓**
Interleukin 1 receptor antagonist	IL1Ra	Anti		**IL1 ↓**
Interleukin 1β	IL1b	Pro	V	**Insulin sensitivity ↓**
Interleukin 6	IL6	Pro	V/S; M1	**Insulin sensitivity ↓**
Interleukin 10	IL10	Anti	M2/$T_{reg(?)}$	**M2↑, Insulin sensitivity ↑**
Interleukin 12	IL-12	Pro	M1/DC	**Th1 ↑**
Transforming growth factor β	TGFβ	Anti		
Angiopoietin-like protein 4	Angptl4	?(Anti)		**Insulin sensitivity ↑, but might induce dyslipidemias**
Serum amyloid A	SAA	Pro		**Insulin sensitivity ↓**
CHEMOKINES				
Monocyte chemoattractant protein-1	MCP1/CCL2	Pro	AC/T_{CD8}	**Macrophage recruitment ↑**
Macrophage migration inhibitory factor	MIF	Pro		**Macrophage activation ↑(?)**
Interleukin 8	IL8	Pro	V	**Neutrophil recruitment ↑**
Osteopontin↑	OPN	Pro		**Macrophage recruitment ↑**

Note that the context of publication has to be considered when cytokines are generally characterized as either pro- or anti-inflammatory. AC, adipocyte; DC, dendritic cell; M1, classically activated, pro-inflammatory macrophage; M2, alternatively activated anti-/non-inflammatory macrophage; PBMC, peripheral blood mononuclear cell; T_{CD8}, CD8+ T-cell; T_{reg}, regulatory T cell; S, subcutaneous adipose tissue; V, visceral adipose tissue.

Interleukin 6

Human studies suggest that IL6 is produced by adipocytes, fibroblasts, endothelial cells and resident macropages in adipose tissue (Maury and Brichard 2010). Adipose tissue might be a major source of circulating IL6, contributing even 15-35% of the systemic IL6 in humans. Circulating levels and adipose tissue production of IL6 are increased in obesity. Moreover, IL-6 is produced 2-3 times more in the visceral than subcutaneous adipose tissue (Cancello and Clement 2006).

IL6 seems to have direct effect on insulin signaling in adipocytes and hepatocytes (Maury and Brichard 2010). It causes insulin resistance by, above all, decreasing the activation of IRS-1 via Janus kinases (JAKs) as intracellular signaling pathways. In mice fed a high-fat diet, the increased production of IL6 by adipose tissue induced hepatic insulin resistance. The effect could be mediated by the increased expression of suppressor of cytokine signaling (SOCS)-3, a protein that binds and inhibits the insulin receptor and also targets IRS proteins for proteosomal degradation. The adipokine can also have indirect effects on insulin action, which are mediated by IL6 stimulation of adipocyte lipolysis.

IS THE INFLAMMATORY STATE IN METABOLIC SYNDROME REVERSIBLE?

Unhealthy diet plays a key role in the development of inflammation (O'Keefe *et al.* 2008). It has been hypothesized that post-prandial dysmetabolism, characterized as repeated high and long-term increases of glucose and lipid concentrations after meals, may result in inflammation, endothelial dysfunction and development of atherosclerotic damage (O'Keefe *et al.* 2008). There is solid evidence that many chronic health problems could be overcome by effective lifestyle modification (Lindström *et al.* 2006, Tuomilehto *et al.* 2001, Knowler *et al.* 2002). In fact, it has been shown that after dramatic weight loss a marked decrease is seen in macrophage accumulation and the production of inflammatory markers in adipose tissue (Cancello and Clement 2006). Adipose tissue responds readily to dietary modifications, indicating that adipose tissue harbours sensory mechanisms for them (Dahlman *et al.* 2005, Kallio *et al.* 2007).

Decreasing energy intake and increasing physical activity may be effective in reducing overall inflammation, cytokine concentrations and expression of linked genetic factors either after moderate weight loss or more dramatically after bariatric surgery (Cancello and Clement 2006). The genetic factors belong to 12 functional families including cytokines, ILs, complement cascade factors, acute phase proteins and some molecules involved in cell-cell contacts or extracellular matrix remodeling. This improvement of the inflammatory profile resulted not only from reduction of the expression of the pro-inflammatory factors but also from increase of the expression of anti-inflammatory factors, such as IL10 or IL1Ra.

Summary Points

- Inflammation is the key pathological factor in the development of complications of obesity, i.e., insulin resistance, metabolic syndrome, type 2 diabetes and cardiovascular diseases.
- Adipose tissue is suggested to be a major player in the development of metabolic syndrome: fatty acid overflow, fat deposition to ectopic deposits, macrophage infiltration and secretion of pro-inflammatory cytokines mediating the pathology.
- The secretory functions of adipocytes and preadipocytes and infiltrating macrophages make adipose tissue an active endocrine organ in which the lean-state homeostasis between anti- and pro-inflammatory cytokines/ hormones is skewed towards pro-inflammatory secretions in obesity, a plausible origin of the metabolic syndrome.
- The core cytokine/hormone imbalances in metabolic syndrome include (1) low concentrations of the anti-inflammatory adipokine adiponectin, (2) elevated levels of pro-inflammatory and cell-mediated immunity-promoting adipokine leptin, (3) production of pro-inflammatory, primarily M1 macrophages and of TNFα, which interferes with the insulin signaling pathway.
- Adipose tissue inflammation in obesity is augmented by macrophage infiltration (especially to the visceral tissue), and transition of the adipose tissue macrophage population from anti-inflammatory, wound-healing (M2) phenotype to pro-inflammatory, classically activated (M1) phenotype.
- Lifestyle modifications are powerful means to overcome and prevent the complications of obesity: weight reduction, increasing the intake of dietary fiber as whole grain, vegetables, fruits and berries, and consumption of fish- and vegetable-based fats as well as increasing habitual physical activity are proven to be effective in prevention of chronic diseases.

Abbreviations

AdipoR1	:	adiponectin receptor 1
AdipoR2	:	adiponectin receptor 2
AMPK	:	AMP-activated protein kinase
CCL2	:	chemokine (C-C motif) ligand 2
CCR2	:	chemokine (C-C motif) ligand receptor 2/ monocyte chemoattractant protein 1 receptor 2
CREBP	:	cAMP response element-binding protein

CRP	:	C-reactive protein
CVD	:	cardiovascular diseases
DC	:	dendritic cells
ER	:	endoplasmic reticulum
GLUT4	:	glucose transporting protein 4
HIF-1α	:	hypoxia inducible factor 1α
HMW	:	high molecular weight
HPA	:	hypothalamo-pituitary-adrenal
ICAM1	:	intracellular cell adhesion molecule 1
IFN	:	interferon
IL	:	interleukin
IL1Ra	:	IL1 receptor antagonist
IR	:	insulin resistance
IRS1	:	insulin receptor substrate 1
JAK	:	Janus kinases
JNK	:	Jun N-terminal kinase
LDL	:	low density lipoprotein
LMW	:	low molecular weight form
LPS	:	lipopolysaccharide
MCP-1	:	monocyte chemoattractant protein 1
MetS	:	metabolic syndrome
MIF	:	macrophage migration inhibitory factor
MMP	:	matrix metalloproteinases
MMW	:	middle molecular weight form
NAD	:	nicotinamide adenine dinucleotide
Nampt	:	nicotinamide phosphoribosyltransferase
NFκB	:	nuclear factor κB
NGF	:	nerve growth factor
OPN	:	osteopontin
PAI1	:	plasminogen activator inhibitor 1
PPAR	:	peroxisome-proliferated receptor
SOCS	:	suppressor of cytokine signaling
T2DM	:	type 2 diabetes
TGFβ	:	transforming growth factor β
Th1	:	T helper cell type 1
TLR	:	toll-like receptor

TNFα	:	tumor necrosis factors α
TNFR1	:	TNFα receptor 1 subtype
VCAM1	:	vascular cell adhesion molecule 1
VEGF	:	vascular endothelial growth factor
VLDL-particle	:	very low density lipoprotein particle

Definition of Terms

Adipokine: Metabolically active cytokine primarily produced and secreted in adipose tissue. Adipokines have various functions in human physiology and they participate in, e.g., energy, glucose and lipid metabolism and immunity.

Cytokine: A secreted small protein molecule affecting the behavior of other cells (cellular communication mediating molecule), the effect of which is usually mediated via specific cytokine receptor. Cytokines are often specified by the cell type by which they are produced, e.g., lymphocyte-produced lymphokines or interleukins and adipocyte-produced adipokines.

Ectopic adipose tissue: Adipose tissue that is located in other organs such as in muscle and liver. These depots have strong negative effect on the normal metabolism of the target tissue, for example, it inhibits insulin signaling in the muscle promoting the development of insulin resistance.

Inflammation: A general term for the state of physiology, which can exist in many different forms. In its classic, acute form, inflammation is a rapidly developing host response intended to remove invading exogenous matter and/or heal injured tissue and is subsequently resolved. It is characterized by accumulation of white blood cells (e.g., macrophages, neutrophils and lymphocytes), fluids and plasma proteins and increased blood flow to tissue, vasodilatation and vessel permeability, which can result in visible redness and swelling and is associated with pain. However, inflammation can also result in absence of obvious exogenous stimulation, injury or autoimmune response and does not resolve by itself. Such inflammation is often described as "chronic", "sub-clinical" and/or "low-grade" as in association of obesity. The inflammation-associated mediators are largely the same but chronic inflammation is typically less severe and macroscopically less obvious but can disturb cellular metabolism, function and turnover rate and result in tissue damage, lesions and fibrosis.

Metabolic syndrome: A constellation of dyslipidemia, abdominal obesity, insulin resistance, inflammation and hypertension that appear together in pathological development more often than would be predicted just by chance.

Subcutaneous adipose tissue: Adipose tissue depot located under the skin. This depot is clearly the largest adipose tissue depot, especially in the state of obesity.

Visceral adipose tissue: Adipose tissue located in abdominal cavity around the organs. This adipose tissue is highly active. For example, lipolytic activity is high and it produces most of the pro-inflammatory molecules, and often in greater amount than the subcutaneous adipose tissue.

References

Al-Attas, O.S. and N.M. Al-Daghri, K. Al-Rubeaan, N.F. da Silva, S.L. Sabico, S. Kumar, P.G. McTernan, and A.L. Harte. 2009. Changes in endotoxin levels in T2DM subjects on anti-diabetic therapies. Cardiovasc. Diabetol. 8: 20.

Alberti, K.G. and R.H. Eckel, S.M. Grundy, P.Z. Zimmet, J.I. Cleeman, K.A. Donato, J.C. Fruchart, W.P. James, C.M. Loria, and S.C. Smith Jr. 2009. Harmonizing the metabolic syndrome: a joint interim statement of the International Diabetes Federation Task Force on Epidemiology and Prevention, National Heart, Lung, and Blood Institute, American Heart Association, World Heart Federation, International Atherosclerosis Society, and International Association for the Study of Obesity. Circulation 120: 1640-1645.

Arner, P. 2007. Introduction: The inflammation orchestra in adipose tissue. J. Int. Med. 262: 404-407.

Cancello, R. and K. Clement. 2006. Is obesity an inflammatory illness? Role of low-grade inflammation and macrophage infiltration in human white adipose tissue. Int. J. Obstetr. Gyn. 113: 1141-1147.

Dahlman, I. and K. Linder, E. Arvidsson-Nordstrom, I. Andersson, J. Liden, C. Verdich, *et al.* 2005. Changes in adipose tissue gene expression with energy-restricted diets in obese women. Am. J. Clin. Nutr. 81: 1275-1285.

Diabetes Prevention Program Research Group. 2005. Intensive lifestyle intervention or metformin on inflammation and coagulation in participants with impaired glucose tolerance. Diabetes 54: 1566-1572.

Garten, A. and S. Petzold, A. Körner, S. Imai, and W. Kiess. 2009. Nampt: linking NAD biology, metabolism and cancer. Trends Endocrinol. Metab. 20(3): 130-138.

Ghoshal, S. and J. Witta, J. Zhong, W. de Villiers, and E. Eckhardt. 2009. Chylomicrons promote intestinal absorption of lipopolysaccharides. J. Lipid Res. 50(1): 90-97.

Grundy, S.M. 2008. Metabolic syndrome pandemic. Arterioscler Thromb Vasc Biol. 28: 629-636.

Herder, C. and M. Peltonen, W. Koenig, K. Sütfels, J. Lindström, S. Martin, P. Ilanne-Parikka, J.G. Eriksson, S. Aunola, S. Keinänen-Kiukaanniemi, T.T. Valle, M. Uusitupa, H. Kolb, J. Tuomilehto, *et al.* for the Finnish Diabetes Prevention Study Group. 2009. Anti-inflammatory effect of lifestyle changes in the Finnish Diabetes Prevention Study. Diabetologia 52: 433-442.

Hotamisligil, G.S. 2006. Inflammation and metabolic disorders. Nature 444: 860-867.

Kallio, P. and M. Kolehmainen, D.E. Laaksonen, J. Kekäläinen, T. Salopuro, K. Sivenius, *et al.* 2007. Dietary carboydrate modification induces marked alterations in gene expression of white adipose tissue in persons with metabolic syndrome the FUNGENUT study. Am. J. Clin. Nutr. 85: 1417-1427.

Karlsson, E.A. and P.A. Sheridan, and M.A. Beck. 2001. Diet-induced obesity impairs the T cell memory response to influenza virus infection. J. Immunol. 184(6): 3127-3133.

Knowler, W.C. and E. Barrett-Connor, S.E. Fowler, R.F. Hamman, J.M. Lachin, E.A. Walker, *et al.* 2002. Reduction in the incidence of type 2 diabetes with lifestyle intervention or metformin. N. Engl. J. Med. 346: 393-403.

La Cava, A. and G. Matarese. 2004. The weight of leptin. Nat. Rev. 4: 371-379.

Lindström, J and P. Ilanne-Parikka, M. Peltonen, S. Aunola, J.G. Eriksson, K. Hemiö, H. Hämäläinen, P. Härkönen, S. Keinänen-Kiukaanniemi, M. Laakso, A. Louheranta, M. Mannelin, M. Paturi, J. Sundvall, T.T. Valle, M. Uusitupa, and J. Tuomilehto. Finnish Diabetes Prevention Study Group. 2006. Sustained reduction in the incidence of type 2 diabetes by lifestyle intervention: follow-up of the Finnish Diabetes Prevention Study. Lancet 368: 1673-1679.

Matarese, G. and E.H. Leiter, and A. La Cava. 2007. Leptin in autoimmunity: many questions, some answers. Tissue Antigens 70: 87-95.

Maury, E. and S.M. Brichard. 2010. Adipokine dysregulation, adipose tissue inflammation and metabolic syndrome. Mol. Cell. Endoc. 314: 1-16.

O'Keefe, J.H. and N.M. Gheewala, and J.O. O'Keefe. 2008. Dietary strategies for improving post-prandial glucose, lipids, inflammation, and cardiovascular health. J. Am. Coll. Cardiol. 51: 249-255.

Tilg, H. and A.R. Moschen. 2006. Adipocytokines: mediators linking adipose tissue, inflammation and immunity. Nat. Rev. 6: 772-783.

Trayhurn, P. and B. Wang, and I.S. Wood. 2008. Hypoxia and the endocrine and signaling role of white adipose tissue. Arch. Physiol. Biochem. 114: 267-276.

Tuomilehto, J. and J. Lindström, J.G. Eriksson, T.T. Valle, H. Hämäläinen, P. Ilanne-Parikka, S. Keinänen-Kiukaanniemi, M. Laakso, A. Louheranta, M. Rastas, V. Salminen, M. Uusitupa. 2001. Finnish Diabetes Prevention Study Group. Prevention of type 2 diabetes mellitus by changes in lifestyle among subjects with impaired glucose tolerance. N. Engl. J. Med. 344: 1343-1350.

Vasquez-Vela, M.E.F. and N. Torres, and A.R. Tovar. 2008. White adipose tissue as an endocrine organ and its role in obesity. Arch. Med. Res. 39: 715-728.

Section VI
Organs and Tissue Systems

CHAPTER 21

Cytokines and Alzheimer's Disease

Lucia Velluto[1], Carla Iarlori[2], Domenico Gambi[3] and Marcella Reale[2,*]

[1]Villa Serena Hospital, V.le Petruzzi 65013, Città Sant'Angelo Pescara, Italy
[2]Department of Oncology and Neurosciences, University "G.d'Annunzio" Chieti-Pescara, Via dei Vestini, 66100 Chieti, Italy
[3]Aging Research Center, Ce.S.I., "G. d'Annunzio" University Foundation, Chieti-Pescara, Italy

ABSTRACT

Alzheimer's disease (AD) is an age-dependent neurodegenerative disorder that results in a progressive loss of cognitive abilities. It is characterized by the accumulation of the amyloid-β peptide into amyloid plaques in the extracellular brain parenchyma as well as by intracellular neurofibrillary tangles. Neuroinflammatory responses in the brain involve different cells of the immune system, resident cells of the central nervous system, many protein components and cytotoxic substances.

Increased levels of pro-inflammatory cytokines demonstrated in the brain and cerebrospinal fluid are able to promote Aβ accumulation, playing a role in the progression of AD.

The risk of AD is substantially influenced by several genetic polymorphisms in the promoter region, or other untranslated regions, of genes encoding inflammatory cytokines. The identification of biomarkers for AD is required to improve the accuracy of clinical diagnosis and to monitor both disease progression and response to treatments. The research on AD biomarkers has been focused on amyloid and tau related species, but cytokines may provide important opportunities as well. Analysis and discussion presented in this report include the studies that have led to the hypothesis that cytokines may be markers to differentiate patients

*Corresponding author

with suspected AD from healthy controls and the possibilities of cytokine-based therapeutic strategies in AD.

INTRODUCTION

Alzheimer's disease (AD) is an age-dependent neurodegenerative disorder that results in a progressive impairment in memory, judgement, decision-making, orientation to physical surroundings and language. AD is characterized by the accumulation of amyloid beta (Aβ) protein into amyloid plaques in the extracellular brain parenchyma as well as by neurofibrillary tangles primarily comprising hyperphosphorylated tau, together with synapses loss and a deficit of presynaptic markers of the cholinergic system in brain regions involved in cognition and mood.

AD has a heterogeneous etiology with a large percentage, termed sporadic AD, arising from unknown causes and a smaller fraction of early onset familial AD, caused by mutations in one of several genes, such as Aβ precursor protein (APP) and presenilins (PS1, PS2). Most investigators believe that Aβ can induce an intracerebral inflammatory response that leads to production of inflammatory cytokines, deposition of complement, generation of free radicals and neuronal death (Fig. 1). Over the last decade, it has become clear that chronic neuroinflammation represents a further contrast feature of AD and it is being increasingly accepted as being associated with the onset of this neurodegenerative disorder. Neuroinflammatory responses in the brain of AD patients involve different cells of

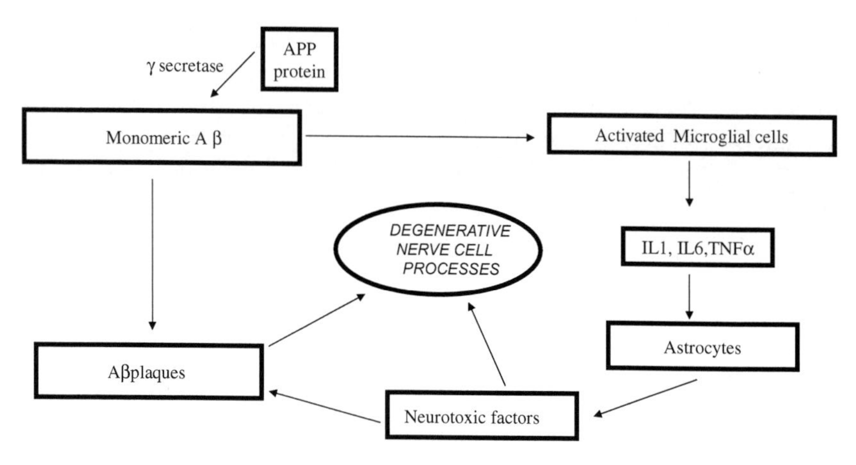

Fig. 1 Pathogenic pathway from amyloid β accumulation to cell death for Alzheimer's disease. The accumulation of amyloid β induces inflammatory reaction with increased cytokines expression by microglial cells. Neurons respond to glial activation and mature amyloid deposits with neuritic degeneration and cell death.

the immune system and resident cells of the central nervous system (CNS). In AD, activated microglia showed increased levels of cytokines that could exacerbate the loss of neurons damaged by the Aβ or by disruption of their cytoskeleton and axonal transport due to accumulation of tau. T cells can cross the blood-brain barrier so they may be responsible for releasing inflammatory mediators in brain during neurodegeneration (Fig. 2). Cytokines released by immunocompetent cells may be responsible for the bidirectional communication between cells of the nervous and immune systems. One hypothesis is that the inflammation starts within the CNS, where inflammatory products are able to modulate cytokine concentration in body fluid. On the other hand, a systemic inflammatory response results in the production of cytokines that circulate in the blood and communicate with neurons in the brain or diffuse from the blood into the brain parenchyma, where they interact with macrophages, or activate the endothelium, which in turn signals to the perivascular macrophages that communicate with the microglia, or by the sensory afferents of the vagus nerve that communicate with neuronal populations in the brain or, finally, by direct active transport across the blood-brain barrier.

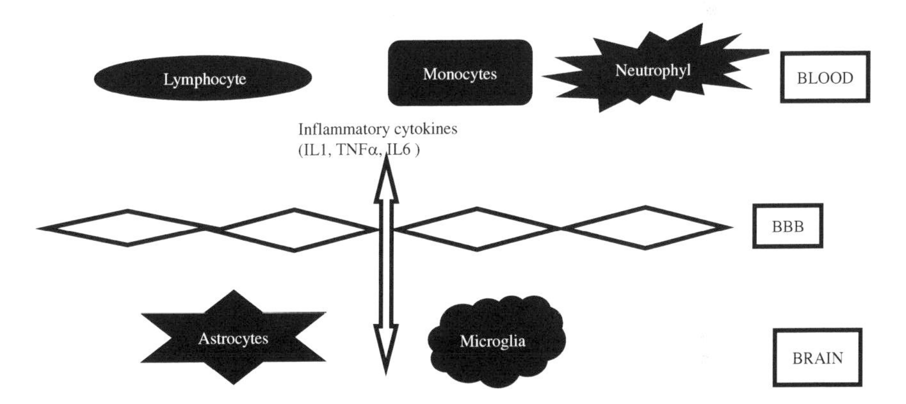

Fig. 2 Bidirectional communication between the brain and periphery: phases of communication between nervous and immune system. Cytokines derived from CNS lead to alteration of "cytokines network" in body fluid. Cytokines released by peripheral immune cells might be responsible for activation of CNS cells. BBB, blood-brain barrier.

Biomarkers may be helpful in the identification of high-risk people, early diagnosis, disease progression and therapeutic approaches. Unfortunately, there is at present no accepted biomarker or combination of biomarkers that can improve the accuracy of clinical diagnosis and monitor both disease progression and response to treatments. Cytokines are proposed for this role. Results of studies evaluating markers of systemic inflammation in AD patients are contradictory since both increased and normal plasma levels of pro-inflammatory cytokines have been observed (Table 1). Several reports have indicated that the risk of AD is influenced by several genetic polymorphisms of genes encoding cytokines (Table 2). The clinical consequence of a CNS dysregulation in this cytokine's balance can lead to cytokine production and synergistic cytokine actions. In this chapter we analyzed the most important cytokines involved in AD.

Table 1 Main cytokines produced in peripheral blood and in brain tissue of AD

	Plasma-PBMC	Brain-CSF
IL1	++	++ in frontal, parietal, temporal, cortex, hippocampus, thalamus, hypothalamus
TNFα	++ in severe AD vs mild AD	++, -, depends on disease stage
IL6	++ greater risk of cognitive decline,	++ production in PBMC ++ in astrocytes
IL4	-- in severe AD	NA
IL12	++ in mild AD	NA
IL18	++ production by stimulated PBMC	++ in frontal lobe
TGFβ	++ in plasma	++ meningeal cells, choroid plexus, glial cells

Cytokine levels in peripheral blood and in different brain regions referred to AD stage or to increased risk of cognitive decline and their expression in different AD brain regions. A peripheral increase of cytokines may represent a step in the pathogenesis of AD. Cytokines are present in AD brain.

Table 2 Cytokine gene polymorphisms associated with risk of disease development and progression

	Polymorphism
IL1	IL1 – 511 C/T
TNFα	TNFα – 850C → T
IL6	IL6 - 174G/C
IL10	IL10 - 1082AG
IL18	IL18 - 607CC
TGFβ	TGFβ

Genetic variation located within cytokine promoter regions influenced the susceptibility or resistance to AD.

IL1

The concentration of IL1β is maximum in those brain regions where AD neuropathology is most prominent, such as the frontal cortex, parietal and temporal cortex, hypothalamus, thalamus and hippocampus. This overall increase in cytokine production might represent an early event in the activation of a neuroimmune cascade leading to cell death and neurodegeneration in brain regions where a primary cause (e.g., genetic, toxic, vascular) facilitates the induction of resting microglia for firing brain immune function (Cacabelos *et al.* 1994).

Increased IL1 expression in reactive microglia surrounding amyloid plaques and in the brains of aged AD mouse models, and plaque-associated microglia provided the initial indication that IL1 may be associated with AD pathogenesis (Griffin *et al.* 1995) (Fig. 3).

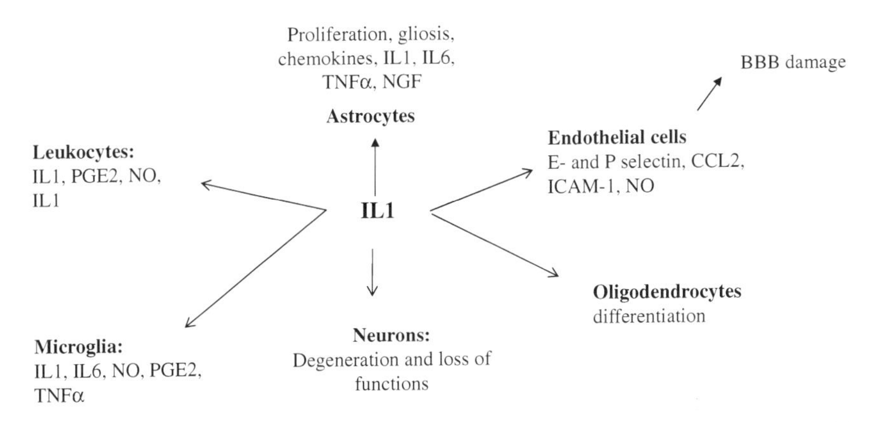

Fig. 3 IL1 pathway between the brain and periphery: IL1 exacerbates neurodegeneration by (1) microglia stimulation to release IL1, IL6, TNFα, NO and PGE1, (2) oligodendrocyte stimulation to differentiation and maturation, (3) lymphocyte stimulation to produce inflammatory mediators, (4) endothelial cell stimulation to release E- and P-selectin, CCL2, NO with subsequent blood-brain barrier (BBB) breakdown, (5) astrocytes stimulation to proliferation, gliosis and production of chemokines and cytokines.

Specific polymorphisms in the IL1α and IL1β genetic loci were shown to be associated with increased disease risk in certain patient populations (McGeer and McGeer 2001). These associations, in addition to observations of IL1 elevations in peripheral blood of AD patients, provided the key evidence for a central role of IL1 in disease pathogenesis (Lombardi *et al.* 1999, Gambi *et al.* 2004). Neuronal injury or insults including amyloid deposition may trigger a self-propagating cytokine cycle, which, if chronically induced, initiates a vicious feedback loop of continuing IL1β elevation promoting further neuronal and synaptic dysfunction and Aβ plaque accumulation. Cultured human monocytes and mouse microglia produce IL1β in response to Aβ exposure or, to a greater extent, to secreted fragments of β-APP. Further support for IL1β-mediated activation of microglia and resultant tau hyperphosphorylation has been indirectly provided in the 3xTg-AD mouse model.

Other studies investigated the effects of some promoter polymorphisms of IL-1 cytokines: the risk of developing AD has been associated with the IL1β (-511 C/T) promoter polymorphism. The IL1α promoter polymorphism -889 C/T has not been associated with the disease, so polymorphism -511 C/T is an independent risk factor for AD.

TNF, IL6 AND IL2

Mild forms of AD compared to severe AD and other dementias such as vascular dementias, showed lower serum levels of TNFα and TNFα/IL1 ratio. Both cytokines displayed significant relationship with age and damage of clinical pattern, suggesting that their profile seems useful to discriminate between mild and severe forms of disease. TNFα may be considered an important agent inducing the pathologic aspects of AD via multiple cellular mechanisms which include the TNFα activation of JNK and p38 kinase, MCP1 chemokine, and γ secretase. Together these elements may modulate the microglial activity, and the production of $A\beta_{1\text{-}42}$ protein.

Increase, decrease, or lack of change of mRNA or soluble TNFα levels in brain tissue and in CSF of AD when compared to non-AD patients was observed (Yamamoto *et al.* 2007). Elevated serum TNFα in dementia subjects appeared to be dependent on disease stage. Available data suggested a significant association between -850 polymorphism and AD risk. Subgroup analysis suggested that the presence of T allele significantly increased the risk of AD associated with carriage of the apolipoprotein E epsilon 4 allele in Caucasian Australians and Northern Europeans. No significant difference in genotype distribution of -308 polymorphism in AD was found. For the -863 and -1031 polymorphisms no association has been found with AD. Current findings support an association between -850 C>T polymorphism and the risk of developing AD (Di Bona *et al.* 2009).

Whereas IL6 has been reported up-regulated in AD brains, other studies have reported no significant increase in IL6 in CSF, and despite these variable results IL6, together with IL1 and TNFα, is considered one of the major mediators of the inflammatory response in AD. Recent data suggested that IL6 is also genetically associated with AD. IL6 has immune-related functions and participates in synaptic activities, as well as IL1 and TNFα. In progressive neurodegenerative disorders IL6 levels are over-expressed, particularly in astrocytes, resulting in severe clinical patterns characterized by ataxia, seizures and cognitive decline.

High plasma levels of IL6 have been associated with an inferior cognitive performance at baseline and with greater risk of cognitive decline. IL6 may permeate across blood-brain barrier and, after reaching brain, may cause or contribute to neuropathology of AD. The production of IL2 and IL6 by peripheral blood mononuclear cells (PBMC) was detected in patients with AD according to the severity of the disease. IL6 gene in humans is located in the short arm of chromosome 7 and has a -174 G/C polymorphism in its promoter region. The C allele at position -174 in the promoter of the IL6 gene has been associated with reduced gene expression and reduced plasma levels of IL6. Given the supposed role of several inflammatory mediators in neurodegeneration, the IL6 -174 G/C promoter polymorphism has been associated with AD (Capurso *et al.* 2004).

Concomitant presence of both mutant TNFa -308 A and IL6 -174 C alleles raised the AD risk three-fold, whereas there was no notable risk for AD afflicted by IL6 -174 polymorphism alone (Vural *et al.* 2009).

IL10 AND IL4

IL10 and IL4 show anti-inflammatory and immuno-suppressive properties; therefore, these cytokines have neuroprotective functions. The IL10 -1082 gene polymorphisms were investigated as susceptibility factors for AD. Heterozygotes (AG) or combined genotype (AG+AA) for IL10 -1082 were associated with approximately double increase in the risk of AD. Carriers of A alleles of both TNFα -308 and IL-10 -1082 had 6.5 times the risk for AD in comparison with non-carriers. IL4 plays an important role in the maintenance of CNS homeostasis limiting microglia activation by pro-inflammatory stimulants. Reduced serum levels were observed in AD patients compared to control subject (Lugaresi *et al.* 2004). In AD patients, treatment with acetylcholinesterase (AchE) inhibitor increases levels of IL4 (Gambi *et al.* 2004, Reale *et al.* 2005).

IL18

A pro-inflammatory cytokine IL18 produced in the brain is implicated in AD and other CNS disorders. Although plasma levels of IL18 showed no correlation with the disease, a significantly increased production of IL18 was obtained from stimulated PBMC of AD patients (Bossu *et al.* 2008).

The expression of IL18 is increased especially in the frontal lobe of AD patients. Immunohistochemistry of AD brain samples detected IL18 in microglia, astrocytes, and surprisingly in neurons, and it is also co-localized with amyloid-beta plaques and tau protein.

In CSF, elevated IL18 level was detected in men and it also correlated with CSF tau in mild cognitive impairment (MCI) and may represent a potential biomarker for AD in men, but not in women (Ojala *et al.* 2009). Aβ may induce the synthesis of IL18 and IL18 kinases involved in tau phosphorylation. IL18 would induce its own synthesis and that of other cytokines with neurotoxic properties and could promote the expression of IL18 receptor as a part of the amyloid-associated inflammatory reaction. The presence of IL18 receptor in an inflammatory setting combined with the induction of IFNγ expression via IL18 signaling suggests a positive feedback loop in the vasculature, which might contribute to the deregulated inflammatory response characterizing cardiovascular disease. While the mechanisms of IL18-mediated neuroinflammation were extensively studied, the exact role of IL18 in terms of beneficial or detrimental effects still needs to be clarified.

Furthermore, a significant correlation between IL18 production and cognitive decline was observed in AD patients. Overall, these data indicate that IL18-related inflammatory pathways, probably also by virtue of polymorphic IL18 gene influence, are exacerbated in AD patients, leading to dementia. The human IL18 gene promoter has two polymorphisms at positions -607 (C/A) and -137 (G/C) on both susceptibility and progression of AD. The genotype distribution of the -607 (C/A) polymorphism was different between patients with AD and control subjects. In particular, carriers of the CC genotype were at increased risk of developing AD; in addition, in a 2 yr follow-up study, the -137 CC genotype was strongly and specifically associated with a faster cognitive decline. As IL18 cytokine promoter gene polymorphisms have been previously described to have functional consequences on IL18 expression, it is possible that individuals with a prevalent IL18 gene variant have a dysregulated immune response, suggesting that IL18-mediated immune mechanisms may play a crucial role in AD.

TGF

TGFβ1, a cytokine part of the TGFβ superfamily, is expressed in meningeal cells, choroid plexus epithelial cells and glial cells. TGFβ1 immunoreactivity is increased in amyloid plaques and around cerebral vessels, and it is increased in CSF and serum as well. The plasma levels of TGFβ1 were significantly elevated in patients with AD compared to non-demented age-matched subjects. There was inverse correlation between the levels of IL18 and TGFβ1 in AD patients, whereas, in the non-demented, age-matched subjects, a positive correlation between IL18 and TGFβ1 levels was observed. TGFβ gene contains single nucleotide polymorphisms (SNPs) at codon +10 (T→C) and +25 (G→C) that appear to influence the level of expression of TGFβ1. It was observed that both the +10 C allele and the CC genotype were over-represented in AD when compared to healthy control. These variants significantly raised the risk of disease. The CC genotype was also over-expressed in MCI. These results suggest that TGFβ1 may be one of the early markers involved in the inflammatory mechanisms underlying the pathogenesis of AD (Arosio *et al.* 2007).

IL12 AND IL16

IL12 is considered important in the early inflammation response, acting to induce IFNγ production from T and NK cells. A positive correlation between IL12 and Mini Mental State examination (MMSE) scores has been reported in mild AD patients (Rentzos *et al.* 2006).

High levels of IL16 mRNA expression were observed in monocyte-macrophages of the AD peripheral blood. The plasma levels of IL12 and IL16 in AD patients at different stages of the disease correlate with the disease progression: they were higher in AD-mild patients and slightly lower in AD-moderate patients, whereas no significant difference was observed between AD-severe patients and non-demented age-matched subjects. These plasma levels of IL12 and IL16 follow the degree of disease, suggesting a gradual decline of immune responsiveness in AD (Motta *et al.* 2007). In other studies the expression of IL16 was found to be higher in patients with cardiovascular disease than in AD patients, whereas in non-demented age-matched subjects IL16 expression was significantly lower. IL16 seems to play a crucial role in the induction of blood-brain barrier breakdown leading to edema formation and subsequent secondary injury, such as activation of humoral and neuroendocrine pathways and resulting in a highly modulated peripheral inflammatory response. The high levels of IL16 observed in AD may be the systemic expression of microglial activity in response to Aβ. In conclusion, IL16, IL18 and TGFβ1 orchestrate many of the complex processes underlying this human disease.

IL15

IL15 levels in CSF of patients with AD or with frontotemporal dementia are high. In AD a significant positive correlation was noted between IL15 levels and age of onset, suggesting that IL15 may be implicated in the pathophysiology of AD and frontotemporal dementia (Rentzos *et al.* 2006).

IFN γ

IFNγ, an important pro-inflammatory cytokine, is encoded by a single gene mapped to chromosome 12, one of the candidate loci of AD. The first intron in the IFNγ gene represents a CA repeat polymorphism that may affect the IFNγ secretion dose. It is observed that the polymorphism may have some effect on the inflammatory process and the pathologic change in AD, so Oda *et al.* analyzed the IFNγ gene polymorphism and concluded that IFNγ gene polymorphism is not associated with development of AD (Oda *et al.* 2004). Yamamoto *et al.* generated a novel Swedish beta-amyloid precursor protein mutant (APP) transgenic mouse in which the IFNγ receptor type I was knocked out (APP/GRKO). IFNγ signaling loss in the APP/GRKO mice reduced gliosis and amyloid plaques at 14 month of age. Aggregated Aβ induced IFNγ production from co-culture of astrocytes and microglia, and IFNγ elicited TNFα secretion in wild-type but not GRKO microglia co-cultured with astrocytes. Both IFNγ and TNFα enhanced Aβ production from APP-expressing astrocytes and cortical neurons. TNFα directly stimulated beta-site

APP-cleaving enzyme (BACE1) expression and enhanced beta-processing of APP in astrocytes. The numbers of reactive astrocytes expressing BACE1 were increased in APP compared with APP/GRKO mice in both cortex and hippocampus. IFNγ and TNFα activation of wild-type microglia suppressed Aβ degradation, whereas GRKO microglia had no changes. These results support the idea that glial IFNγ and TNFα enhance Aβ deposition through BACE1 expression and suppression of Aβ clearance (Yamamoto *et al.* 2007).

Taken together, these observations suggest that pro-inflammatory cytokines are directly linked to AD pathogenesis and a link can be logically proposed between the cytokine profiles in the brain and systemic circulation.

CLINICAL APPLICATIONS

The detection of a risk profile that will potentially allow both the early identification of individuals at risk for disease and the possible discovery of potential targets for medication is the aim of modern research in AD.

One therapeutic approach is based on the fact that different single nucleotide polymorphisms have already been associated with substantial changes in the effects of drugs, and some are now being used to predict clinical response. Since most of the drug effects are determined by the relationship of different gene products that influence the response to therapy, polygenic determinants of drug effects have become increasingly important in pharmacogenomics. The occurrence of a high-risk genetic profile linked to the presence of high responder alleles of pro-inflammatory cytokines or of low responder alleles of anti-inflammatory cytokines might suggest the treatment with biologics as monoclonal antibodies directed versus the pro-inflammatory cytokines.

A strategy for treatment of AD has been to use AChE inhibitors (e.g., Donepezil) to increase the levels of acetylcholine and enhance cholinergic activity in the affected regions of the brain, not only for activation of cholinergic transmission but also for a neuroprotective effect, by reducing the amount of the toxic form of Aβ fibrils, by attenuating the production of IL1β in hippocampus and the production of other functionally important inflammatory cytokines that may contribute to the immunopathology associated with AD (Reale *et al.* 2005).

Even the TNFα antagonists have been considered in therapeutic programs of AD. Tobinick and co-workers have demonstrated that the use of a biologic TNFα inhibitor, Etarnecept, supplied by perispinal extrathecal administration in a pilot study of AD patients, leads to significant improvement of MMSE score (Tobinick 2009).

CONCLUSIONS

The evaluation of cytokine expression, production and gene polymorphisms in AD patients has contributed to clarify the pathogenesis of this disease. The Aβ deposition, facilitation of neurotoxicity, abnormal tau phosphorylation, resulting in the neurodegenerative pattern and loss of function in AD patients, could be affected by CNS and peripheral cytokine levels. The aim of the clinical research is to identify markers that may open the way to new diagnostic approaches for the early diagnosis of relevant preclinical states of AD (Table 3). In fact, a major goal of clinical research is to improve early detection of AD before the clinical manifestation of the illness, where anti-inflammatory treatment might play a decisive role in preventing or significantly retarding the manifestation of the disease. In addition, longitudinal investigations focused upon the association of genetic background of immune factors such as cytokines, with the manifestation of cognitive impairment and mental performance deterioration in the elderly population.

Table 3 Mechanisms of AD and biomarkers

Mechanism	*Biomarker*
Aβ deposition	Tau and Aβ (CSF)
Microglial activation	IL1 expression
Peripheral inflammation	Cytokine production from PBMCs
Oxidative stress	NO
Neuronal dysfunction	Brain atrophy and cognitive decline

Identification of biomarkers may open the way to new diagnostic approaches for early diagnosis of AD.

Summary Points

- Neuroinflammation represents a contrast feature of AD and it is being increasingly accepted as being associated with the onset of this neurodegenerative disorder.
- Cytokines are expressed in different regions of AD brain and produced in peripheral blood by T lymphocytes in an abnormal way.
- Cytokines are modulators of Aβ accumulation and thus play an important role in the amplification of detrimental mechanisms likely to be involved in progression of AD.
- Cytokine gene polymorphisms are linked to greater risk to cognitive decline.
- Cytokines may represent additional peripheral biomarkers of AD.
- Inflammatory genetic variation, like cytokine polymorphisms, may contribute to AD susceptibility and represents the basis for future recognition of individuals at risk and for the pharmacogenomic driving of drug responsiveness.

Abbreviations

AchE	:	acetilcholinesterase
AD	:	Alzheimer's disease
APP	:	amyloid precursor protein
Aβ	:	amyloid beta
CNS	:	central nervous system
IL	:	interleukin
MCI	:	mild cognitive impairment
NFT	:	neurofibrillary tangle
NSAD	:	non-steroidal anti-inflammatory drug
PBMC	:	peripheral blood mononuclear cell
TGFβ	:	transforming growth factor beta
TNFα	:	tumor necrosis factor alpha

Definition of Terms

Amyloid β plaque: Aβ is the main constituent of amyloid plaques in AD brains. Plaques are composed of amyloid fibrillar.

Amyloid precursor protein: Aβ is formed after sequential cleavage of the amyloid precursor protein by α-, β- and γ-secretases.

Anti-inflammatory cytokines: Cytokines involved in the reduction of inflammatory reactions.

Neurofibrillary tangles: Formed by hyperphosphorylation of a microtubule-associated protein known as tau, causing it to aggregate in an insoluble form.

Neuroinflammation: Release of endogenous damage/alarm signals in response to accumulated cell distress is the earliest triggering event in AD, leading to activation of innate immunity and inflammatory cascade in the brain tissue.

Pro-inflammatory cytokines: Cytokines produced by activated immune cells such as microglia involved in the amplification of inflammatory reactions.

References

Arosio, B. and L. Bergamaschini, L. Galimberti, C. La Porta, M. Zanetti, C. Calabresi, E. Scarpini, G. Annoni, and C. Vergani. 2007. +10 T/C polymorphisms in the gene of transforming growth factor-beta1 are associated with neurodegeneration and its clinical evolution. Mech. Ageing Dev. 128: 553-557.

Bossu, P. and A. Ciaramella, F. Salani, F. Bizzoni, E. Varsi, F. Di Iulio, F. Giubilei, W. Gianni, A. Trequattrini, M.L. Moro, S. Bernardini, C. Caltagirone, and G. Spalletta. 2008. Interleukin-18 produced by peripheral blood cells is increased in Alzheimer's disease and correlates with cognitive impairment. Brain Behav. Immun. 22: 487-492.

Cacabelos, R. and X.A. Alvarez, L. Fernandez-Novo, A. Franco, R. Mangues, A. Pellicer, and T. Nishimura. 1994. Brain interleukin-1 beta in Alzheimer's disease and vascular dementia. Methods Find Exp. Clin. Pharmacol. 16: 141-151.

Capurso, C. and V. Solfrizzi, A. D'Introno, A.M. Colacicco, S.A. Capurso, A. Capurso, and F. Panza. 2004. Interleukin 6-174 G/C promoter gene polymorphism and sporadic Alzheimer's disease: geographic allele and genotype variations in Europe. Exp. Gerontol. 39: 1567-1573.

Di Bona, D. and G. Candore, C. Franceschi, F. Licastro, G. Colonna-Romano, C. Cammà, D. Lio and C. Caruso. 2009. Systematic review by meta-analyses on the possible role of TNF-alpha polymorphisms in association with Alzheimer's disease. Brain Res. Rev. 61: 60-68.

Gambi, F. and M. Reale, C. Iarlori, A. Salone, L. Toma, C. Paladini, G. De Luca, C. Feliciani, M. Salvatore, R.M. Salerno, T.C. Theoharides, P. Conti, M. Exton, and D. Gambi. 2004. Alzheimer patients treated with an AchE inhibitor show higher IL-4 and lower IL-1 beta levels and expression in peripheral blood mononuclear cells. J. Clin. Psychopharmacol. 24: 314-321.

Griffin, W.S. and J.G. Sheng, G.W. Roberts, and R.E. Mrak. 1995. Interleukin-1 expression in different plaque types in Alzheimer's disease: significance in plaque evolution. J. Neuropathol. Exp. Neurol. 54: 276-281.

Lombardi, V.R.M. and M. Garcia, L. Rey, and R. Cacabelos. 1999. Characterization of cytokine production, screening of lymphocyte subset patterns and in vitro apoptosis in healthy an Alzheimer disease (AD) individuals. J. Neuroimmunol. 97: 163-171.

Lugaresi, A. and A. Di Iorio, C. Iarlori, M. Reale, G. De Luca, E. Sparvieri, A. Michetti, P. Conti, D. Gambi, G. Abate, and R. Paganelli. 2004. IL4 in vitro production is up-regulated in Alzheimer's disease patients treated with acetylcholinesterase inhibitors. Exp. Gerontol. 39: 653-657.

McGeer, P.L. and E.G. McGeer. 2001. Polymorphisms in inflammatory genes and the risk of Alzheimer disease. Arch. Neurol. 58: 1790-1792.

Motta, M. and R. Imbesi, M. Di Rosa, F. Stivala, and L. Malaguarnera. 2007. Altered plasma cytokine levels in Alzheimer's disease: correlation with the disease progression. Immunol. Lett. 114: 46-51.

Oda, M. and H. Maruyama, Y. Izumi, H. Morino, T. Torii, S. Nakamura, and H. Kawakami. 2004. Dinucleotide repeat polymorphism in interferon-gamma gene is not associated with sporadic Alzheimer's disease. Am. J. Med. Genet. B Neuropsychiatr. Genet. 124B: 48-49.

Ojala, J. and I. Alafuzoff, S.K. Herukka, T. van Groen, H. Tanila, and T. Pirttilä. 2009. Expression of interleukin-18 is increased in the brains of Alzheimer's disease patients. Neurobiol. Aging 30: 198-209.

Reale, M. and C. Iarlori, F. Gambi, C. Feliciani, I. Lucci, and D. Gambi. 2005. The acetylcholinesterase inhibitor, Donepezil, regulates a Th2 bias in Alzheimer's disease patients. Neuropharmacology 20: 1-8.

Rentzos, M. and G.P. Paraskevas, E. Kapaki, C. Nikolaou, M. Zoga, A. Rombos, A. Tsoutsou, and D. Vassilopoulos. 2006. Interleukin-12 is reduced in cerebrospinal fluid of patients with Alzheimer's disease and frontotemporal dementia. J. Neurol. Sci. 249: 110-114.

Rentzos, M. and M. Zoga, G.P. Paraskevas, E. Kapaki, A. Rombos, C. Nikolaou, A. Tsoutson, and D. Vassilopoulos. 2006. IL-15 is elevated in cerebrospinal fluid of patients with Alzheimer's disease and frontotemporal dementia. M.J. Geriatr. Psychiatry Neurol. 19: 114-117.

Tobinick, E. 2009. Tumour necrosis factor modulation for treatment of Alzheimer's disease: rationale and current evidence. CNS Drugs 23: 713-725.

Vural, P. and S. Degirmencioglu, H. Parıldar-Karpuzoglu, S. Dogru-Abbasoglu, H.A. Hanagasi, B. Karadag, H. Gurvit, M. Emre, and M. Uysal. 2009. The combinations of TNFa -308 and IL-6-174 or IL-10 -1082 genes polymorphisms suggest an association with susceptibility to sporadic late-onset Alzheimer's disease. Acta Neurol. Scand. 120: 396-401.

Yamamoto, M. and T. Kiyota, M. Horiba, J.L. Buescher, S.M. Walsh, H.E. Gendelmann, and T. Ikezu. 2007. Interferon-gamma and tumor necrosis factor-alpha regulate amyloid-beta plaque deposition and beta-secretase expression in Swedish mutant APP transgenic mice. Am. J. Pathol. 170: 680-692.

CHAPTER 22

Parkinson's Disease and Cytokines

Carla Iarlori[1], Domenico Gambi[2] and Marcella Reale[1,*]

[1]Department Oncology and Neurosciences, University "G.d'Annunzio" Chieti-Pescara, Via dei Vestini, 66100 Chieti, Italy

[2]Aging Research Center, Ce.S.I., "G.d'Annunzio" University Foundation Chieti-Pescara, Italy

ABSTRACT

Parkinson's disease (PD) is characterized by tremor, rigidity, and slowness of movements and is associated with progressive neuronal loss of the substantia nigra (SN) and other brain structures. Ample evidence of chronic inflammatory reactions in the brain of PD patients is shown. Neuroinflammatory responses in the brain involve different cells of the immune system (e.g., macrophages, mast cells, T and B lymphocytes, dendritic cells), resident cells of the central nervous system (CNS) (e.g., microglia, astrocytes, neurons), many protein components (e.g., complement, adhesion molecules, chemokines, cytokines) and cytotoxic substances (e.g., reactive oxygen and nitrogen species). Soluble molecules or cytokines released by immuno-competent cells may be responsible for the bidirectional communication between cells of the nervous and immune systems. One hypothesis is that inflammation starts within the CNS, where several inflammatory products are formed and are quickly removed into the bloodstream. On the other hand, it has also been proposed that inflammation develops at first in the periphery and then will contribute to brain damage and finally neurodegeneration. Elevated levels of pro-inflammatory cytokines, such as tumour necrosis factor α (TNF-α), interleukin (IL) 1β, IL-6 and the colony-stimulating factor, have been demonstrated in the brain and cerebrospinal fluid (CSF) as well as basal ganglia of PD patients. Peripheral blood mononuclear cells deriving from patients with PD have been reported to have an altered production of TNFα- as well as IL-1α and -1β compared to healthy controls, while IL-2,

*Corresponding author

IFN-γ, IL-6 and plasma-soluble interleukin-2 receptor (sIL-2R) are no different from healthy controls.

The identification of biomarkers for neurodegenerative diseases such as PD is required to improve the accuracy of clinical diagnosis and monitor both disease progression and response to treatments.

INTRODUCTION

Parkinson's disease (PD), one of the most common progressive neurodegenerative disorders, is characterized by the progressive loss of dopaminergic neurons (DN), neuronal loss is extremely severe in the substantia nigra (SN) pars compacta, moderate in the ventral tegmental area and catecholaminergic cell group A8, and almost nil in the central gray substance.

Various mechanisms of neuronal degeneration in PD have been proposed. These include formation of free radicals and oxidative stress, mitochondrial dysfunction, excitotoxicity, calcium cytotoxicity, trophic factor deficiency, inflammatory processes, genetic factors, environmental impact factors (Table 1), toxic action of nitric oxide (NO) and apoptosis, all of which may interact and amplify each other in a vicious cycle of toxicity leading to neuronal dysfunction, atrophy, and finally cell death (Fig. 1)

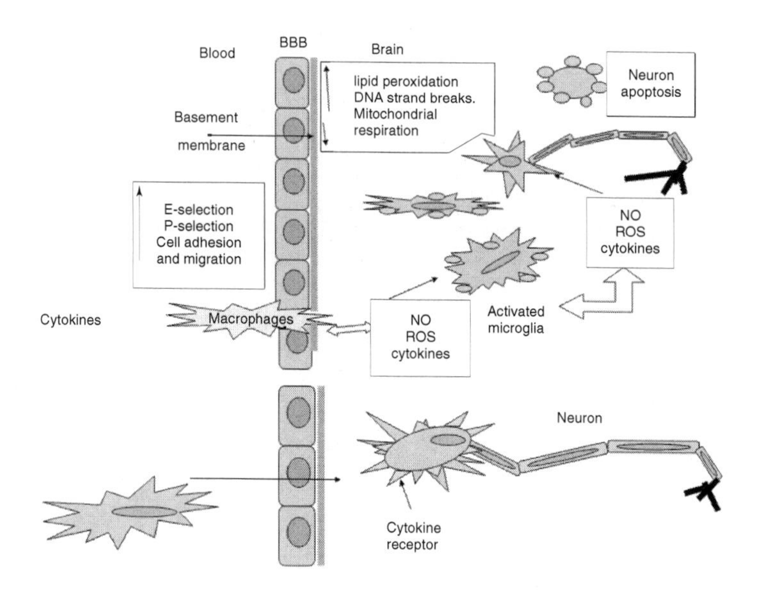

Fig. 1 Key features of cytokine localization in the brain of Parkinson's disease patients. Neuroinflammation is an indisputable neuropathological feature of PD. A marked increase of pro-inflammatory cytokines has been observed in the CNS of Parkinson's patients. The increase in the level of cytokines was demonstrated in nigrostriatal regions, lateral ventricles and substantia nigra.

Table 1 Environmental impact factors implicated in the increased and decreased development of sporadic Parkinson's disease

Putative agent	*Risk factor*
	Increased risk
Paraquat, organochlorines, carbamates	Pesticide exposure
Lead with iron, iron with copper, manganese	Metal exposure
	Decreased risk
Nicotine, MAO-B,	CYP2D6 Smoking
Caffeine	Coffee consumption

Pesticide exposure may play a role in the development of sporadic PD. Metals such as iron, copper and manganese interact to α-synuclein and promote its fibrillation. Caffeine consumption and cigarette smoking may decrease the risk of PD.

In both animal (Depino *et al.* 2003) and human models of sporadic PD (Langston 1999), a neuroinflammatory process mediated by microglial activation from toxic exposure to substances such as 1-methy 1-4-phenyl 1-1,2,3,6-tetrahydropyridine (MPTP) or the bacterial toxin lipopolysaccharide (LPS) has been implicated in dopaminergic nigral cell loss and the degenerative process. Specifically, several studies have reported an increased expression of pro-inflammatory cytokines by glial cells, a reduction in glial-derived neurotrophic factors, and an increased density of glial cells, associated with PD (Table 2) (Mogi *et al.* 1996a).

Table 2 Pro-inflammatory cytokines in various Parkinson's disease animal models

Cytokines	*MPTP Model*	*6-OHDA Model*	*LPS Model*
IL-1β	Yes	Yes	Yes
IL-2	NA	NA	NA
IL-4	NA	NA	NA
IL-6	Yes	NA	Yes
TNFα	Yes	Yes	Yes
IFNγ	Yes	NA	NA

The presence of cytokines in PD models suggests the role of these cytokines in the DA neurodegeneration. MPTP (1-methyl-4-phenyl-1,2,3,6-tetrahydropyridine) has become the most widely used agent to reproduce PD pathologies. Microglial activation and IL-1β and TNF-α expression was noticed in 6-hydroxydopamine (6-OHDA) PD model. The LPS (lipopolysaccharide) PD model has to decipher the role of glial cells, especially that of microglia in the DA neurodegeneration process. NA: not available.

Increased levels of interleukin (IL)-1β, IL-2, IL-4, IL-6, transforming growth factor (TGF)-α, TGF-β1 and TGF-β2 have been demonstrated in nigrostriatal dopamine regions, and ventricular and lumbar CSF (Table 3) (Nagatsu *et al.* 2000). Two mechanisms that are not mutually exclusive may explain the deleterious role of cytokines in parkinsonian SN. First, pro-inflammatory cytokines increase leukocyte adhesion and migration, disrupt the blood-brain barrier and can induce the product of NO in glial cells. NO may diffuse towards dopaminergic neurons where it drive an oxidant stress specific for vulnerable SNC dopaminergic neurons. That may play a deleterious role by inducing lipid peroxidation and DNA strand

Table 3 Cytokines that contribute significantly to DA neurodegeneration

Cytokines	Human PD Brain
IL-1β	Yes
IL-2	Yes
IL-4	Yes
IL-6	Yes
TNF-α	Yes
IFN-γ	Yes

Cytokines released by activated microglial cells induce neuronal apoptosis and contribute significantly to DA neurodegeneration.

breaks, inhibiting mitochondrial respiration and energy metabolism, and inducing apoptosis and specific cytolysis caused by the activation of the complement system. The second mechanism by which pro-inflammatory cytokines may play a deleterious role in PD may be more direct. In fact, the recent evidences that neurons and glial cells express cytokines and their receptors confirm the likelihood of a cross-talk between the immune and nervous system. After receptor binding a specific receptor-signaling pathway drives the apoptotic cell death (Fig. 2).

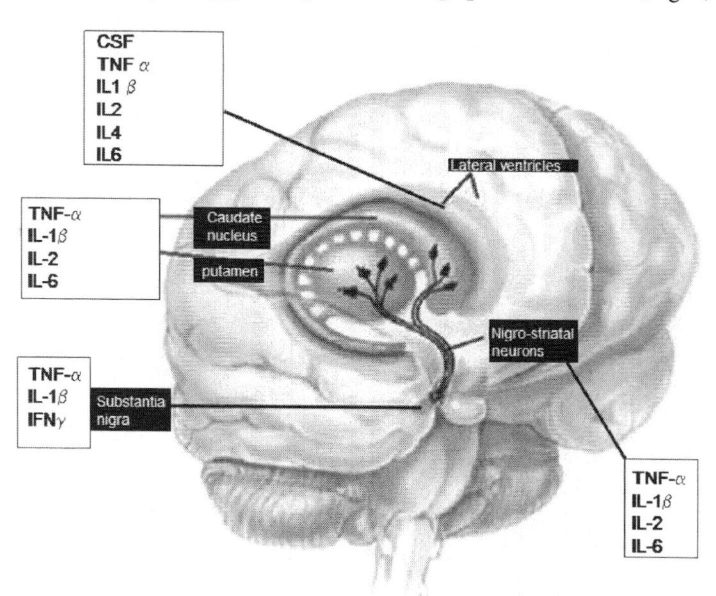

Fig. 2 The hypothesized mechanism of neuronal cells death mediated by cytokines in Parkinson's disease. **(A)** (1) Pro-inflammatory cytokines released by activated microglial cells induce recruitment of macrophages. (2) Macrophages cross the blood-brain barrier and generate oxygen-free radicals, nitric oxide and cytokines. (3) NO may diffuse towards dopaminergic neuron inducing lipid peroxidation and DNA strand breaks, inhibiting mitochondrial respiration and energy metabolism. (4) Neuronal dysfunction and apoptosis. **(B)** (1) Dopaminergic neurons express specific receptors for proinflammatory cytokines. (2) Cytokine receptor binding directly leads to apoptotic neuronal death.

TUMOR NECROSIS FACTOR (TNF)

Combined evidence from histopathologic, epidemiologic, and pharmacological studies supports a role for TNF, a potent pro-inflammatory cytokine, in eliciting dopaminergic neuron loss and nigrostriatal degeneration. Multiple studies indicate that TNF is highly toxic to dopaminergic neurons in both *in vitro* primary cultures (Clarke and Branton 2002) and *in vivo* (Carvey *et al.* 2005). TNF and soluble circulating trimer (solTNFR1) levels are elevated in cerebrospinal fluid and tissues of PD patients as well as in postmortem PD brains with the highest TNF levels present in areas that have the greatest loss of dopaminergic neurons (Boka *et al.* 1994).

INTERFERON-γ(INFγ)/CD23

Also, *in vitro*, interferon-γ (IFN-γ) together with IL-1β and TNF-α can induce the expression of CD23 in glial cells. CD23 is a low-affinity IgE receptor expressed at the cell surface of various cell types after stimulation by different cytokines. Ligation of CD23 with specific antibodies resulted in the induction of inducible nitric oxide synthases (iNOS) and the subsequent release of NO. The activation of CD23 also led to an up-regulation of TNF-α production, which was dependent on NO release. In the SN of PD patients, a significant increase in the density of glial cells expressing TNF-α, IL-1β, and IFN-γ was observed. Furthermore, although CD23 was not detectable in the SN of control subjects, it was found in both astroglial and microglial cells in parkinsonian patients.

INTERLEUKIN-1 (IL-1)

The IL-1 family comprises three proteins, the pro-inflammatory IL-1α and IL1-β and their inhibitor IL-1 receptor antagonist (IL-1Ra), which are encoded by the *IL-1A, IL-1B*, and *IL-1RN* genes respectively. Variation in the *IL-1A*, *IL-1B*, and *IL-1RN* genes may be of importance in the development of these disorders.

Pott Godoy *et al.* have shown that central LPS injection in a degenerating SN exacerbated neurodegeneration, accelerated and increased motor signs and shifted microglial activation towards a pro-inflammatory phenotype with increased IL-1β secretion. Importantly, chronic systemic expression of IL-1 also amplified neurodegeneration and microglial activation, and specific IL-1 inhibition reversed these effects. Thus, IL-1 exerts its exacerbating effect on degenerating dopaminergic neurons by direct and indirect mechanisms (Pott Godoy *et al.* 2008).

INTERLEUKIN-6 (IL-6)

As a member of pro-inflammatory cytokines, IL-6 is known to play a key role in this interaction between nervous and immune system (Gruol and Nelson 1997). Its actions are manifested in an opposite fashion, either supporting cell survival or inducing cell death. Concentration of IL-6 increases in nigrostriatal region of postmortem brain and in CSF of PD patients and its levels in CSF inversely correlate with the severity of PD. Some studies have shown no difference in plasmatic levels of IL-6 (Hofman *et al.* 2009), while others found elevated IL-6 levels in PD patients with severe depression (Selikhova *et al.* 2002). Bessler *et al.* reported that IL-6 was at higher plasmatic levels in patients with a rapidly progressing disease when compared to patients with usual progression. It was also reported that levodopa, drug used for treatment of PD patients, in physiological concentrations, presented an immunomodulatory effect on cells from both PD patients and controls and caused a stimulation of IL-6 production (Bessler *et al.* 1999). In patients with sporadic PD, Nagatsu and Sawada (2005), reported increased levels of TNF-α, IL-1β and IL-6 in nigrostriatal tissues, ventricular and lumbar CSF compared to controls (Nagatsu and Sawada 2005). Additionally, Gribova *et al.* reported that IFN-γ and TNF-α were elevated in serum from PD patients (Gribova *et al.* 2003), IL-1β and IL-6 were elevated in lumbar CSF of PD patients (Blum-Degen *et al.* 1995) and TNFα and IL-6 were present in microglia of PD brain (Imamura *et al.* 2003). Despite these findings, controversy remains regarding the role of these inflammation-associated compounds in the initiation and/or progression of the disease. It has thus been hypothesized that a continuous exposure to cytokines that mimic mild but persistent infectious/inflammatory responses, or high-level exposures that mimic more serious infectious/inflammatory responses, are necessary to initiate a neurodegenerative process, or to predispose glial cells to initiate cell death pathways. There are several disparate reports demonstrating both neuroprotective and deleterious effects of cytokines in neurodegenerative disorders, such as the protective role proposed for TNF-α and IL-6 (Imamura *et al.* 2003). Using a mesencephalic cell culture model of tyrosine hydroxylase immunoreactive neurons (THir), Gayle *et al.* demonstrated that cultures exposed to LPS decreased their THir cells and increased IL-1β and TNF-α synthesis, while neutralizing antibodies to IL-1β and TNF-α reduced the LPS-induced THir cell loss (Gaile *et al.* 2002). Since THir cells are thought to model the dopamine-containing neurons in the SN, this compelling study provided critical support for modulating the neuroinflammatory process to prevent, or even protect the brain from, the poorly defined degenerative process in PD. Additionally, there is evidence that different neuronal cell types (Recchia *et al.* 2004, Braak *et al.* 2003), glial cells (Piao *et al.* 2001) and brain areas exhibit both cytokine-related changes and alterations in the PD-associated proteins α-synuclein, tau and ubiquitin (Zhang *et al.* 2005).

CYTOKINES/APOPTOSIS

Decreased levels of neurotrophins such as *brain-derived neurotrophic factor* (BDNF) and *nerve growth factor* (NGF), which changes are known to trigger the process of apoptosis associated with increased levels of pro-inflammatory cytokines, strongly suggest a pro-apoptotic environment in the striatum in PD. In fact, the levels of apoptosis-related factors such as Bcl-2 (Mogi *et al.* 1996), soluble FAS, TNF R1 (p55), caspase 1 (IL-1-beta converting enzyme), and caspase 3 are increased in the PD brain (Mogi *et al.* 2000). Fas antigen and two TNF receptors, p55 and p75, are implicated in triggering cell death upon stimulation by their natural ligands, i.e., TNF-α and Fas ligands. Since TNF R1 and caspases 1 and 3 have been implicated as mediators of apoptotic cell death, their increased levels support the presence of pro-apoptotic environment in the striatum in the PD brain.

INFLAMMATORY CYTOKINE GENE POLYMORPHISMS

Because inflammation plays an important role in the pathogenesis of PD, it has been hypothesized that genetically determined differences in the immune response might influence the risk of developing neurodegenerative disorders or affect the rate of disease progression (Yucesoy *et al.* 2007). Genes that code for cytokines are highly polymorphic, and some of these polymorphisms directly or indirectly influence cytokine expression. The most frequent types of mutations are characterized by a change in a single nucleotide base pair and are called single nucleotide polymorphisms (SNPs). In a Polish population, IL-10 polymorphism has been studied as a risk factor of sporadic PD. No statistically significant differences between PD patients and controls were found in the frequency of a single locus (−1082, −519) of IL-10 promoter (Bialecka *et al.* 2007). Wahner *et al.* have investigated whether functional DNA polymorphisms of the TNF-α and IL-1 β genes affect the risk of PD. A smaller magnitude of PD risk increase among carriers of the heterozygous genotype for either or both polymorphisms suggests a gene-dosing effect (Wahner *et al.* 2007). No association was observed for single nucleotide polymorphism in the promoter regions of the IL-2, IL-6, and TNF-α genes. The single nucleotide polymorphism in the chemokine IL-8 gene was observed to associate with PD and appeared to be independent of age at onset (Owen *et al.* 2004).

PERIPHERAL CYTOKINE LEVELS

The question of the existence of a systemic inflammation in PD, and of which cytokines may be possibly involved in this phenomenon, might be relevant to clinicians in order to better understand the pathophysiology of PD. Interestingly, abnormalities in peripheral immune functions in patients with PD, such as changes in lymphocytic subpopulations in the blood and cerebrospinal fluid, deviation of T lymphocyte subset, impaired production of IL-2 and higher production of IFN-γ by PBMC as well as plasma elevations of some cytokines, have been reported by several authors (Table 4). PBMCs deriving from patients with PD have been reported to induce an altered (often reduced) production of TNF-α as well as IL-1α and IL-1β compared to healthy controls, and their monocyte/macrophages induce less TNF-α (Tweedie *et al.* 2007). This picture remains confusing as Reale *et al.* (2008) observed a higher expression and production of IL-1β from PBMCs isolated from PD subjects compared to controls. Production of IFN-γ and IL-2 from LPS-stimulated PBMCs from PD patients has been determined to be significantly lower than in controls, and a lack of difference in IL-10 production has been described. A significant negative correlation was noted in IL-6, TNF-α, IL-1β and IL-1α levels produced by both LPS-stimulated and unstimulated PBMCs from PD patients versus their Hoehn Yahr disability score, signifying that impaired cytokine production may progress with advancing disease.

Table 4 Cytokines released by peripheral blood mononuclear cells (PBMC) isolated from Parkinson's disease patients

Cytokine	*Level* [a]	*References*
IL-1β	+	Reale *et al.* 2009
	–	Hosegawa *et al.* 2000
IL-2		Stypula *et al.* 1996, Gangemi *et al.* 2003, Rentzos *et al.* 2007, Bogdan *et al.*2008
	–	Kluter *et al.* 1995
IL-4		Bogdan *et al.* 2008
IL-6	+	Selikhova *et al.* 2002, Bogdan *et al.* 2008
IL-8	+	Reale *et al.* 2009
	–	Hosegawa *et al.* 2000
IL-10		Bogdan *et al.* 2008
IL-15	+	Gangemi *et al.* 2003, Rentzos *et al.* 2007, Stypula *et al.* 1996
IFN-γ	+	Bogdan *et al.* 2008, Reale *et al.* 2009
TNF-α	+	Mogi *et al.* 1994, Bogdan *et al.* 2008, Reale *et al.* 2009
	–	Hosegawa *et al.* 2000
TGF-α	+	Mogi *et al.* 1996
TGFβ1-β2	+	Mogi *et al.* 1996

A peripheral increase of chemokines may represent a step in the pathogenesis of PD. Cytokines released by immunocompetent cells may be responsible for the bidirectional communication between cells of the nervous and immune systems. Measure and profile of peripheral levels of cytokines may used alone or in combination with other biomarkers in PD and similar diseases involving neuroinflammation. [a]+: Increased levels were observed compared to healthy controls. -: Decreased levels were observed compared to healthy controls.

APPLICATION

Increased rate of cell death in the nervous system underlies neurodegenerative disease and is a hallmark of PD. Pro-inflammatory cytokines increasing leukocyte adhesion and migration induces injury to the blood-brain barrier and production of NO that causes apoptosis. Therapeutic agents are being developed to interfere with these events, thus conferring the potential to be neuroprotective. Furthermore, it is necessary to take into account that cytokines not only exert numerous primary effects, but also induce a cascade of secondary effects. Another difficulty in cytokine therapy is that cytokines do not act independently, but within a complex network dosage is another problem. In fact, too high a dosage may suppress a desired response and too low a dosage may fail to invoke it. In conclusion, at the moment, there are difficulties in designing and conducting studies on cytokine therapy, but the financial aspect and medical care systems must also be taken into account (Table 5).

Table 5 Principles for preclinical cytokines therapy

- Obtain *in vitro* data allowing conclusion on possible *in vivo* effects.
- Conduct animal study to obtain data on toxicity and pharmacokinetics.
- Identify clinically relevant hypothesis predicted from *in vitro* research and animal study.

CONCLUSION

The data reviewed suggest that pro-inflammatory cytokines may participate in the pathophysiology of PD either by producing NO, which may subsequently have a deleterious effect on dopaminergic neurons, or by activating specific transduction pathways coupled to receptors for cytokines.

Biological therapy with cytokines might represent a completely new approach in PD therapy. Furthermore, the studies of different cytokine profiles might help to better understand the progression of disease in PD patients, and the cytokines network might have important implications for clinical monitoring of PD and serve as markers for therapy efficacy.

Summary Points

- Parkinson's disease (PD) is a complex disorder characterized by the progressive degeneration of dopaminergic neurons in the midbrain.
- Various mechanisms of neuronal degeneration in PD have been proposed. These include formation of free radicals and oxidative stress, mitochondrial dysfunction, excitotoxicity, calcium excitotoxicity, trophic factor deficiency, inflammatory processes, genetic factors, environmental impact factors, toxic action of nitric oxide and apoptosis.
- One of the first features of the inflammation-associated modification described in PD is the up-regulated expression of major histocompatibility complex (MHC) molecules in the striatum and SNpc of mice intoxicated with 1-methyl-4-phenyl-1,2,3,6-tetrahydropyridine (MPTP).

- A marked increase in cytokines levels in the striatum and cerebrospinal fluid of parkinsonian patients compared to control subjects has been observed.
- The frequency of polymorphism located in the critical promoter region of the cytokine genes has been investigated to verify whether these polymorphisms are associated with the susceptibility to develop PD.
- The cytokines data contributes to a growing body of evidence suggesting that neuroinflammation has an important role in pathophysiology of PD.
- Further *in vitro* and *in vivo* population-based studies are warranted to corroborate these findings.
- An improved understanding of the relationship between cytokines and PD could have major therapeutic implications, targeting the role of cytokines in specific neuroinflammation-associated deleterious mechanisms may prove effective in slowing or even halting the disease course.

Abbreviations

6-OHDA	:	6-hydroxydopamine
CSF	:	cerebrospinal fluid
DN	:	dopaminergic neurons
IFN-γ	:	interferon-γ
IL	:	interleukin
iNOS	:	nitric oxide synthases
LPS	:	lipopolysaccharide
MPTP	:	1-methyl-4-phenyl-1,2,3,6-tetrahydropyridine
NO	:	nitric oxide
PBMC	:	peripheral blood mononuclear cells
PD	:	Parkinson's disease
SN	:	substantia nigra
TNF	:	tumor necrosis factor

Definition of Terms

Alpha-synuclein: A protein expressed in the neocortex, hippocampus, substantia nigra, thalamus, and cerebellum. It is predominantly a neuronal protein but can also be found in glial cells. Alpha-synuclein interacts with tubulin and may have an activity as potential microtubule-associated protein like tau. Alpha-synuclein is the primary structural component of Lewy body fibrils.

Apoptosis: Morphological processes leading to controlled cellular self-destruction. Apoptosis can be triggered by various stimuli from outside or inside the cell, e.g., by ligation of cell surface receptors, DNA damage as a cause of defects in DNA repair mechanisms, contradictory cell cycle signalling or developmental death signals.

Cytokines: Polypeptides produced throughout the body by cells of diverse embryological origin, which carry signals locally between cells and thus have an effect on other cells. They act by binding to specific membrane receptors, which then signal the cell via second messengers, often tyrosine kinases, to alter its behaviour (gene expression). Responses to cytokines include increasing or decreasing expression of membrane proteins (including cytokine receptors), proliferation, and secretion of effector molecules. T cells are initially activated as Th0 cells, which produce IL-2, IL-4 and IFN-γ. The nearby cytokine environment then influences differentiation into Th1 or Th2 cells. IL-4 stimulates Th2 activity and suppresses Th1 activity, while IL-12 promotes Th1 activities. Th1 and Th2 cytokines are antagonistic in activity.

Dopaminergic neurons: Neurons of the midbrain are the main source of dopamine (DA) in the mammalian central nervous system. Their loss is associated with Parkinson's disease. Dopaminergic neurons are found in a "harsh" region of the brain, the substantia nigra pars compacta. Although their numbers are small, dopaminergic neurons play an important role in the control of multiple brain functions including voluntary movement and a broad array of behavioural processes such as mood, reward, addiction, and stress.

Lewy bodies: Composed of the protein alpha-synuclein associated with other proteins such as ubiquitin and neurofilament protein. Lewy body proteins are found in an area of the brain stem where they deplete the neurotransmitter dopamine, causing Parkinsonian symptoms.

Neuronal degeneration: Loss of functional activity and trophic degeneration of nerve axons and their terminal arborizations following the destruction of their cells of origin or interruption of their continuity with these cells.

Polymorphisms: Proteins are gene products and so polymorphic versions are simply reflections of allelic differences in the gene; that is, allelic differences in Danto or more variants of an enzyme may be encoded by a single locus. The variants differ slightly in their amino acid sequence and often this causes them to migrate differently under electrophoresis. Alleles whose sequence reveals only a single changed nucleotide are called single nucleotide polymorphisms or SNPs.

References

Bessler, H. and R. Djaldetti, H. Salman, M. Bergman, and M. Djaldetti. 1999. IL-1 beta, IL-2, IL-6 and TNF-alpha production by peripheral blood mononuclear cells from patients with Parkinson's disease. Biomed. Pharmacother. 53: 141-145.

Bialecka, M. and G. Klodowska-Duda, M. Kurzawski, J. Slawek, G. Opala, P. Bialecki, K. Safranow, and M. Droździk. 2007. Interleukin-10 gene polymorphism in Parkinson's disease patients. Arch. Med. Res. 38: 858-863.

Boka, G. and P. Anglade, D. Wallach, F. Javoy-Agid, Y. Agid, and E.C. Hirsch. 1994. Immunocytochemical analysis of tumor necrosis factor and its receptors in Parkinson's disease. Neurosci. Lett. 172: 151-154.

Carvey, P.M. and E.Y. Chen, J.W. Lipton, C.W. Tong, Q.A. Chang, and Z.D. Ling. 2005. Intra-parenchymal injection of tumor necrosis factor-alpha and interleukin 1-beta produces dopamine neuron loss in the rat. J. Neural. Transm. 112: 601-612.

Clarke, D.J. and R.L. Branton. 2002. A role for tumor necrosis factor alpha in death of dopaminergic neurons following neural transplantation. Exp. Neurol. 176: 154-162.

Depino, A.M. and C. Earl, E. Kaczmarczyk, C. Ferrari, H. Besedovsky, A. del Rey, F.J. Pitossi, and W.H. Oertel. 2003. Microglial activation with atypical proinflammatory cytokine expression in a rat model of Parkinson's disease. Eur. J. Neurosci. 18: 2731-2742.

Gayle, D.A. and Z. Ling, C. Tong, T. Landers, J.W. Lipton, and P.M. Carvey. 2002. Lipopolysaccharide (LPS)-induced dopamine cell loss in culture: roles of tumor necrosis factor-alpha, interleukin-1beta, and nitric oxide. Brain Res. Dev. Brain Res. 133: 27-35.

Gribova, I.E. and B.B. Gnedenko, V.V. Poleshchuk, and S.G. Morozov. 2003. Level of interferon-gamma, tumor necrosis factor alpha, and antibodies to them in blood serum from Parkinson disease patients. Biomed. Khim. 49: 208-212.

Gruol, D.L. and T.E. Nelson. 1997. Physiological and pathological roles of interleukin-6 in the central nervous system. Mol. Neurobiol. 15: 307-339.

Hofmann, K.W. and A.F. Schumacher Schuh, J. Saute, R. Townsend, D. Fricke, R. Leke, D.O. Souza, L.V. Portela, M.L.F. Chaves, and C.R.M. Rieder. 2009. Interleukin-6 serum levels in patients with Parkinson's disease. Neurochem. Res. 34: 1401-1404.

Imamura, K. and N. Hishikawa, M. Sawada, T. Nagatsu, M. Yoshida, and Y. Hashizume. 2003. Distribution of major histocompatibility complex class II-positive microglia and cytokine profile of Parkinson's disease brains. Acta Neuropathol. 106: 518-526.

Mogi, M. and A. Togari, T. Kondo, Y. Mizuno, O. Komure, S. Kuno, H. Ichinose, and T. Nagatsu. 2000. Caspase activities and tumor necrosis factor receptor R1 level are elevated in the substantia nigra in Parkinson's disease. J. Neural. Transm. 107: 335-341.

Mogi, M. and M. Harada, T. Kondo, P. Riederer, and T. Nagatsu. 1996a. Interleukin-2 but not basic fibroblast growth factor is elevated in parkinsonian brain. J. Neural. Transm. 103: 1077-1081.

Mogi, M. and M. Harada, T. Kondo, Y. Mizuno, H. Narabayashi, P. Riederer, and T. Nagatsu. 1996. bcl-2 Protein is increased in the brain from parkinsonian patients. Neurosci. Lett. 215: 137-139.

Nagatsu, T. and M. Mogi, H. Ichinose, and A. Togari. 2000. Cytokines in Parkinson's disease. J. Neural. Transm. Suppl. 58: 143-151.

Nagatsu, T. and M. Sawada. 2005. Inflammatory process in Parkinson's disease: role for cytokines. Curr. Pharm. Des. 11: 999-1016.

Owen, A. and R.C. O'Neill, I. Maeve Rea, T. Lynch, D. Gosal, A. Wallace, P.M.D. Curran, D. Middleton, and J.M. Gibson. 2004. Functional promoter region polymorphism of the proinflammatory chemokine IL-8 gene associates with Parkinson's disease in the Irish. Human Immunol. 65: 340-346.

Pott Godoy, M.C. and R. Tarelli, C.C. Ferrari, M.I. Sarchi, and J. Fernando. 2008. Central and systemic IL-1 exacerbates neurodegeneration and motor symptoms in a model of Parkinson's disease. Brain 131: 1880-1894.

Reale, M. and C. Iarlori, A. Thomas, D. Gambi, B. Perfetti, M. Di Nicola, and M. Onofrj. 2008. Peripheral cytokines profile in Parkinson's disease. Brain Behav. Immun. 23: 55-63.

Recchia, A. and P. Debetto, A. Negro, D. Guidolin, S.D. Skaper, and P. Giusti. 2004. Alpha-synuclein and Parkinson's disease. FASEB J. 18: 617-626.

Selikhova, M.V. and N.E. Kushlinskii, N.V. Lyubimova, and E.I. Gusev. 2002. Impaired production of plasma interleukin-6 in patients with Parkinson's disease. Bull. Exp. Biol. Med. 133: 81-83.

Tweedie, D. and K. Sambamurti, and N.H. Greig. 2007. TNF-alpha inhibition as a treatment strategy for neurodegenerative disorders: new drug candidates and targets. Curr. Alzheimer Res. 4: 378-385.

Wahner, A.D. and J.S. Sinsheimer, J.M. Bronstein, and B. Ritz. 2007. Inflammatory cytokine gene polymorphisms and increased risk of parkinson disease. Arch. Neurol. 64: 836-840.

Yucesoy, B. and V.J. Johnson, Mi.L. Kashon, and M.I. Luster. 2007. Cytokine polymorphisms and relationship to disease. pp. 113-132. In: Cytokines in Human Health. Immunotoxicology, Pathology, and Therapeutic Applications. **Series:** Methods in Pharmacology and Toxicology, Robert V. House and Jacques Descotes (eds.): **DOI:** 10.1007/978-1-59745-350-9-

Zhang, Y. and M. James, F.A. Middleton, and R.L. Davis. 2005. Transcriptional analysis of multiple brain regions in Parkinson's disease supports the involvement of specific protein processing, energy metabolism, and signaling pathways, and suggests novel disease mechanisms. Am. J. Med. Genet. B Neuropsychiatr. Genet. 137: 5-16.

CHAPTER 23

Cerebrospinal Fluid Cytokines and their Change in Herpes Simplex Virus Encephalitis

Satoshi Kamei
Division of Neurology, Department of Medicine
Nihon University School of Medicine, Tokyo, Japan
30-1 Oyaguchi-kamichou, Itabashi-ku, Tokyo 173-8610, Japan

ABSTRACT

Herpes simplex virus encephalitis (HSVE) is associated with a significant morbidity and mortality, even when appropriate antiviral therapy is administered. A recent trial suggested that corticosteroid was beneficial in HSVE, but that precise role remains unclear. Previous studies indicated that several cerebrospinal fluid (CSF) cytokines including IFN-γ and IL-6 were elevated at the acute stage of HSVE.

It was also found that the values for monocyte chemotactic protein (MCP)-1 in the CSF were reciprocally correlated with the Modified Barthel Index at the time of CSF sampling. These findings have suggested that a host-side immune response associated with HSVE, such as that involving cytokines, could play some role in the outcome of HSVE. We recently reported the differences of CSF cytokine changes between different outcomes and between patients with and without corticosteroid administration at the acute stage of HSVE. The initial IFN-γ and maximum IL-6 with a poor outcome were higher than those with a good outcome. The decline rate of IL-6 in patients with corticosteroid was higher than that without corticosteroid. The initial IFN-γ and maximum IL-6 CSF values represented prognostic biomarkers in HSVE. One pharmacological mechanism related to corticosteroid in HSVE is apparently inhibition of pro-inflammatory cytokines such as IL-6.

INTRODUCTION

Herpes simplex virus (HSV) is a human herpes virus that can cause HSV encephalitis (HSVE), the most common, serious, sporadic, viral encephalitis in humans. HSVE patients who do not receive antiviral treatment have an extremely high mortality rate (about 70%), and fewer than 3% of survivors return to normal function (Sköldenberg *et al.* 1984, Whitley *et al.* 1986, Whitley and Lakeman 1995). Of the common central nervous system (CNS) viral infections, HSVE has a disproportionately high mortality compared with encephalitis due to other viruses. The introduction of acyclovir (ACV) has dramatically reduced mortality and morbidity for patients with HSVE; mortality rates for HSVE have decreased to 19-28% (Sköldenberg *et al.* 1984, Whitley *et al.* 1986, Whitley and Lakeman 1995). Although ACV treatment for HSVE is highly effective, the rate of poor outcome including advanced sequelae remains high, at 30-50%, and the rate of return to normal living is less than 50%. Thus, the morbidity and mortality remain significantly high for HSVE despite standard ACV treatment at the acute stage (Sköldenberg *et al.* 1984, Whitley *et al.* 1986, Whitley and Lakeman 1995). This finding indicates a need to develop further improved therapeutic regimens for HSVE.

We recently reported that combination therapy with corticosteroid and ACV appeared to give a better outcome in adult patients with HSVE (Kamei *et al.* 2005). The pharmacological mechanism of corticosteroid in HSVE, except for improvement of brain edema, remains unclear. In a clinical guide (Solomon *et al.* 2007) and in a clinical review (Fitch and van de Beek D 2008) concerning HSVE, it was also noted that "a recent trial suggests that steroid may be beneficial even in patients without marked swelling. Their role in HSVE merits further study."

SERIAL CSF CYTOKINE CHANGES IN HSVE

There have been only four previous detailed clinical studies on serial changes in intrathecal cytokines and chemokines in patients with HSVE (Aurelius *et al.* 1994, Rösler *et al.* 1998, Asaoka *et al.* 2004, Ichiyama *et al.* 2008). One of these earlier studies described the serial intrathecal cytokine changes in nine patients with HSVE, demonstrating that IFN-γ and IL-6 were first increased during the first week of illness, and TNF-α, IL-2, and soluble CD8 antigen then became elevated at 2 to 6 wk, but IL-1 β could not be detected (Aurelius *et al.* 1994). Previous studies (Asaoka *et al.* 2004, Ichiyama *et al.* 2008) have also noted that the elevation of intrathecal IL-6 values in six patients with HSVE was greater than that in patients with limbic encephalitis and that the CSF concentrations of IFN-γ and soluble tumor necrosis factor receptor 1 in HSVE was higher than those with non-herpetic acute limbic encephalitis. The results obtained for the elevation of intrathecal IFN-γ and IL-6 values at the acute stage of HSVE in the present study confirmed these earlier findings.

There have been several previous fundamental and basic investigations of serial changes in cytokines employing experimental animal HSVE models (Geiger *et al.* 1997, Shimeld *et al.* 1997, Dvorak *et al.* 2004, Sellner *et al.* 2005). In these experimental studies, the acute neuroinflammatory reaction in HSVE was found to comprise a rapid secretion of IFN-γ, pro-inflammatory cytokines including IL-6, and anti-inflammatory cytokines such as IL-10, followed by a specific synthesis of chemokines in order to attack inflammatory cells (Geiger *et al.* 1997, Shimeld *et al.* 1997, Sellner *et al.* 2005). Moreover, a recent investigation employing an experimental HSVE mouse model has indicated that IL-6 acted as a potent mediator of neuronal injury at the acute stage of HSVE (Dvorak *et al.* 2004). Furthermore, it has been suggested that a host-side immune response associated with HSVE, such as that involving cytokines, could play some role in the outcome of HSVE based on the findings of an *in vitro* experimental study on infection with HSV (Oshima *et al.* 2001).

ALTERATION OF CSF CYTOKINES AMONG DIFFERENT OUTCOMES AND BETWEEN THE ANTIVIRAL THERAPY WITH AND WITHOUT CORTICOSTEROID ADMINISTRATION

It was found that the values for monocyte chemotactic protein (MCP)-1 in the CSF were reciprocally correlated with the Modified Barthel Index at the time of CSF sampling (Rösler *et al.* 1998). However, the differences in serial intrathecal cytokine changes between different clinical outcomes and between patients with and without additional administration of corticosteroid under ACV treatment at the acute stage of HSVE remain unclear. Based on the above-mentioned background, we recently provided a first assessment of the differences in serial CSF cytokine changes among different clinical outcomes, and also an assessment of the differences between treatment with and without corticosteroid administration under ACV at the acute stage in adult patients with HSVE (Kamei *et al.* 2009).

The clinical characteristics of the 20 patients with HSVE enrolled in this study were as follows (Kamei *et al.* 2009). The numbers of good and poor outcomes were 13 and 7. The numbers of patients with and without corticosteroid administration amounted to 10 in each group. The baseline clinical characteristics of the groups with different outcomes and of the patients with and without corticosteroid administration are summarized in Table 1. The age of the patients with a poor outcome was greater than that of the patients with a good outcome. All factors among the clinical characteristics of the patients with and without corticosteroid administration showed no significant differences.

Table 1 Baseline clinical characteristics of the patient groups

	Outcome			*Corticosteroid treatment under acyclovir therapy*		
	Good (*N* = 13)	*Poor* (*N* = 7)	*Difference between the two groups**	*Administration* (*N* = 10)	*No treatment* (*N* = 10)	*Difference between the two groups**
(1) Male (%)	69	71	NS	20	40	NS
(2) Age at onset (M <SD>; years)	39.2 <21.6>	57.6 <12.9>	P=0.027	48.8 <23.3>	42.4 <18.4>	NS
(3) Days after onset at initiation of acyclovir (M <SD>)	4.7 <3.3>	5.1 <3.4>	NS	4.3 <3.2>	5.4 <3.4>	NS
(4) GCS score at initiation of acyclovir (M <SD>)	7.5 <3.2>	5.0 <2.8>	NS	6.5 <3.0>	6.7 <3.6>	NS
(5) Number of cells in initial CSF (M <SD>; /μl)	267 <386>	229 <339>	NS	197 <298>	310 <424>	NS
(6) Concentration of protein in initial CSF (M <SD>; mg/dl)	143 <143>	124 <130>	NS	115 <107>	158 <163>	NS
(7) Detection of lesion by initial CT (%)	31	71	NS	60	50	NS
(8) Detection of lesion by initial MRI (%)	85	100	NS	80	100	NS

M = mean. SD = standard deviation. GCS = Glasgow coma scale. CSF = cerebrospinal fluid. CT = cranial computed tomography; MRI = magnetic resonance imaging.

NS = not significant ($p>0.05$); *statistical differences among the two groups (Fisher's exact probability test or the Mann-Whitney U test). (With permission, from the original manuscript.)

Among the measurements of the CSF cytokine values in the present 20 patients with HSVE, IL-6 was elevated in all patients. Also, IL-10 and IFN-γ were elevated in 17 out of the 20 patients. Other cytokines revealed less than the minimum detection limit in all the CSF samples at the acute stage of HSVE. The initial value of IL-6 represented the maximum value among the serial CSFs in 17 out of the 20 patients. The IL-6 values in the subsequent CSF samples declined in these 17 patients. However, in the remaining three patients, the value of IL-6 in the second CSF sample was higher than that in the initial CSF sample. The initial values of IL-10 and IFN-γ represented the maximum values among the serial CSF samples in all 17 patients who revealed detectable values, and the values in the subsequent CSF samples declined to the range of less than the minimum detection limit. The results for the CSF cytokines including the statistical differences among different clinical outcomes are illustrated in Fig. 1.

The mean, median, and range of initial IFN-γ values in the 13 patients with a good outcome revealed 6, 3, and 0 to 23 IU/ml. Those of initial IFN-γ values in the seven patients with a poor outcome revealed 25, 24, and 0 to 53 IU/ml. The mean, median, and rang of maximum IL-6 values in the 13 patients with a good outcome revealed 346, 221, and 45 to 1285 pg/ml. Those of maximum IL-6 values in the seven patients with a poor outcome revealed 2617, 2390, and 134 to 8340 pg/ml. The mean values of initial IFN-γ and the maximum value of IL-6 in the patients with a poor outcome were significantly higher than those in the patients with a good outcome (p=0.019 for the initial value of INF-γ, and p=0.013 for the maximum value of IL-6; Mann-Whitney *U* test). The initial values of IL-6 and IL-10 among the different clinical outcomes were not significantly different.

Another notable finding of our study was that the rate of decline of IL-6 in patients with corticosteroid administration at the acute stage of HSVE was significantly higher than that without corticosteroid administration (p=0.034, Mann-Whitney *U* test). The results for the changes in IFN-γ, IL-10, and IL-6 values of the serial CSF samples at the acute stage in each HSVE patient with and without corticosteroid administration are illustrated in Figs. 2 to 4.

The IL-6 CSF values in the patients with corticosteroid administration declined rapidly to less than the minimum detection limit as compared to those in the patients without corticosteroid administration. The mean, median, and range of the IL-6 decline rate in the patients with corticosteroid administration revealed 173, 73, and 29 to 637 pg/ml/day. On the other hand, those values of the IL-6 decline rate in the patients without corticosteroid administration revealed -996 (increase), 11, and -8293 (increase) to 459 pg/ml/day. The differences in decline rates of IFN-γ and IL-10 between the patients with and without administration of corticosteroids were not significant. The differences in decline rates of IL-6, IFN-γ, and IL-10 between the different outcomes were also not significant.

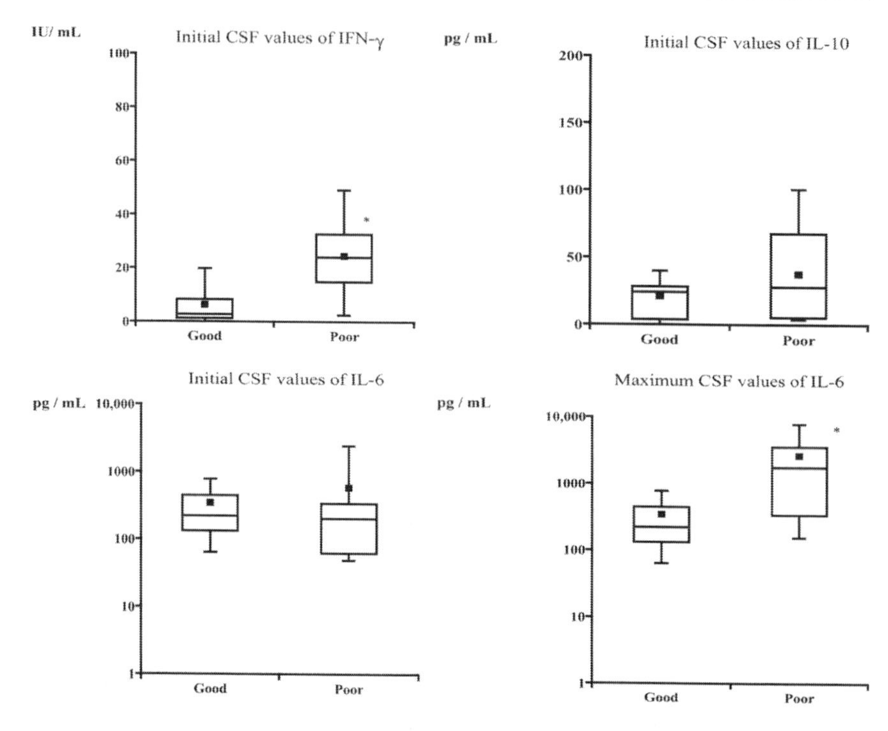

Fig. 1 Results for intrathecal cytokine concentrations among different outcomes. CSF = cerebrospinal fluid, IFN = interferon, IL = interleukin, IU = international units. In the box plots, the top of the box is placed at the first quartile, the bottom at the third quartile, and a horizontal line is also placed at the median. The vertical lines extending from the top and bottom of each box indicate minimum and maximum values. Symbols: ▪ indicates mean value; * indicates a significant difference among different outcomes ($P < 0.05$).

The differences in mean values of the initial values of INF-γ (left upper graph) and the maximum values of IL-6 (right lower graph) among different outcomes were significant. Those of the initial values of IL-10 and IL-6 among different outcomes were not significant. The initial CSF values of INF-γ and IL-10 in all patients represented the maximum values of the serial CSFs. Since all initial CSF samples at admission were taken at the time just before treatment, a comparison of CSF cytokine concentrations between the patients with corticosteroids and those without corticosteroids was not performed. (With permission, from the original manuscript.)

PHARMACOLOGICAL MECHANISM IN HSVE BASED ON THE ALTERATION OF CSF CYTOKINES

The above-mentioned findings appeared to represent appropriate clinical findings for the acute stage of HSVE based on the descriptions given in recent experimental studies (Oshima *et al.* 2001, Dvorak *et al.* 2004, Sellner *et al.* 2005); namely, the initial CSF value of IFN-γ and the maximum CSF value of IL-6 in our patients

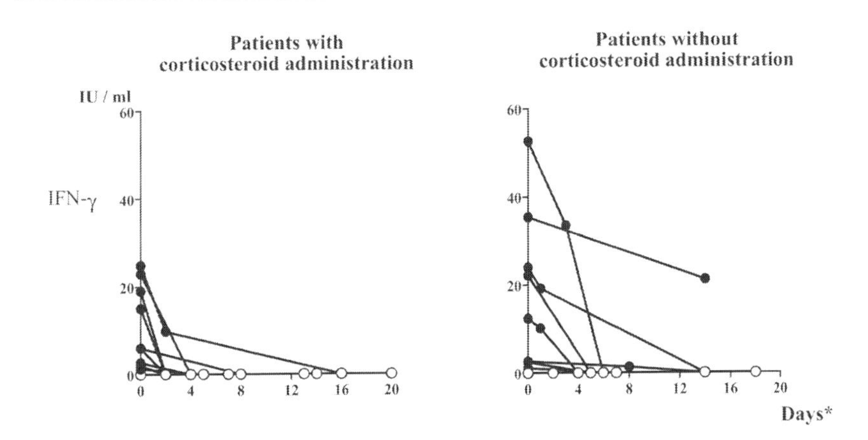

Fig. 2 Serial changes of intrathecal IFN-γ concentrations in each patient with and without corticosteroid administration. IU = international units. ○ indicates less than the minimum detection limit; Days* indicates days after initiation of treatment. The data for the IFN-γ serial concentrations between the patients with corticosteroid administration and those without corticosteroid administration indicated that the rates of decline in the patients with corticosteroid administration were higher than those in the patients without corticosteroid, but such differences in decline rates among the different treatments were not significant. (With permission, from the original manuscript.)

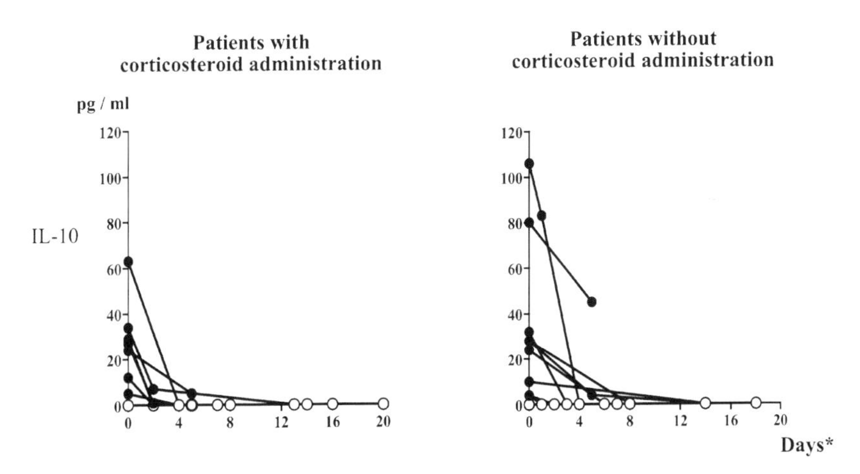

Fig. 3 Serial changes of intrathecal IL-10 concentrations in each patient with and without corticosteroid administration. ○ indicates less than the minimum detection limit; Days* indicates days after initiation of treatment. The data for IL-10 serial concentrations between the patients with corticosteroid administration and those without corticosteroid administration indicated that the rates of decline in the patients with corticosteroid administration were higher than those in the patients without corticosteroid, but such differences in decline rates among the different treatments were not significant. (With permission, from the original manuscript.)

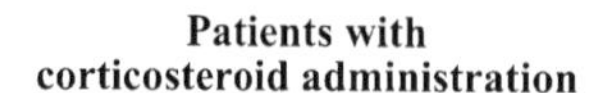

Patients without
corticosteroid administration

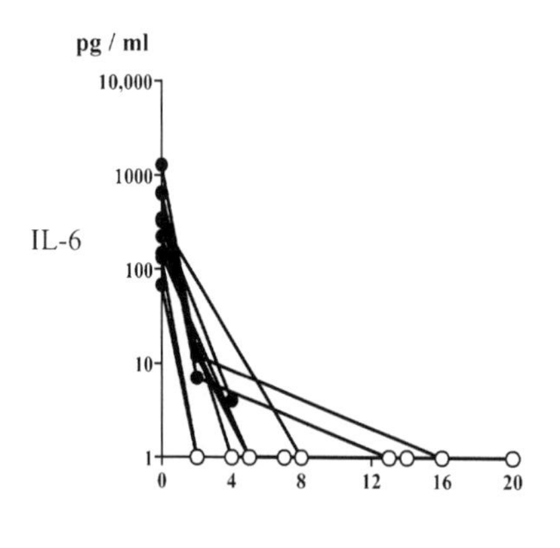

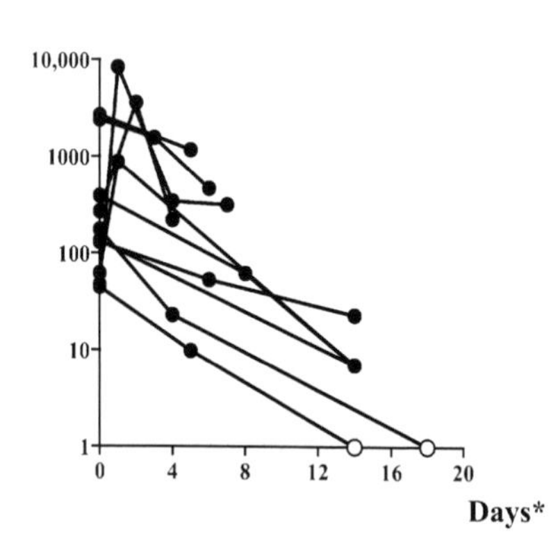

Fig. 4 Serial changes of intrathecal IL-6 concentrations in each patient with and without corticosteroid administration. ○ indicates less than the minimum detection limit; Days* indicates days after initiation of treatment. The IL-6 concentrations in the initial CSF samples, which were taken just before treatment, demonstrated high values in all patients with HSVE. The initial CSF values for the IL-6 concentration revealed no significant difference between the patients with corticosteroid administration (left graph) and those without corticosteroid administration (right graph). The serial changes of IL-6 CSF values in 9 out of the 10 patients with corticosteroid administration revealed a marked decline to less than the minimum detection limit in the interval from the 2nd to 16th day after initiation of the treatment with acyclovir and corticosteroid. However, the ongoing IL-6 values in the CSF of only 2 out of the 10 patients without corticosteroid administration declined to less than the minimum detection limit by the 20th day. All of the three patients who exhibited a transient increase in IL-6 values at the second CSF sample did not receive corticosteroid. (With permission, from the original manuscript.)

with a poor outcome were significantly higher than those in patients with a good outcome. Moreover, a recent investigation of the levels of CSF cytokines and chemokines in patients with Japanese encephalitis has yielded similar findings in which the IFN-α, IL-6, and IL-8 values of non-survivors were higher than those of survivors (Winter *et al.* 2004). Based on the notion that IFN-γ is an inhibitor of viral replication (Samuel 1991, Finke *et al.* 1995) and IL-6 is one of the pro-inflammatory cytokines, the results of the present study indicate that the initial IFN-γ and maximum IL-6 values at the acute stage of HSVE could be appropriate for use as prognostic biomarkers in adult patients with HSVE.

Another notable finding of the present study was that the rate of decline in CSF IL-6 values at the acute stage of HSVE among the patients receiving corticosteroid administration was significantly higher than that without corticosteroid. Some reports have examined the effects of corticosteroid administration with ACV treatment in animal models of HSVE (Thompson *et al.* 2000, Meyding-Lamade *et al.* 2003). The data obtained indicated that the HSV viral load of the brain tissue in animals treated with both ACV and corticosteroid was similar to that of the brain tissue in animals treated with ACV alone. These studies also revealed that the corticosteroid did not inhibit the antiviral action of ACV and might decrease the extent of HSVE infection (Thompson *et al.* 2000, Meyding-Lamade *et al.* 2003). Seven out of nine patients in the previous detailed clinical report on serial cytokine values in HSVE (Aurelius *et al.* 1994) were administered corticosteroid with ACV, so that the above-mentioned observations made it difficult to assess the differences in intrathecal cytokine changes between the patients with and without corticosteroid administration. In the Discussion section of this previous clinical study (Aurelius *et al.* 1994), it was speculated that a decline of intrathecal cytokines might accompany the therapeutic use of corticosteroid. The marked decline rate of IL-6 observed in the adult HSVE patients with therapeutic use of corticosteroid in the present study could provide the first supportive finding for this earlier speculation (Aurelius *et al.* 1994). This finding thus suggests that the pharmacological mechanism of corticosteroid treatment in the acute stage of adult HSVE patients may involve not only improvement of brain edema, but also inhibition of the pro-inflammatory cytokine cascade based on the host-side immune response associated with the acute stage of HSVE.

Summary Points

- Herpes simplex virus encephalitis (HSVE) is associated with a significant morbidity and mortality, even when appropriate antiviral therapy is administered.
- A recent trial suggested that corticosteroid was beneficial in HSVE, but that precise role remains unclear.
- Previous studies indicated that several cerebrospinal fluid (CSF) cytokines including IFN-γ and IL-6 were elevated at the acute stage of HSVE.
- It was also found that the values for monocyte chemotactic protein (MCP)-1 in the CSF were reciprocally correlated with the Modified Barthel Index at the time of CSF sampling.
- We recently reported the differences of CSF cytokine changes between different outcomes and between patients with and without corticosteroid administration at the acute stage of HSVE.
- The initial IFN-γ and maximum IL-6 with a poor outcome were higher than those with a good outcome.
- The decline rate of IL-6 in patients with corticosteroid was higher than that without corticosteroid.
- The initial IFN-γ and maximum IL-6 CSF values represented prognostic biomarkers in HSVE.

- One pharmacological mechanism related to corticosteroid in HSVE is apparently inhibition of pro-inflammatory cytokines such as IL-6.

Abbreviations

ACV	:	acyclovir
CLEIA	:	chemiluminescent enzyme immunoassay
CSF	:	cerebrospinal fluid
CT	:	cranial computed tomography
EASIA	:	enzyme amplified sensitivity immunoassay
GCS	:	Glasgow coma scale
HSV	:	herpes simplex virus
HSVE	:	herpes simplex virus encephalitis
IFN	:	interferon
IL	:	interleukin
IU	:	international units
M	:	mean
MCP	:	monocyte chemotactic protein
MRI	:	magnetic resonance imaging
NS	:	not significant
SD	:	standard deviation
TNF	:	tumor necrosis factor

Definition of Terms

Acyclovir: Anti-herpes simplex virus drug.

Biomarker: A biological product that can be objectively measured and that indicates pathological processes or outcome.

Chemokines: A class of cytokines that chemoattracts, or recruits, specific cells that express the appropriate chemokine receptor.

Glasgow coma scale: Assessment for the disturbance of consciousness; lower score indicates severe impairment.

Modified Barthel Index: Scale of daily activity.

References

Asaoka, K. and H. Shoji, S. Nishizaka, M. Ayabe, T. Abe, N. Ohori, *et al.* 2004. Non-herpetic acute limbic encephalitis: cerebrospinal fluid cytokines and magnetic resonance imaging findings. Intern. Med. 43: 42-48.

Aurelius, E. and B. Andersson, M. Forsgren, B. Sköldenberg, and O. Strannegård. 1994. Cytokines and other markers of intrathecal immune response in patients with herpes simplex encephalitis. J. Infect. Dis. 170: 678-681.

Dvorak, F. and F. Martinez-Torres, J. Sellner, J. Haas, P.D. Schellinger, M. Schwaninger, *et al.* 2004. Experimental herpes simplex virus encephalitis: a long-term study of interleukin-6 expression in mouse brain tissue. Neurosci. Lett. 367: 289-292.

Finke, D. and U.G. Brinckmann, V. ter Meulen, and U.G. Liebert. 1995. Gamma interferon is a major mediator of antiviral defense in experimental measles virus-induced encephalitis. J. Virol. 69: 5469-5474.

Fitch, M.T. and D. van de Beek. 2008. Drug insight: steroids in CNS infectious diseases–new indications for an old therapy. Nat. Clin. Pract. Neurol. 4: 97-104.

Geiger, K.D. and T.C. Nash, S. Sawyer, T. Krahl, G. Patstone, J.C. Reed, *et al.* 1997. Interferon-gamma protects against herpes simplex virus type 1-mediated neuronal death. Virology 238: 189-197.

Ichiyama, T. and H. Shoji, Y. Takahashi, T. Matsushige, M. Kajimoto, T. Inuzuka, *et al.* 2008. Cerebrospinal fluid levels of cytokines in non-herpetic acute limbic encephalitis: Comparison with herpes simplex encephalitis. Cytokine 44: 149-153.

Kamei, S. and T. Sekizawa, H. Shiota, T. Mizutani, Y. Itoyama, T. Takasu, *et al.* 2005. Evaluation of combination therapy using aciclovir and corticosteroid in adult patients with herpes simplex virus encephalitis. J. Neurol. Neurosurg. Psychiatr. 76: 1544-1549.

Kamei, S. and N. Taira, M. Ishihara, T. Sekizawa, A. Morita, K. Miki, *et al.* 2009. Prognostic value of cerebrospinal fluid cytokine changes in herpes simplex virus encephalitis. Cytokine 46: 187-193.

Meyding-Lamade, U.K. and C. Oberlinner, P.R. Rau, S. Seyfer, S. Heiland, J. Sellner, *et al.* 2003. Experimental herpes simplex virus encephalitis: a combination therapy of acyclovir and glucocorticoids reduces long-term magnetic resonance imaging abnormalities. J. Neurovirol. 9: 118-125.

Oshima, M. and H. Azuma, T. Suzutani, H. Ikeda, and A. Okuno. 2001. Direct and mononuclear cell mediated effects on interleukin 6 production by glioma cells in infection with herpes simplex virus type 1. J. Med. Virol. 63: 252-258.

Rösler, A. and M. Pohl, H.J. Braune, W.H. Oertel, D. Gemsa, and H. Sprenger. 1998. Time course of chemokines in the cerebrospinal fluid and serum during herpes simplex type 1 encephalitis. J. Neurol. Sci. 157: 82-89.

Samuel, C.E. 1991. Antiviral actions of interferon: interferon-regulated cellular proteins and their surprisingly selective antiviral activities. Virology 183: 1-11.

Sellner, J. and F. Dvorak, Y. Zhou, J. Haas, R. Kehm, B. Wildemann, *et al.* 2005. Acute and long-term alteration of chemokine mRNA expression after anti-viral and anti-inflammatory treatment in herpes simplex virus encephalitis. Neurosci. Lett. 374: 197-202.

Shimeld, C. and J.L.Whiteland, N.A. Williams, D.L. Easty, and T.J. Hill. 1997. Cytokine production in the nervous system of mice during acute and latent infection with herpes simplex virus type 1. J. Gen. Virol. 78 (Pt 12): 3317-3325.

Sköldenberg, B. and M. Forsgren, K. Alestig, T. Bergström, L. Burman, E. Dahlqvist, *et al.* 1984. Acyclovir versus vidarabine in herpes simplex encephalitis: randomised multicentre study in consecutive Swedish patients. Lancet 2: 707-711.

Solomon, T. and I.J. Hart, and N.J. Beeching. 2007. Viral encephalitis: a clinician's guide. Pract. Neurol. 7: 288-305.

Thompson, K.A. and W.W. Blessing, and S.L. Wesselingh. 2000. Herpes simplex replication and dissemination is not increased by corticosteroid treatment in a rat model of focal herpes encephalitis. J. Neurovirol. 6: 25-32.

Winter, P.M. and N.M. Dung, H.T. Loan, R. Kneen, B. Wills, T. Thu le, *et al.* 2004. Proinflammatory cytokines and chemokines in humans with Japanese encephalitis. J. Infect. Dis. 190: 1618-1626.

Whitley, R.J. and C.A. Alford, M.S. Hirsh, R.T. Schooley, J.P. Luby, F.Y. Aoki, *et al.* 1986. Vidarabine versus acyclovir therapy in herpes simplex encephalitis. N. Engl. J. Med. 314: 144-149.

Whitley, R.J. and C.G. Cobbs, C.A. Alford, S.J. Soong, M.S. Hirsch, J.D. Connor, *et al.* 1989. Diseases that mimic herpes simplex encephalitis: diagnosis, presentation and outcome. JAMA 262: 234-239.

Whitley, R.J. and F. Lakeman. 1995. Herpes simplex virus infections of the central nervous system: therapeutic and diagnostic considerations. Clin. Infect. Dis. 20: 414-420.

CHAPTER **24**

Cytokines in Skeletal Muscle Insulin Resistance

Adam Whaley-Connell[1,3,4], Vincent G. DeMarco[1,2,3] and James R. Sowers[1,2,3,4,*]

University of Missouri-Columbia School of Medicine, Departments of [1]Internal Medicine [2]Medical Pharmacology & Physiology, [3]Diabetes and Cardiovascular Research Center, and [4] Harry S. Truman VA Hospital, Columbia, MO, USA

ABSTRACT

Insulin resistance is the linchpin for development of the cardiometabolic syndrome and type 2 diabetes. Resistance to the actions of insulin in skeletal muscle is important in development of systemic insulin resistance given that skeletal muscle normally accounts for approximately 75% of all insulin-mediated glucose disposal. However, the molecular mechanisms responsible for skeletal muscle insulin resistance remain poorly defined. Understanding the mechanisms by which skeletal muscle tissue develops resistance to insulin could provide attractive targets for therapeutic interventions. There is emerging evidence of an integral relationship between chronic inflammation, oxidative stress, and skeletal muscle insulin resistance due to circulating inflammatory cytokines derived from adipose tissue (e.g., adipokines) or from local autocrine/paracrine effects of skeletal muscle–derived cytokines (e.g., myokines). This chapter is focused on the effects of inflammatory cytokines and oxidative stress on insulin signaling in skeletal muscle and consequent development of systemic insulin resistance.

INTRODUCTION

The ability to maintain normal blood glucose levels involves a complex interplay between insulin secretion by pancreatic beta cells and insulin metabolic

Corresponding author

responsiveness in skeletal muscle, liver, adipose and other tissues (Shoelson *et al.* 2006, Wellen and Hotamisligil 2005). Skeletal muscle is particularly important as it comprises 40-50% of body mass and mediates over 75% of all insulin-mediated glucose disposal under normal physiological conditions (Wellen and Hotamisligil 2005, Cooper *et al.* 2007). Increasing evidence suggests that immune-mediating inflammatory products are integral in regulating glucose metabolism and the excessive activation of inflammatory pathways may represent a fundamental step in the development of systemic insulin resistance (Shoelson *et al.* 2006, Wellen and Hotamisligil 2005, Cooper *et al.* 2007). Chronic low-grade inflammation and oxidative stress due to increased reactive oxygen species (ROS) generation and/or compromised anti-oxidant systems represent important factors in the progression of insulin resistance (Wellen and Hotamisligil 2005, Cooper *et al.* 2007). Despite significant progress in understanding the development of insulin resistance, the precise molecular mechanisms responsible for insulin resistance, particularly in skeletal muscle, still remain incompletely understood.

THE IMPORTANCE OF INSULIN RESISTANCE IN SKELETAL MUSCLE

Insulin resistance denotes an impaired biological response by target organs to insulin. The signaling pathway defect implicated in impaired insulin metabolic signaling involves the phosphatidylinositide 3-kinase (PI3-K) and protein kinase B (Akt) pathway. Upon binding to the insulin receptor (IR), in the insulin-sensitive state, there is auto-phosphorylation of the beta-subunit, which mediates non-covalent but stable interaction between the receptor and cellular proteins. Several proteins are then rapidly phosphorylated on tyrosine residues by ligand-bound insulin receptors, including insulin receptor substrate 1 (IRS-1). Tyrosine phosphorylation of IRS-1 results in engagement of the p85 regulatory subunit of PI3-K and activates the p110 catalytic subunit, which increases phosphoinositides such as phosphatidylinositol 3,4,5-trisphosphate (Fig. 1). This leads to the activation of phosphoinositide-dependent protein kinase (PDK) and downstream Akt and/or atypical protein kinase C (PKC). Phosphorylation of Akt substrate 160 (AS160) facilitates translocation of glucose transport 4 (GLUT-4) to the sarcolemma to facilitate glucose entry into the cell. The interruption of this signaling cascade results in resistance to the actions of insulin and impaired insulin-mediated glucose metabolism. There are several putative mechanisms that mitigate insulin actions in skeletal muscle glucose metabolism and cross-talk with the insulin metabolic signaling via the IRS-PI3-K-Akt pathway that contribute to reduced insulin resistance including low-grade inflammation and oxidative stress (Fig. 1).

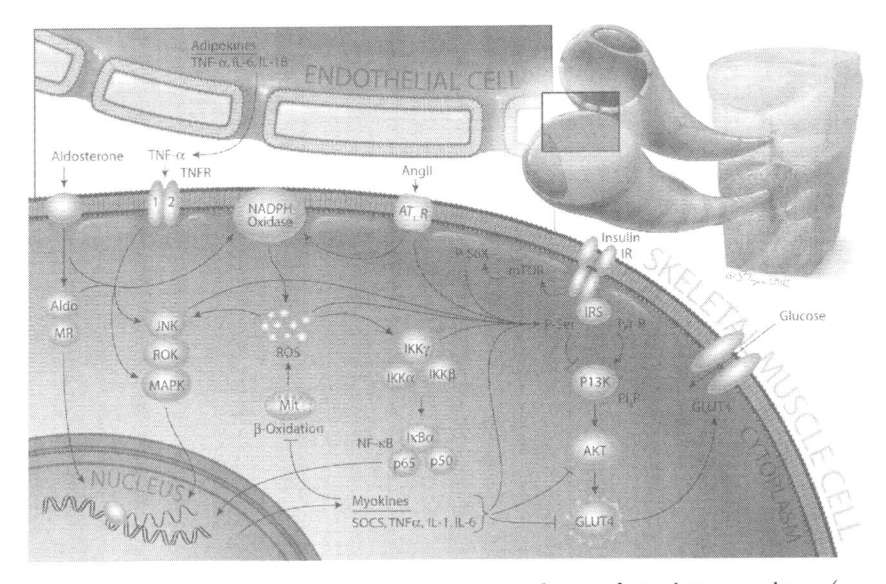

Fig. 1 Visceral adipose tissue is an important contributor of circulating cytokines (e.g., adipokines) that, in concert with local production of cytokines in skeletal muscle (e.g., myokines), contribute to inflammatory pathways that react with oxidative stress and activation of the RAAS to impair insulin metabolic signaling in skeletal muscle tissue. The consequent impairment in insulin-mediated glucose utilization in skeletal muscle is an important contributor to skeletal muscle insulin resistance as well as systemic insulin resistance. (Original artwork by Stacy Turpin.) (→ enhancement; ⊣ inhibition)

Color image of this figure appears in the color plate section at the end of the book.

THE ROLE OF INFLAMMATION IN SKELETAL MUSCLE INSULIN RESISTANCE

The relationship between chronic inflammation and skeletal muscle insulin resistance is thought to be mediated by either circulating inflammatory cytokines derived from adipose tissue (e.g., adipokines) or skeletal muscle–derived cytokines (e.g., myokines). The evidence that skeletal muscle is susceptible to either circulating or local cytokine effects is demonstrated by increased numbers of inflammatory macrophages and CD154 (T cell membrane marker) levels in muscle biopsies from diabetic patients (Wellen and Hotamisligil 2005) as well as increased levels of tumor necrosis factor alpha (TNF-α), IL-6, inducible nitric oxide synthase (iNOS), fibrinogen, C-reactive protein (CRP), plasminogen activator inhibitor 1 (PAI-1), and sialic acid (Perreault and Marette 2001). In obese states, skeletal muscle is susceptible to both circulating adipokines, such as TNF-α, IL-6, and IL-1β, and locally generated inflammatory myokines (Wellen and Hotamisligil 2005, Cooper *et al.* 2007, Perreault and Marette 2001, Pedersen *et al.* 2007, Hotamisligil *et al.* 1993). This notion is further supported by the fact that skeletal muscle possesses many of the components of the innate immune system, including cytokine receptors and toll-like receptors (TLRs) (Wellen and Hotamisligil 2005, Cooper *et al.* 2007) (Fig. 1). Indeed, the list of adipokines and myokines is growing. Herein, we will summarize recent evidence that myokines and chemokines, in conjunction with an altered innate immune response in skeletal muscle, collectively contribute to skeletal muscle insulin resistance.

Tumor Necrosis Factor α

TNF-α is a pleiotropic cytokine that can produce systemic inflammation as well as stimulate cellular responses such as apoptosis, proliferation, and even local production of inflammatory molecules. TNF-α is largely generated by macrophages but also skeletal muscle and is the first cytokine recognized to have a direct role in promoting insulin resistance (Hotamisligil *et al.* 1993, 1995). In insulin-resistant animal models and humans, there are demonstrable increases in skeletal muscle TNF-α and cultured skeletal muscle cells (Hotamisligil *et al.* 1993, 1995, Togashi *et al.* 2000, Uysal *et al.* 1997, Hotamisligil *et al.* 1993, Zhang *et al.* 2000). Indeed, mice lacking TNF-α or its receptors 1 and/or 2 (TNF-R1 and TNF-R2) are protected from obesity-induced insulin resistance (Uysal *et al.* 1997), while suppression of TNF-α by anti-TNF-α antibodies or TNF-α converting enzyme inhibitors improves insulin sensitivity in either obese or non-obese insulin-resistant models (Hotamisligil *et al.* 1993). Thereby, targeting TNF-α is essential in developing our understanding of insulin resistance as skeletal muscle accounts for the bulk of *in vivo* glucose disposal.

TNF-α elicits various cellular responses in skeletal muscle through binding to specific receptors, TNF-R1 and TNF-R2 (Uysal *et al.* 1997, Zhang *et al.* 2000), that promote a complex array of post-receptor signaling events. TNF-α decreases tyrosine phosphorylation and increases serine phosphorylation of IRS-1 that contribute to impairments in insulin metabolic signaling (Hotamisligil *et al.* 1995, Cooper *et al.* 2007). Collectively, the relative increase in serine to tyrosine phosphorylation contributes to increased ubiquinization/proteosomal degradation of IRS-1, or decreased ability of IRS-1 to engage the p85 subunit of PI3-K leading to decreased insulin metabolic signaling. Alternatively, targeting TNF-α through infusion of anti-TNF-α antibody leads to improvements in insulin receptor phosphorylation. TNF-α also impairs insulin signal transduction via reductions in Akt and the AS160 that ultimately contribute to reductions in skeletal muscle insulin-stimulated glucose uptake. Furthermore, the reductions in glucose utilization occur through decreased IRS tyrosine phosphorylation and Akt activation in a p38 MAP kinase (MAPK)-dependent manner (de Alvaro *et al.* 2004).

TNF-α signaling through TNF-R1 suppresses AMP-activated protein kinase (AMPK) activity through transcriptional up-regulation of protein phosphatase 2C. This, in turn, reduces acetyl CoA carboxylase (ACC) phosphorylation, suppresses fatty-acid oxidation, and increases intramuscular diacylglycerol accumulation. MAPK is an important target in understanding the link between TNF-α and insulin metabolic signaling. The MAPK isoform 4 (MAP4K4) links extracellular signal–regulated kinase 1/2 (ERK1/2) and N-terminal kinase (JNK) signaling pathways to TNF-α in human skeletal muscle cells (Wellen and Hotamisligil 2005, Cooper *et al.* 2007). TNF-α infusion increases phosphorylation of p70 S6 kinase, ERK1/2 and JNK, concomitant with increased serine and reduced tyrosine phosphorylation of IRS-1. Effects that are then associated with impaired PI3-K activation and phosphorylation of AS160 (Plomgaard *et al.* 2005), a critical link between the PI3-K and AMPK pathways for stimulating glucose transport (Fig. 1).

Interleukin-6 (IL-6)

The cytokine IL-6 modulates immune responses and exhibits both pro- and anti-inflammatory effects in skeletal muscle (Adam *et al.* 2000, Lopez-Soriano *et al.* 2006) that inhibit or promote insulin actions on glucose metabolism in skeletal muscle, respectively. These apparently contradictory roles of IL-6 in modulating skeletal muscle insulin resistance remain controversial and further study is needed for clarification.

In humans, IL-6 exerts an insulin-sensitizing effect and enhances insulin-stimulated glucose disposal in skeletal muscle (Carey *et al.* 2006, Ruderman *et al.* 2006). In this context, recent data suggest that exercise, a potent insulin sensitizer, contributes to release of IL-6 from muscle (Lopez-Soriano *et al.* 2006). Further, transgenic mice that overexpress IL-6 are protected from obesity induced by a high-fat diet and from insulin resistance compared with control wild type mice (Ruderman *et al.* 2006). Evidence that acute IL-6 administration does not impair muscle glucose uptake or whole body glucose disposal in healthy humans supports a role for IL-6 in improving glucose handling.

Alternatively, there is ample evidence that IL-6 heightens the deleterious effects of insulin on glucose homeostasis in the insulin-resistant state. There are elevations in circulating IL-6 in various insulin-resistant states including type 2 diabetes mellitus (T2DM) and obesity. In mice subjected to hyperinsulinemic-euglycemic clamps, infusion of IL-6 reduces insulin-stimulated skeletal muscle glucose uptake and IRS-1/PI3-K activity and increases fatty acyl-CoA levels (Kim *et al.* 2004). Under the same conditions, IL-6 inhibits gene transcription of IRS-1, GLUT-4, and peroxisome proliferator-activated receptor gamma (PPAR-γ) (Cai *et al.* 2005) as well as a rapid, transient IRS-1 serine phosphorylation and IRS-1 ubiquinization.

The discordance in understanding how IL-6 contributes to skeletal muscle insulin resistance has yet to be fully elucidated. Factors that may contribute to this gap in our knowledge include disparate effects of IL-6 in the models used (e.g., mouse, rat, or human), the method of administration (acute vs. chronic), or the health status of the test subjects (e.g., normal or diseased). However, it is clear that IL-6 contributes to glucose homeostasis in some manner at the level of skeletal muscle.

Chemotactic Cytokines (Chemokines)

Chemokines are small molecular weight cytokines that regulate leukocyte trafficking, infiltration, and activation and are integral in mediating insulin actions on glucose metabolism as well as insulin resistance (Kanda *et al.* 2006). Skeletal muscle produces several chemokines such as monocyte chemotactic protein (MCP)-1 and IL-8, and express chemokine receptors (CXCR1,2 and CCR1,2,4,5,10) (Sell *et al.* 2006). MCP-1 contributes to macrophage infiltration of skeletal muscle that leads to low-grade inflammation. Macrophage infiltration is markedly increased in insulin-resistant human skeletal muscle (Cooper *et al.* 2007, Perreault and Marette 2001). MCP-1 exerts direct inhibitory effects on insulin signaling and reduces glucose uptake in skeletal muscle cells. In contrast, macrophage inflammatory protein (MIP)-1β only impairs insulin metabolic signaling at very high concentrations. MCP-1 may represent the critical molecular link in skeletal muscle leading to insulin resistance (Sell *et al.* 2006).

THE ROLE OF INNATE IMMUNE-RESPONSE IN SKELETAL MUSCLE INSULIN RESISTANCE

Toll-like Receptors

Skeletal muscle expresses multiple toll-like receptors (TLRs) including TLR4, TLR1-7, TLR9, and TLR2 (Akira *et al.* 2006) that activate downstream signal transduction pathways such as the MAPK pathway that eventually results in the activation of transcription factors, including nuclear factor-κB (NF-κB), AP-1, and Interferon Regulatory Factor, which in turn lead to the transcription of multiple pro-inflammatory cytokines (Cooper *et al.* 2007). Recent data suggest that TLR2 is essential for the development of insulin resistance in myotubes via inhibition of tyrosine phosphorylation of the IR and the phosphorylation of Akt. TLR4 has been implicated in diet-induced obesity, inflammation, insulin resistance, and overt diabetes (Tschop and Thomas 2006). Signaling through TLR2 or TLR4 elicits robust increases in MCP-1 expression and activates the NF-κB pathway. The precise role of TLRs in skeletal muscle insulin resistance requires further investigation (Tschop and Thomas 2006).

Suppressor of Cytokine Signaling (SOCS)

Members of the SOCS family are important mitigators of cytokine signal transduction. SOCS signaling has been shown to suppress the cytokine-activated Janus kinase/signal transducers and activators of transcription (JAK/STAT) signaling pathways. SOCS1, 3, and 6, are involved in cytokine-mediated inhibition of insulin signaling in adipose tissue, liver and brain (Jamieson *et al.* 2005, Mori *et al.* 2004). Further, SOCS3 is up-regulated in skeletal muscle after high-fat feeding or by IL-6 stimulation and overexpression via adenovirus-transfection prevents leptin activation of AMPK signaling. Thereby, elevated expression of SOCS3 in the skeletal muscle may impair AMPK modulation of insulin-mediated glucose uptake. Furthermore, SOCS3 expression in skeletal muscle may contribute to the exercise-induced increase in IL-6 expression through NF-κB activation. This is notable when considering both SOCS1 and SOCS3 can bind directly to the IR, where SOCS3 inhibits tyrosine phosphorylation of IRS-1 and -2 and SOCS1 inhibits tyrosine phosphorylation of IRS-1. Collectively, this would suggest that SOCS and especially SOC3 is important in regulating skeletal muscle insulin resistance.

Nuclear Factor κB (NF-κB)

The role of NF-κB in skeletal muscle insulin resistance has been demonstrated in several models. Recently, it was shown that lipid-induced insulin resistance in L6 myotubes and muscle from rodents and humans is associated with activation of the inhibitor of κB (IκB)/NF-κB pathway (Shoelson *et al.* 2006). Intralipid infusion has been shown to activate PKC-θ and NF-κB kinase complex (IKKβ) promoting insulin resistance in rodent models. Further, decreased IκBβ content and enhanced IκB/NF-κB signaling is noted in skeletal muscle from patients with insulin resistance. It is thought lipid-induced activation of IκB/NF-κB signaling is critical in this process. Moreover, skeletal muscle of insulin-resistant subjects is characterized by increases in fatty acyl CoA and ceramides.

Further evidence that NFκB activation promotes insulin resistance is strengthened by the notion that inhibition improves insulin sensitivity. Recent data suggest that diet-induced obesity in rats leads to a decrease in muscle IκB content (Bhatt *et al.* 2006). The same data suggest metabolites of fatty acids (i.e., fatty acyl CoAs, diacylglycerol, and ceramides) are believed to be responsible for this insulin resistance. Thereby, an activated IκB/NFκB pathway is critical in the low-grade chronic inflammation associated with insulin resistance (Fig. 1).

THE CROSS-TALK BETWEEN OXIDATIVE STRESS AND INFLAMMATION IN SKELETAL MUSCLE INSULIN RESISTANCE

ROS have evolved to play an important role in routine physiological and cellular processes (Wellen and Hotamisligil 2005, Cooper *et al.* 2007). ROS not only enable immune cells to kill invading pathogens but also are an essential mediator of inflammatory signaling. This latter role is not limited to inflammation; rather, it is now known that ROS serve as essential mediators of a wide array of signaling mechanisms such as cell cycle progression, growth, and proliferation as well as insulin signaling. In addition to the contributions from metabolic oxidases such as NAD(P)H oxidase and xanthine oxidase, ROS are produced as byproducts of mitochondrial aerobic metabolism, with ~2% of oxygen (O_2) being converted to $O_2^{\bullet -}$ at any given time. Excess ROS, redox imbalance, and oxidative stress are thought to be the main cause underlying the aging processes and chronic diseases as their electrophilic character allows them to oxidize cell constituents such as proteins, lipids, and DNA. The paradoxical nature of ROS can be resolved by understanding that only excessive production of these radicals results in damage, while their roles as mediators of cell signaling are temporally and spatially controlled. As an example, ROS are known to oxidize/reduce cysteine residues within proteins, a mechanism particularly active with MAPK, protein tyrosine phosphatases (PTP), protein tyrosine kinase (PTK), transcription factors, and even other enzymes being redox-regulated via their cysteine residues. This redox regulation allows the cell to activate/inhibit signaling proteins and hence dynamically change gene expression according to external stimuli. NF-κB is a redox-regulated transcription factor involved in the activation of inflammatory signaling and its constitutive activation may underlie its role as a chronic inflammatory mediator in the pathogenesis of insulin resistance.

Excess ROS overwhelm anti-oxidant defenses that lead to oxidative stress, and this, in turn, plays an important role in promoting the pro-inflammatory pathogenesis and progression of many disorders including insulin resistance (Wellen and Hotamisligil 2005, Cooper *et al.* 2007). Increased O_2^- content has been demonstrated in skeletal muscle from insulin-resistant animal models, e.g., KK-Ay mice and transgenic Ren2 rats. Further evidence of a role for oxidative stress in promoting insulin resistance comes from *in vitro* studies showing that exposure of insulin-sensitive cell lines such as 3T3-1 adipocytes and L6 myocytes to hydrogen peroxide induces insulin resistance, while the addition of anti-oxidants such as α-lipoic acid and apocynin to the cells abrogates this effect (Henriksen 2006). Collectively, these data suggest that oxidative stress plays a causal role in insulin resistance.

ROS have also been shown to promote various signaling pathways involving Foxo, MAPK, JAK/STAT, p53, phospholipase C, and PI3-K, which depend on the magnitude and type of ROS, the cell type, the duration of exposure, and other factors. ROS activate transcription factors (e.g., NF-κB and AP-1) and up-regulate expression of pro-inflammatory genes such as TNF-α, IL-6, MCP-1, CRP (Brasier *et al.* 2002, Nathan 2003), which are involved in the pathogenesis of inflammation, obesity, and insulin resistance (Wellen and Hotamisligil 2005, Cooper *et al.* 2007). ROS also simulate the IκB/NFκB pathway, and hyperglycemia-induced ROS generation could contribute to the lower IκB seen in the T2DM patients. ROS induce JNK activation, which, in turn, may result in insulin resistance (Hirosumi *et al.* 2002, Kamata *et al.* 2005).

RENIN ANGIOTENSIN ALDOSTERONE SYSTEM (RAAS)-INDUCED SKELETAL MUSCLE INSULIN RESISTANCE

Recent studies have used several transgenic and knockout strategies in small animal models to investigate the relationship between RAAS activation and insulin metabolic signaling. One model that has been explored is the transgenic Ren2 rat with over-expression of the mouse renin transgene that manifests increased tissue ROS and circulating aldosterone in various tissues, including the vasculature, heart, skeletal muscle, pancreas, and liver. In measuring glucose uptake *ex vivo* of skeletal muscle, there is a diminution in deoxyglucose uptake under the auspices of insulin in the soleus skeletal muscle, slow-twitch muscle tissue from the Ren2 rat. These results show concordance of the increase in ROS and inflammatory markers, such as TNF-α, in skeletal muscle tissue. By treating *in vivo* with an ARB or tempol, a superoxide dismutase catalase mimetic, during this period of development of metabolic abnormalities, insulin resistance is improved.

Further, elevations in angiotensin II (Ang II) levels may contribute to skeletal muscle insulin resistance through various pro-inflammatory and pro-oxidant pathways (Cooper *et al.* 2007). Systemic or local infusion of Ang II causes insulin resistance in skeletal muscle independent of hemodynamic influences, which supports the idea that Ang II can directly and negatively modulate the muscle glucose transport system (Cooper *et al.* 2007). Data from our group further suggests that Ang II-induced skeletal muscle insulin resistance is mediated by oxidative stress and inflammation (Cooper *et al.* 2007). For example, Ang II causes ROS generation in skeletal muscle and impairs insulin-mediated IRS-1 tyrosine phosphorylation, Akt activation, GLUT-4 plasma membrane translocation, and skeletal muscle glucose uptake, all of which are significantly attenuated by angiotensin type-1 receptor (AT_1R) blockage or anti-oxidant treatment. Furthermore, we found a significant linear relationship between oxidative stress and NF-κB activity. Specifically we reported that Ang II increases ROS generation and subsequently mediates Ang II-induced increases of NF-κB activation and TNF-α expression in the soleus muscles from insulin-resistant TG(mRen2)27 (Ren2) rats. We also observed that L6 myotubes treated Ang II, exhibited impairments in insulin-stimulated Akt activation, GLUT-4 plasma membrane translocation, and glucose uptake. High-fructose diets induce skeletal muscle insulin resistance with increased TNF-α expression, and AT_1R antagonist treatment improves insulin sensitivity with reduced TNF-α expression in skeletal muscle (Wellen and Hotamisligil 2005, Cooper *et al.* 2007). Like Ang II, excess aldosterone promotes oxidative stress and inflammation, which results in impaired insulin signaling, which in turn contributes to insulin resistance in many human and animal tissues (Cooper *et al.* 2007). Inhibiting aldosterone with the mineralocorticoid receptor antagonist, spironolactone, substantially improves insulin-mediated glucose uptake in skeletal muscle from hypertensive, insulin-resistant Ren2 rats due to attenuation of ROS generation and NAD(P)H oxidase activity. These studies suggest that oxidative stress and inflammatory cytokines are critical mediators in Ang II-induced insulin resistance in skeletal muscle. However, the precise mechanisms that link RAAS and skeletal muscle insulin resistance need further investigation (Fig. 1).

Table 1 Key features of skeletal muscle

- Skeletal muscle is responsible for human movement and requires glucose for energy metabolism.
- Skeletal muscle is the primary tissue responsible for the bulk of glucose disposal in the body and requires the hormone insulin for uptake.
- Alterations in metabolic signaling pathways contribute to impaired glucose utilization that contributes to impaired skeletal muscle insulin sensitivity and a condition known as insulin resistance.
- Insulin resistance is a precursor to diabetes mellitus and other metabolic derangements such as hypertension that predict a very high cardiovascular morbidity and mortality.

SUMMARY

Significant progress in understanding the development of insulin resistance has been made in the last decade; however, the precise molecular mechanisms responsible for insulin resistance, particularly in skeletal muscle, still remain incompletely understood. Disturbances in IR/IRS-1/Akt signaling and glucose uptake and disposal are mediated by a complex interaction of oxidants, cytokines, receptors and signaling pathways that promote states of chronic low-grade inflammation and oxidative stress. The presence of hypertension, obesity, and lipid abnormalities are often contributing factors in the development of the pro-inflammatory state associated with skeletal muscle insulin resistance. These patients often exhibit an activated RAAS, which is one of the principal drivers of oxidative stress and inflammation that promotes insulin resistance.

Summary Points

- Resistance to the actions of insulin in skeletal muscle is important in development of systemic insulin resistance given that skeletal muscle normally accounts for approximately 75% of all insulin-mediated glucose disposal.
- There is emerging evidence of an integral relationship between chronic inflammation, oxidative stress, and skeletal muscle insulin resistance due to circulating inflammatory cytokines derived from adipose tissue (e.g., adipokines) or from local autocrine/paracrine effects of skeletal muscle–derived cytokines (e.g., myokines).
- Inflammatory cytokines and oxidative stress contribute to impairments in insulin signaling and glucose utilization in skeletal muscle and consequent development of systemic insulin resistance.
- Visceral adipose tissue is an important contributor of circulating cytokines (e.g., adipokines) that, in concert with local production of cytokines in skeletal muscle (e.g., myokines), contribute to inflammatory pathways that react with oxidative stress and activation of the RAAS to impair insulin metabolic signaling in skeletal muscle tissue.
- There is a relative cross-talk between inflammation and oxidative stress that contribute to alterations in innate immunity and promotion of local production of myokines in skeletal muscle that contribute to impaired insulin sensitivity.

Abbreviations

ACC	:	acetyl CoA carboxylase
Akt	:	protein kinase B
AMPK	:	AMP-activated protein kinase
Ang II	:	angiotensin II
AS160	:	Akt substrate 160
AT_1R	:	angiotensin type-1 receptor
Chemokines	:	chemotactic cytokines
CRP	:	C-reactive protein
ERK1/2	:	extracellular signal–regulated kinase 1/2
GLUT-4	:	glucose transport 4
IKK	:	NF-κB kinase complex

IL	:	interleukin
iNOS	:	inducible nitric oxide synthase
IR	:	insulin receptor
IRS	:	insulin receptor substrate
IκB	:	inhibitor of κB
JAK/STAT	:	Janus kinase/signal transducers and activators of transcription
JNK	:	N-terminal kinase
MAP4K4	:	mitogen-activated protein kinase isoform 4
MAPK	:	mitogen-activated protein kinase
MCP	:	monocyte chemotactic protein
MIP	:	macrophage inflammatory protein
MYD88	:	myeloid differentiation primary response gene (88)
NF-κB	:	nuclear factor κB
PAI-1	:	plasminogen activator inhibitor 1
PDK	:	phosphoinositide-dependent protein kinase
PI3-K	:	phosphoinositide-3 kinase
PKC-θ	:	protein kinase C
PPAR-γ	:	peroxisome proliferator-activated receptor *gamma*
PTK	:	protein tyrosine kinase
PTP	:	protein tyrosine phosphatase
RAAS	:	renin angiotensin aldosterone system
RANK	:	receptor activator of nuclear factor κB
ROS	:	reactive oxygen species
SOCS	:	suppressor of cytokine signaling
T2DM	:	type 2 diabetes mellitus
TIR	:	toll-IL-1 receptor
TLR	:	toll-like receptors
TNF-α	:	tumor necrosis factor alpha

Definition of Terms

Adipokines: Inflammatory cytokines produced by visceral adipose tissue.

Innate immunity: Immunity that is naturally present and not due to prior sensitization to an antigen from a pathogen.

Myokines: Production of inflammatory cytokines by skeletal muscle.

References

Adam, S. and G. van Hall, T. Osada, M. Sacchetti, B. Saltin, and B.K. Pedersen. 2000. Production of interleukin-6 in contracting human skeletal muscles can account for the exercise-induced increase in plasma interleukin-6. J. Physiol. (Lond). 529: 237-242.

Akira, S. and S. Uematsu, and O. Takeuchi. 2006. Pathogen recognition and innate immunity. Cell. 124: 783-801.

Bhatt, B.A. and J.J. Dube, N. Dedousis, J.A. Reider, and R.M. O'Doherty. 2006. Diet-induced obesity and acute hyperlipidemia reduce IkappaBalpha levels in rat skeletal muscle in a fiber-type dependent manner. Am. J. Physiol. Regul. Integr. Comp. Physiol. 290: R233-R240.

Brasier, A.R. and A. Recinos III, and M.S. Eledrisi. 2002. Vascular inflammation and the renin-angiotensin system. Arterioscler. Thromb. Vasc. Biol. 22: 1257-1266.

Cai, D. and M. Yuan, D.F. Frantz, P.A. Melendez, L. Hansen, J. Lee, and S.E. Shoelson. 2005. Local and systemic insulin resistance resulting from hepatic activation of IKK-beta and NF-kappaB. Nat. Med. 11: 183-190.

Carey, A.L. and G.R. Steinberg, S.L. Macaulay, W.G. Thomas, A.G. Holmes, G. Ramm, O. Prelovsek, C. Hohnen-Behrens, M.J. Watt, D.E. James, B.E. Kemp, B.K. Pedersen, and M.A. Febbraio. 2006. Interleukin-6 increases insulin-stimulated glucose disposal in humans and glucose uptake and fatty acid oxidation in vitro via AMP-activated protein kinase. Diabetes 55: 2688-2697.

Cooper, S.A. and A.Whaley-Connell, J. Habibi, Y. Wei, G. Lastra, C.M. Manrique, S. Stas, and J.R. Sowers. 2007. Renin-Angiotensin-Aldosterone system and oxidative stress in cardiovascular insulin resistance. Am. J. Physiol. Heart Circ. Physiol. 293: H2009-H2023.

de Alvaro, C. and T. Teruel, R. Hernandez, and M. Lorenzo. 2004. Tumor necrosis factor alpha produces insulin resistance in skeletal muscle by activation of inhibitor kappaB kinase in a p38 MAPK-dependent manner. J. Biol. Chem. 279: 17070-17078.

Henriksen, E.J. 2006. Exercise training and the antioxidant alpha-lipoic acid in the treatment of insulin resistance and type 2 diabetes. Free Radic. Biol. Med. 40: 3-12.

Hirosumi, J. and G. Tuncman, L. Chang, C.Z. Gorgun, K.T. Uysal, K. Maeda, M. Karin, and G.S. Hotamisligil. 2002. A central role for JNK in obesity and insulin resistance. Nature 420: 333-336.

Hotamisligil, G.S. and P. Arner, J.F. Caro, R.L. Atkinson, and B.M. Spiegelman. 1995. Increased adipose tissue expression of tumor necrosis factor-alpha in human obesity and insulin resistance. J. Clin. Invest. 95: 2409-2415.

Hotamisligil, G.S. and N.S. Shargill, and B.M. Spiegelman. 1993. Adipose expression of tumor necrosis factor-alpha: direct role in obesity-linked insulin resistance. Science 259: 87-91.

Jamieson, E. and M.M. Chong, G.R. Steinberg, V. Jovanovska, B.C. Fam, D.V. Bullen, Y. Chen, B.E. Kemp, J. Proietto, T.W. Kay, and S. Andrikopoulos. 2005. Socs1 deficiency enhances hepatic insulin signaling. J. Biol. Chem. 280: 31516-31521.

Kamata, H. and S. Honda, S. Maeda, L. Chang, H. Hirata, and M. Karin. 2005. Reactive oxygen species promote TNFalpha-induced death and sustained JNK activation by inhibiting MAP kinase phosphatases. Cell 120: 649-661.

Kanda, H. and S. Tateya, Y. Tamori, K. Kotani, K. Hiasa, R. Kitazawa, S. Kitazawa, H. Miyachi, S. Maeda, K. Egashira, and M. Kasuga. 2006. MCP-1 contributes to macrophage infiltration into adipose tissue, insulin resistance, and hepatic steatosis in obesity. J. Clin. Invest. 116: 1494-1505.

Kim, H.J. and T. Higashimori, S.Y. Park, H. Choi, J. Dong, Y.J. Kim, H.L. Noh, Y.R. Cho, G. Cline, Y.B. Kim, and J.K. Kim. 2004. Differential effects of interleukin-6 and -10 on skeletal muscle and liver insulin action in vivo. Diabetes 53: 1060-1067.

Lopez-Soriano, J. and C. Chiellini, M. Maffei, P.A. Grimaldi, and J.M.Argiles. 2006. Roles of skeletal muscle and peroxisome proliferator-activated receptors in the development and treatment of obesity. Endocr. Rev. 27: 318-329.

Mori, H. and R. Hanada, T. Hanada, D. Aki, R. Mashima, H. Nishinakamura, T. Torisu, K.R. Chien, H. Yasukawa, and A.Yoshimura. 2004. Socs3 deficiency in the brain elevates leptin sensitivity and confers resistance to diet-induced obesity. Nat. Med. 10: 739-743.

Nathan, C. 2003. Specificity of a third kind: reactive oxygen and nitrogen intermediates in cell signaling. J. Clin. Invest. 111: 769-778.

Pedersen, B.K. and T.C.A. Akerstrom, A.R. Nielsen, and C.P. Fischer. 2007. Role of myokines in exercise and metabolism. J. Appl. Physiol. 103: 1093-1098.

Perreault, M. and A. Marette. 2001. Targeted disruption of inducible nitric oxide synthase protects against obesity-linked insulin resistance in muscle. Nat. Med. 7: 1138-1143.

Plomgaard, P. and K. Bouzakri, R. Krogh-Madsen, B. Mittendorfer, J.R. Zierath, and B.K. Pedersen. 2005. Tumor necrosis factor-{alpha} induces skeletal muscle insulin resistance in healthy human subjects via inhibition of Akt substrate 160 phosphorylation. Diabetes 54: 2939-2945.

Ruderman, N.B. and C. Keller, A.M. Richard, A.K. Saha, Z. Luo, X. Xiang, M. Giralt, V.B. Ritov, E.V. Menshikova, D.E. Kelley, J. Hidalgo, B.K. Pedersen, and M. Kelly. 2006. Interleukin-6 regulation of AMP-Activated protein kinase: potential role in the systemic response to exercise and prevention of the metabolic syndrome. Diabetes 55: S48-S54.

Sell, H. and D. Dietze-Schroeder, U. Kaiser, and J. Eckel. 2006. Monocyte chemotactic protein-1 is a potential player in the negative cross-talk between adipose tissue and skeletal muscle. Endocrinology 147: 2458-2467.

Shoelson, S.E. and J. Lee, and A.B. Goldfine. 2006. Inflammation and insulin resistance. J. Clin. Invest. 116: 1793-1801.

Togashi, N. and N. Ura, K. Higashiura, H. Murakami, and K. Shimamoto. 2000. The contribution of skeletal muscle tumor necrosis factor-alpha to insulin resistance and hypertension in fructose-fed rats. J. Hypertens. 18: 1605-1610.

Tschop, M. and G. Thomas. 2006. Fat fuels insulin resistance through Toll-like receptors. Nat. Med. 12: 1359-1361.

Uysal, K.T. and S.M. Wiesbrock, M.W. Marino, and G.S. Hotamisligil. 1997. Protection from obesity-induced insulin resistance in mice lacking TNF-alpha function. Nature 389: 610-614.

Wellen, K.E. and G.S. Hotamisligil. 2005. Inflammation, stress, and diabetes. J. Clin. Invest. 115: 1111-1119.

Zhang, Y. and G. Pilon, A. Marette, and V.E. Baracos. 2000. Cytokines and endotoxin induce cytokine receptors in skeletal muscle. Am. J. Physiol. Endocrinol. Metab. 279: E196-E205.

CHAPTER 25

Cytokines and Bone

Patrizia D'Amelio
Department of Surgical and Medical Disciplines Section of Gerontology
Corso Bramante 88/90, 10126 Torino, Italy

ABSTRACT

Bone turnover is due to cyclic bone resorption followed by bone apposition; these processes are due to the coordinated actions of osteoclasts (OCs) and osteoblasts (OBs). The actions of these two cellular types are orchestrated by osteocytes (OSs) that differentiate from osteoblasts and are the most abundant cells in bone. OCs are formed by the attraction of myelomonocytic precursors to the resorption site; the fusion of these cells generates a multinucleated cell attached to the bone surface. OBs derive from a mesenchymal stem cell precursor shared with adipocytes. OSs are thought to be the cells primarily responsible for mechanosensing in bone. Numerous cytokines are thought to be responsible for the regulation of bone turnover; most of them have pleiotropic actions and are involved in the regulation of systems other than skeleton. OC formation and function are mainly regulated by the essential factor RANKL, whereas other cytokines increased during inflammation up-regulate OCs and are involved in inflammation-induced bone loss. OB formation and activity are believed to be mainly regulated by the Wnt and BMP signaling pathways.

This review will focus on the main cytokines involved in the regulation of osteoclastogenesis, osteoblastogenesis, and coupling of osteoclasts and osteoblasts under physiological and pathological conditions.

INTRODUCTION

Bone is a dynamic tissue that is constantly formed and resorbed in response to changes in mechanical loading, altered serum calcium and pH levels. These changes are mediated by paracrine and endocrine factors. Bone turnover is due to continuous and cyclic bone resorption followed by apposition; these processes are due to the coordinated actions of osteoclasts (OCs, which destroy bone) and osteoblasts (OBs, which form bone). The actions of these two cellular types may be orchestrated to some extent by osteocytes (OSs). In physiological conditions, OB and OC activity is coupled, so that the amount of resorption is equal to the amount of formation. However, in pathological conditions or during senescence, resorption is higher than formation, leading to bone loss. OC and OB formation, as well as the coupling between these two cell types, are mediated mainly through cytokines.

The aim of this review is to describe the complex cytokine network involved in the regulation of bone cell function and to illustrate the close relationship between the skeleton and other systems, such as the immune system (D'Amelio and Isaia 2009), cardiovascular system (D'Amelio *et al.* 2009), and adipose tissue (Isaia *et al.* 2005).

OSTEOCLASTS

OCs are formed by the attraction of myelomonocytic precursors to the resorption site, followed by their fusion, and attachment of the subsequent multinucleated cell to the bone surface (Fig. 1A).

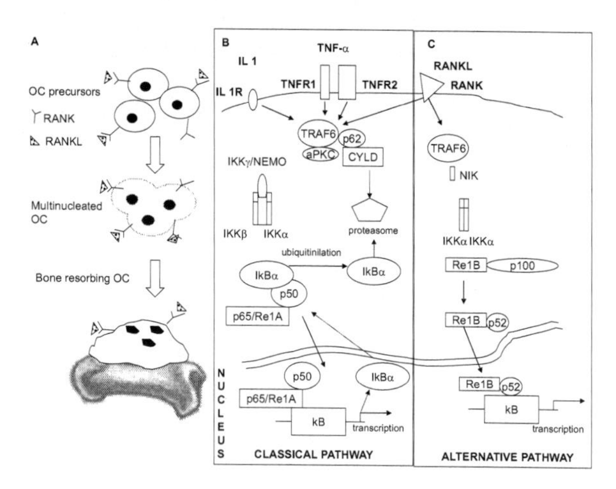

Fig. 1 (A) Key steps of OC formation and activity. (B) RANKL, IL-1 and TNF-α activate classical pathways of NF-kB in osteoclast-like cells. (C) RANKL activates non-classical pathways of NF-kB in osteoclast-like cells (data from Xu *et al.* 2009).

RANKL/RANK/OPG System

A central role in OC biology is played by the receptor activator of NF-kB ligand (RANKL), which is essential for osteoclastogenesis and bone resorption (Leibbrandt and Penninger 2008). Mice and humans deficient in the RANKL gene completely lack OC and exhibit variable forms of osteopetrosis. RANKL has also been implicated in regulation of immune response and in arterial wall calcification (D'Amelio and Isaia 2009, D'Amelio *et al.* 2009). The functional receptor for RANKL, RANK is encoded by a tumour necrosis factor receptor (TNFR) superfamily gene (TNFGS11A) and is expressed on OC precursors. Mice lacking TNFGS11A have a profound defect in bone resorption and in the development of cartilaginous growth plates. One of the key steps upon activation of the RANK pathway is the binding of TNFR-associated cytoplasmic factors (TRAFs) to specific domains within the cytoplasmic domain of RANK. The TRAF family proteins are cytoplasmic adapter proteins involved in the mediation of several cytokine-signaling pathways. Different members of the family activate different transcriptional pathways: TRAF2, 5 and 6 are involved in the activation of NF-kB through IkB kinase (IKK) activation and AP-1 through activation of mitogen-activated protein kinases (MAPKs), including Jun-N-terminal kinase (JNK), p38 and extracellular signal-regulated kinase (ERK). Moreover, TRAF6 functions as an ubiquitin ligase, which catalyzes the formation of a polyUb chain. This leads to the activation of IKK and JNK through a proteasome-independent mechanism (Xu *et al.* 2009) (Fig.1BC).

RANKL/RANK signaling promotes the differentiation of OC precursors into mature multinucleated OCs, stimulates their capacity to resorb bone, and decreases OC apoptosis. RANKL is present as both a transmembrane molecule and a secreted form; its interaction with RANK is opposed by osteoprotegerin (OPG), a neutralizing soluble decoy receptor, produced by marrow stromal cells and OBs (Grundt *et al.* 2009). The imbalance between RANKL and OPG has been indicated as the pivotal mechanism responsible for bone loss in case of estrogen deficiency (D'Amelio *et al.* 2008), inflammation (Leibbrandt and Penninger 2008), and cancer-induced bone loss.

M-CSF

M-CSF induces the proliferation of OC precursors and their differentiation and increases the survival of mature OCs; OC formation occurs when monocytes are co-stimulated by the essential osteoclastogenic factors RANKL and M-CSF. In addition to RANKL, TNF-α and IL-1 directly regulate NF-kB signaling in OC. As shown in Fig. 1B-C, RANKL activates both classical and alternative pathways of NF-kB, whereas TNF-α and IL-1 activate the classical pathways.

TNF-α

TNF-α enhances OC formation by up-regulating stromal cell production of RANKL and M-CSF, and by augmenting the responsiveness of OC precursors to RANKL. TNF directly induces marrow precursor differentiation into OCs, although according to some studies it is not osteoclastogenetic in cells not primed by RANKL. The ability of TNF to increase the osteoclastogenic activity of RANKL is due to synergistic interactions at the level of NFkB and AP-1 signaling (Fig.1B). In addition, TNF and RANKL synergistically up-regulate RANK expression. *In vivo* blockade of TNF in postmenopausal osteoporosis reduces bone resorption (Charatcharoenwitthaya *et al.* 2007); this suggests that TNF-α increase could be one of the mechanisms responsible for postmenopausal bone loss. TNF is mainly produced by activated T cells and it is also involved in inflammation and cancer-induced bone loss both systemically and locally.

IL-1

IL-1 plays an important role in bone loss induced by estrogen deficiency; its level increases after menopause and is reversed by estrogen replacement. Bone loss does not occur after ovariectomy in mice deficient in receptors for IL-1, and treatment with IL-1 receptor antagonist decreases OC formation and activity. A recent study demonstrates that the blockade of IL-1 reduces bone resorption in postmenopausal osteoporosis (Charatcharoenwitthaya *et al.* 2007). IL-1 acts by increasing RANKL expression by bone marrow stromal cells and directly targets OC precursors, promoting OC differentiation in the presence of permissive levels of RANKL (Fig. 1A). The effect of TNF-α on osteoclastogenesis is up-regulated by IL-1.

Additional cytokines are responsible for the up-regulation of OC formation in diseases such as estrogen deficiency, local and systemic inflammation and bone metastasis. Some of these molecules have a well-established role in controlling osteoclastogenesis, while others do not. Among these the most involved in OC formation are IL-7, IL-17, IL-23, IL-6, TGFβ and IFNγ. Most of these molecules are also involved in the regulation of immune system and this may explain part of the relationship between immune and bone cells.

IL-7

IL-7 is known for its ability to stimulate T and B cell number and the reaction to antigenic stimuli. A role for IL-7 has also been postulated in bone remodelling. We have demonstrated that IL-7 promotes osteoclastogenesis by up-regulating T and B cell-derived RANKL (D'Amelio *et al.* 2006) and that the production of IL-7 is down-regulated by estrogen.

In humans it has been suggested that IL-7 is osteoclastogenic in psoriatic arthritis and in solid tumors; also, in healthy volunteers the expression of IL-7 receptor on T lymphocytes correlates with their ability to induce osteoclastogenesis from human monocytes.

IL-17 AND IL-23

IL-17 family members are mainly expressed by a type of human T helper cell (Th17) (Yao *et al.* 1995). It is now believed that this cytokine plays a crucial role in inflammation and the development of autoimmune diseases such as rheumatoid arthritis; however, its mechanism of action in the development of bone erosions, especially in relation to other known key cytokines such as IL-1, TNF-α and RANKL, remains unclear. Recently, IL-17 has been suggested to be involved in the up-regulation of OC formation in inflammation by increasing the release of RANKL, which may synergize with IL-1 and TNF (Lubberts *et al.* 2005).

One of the stimuli to IL-17 production is IL-23 produced by activated dendritic cells and macrophages. IL-23 drives the T helper 1 response and is implicated in autoimmune diseases; hence, it has been suggested that the IL-23/IL-17 axis is critical for controlling inflammatory bone loss. However, in contrast to IL-17-deficient mice, IL-23 knockout mice were completely protected from bone and joint destruction in the collagen-induced arthritis model, indicating that the IL-23-induced bone loss may not be entirely mediated by IL-17, and raising the question whether IL-23 can directly stimulate OCs. Recent work supports this hypothesis, suggesting that IL-23 promotes OC formation (Yago *et al.* 2007). Other *in vivo* studies suggest that IL-23 inhibits OC formation via T cells (Quinn *et al.* 2008). In physiological conditions (unlike inflammation), IL-23 favours higher bone mass in long bones by limiting resorption of immature bone forming below the growth plate (Quinn *et al.* 2008). These contrasting data suggest different roles of this cytokine in the control of physiological or inflammatory bone turnover.

IL-6

Activation of the signaling pathway mediated by glycoprotein (gp)-130 by IL-6 and its soluble receptor has been regarded as a pivotal mechanism for the regulation of osteoclastogenesis (Roodman 1992). Nevertheless, in IL-6 knockout mice (IL6KO), as well as in gp-130-deficient mice, no decrease in OC formation and function was found. These data may suggest that IL-6 is not essential for bone resorption. However, IL6KO mice were protected against ovariectomy-induced bone loss, and this finding, together with the observation of increased level of IL-6 after menopause in women, may suggest a peculiar role for IL-6 in bone loss due to estrogen deprivation. IL-6 was also shown to be involved in other diseases associated with accelerated bone turnover such as Paget's disease of bone, multiple myeloma, rheumatoid arthritis and renal osteodystrophy.

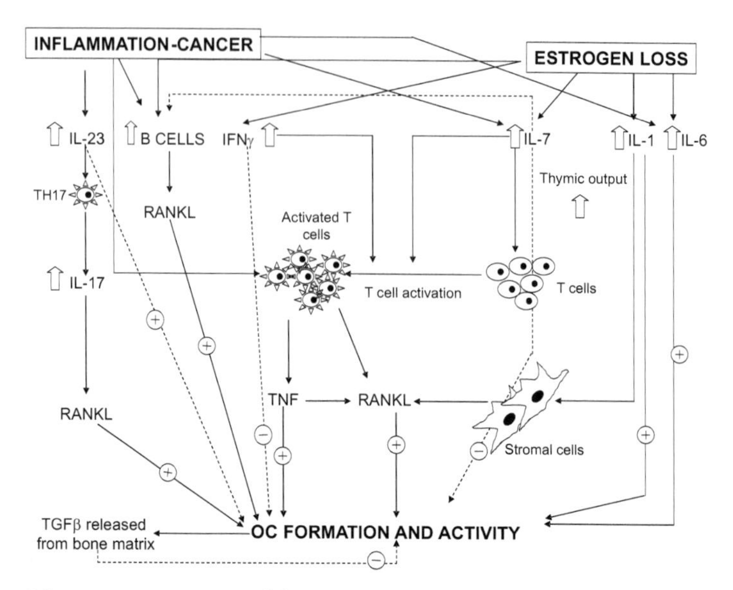

Fig. 2 Schematic representation of the main cytokines involved in OC formation and activity. The dotted lines represent the more controversial pathways, while the solid lines represent the pathways with more concordant data (unpublished).

IFNγ

The effect of IFNγ on OC formation and activity is controversial. IFNγ behaves like an anti-osteoclastogenic cytokine *in vitro* (Takayanagi *et al.* 2000), *in vivo* in nude mice (Sato *et al.* 1992), and in knockout models in which the onset of collagen-induced arthritis is more rapid, as compared with wild type controls. These data are not confirmed by studies in humans and in experimental models of diseases that indicate an increased level of IFNγ during estrogen deficiency.

In humans IFNγ is positively correlated with bone erosions in leprosy and rheumatoid arthritis. Data from randomized controlled trials have shown that IFNγ does not prevent bone loss in rheumatoid arthritis. The use of IFNγ in humans has been suggested to employ IFNγ for the treatment of osteopetrosis, in which condition IFNγ is able to restore bone resorption.

Taken together, the data in humans suggest that, in some conditions, IFNγ stimulates bone resorption. These discrepancies could be explained by the fact that IFNγ directly blocks OC formation targeting maturing OC and induces antigen presentation and thus T cell activation *in vivo*. Therefore, when IFNγ levels are increased *in vivo*, activated T cells secrete pro-osteoclastogenic factors and this activity offsets its anti-osteoclastogenic effect.

TGFβ

TGFβ plays a complex role in osteoclastogenesis. It has wide-ranging effects and it has been suggested that it may play a pivotal role in the growing skeleton, contributing to coupling between OB and OC (Massague 1990). Three isoforms of TGFβ have been described (TGFβ1–3), which all interact with the same receptor complex. TGFβ1 is mainly expressed in lymphoid organs and in serum. TGFβ2 and TGFβ3 are predominantly expressed in mesenchymal tissues and bone. TGFβ is produced by many cell types, including bone marrow cells, OBs and stromal cells, and is secreted in a latent form that must be activated to mediate its effects. Although several mechanisms of activation *in vivo* have been proposed, the precise mechanism of this process is not known. Both *in vitro* and *in vivo* studies have shown that TGFβ1–3 have complex effects on bone. They stimulate or repress proliferation or formation of OBs and OCs, depending on cell types and culture conditions used. Mice with OB-specific over-expression of TGFβ2 develop high turnover osteoporosis.

TGFβ has also been implicated in the pathogenesis of ovariectomy-induced bone loss because local injection of TGFβ1 and TGFβ2 prevents bone loss at the site of the injection in ovariectomized rats. Furthermore, estrogen is known to up-regulate the expression of TGFβ in murine OBs, bone extracts and bone marrow cells and long-term *in vivo* estrogen treatment has been shown to increase serum TGFβ1 and TGFβ2 levels in humans. Latent TGFβ is abundantly present in the bone matrix and is released and activated during bone resorption, and it feeds back to modulate OB and OC activity. In particular, TGFβ is believed to induce OC apoptosis that follows bone resorption *in vivo*.

OSTEOBLASTS

Osteoblasts are the cells responsible for the production of bone matrix components such as type I collagen. OBs and adipocytes differentiate from a common precursor, the pluripotent mesenchymal stem cell (MSC) found in bone marrow and adipose tissue. Numerous transcription factors and multiple extracellular and intracellular signals regulating adipogenesis and osteoblastogenesis have been identified and analyzed in recent years. Over the last two decades, many factors have been identified that regulate differentiation. Runt-related transcription factor 2 (Runx2), Osterix (Osx), Msh homeobox 2 (Msx2), bone morphogenetic proteins (BMPs), Wnt and Hedgehog (Hh) have been shown to be critical in osteoblastogenesis (Marie 2008) (Fig. 3).

Differentiation factors controlling osteoblastogenesis inhibit adipogenesis, and vice versa. In MSC, the Wnt signaling pathways induce osteoblastogenesis and stimulate OB proliferation and maturation. PPAR-γ, a member of the nuclear hormone receptor superfamily, induces MSCs to differentiate into adipocytes

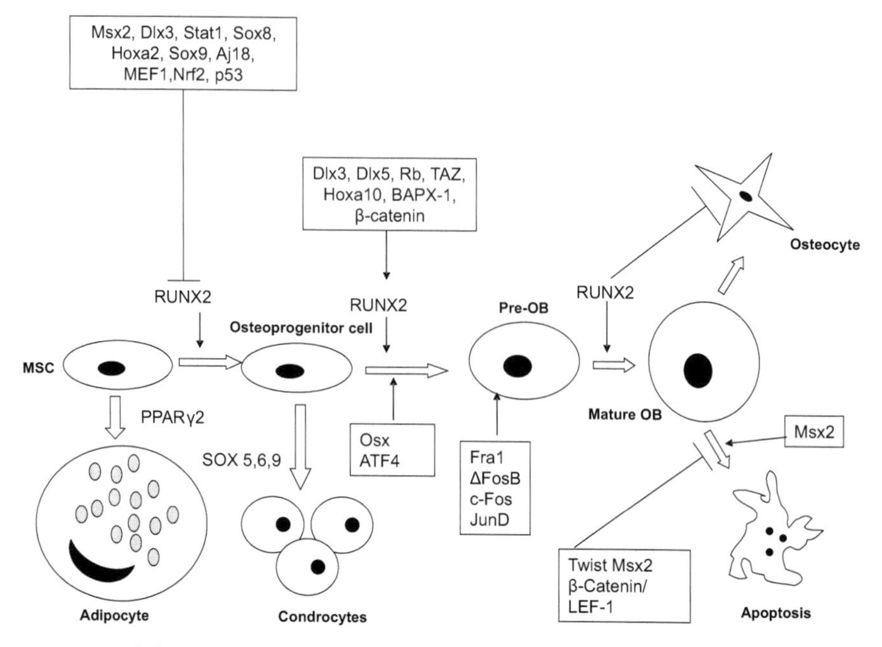

Fig. 3 OB differentiation from the MSC, with the main transcription factors control involved in each step (data from Marie 2008).

(Fig. 3). The actions of Wnt and PPAR-γ in osteoblastogenesis and adipogenesis have been extensively studied; however, their molecular interactions and the effects of this interaction have remained unclear.

Wnts

Wnts are a family of secreted lipid-modified proteins that bind to a receptor complex comprising frizzled (fz) and the low-density lipoprotein receptor-related proteins 5 or 6 (LRP5 or LRP6). Activation of this receptor leads to inactivation of glycogen synthase kinase 3β (GSK-3β), which prevents the proteosomal degradation of the transcriptional co-activator β-catenin and, thereby, promotes its accumulation in the cytoplasm. β-catenin translocates into the nucleus and regulates the expression of Wnt target genes (Fig. 4). Several molecules serve as antagonists of Wnt signaling: secreted frizzled-related proteins (SFRPs) act as soluble decoy fz receptors by preventing binding of Wnt to fz. Dickkopfs (Dkks) and sclerostin (Sost) bind to and inactivate signaling from LRP5/LRP6 through Kremen (Krm) (Fig. 4) (Macsai *et al.* 2008). Besides this so-called canonical pathway, Wnts can signal via the protein kinase C (PKC), Rho- or c-Jun N-terminal kinase (JNK). The role of this pathway in bone biology is still unclear. Wnt canonical signaling is a critical determinant of bone mass; indeed, loss or gain of function mutations in

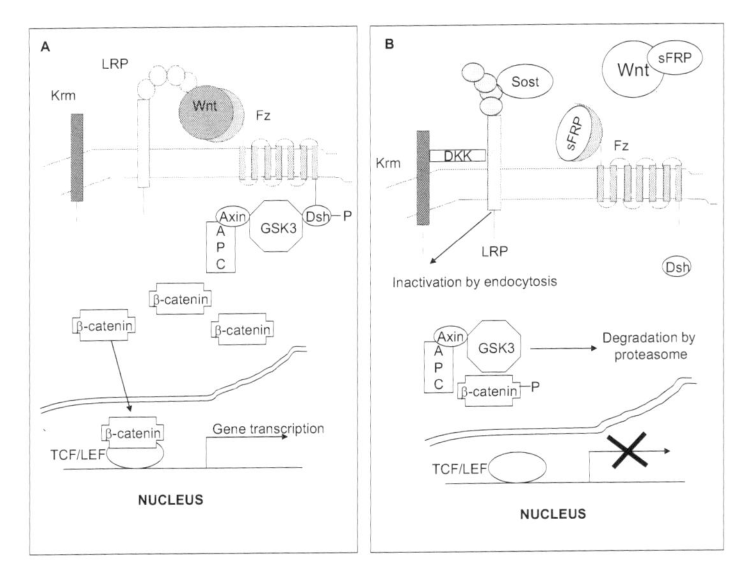

Fig. 4 A: Activation of the canonical pathway is initiated when Wnt binds to Fz receptors and low LRP-5/6 co-receptors. B: Inactivation of canonical pathway through a variety of inhibitors mediates the proteasomal degradation of β-catenin via its phosphorylation (data from Macsai *et al.* 2008).

LRP5 leads to osteoporosis-pseudoglioma syndrome or the hereditary high bone mass trait, respectively.

BMPs

The bone morphogenetic proteins (BMPs) are cytokines belonging to the TGFβ family. More than 20 BMPs have been identified to date and these are divided into classes based on their sequences and functions. Some BMPs (-2, -3, -4, -6, -7, -8 and -8b) are produced by preosteoblasts, while others are secreted by non-parenchymal cells of the liver (BMP-9) or by human endochondral and intramembranous bone. BMP-2, -6, -7 and -9 regulate the proliferation, differentiation and apoptosis of bone cells.

BMPs act on the cells involved in bone ossification via two pathways: the canonical Smad pathway and a pathway involving the mitogen-activated protein kinase (MAPK) (Senta *et al.* 2009) (Fig. 5). The Smad pathway is activated with the binding of BMPs to their heterotetrameric receptors. Smads form a complex that translocates into the nucleus and binds to a specific DNA sequence and to proteins involved in target gene expression, and activates OB-specific genes. The Smad pathway is regulated by the breakdown of Smad by the ubiquitination of the protein via the Smad-ubiquitination regulatory factor (Smurf1), and the

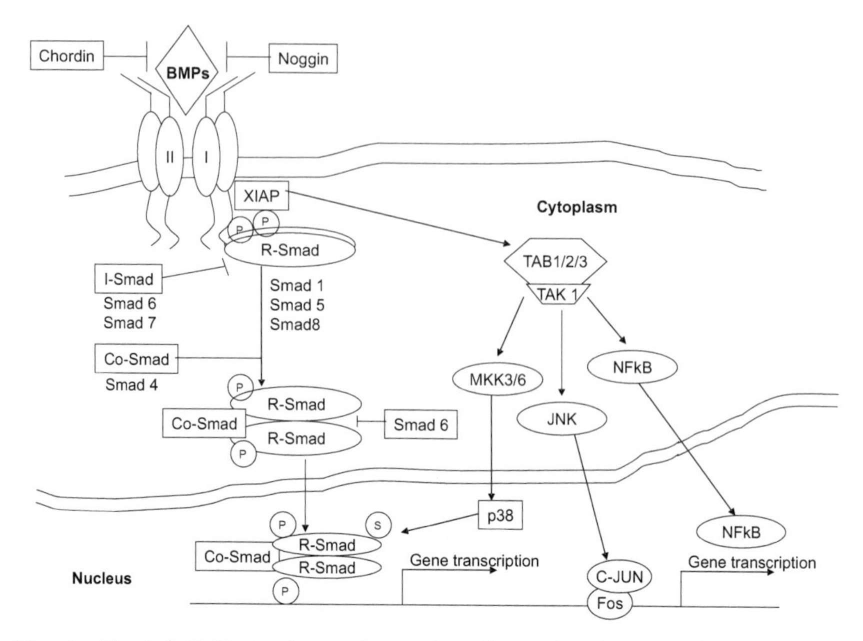

Fig. 5 Classical BMP signaling pathways through Smad and MAPK (data from Senta *et al.* 2009).

subsequent proteasome proteolytic pathway. It can also be inhibited by extracellular antagonists of BMPs such as noggin and chordin. Intracellular I-Smad can interact with the type I receptor to block R-Smad phosphorylation. Smad 6 also competes with Smad 4, which inhibits the formation of R-Smad/Smad 4 complexes. The Wnt/β-catenin pathway is also involved in the regulation of the Smad cascade by inhibiting GSK3.

The non-canonical MAPK pathway involves three different cascades: an extracellular signal-regulated kinase (ERK), JNK and p38 MAPK. Some subgroups of BMPs can activate the TGFβ activated kinase 1 (TAK1)/TAK1 binding protein 1 (TAB1) complex, and thus activate p38. Activation of this pathway also increases the degradation of Smad.

TGFβ

TGFβ can act on OB, promoting osteoprogenitor proliferation and inhibiting terminal differentiation, by repressing the function of Runx2 (Alliston *et al.* 2001). TGFβ also regulates osteoblast expression of osteoclast regulatory factors M-CSF, RANKL and OPG. Currently, TGF is thought to be involved mainly in coupling the actions of OC and OB and to play a fundamental role in the developing skeleton, whereas its role in physiological bone remodelling remains unclear.

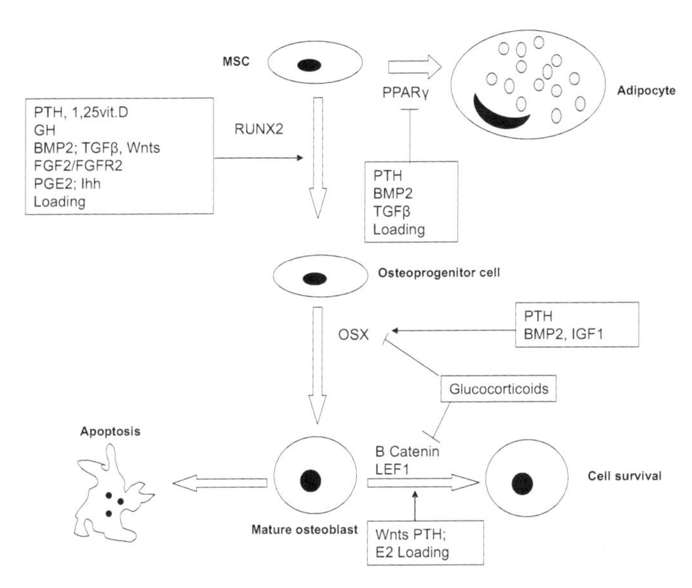

Fig. 6 Schematic representation of the signals involved in OB differentiation (data from Marie 2008).

PPAR-γ

PPAR-γ is a member of the nuclear receptor gene superfamily of ligand-activated transcription factors. Two known isoforms of PPAR-γ (1 and 2) are generated by alternative promoter usage and splicing, with PPAR-γ2 being predominant in adipose tissue. PPAR-γ1 is expressed at lower levels in adipose tissue and in other cell types. The expression of both PPAR-γ1 and PPAR-γ2 is elevated during adipogenesis, and both isoforms can induce adipocyte differentiation and are antiosteoblastogenic. In particular, treatment with TNF-α or IL-1 inhibited Troglitazone (Tro)-induced transcriptional activity of PPAR-γ, demonstrating that cytokines may interfere with differentiation of MSC (Suzawa *et al.* 2003). Canonical Wnt signaling seems to stimulate osteoblastic differentiation without affecting adipo-/osteoprogenitor cells maturing into adipocytes. However, subsequent studies have shown that non-canonical Wnt signaling may directly regulate the transactivation function of PPAR-γ in MSCs (Takada *et al.* 2007)

OSTEOCYTES

Osteocytes are derived from OBs. It is not clear why some OBs are designated to become OSs, but they make connections with existing embedded cells and then are engulfed in osteoid. OSs make up more than 95% of all bone cells in the adult skeleton, whereas OBs compose less than 5% and OCs less than 1%. They are viable for years, even decades, whereas OBs live for weeks and OCs for days. The unique feature of OSs is the formation of long dendritic processes that connect OSs within their lacunae with cells on the bone surface and into the bone marrow. OSs send signals modulating both bone resorption and formation, but how this happens is still unclear. It has been suggested that dying OSs send signals of resorption (Verborgt *et al.* 2002, Kurata *et al.* 2006) and recently it has been shown that OSs can target OB through Sost and thus inhibit bone formation (van Bezooijen *et al.* 2005). OSs are thought to drive bone remodelling by sensing mechanical strain along their dendritic processes, the cilia and the cell body.

Osteocytes influence bone remodelling through cytokines as Sost, M-CSF and RANKL. Sost, a Wnt antagonist, is a marker for late embedded osteocytes and its mutation in humans results in sclerostosis (Fig. 4). There are also three key molecules expressed in OSs that play a role in phosphate homeostasis: dentine matrix protein 1 (Dmp1), phosphate regulating neutral endopeptidase on chromosome X (Pex), and FGF23. Deletion or mutation of either Pex or Dmp1 results in hypophosphatemic rickets with elevation of FGF23. FGF23 is a phosphaturic factor that prevents reabsorption of Pi by the kidney leading to hypophosphatemia (Fig. 7). Some data suggest that Dmp1 could function intracellularly to regulate transcription and extracellularly to regulate mineralization of osteoid.

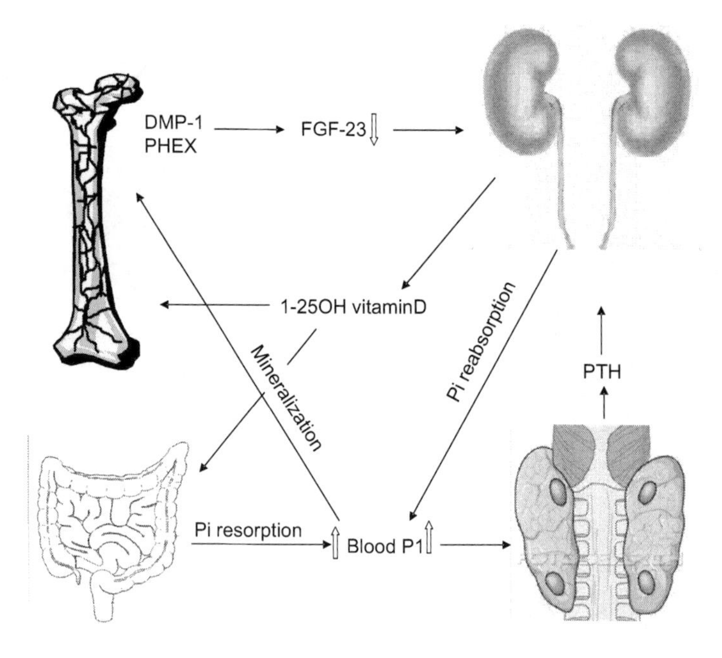

Fig. 7 Interactions between Dmp1, Phex, and FGF23 (data from Bonewald 2007).

On the basis of these observations it has been proposed that the OS network can function as an endocrine gland to regulate phosphate homeostasis. As both Dmp1 and Pex are regulated by mechanical loading, it will be important to determine if skeletal loading can play a role in mineral and phosphate metabolism (Bonewald 2007).

CONCLUSION

In summary, bone is a living tissue in which three main cells types act in response to multiple signals. The hormones and the cytokines involved in bone turnover have pleiotropic functions that may help explain the complex relationships between skeleton and immune and cardiovascular systems. Future studies will allow us to clarify these relations and to deeply understand signals involved in bone metabolism.

Table 1 Key features of bone

- Bone is a living tissue subjected to continuous remodelling.
- Bone resorption is due to osteoclasts and bone apposition to osteoblasts; these two cellular types are coordinated by osteocytes.
- Bone turnover is important to keep calcium, phosphate and pH homeostasis and it is regulated by multiple factors.
- Cytokines involved in the regulation of bone turnover have pleiotropic functions that explain the deep interactions between bone cells and other organs and systems (Fig. 8).

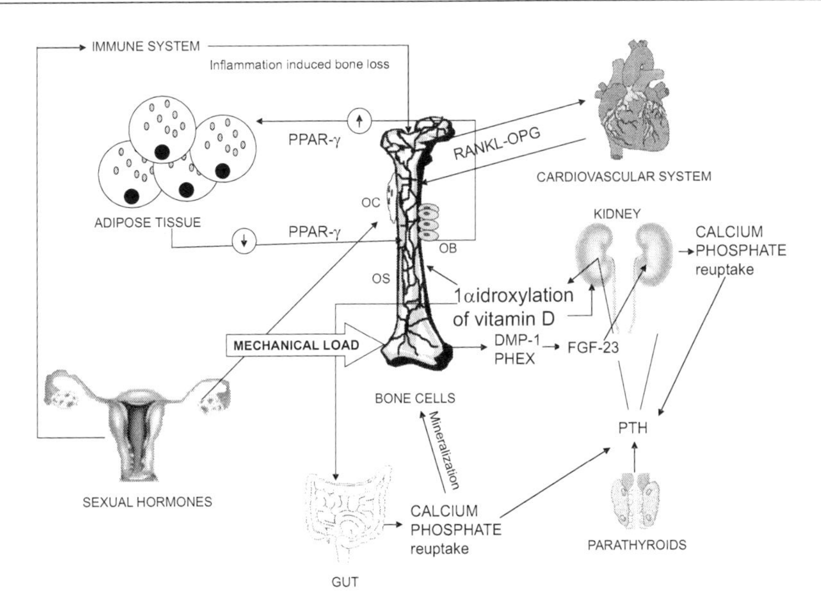

Fig. 8 Simplified diagram of the interactions between bone cells and other organs and systems (unpublished).

Summary Points

- Bone is a dynamic tissue and its turnover is due to cyclic bone resorption followed by apposition; these processes are due to the coordinated actions of osteoclasts and osteoblasts, orchestrated by osteocytes.
- Osteoclast and osteoblast action is normally coupled in physiological conditions, whereas in pathological conditions their imbalance leads to bone loss.
- Osteoclasts are formed from the fusion of myelomonocytic precursors. Their recruitment, fusion and activation are mainly regulated by the RANK/RANKL/OPG and the NFκB pathway.
- Osteoclast formation is up-regulated by estrogen deficiency, inflammation and cancer and hypoxia.
- Osteoblasts differentiate from MSC as well as adipocytes: a central role in osteoblast biology is played by the transcriptional factors that influence the fate of MSC.
- The main cytokines involved in osteoblastogenesis and activity are Wnts and BMPs.
- Osteocytes are the most abundant cells in bone, and they are thought to regulate bone remodelling mainly by sensing the mechanical load.
- Several cytokines produced by osteocytes affect bone formation and bone resorption. Recently, it has been proposed that these cells regulate mineral homeostasis by specific cytokines.

Abbreviations

BMP	:	bone morphogenetic protein
Dkks	:	Dickkopfs
Dmp1	:	dentine matrix protein 1
FGF	:	fibroblast growth factor
fz	:	frizzled
gp-130	:	glycoprotein 130
GSK-3β	:	glycogen synthase kinase 3β
Hh	:	hedgehog
IFN	:	interferon
IL	:	interleukin
JNK	:	c-Jun N-terminal kinase
Krm	:	kremen
LRP5	:	low-density lipoprotein receptor-related protein 5

M-CSF	:	macrophage colony stimulating factor
MSC	:	mesenchymal stem cell
Msx2	:	Msh homeobox 2
OB	:	osteoblasts
OC	:	osteoclasts
OPG	:	osteoprotegerin
OS	:	osteocytes
Osx	:	osterix
Pex	:	phosphate regulating neutral endopeptidase on chromosome X
PKC	:	protein kinase C
PPAR-γ	:	peroxisome proliferator-activated receptor gamma
RANK	:	receptor activator of nuclear factor κB
RANK	:	receptor activator of nuclear factor κB ligand
Runx2	:	Runt-related transcription factor 2
SFRPs	:	secreted frizzled-related proteins
Sost	:	sclerostin
TAB1	:	TAK1 binding protein 1
TAK1	:	TGFβ activated kinase 1
TGF	:	transforming growth factor
Th17	:	T helper 17
TNF	:	tumor necrosis factor
TRAFs	:	TNFR-associated cytoplasmic factors
Wnt	:	ingless

Definition of Terms

BMPs signaling: Together with Wnt, the key system controlling osteoblasts; it acts on OB commitment and activity.

Mechanotransduction: The feature that allows OS to sense mechanical load and to orchestrate bone remodelling and mineral homeostasis according to its variations.

Osteoblast: Cells that form new bone and are derived from the MSC.

Osteoclast: Cells that resorb bone and are derived from myeloid precursors.

Osteocyte: The most abundant cells in bone that orchestrate physiological bone remodelling; they are the mature form of osteoblasts.

RANKL/RANK/OPG system: The key system controlling osteoclasts; OPG is the decoy receptor of RANKL and inhibits osteoclastogenesis and activity.

Wnt signaling: Together with BMPs, the key system controlling osteoblasts; it acts on OB commitment and activity.

References

Alliston, T. and L. Choy, P. Ducy, G. Karsenty, and R. Derynck. 2001. TGF-beta-induced repression of CBFA1 by Smad3 decreases cbfa1 and osteocalcin expression and inhibits osteoblast differentiation. EMBO J. 20: 2254-2272.

Bonewald, L.F. 2007. Osteocytes as dynamic multifunctional cells. Ann. NY Acad. Sci. 1116: 281-290.

Charatcharoenwitthaya, N. and S. Khosla, E.J. Atkinson, L.K. McCready, and B.L. Riggs. 2007. Effect of blockade of TNF-alpha and interleukin-1 action on bone resorption in early postmenopausal women. J. Bone Miner. Res. 22: 724-729.

D'Amelio, P. and G. Isaia. 2009. Immune system and postmenopausal bone loss. CRBMM 7: 261-267.

D'Amelio, P. and A. Grimaldi, S. Di Bella, S.Z. Brianza, M.A. Cristofaro, C. Tamone, G. Giribaldi, D. Ulliers, G.P. Pescarmona, and G. Isaia. 2008. Estrogen deficiency increases osteoclastogenesis up-regulating T cells activity: a key mechanism in osteoporosis. Bone 43: 92-100.

D'Amelio, P. and G. Isaia, and G.C. Isaia. 2009. The osteoprotegerin/RANK/RANKL system: a bone key to vascular disease. JEI 32: 6-9.

Grundt, A. and I.A. Grafe, U. Liegibel, U. Sommer, P. Nawroth, and C. Kasperk. 2009. Direct effects of osteoprotegerin on human bone cell metabolism. Biochem. Biophys. Res. Commun. 389: 550-555.

Isaia, G.C. and P. D'Amelio, S. Di Bella, and C. Tamone. 2005. Is leptin the link between fat and bone mass? J. Endocrinol. Invest. 28: 61-65.

Kurata, K. and T.J. Heino, H. Higaki, and H.K. Vaananen. 2006. Bone marrow cell differentiation induced by mechanically damaged osteocytes in 3D gel-embedded culture. J. Bone Miner. Res. 21: 616-625.

Leibbrandt, A. and J.M. Penninger. 2008. RANK/RANKL: regulators of immune responses and bone physiology. Ann. NY Acad. Sci. 1143: 123-150.

Lubberts, E. and M.I. Koenders, and W.B. van den Berg. 2005. The role of T-cell interleukin-17 in conducting destructive arthritis: lessons from animal models. Arthritis Res. Ther. 7: 29-37.

Macsai, C.E. and B.K. Foster, and C.J. Xian. 2008. Roles of Wnt signalling in bone growth, remodelling, skeletal disorders and fracture repair. J Cell Physiol. 215: 578-587.

Marie, P.J. 2008. Transcription factors controlling osteoblastogenesis. Arch. Biochem. Biophys. 473: 98-105.

Massague, J. 1990. The transforming growth factor-beta family. Annu. Rev. Cell. Biol. 6: 597-641.

Quinn, J.M. and N.A. Sims, H. Saleh, D. Mirosa, K. Thompson, S. Bouralexis, E.C. Walker, T.J. Martin, and M.T. Gillespie. 2008. IL-23 inhibits osteoclastogenesis indirectly through lymphocytes and is required for the maintenance of bone mass in mice. J. Immunol. 181: 5720-5729.

Roodman, G.D. 1992. Interleukin-6: an osteotropic factor? J. Bone Miner. Res. 7: 475-478.

Sato, K. and T. Satoh, K. Shizume, Y. Yamakawa, Y. Ono, H. Demura, T. Akatsu, N. Takahashi, and T. Suda. 1992. Prolonged decrease of serum calcium concentration by murine gamma-interferon in hypercalcemic, human tumor (EC-GI)-bearing nude mice. Cancer Res. 52: 444-449.

Senta, H. and H. Park, E. Bergeron, O. Drevelle, D. Fong, E. Leblanc, F. Cabana, S. Roux, G. Grenier, and N. Faucheux. 2009. Cell responses to bone morphogenetic proteins and peptides derived from them: biomedical applications and limitations. Cytokine Growth Factor Rev. 20: 213-222.

Suzawa, M. and I. Takada, J. Yanagisawa, F. Ohtake, S. Ogawa, T. Yamauchi, T. Kadowaki, Y. Takeuchi, H. Shibuya, Y. Gotoh, K. Matsumoto, and S. Kato. 2003. Cytokines suppress adipogenesis and PPAR-gamma function through the TAK1/TAB1/NIK cascade. Nat. Cell. Biol. 5: 224-230.

Takada, I. and M. Mihara, M. Suzawa, F. Ohtake, S. Kobayashi, M. Igarashi, M.Y. Youn, K. Takeyama, T. Nakamura, Y. Mezaki, S. Takezawa, Y. Yogiashi, H. Kitagawa, G. Yamada, S. Takada, Y. Minami, H. Shibuya, K. Matsumoto, and S. Kato. 2007. A histone lysine methyltransferase activated by non-canonical Wnt signalling suppresses PPAR-gamma transactivation. Nat. Cell. Biol. 9: 1273-1285.

Takayanagi, H. and K. Ogasawara, S. Hida, T. Chiba, S. Murata, K. Sato, A. Takaoka, T. Yokochi, H. Oda, K. Tanaka, K. Nakamura, and T. Taniguchi. 2000. T-cell-mediated regulation of osteoclastogenesis by signalling cross-talk between RANKL and IFN-gamma. Nature 408: 600-605.

van Bezooijen, R.L. and P. ten Dijke, S.E. Papapoulos, and C.W. Lowik. 2005. SOST/sclerostin, an osteocyte-derived negative regulator of bone formation. Cytokine Growth Factor Rev. 16: 319-327.

Verborgt, O. and N.A. Tatton, R.J. Majeska, and M.B. Schaffler. 2002. Spatial distribution of Bax and Bcl-2 in osteocytes after bone fatigue: complementary roles in bone remodeling regulation? J. Bone Miner. Res. 17: 907-914.

Xu, J. and H.F. Wu, E.S. Ang, K. Yip, M. Woloszyn, M.H. Zheng, and R.X. Tan. 2009. NF-kappaB modulators in osteolytic bone diseases. Cytokine Growth Factor Rev. 20: 7-17.

Yago, T. and Y. Nanke, M. Kawamoto, T. Furuya, T. Kobashigawa, N. Kamatani, and S. Kotake. 2007. IL-23 induces human osteoclastogenesis via IL-17 in vitro, and anti-IL-23 antibody attenuates collagen-induced arthritis in rats. Arthritis Res. Ther. 9: R96.

Yao, Z. and S.L. Painter, W.C. Fanslow, D. Ulrich, B.M. Macduff, M.K. Spriggs, and R.J. Armitage. 1995. Human IL-17: a novel cytokine derived from T cells. J. Immunol. 155: 5483-5486.

CHAPTER 26

Cytokines in Rheumatoid Arthritis

Ivan V. Shirinsky* **and Valery S. Shirinsky**
Laboratory of Clinical Immunopharmacology
Institute of Clinical Immunology RAMS
6 Zalesskogo str. 630047 Novosibirsk, Russia

ABSTRACT

Rheumatoid arthritis (RA) is a common chronic autoimmune disease leading to significant morbidity and mortality. Various cytokines are involved in RA pathogenesis forming a complex network with multidirectional relationships at different levels. Cytokines probably have a role in all stages of RA development from loss of tolerance to joint localized inflammation and its systemic consequences such as accelerated atherosclerosis. In this chapter, cytokines of different families, such as the TNF superfamily, IL-6 family, IL-10 superfamily, IL-2/15 superfamily, and some other cytokines are reviewed. Based on preclinical data and animal models, as well as findings on local expression of cytokines in inflamed joints, many cytokines have been shown to play a role in RA. The most overwhelming evidence comes from clinical studies exploring the efficacy of therapeutic disruption of particular cytokine pathways. To date, the most efficient strategies of anti-cytokine treatment have been the blocking of either TNF-α or IL-6. Therefore, these cytokines are currently considered of paramount importance in inflamed synovium. However, the above-mentioned therapeutic approaches are not universally effective, suggesting alternative pathways of regulation in synovial inflammation. Many newer cytokines are being studied in both experimental and clinical settings and, in the future, we hope to gain a greater understanding of the cytokine regulatory network and its interactions in RA.

*Corresponding author

INTRODUCTION AND OVERVIEW OF RA PATHOGENESIS

Rheumatoid arthritis is a chronic autoimmune inflammatory disease primarily affecting synovial tissue, cartilage and subchondral bone. Its prevalence ranges from 0.5 to 1% in European and North American adults and it is associated with increased morbidity and mortality. RA represents a natural model of chronic inflammation, which essentially shares common features with many prevalent inflammatory diseases including atherosclerosis, psoriasis, and inflammatory bowel disease. Therefore, investigating molecular mechanisms of RA may provide clues for understanding the pathogenesis of other inflammatory conditions.

There is significant evidence that various cytokines play an important role in the pathogenesis of RA (Table 1). The most compelling evidence comes from clinical studies showing that blocking of several cytokine pathways, such as TNF-α, in most patients, leads to amelioration of the clinical course of RA and a reduction of structural damage. In this chapter, we briefly describe the major cytokines involved in RA pathogenesis as well as some newer cytokines whose role in RA has recently become recognized.

Table 1 Key facts about cytokines in rheumatoid arthritis

- Rheumatoid arthritis is a chronic autoimmune inflammatory disease affecting approximately 1% of the population.
- RA is characterized by synovial hyperplasia and joint destruction.
- Cells infiltrating RA synovium produce a number of cytokines that form a complex network.
- The most studied cytokines with a well-established role in RA are pro-inflammatory TNF-α, IL-1, and IL-6.
- Many other cytokines appear to be important in RA pathogenesis. Some of them are promising targets for RA treatment.

There is a widely accepted hypothesis that RA is initiated and sustained by autoreactive T cells in genetically predisposed individuals. T cell tolerance loss is due to poorly defined mechanisms that probably arise at a systemic immune regulatory level and may include aberrant thymic selection or peripheral tolerance.

In the next phase, some not yet well-defined factors trigger autoimmune inflammation localized in the joints. After the start of inflammation, normally thin synovial membrane (typically one to three cells thick) becomes hypercellular. Subsequently, a pannus develops that invades articular cartilage and periarticular bone (Fig. 1).

Histologically, RA is characterized by infiltration of the synovium by fibroblast-like synoviocytes (FLS), macrophages, mast cells, CD4+ T cells, CD8+ T cells, natural killer (NK) cells, NKT cells, B cells and plasma cells. Any of these cells

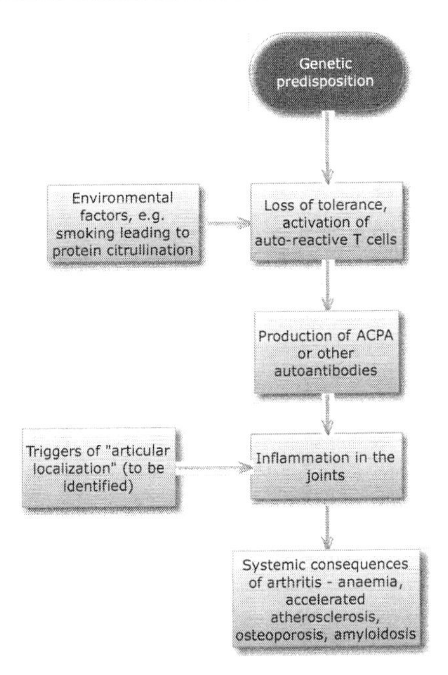

Fig. 1 A schematic model of RA pathogenesis. This figure illustrates key events in RA pathogenesis. ACPA, anti-citrullinated protein antibodies.

may be a source of various cytokines involved in perpetuating joint inflammation (Fig. 2).

The contribution of particular cytokines and cytokine families to joint inflammation and damage is discussed below.

CYTOKINES OF THE TNF SUPERFAMILY IN RA

The TNF superfamily consists of over 20 members that can interact with more than 30 receptors. All members of the TNF superfamily are type II transmembrane proteins and can be expressed in both a membrane-bound and a soluble form. TNF superfamily ligands have both a traditional name and a recently proposed systematic name (Table 2).

A prototypical member of this ligand family, TNF-α, has various biological functions, which are reviewed in detail elsewhere (Dinarello and Moldawer 2002). TNF-α is considered a pivotal cytokine implicated in the pathogenesis of RA. TNF is detected in large quantities in synovial fluid (SF) and is expressed in synovial tissue. Therapeutic disruption of the TNF pathway leads to significant clinical effect in approximately 70% of patients with RA. This effect is accompanied by a decrease in plasma levels of IL-6 and acute-phase reactants, suppression of

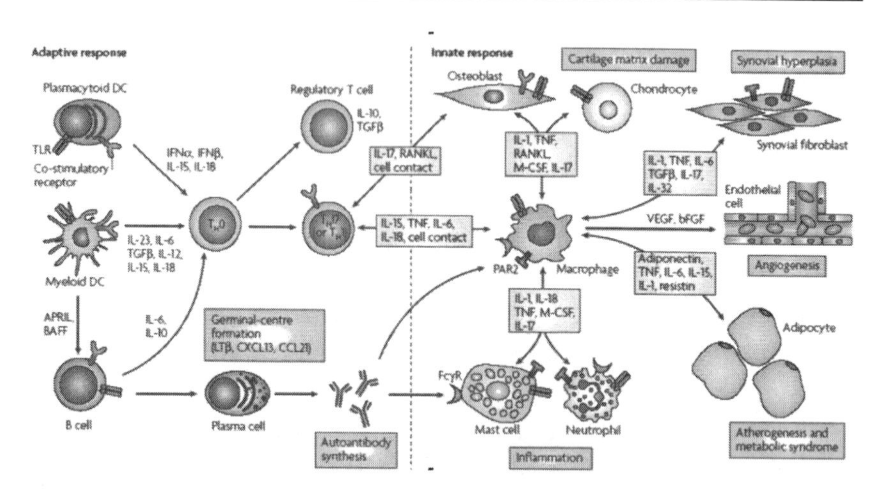

Fig. 2 An overview of the cytokine-mediated regulation of synovial interactions. The component cells of the inflamed rheumatoid synovial membrane are depicted in innate and adaptive predominant compartments of the inflammatory response.Pivotal cytokine pathways are depicted in which activation of dendritic cells (DCs), T cells, B cells and macrophages underpin the dysregulated expression of cytokines that in turn drive activation of effector cells, including neutrophils, mast cells, endothelial cells and synovial fibroblasts. The clinical manifestations of such effects are highlighted. Reprinted with permission of Nature Publishing Group from McInnes and Schett (2007). APRIL, a proliferation-inducing ligand; BAFF, B-cell activating factor; bFGF, basic fibroblast growth factor; CCL21, CC-chemokine ligand 21; CXCL13, CXC-chemokine ligand 13; FcγR, Fc receptor for IgG; IFN, interferon; IL, interleukin; LTβ, lymphotoxin-β; M-CSF, macrophage colony-stimulating factor; PAR2, protease-activated receptor 2; RANKL, receptor activator of nuclear factor-κB (RANK) ligand; TGFβ, transforming growth factor-β; TH, T helper; TLR, Toll-like receptor; TNF, tumour-necrosis factor; VEGF, vascular endothelial growth factor.

Color image of this figure appears in the color plate section at the end of the book.

leukocyte migration, endothelial-cell deactivation, and restoration of Tregs population (McInnes and Schett 2007).

The role of other TNF superfamily ligand members in the pathogenesis of RA is less well defined. Lymphotoxin α (LTα) has a crucial role in lymphoid organ development. LTα also stimulates proliferation and pro-inflammatory cytokine secretion in RA FLS as effectively as TNF-α. Lymphotoxin β (LTβ), which is produced by T cells and B cells, plays a central role in the normal development of lymph nodes and is crucial in the formation of ectopic germinal centre reactions in rheumatoid synovitis (McInnes and Schett 2007).

OX40 ligand (OX40L) binding to OX40 (CD134) is thought to be important in T cell activation through T cell/antigen-presenting cell interaction. Administration of anti-OX40L in mice with CIA ameliorates disease severity. OX40L is expressed on sublining cells in the synovial tissue of RA patients (Hsu *et al.* 2006).

Table 2 TNF superfamily members: role in inflammation

Systematic name	*Functional name*	*Effects on inflammation*
TNFSF1	LT-α/TNF-β	↑
TNFSF2	TNF-α	↑
TNFSF3	Lymphotoxin-β, TNF-γ	↑
TNFSF4	OX40L	↑
TNFSF5	CD40L/CD154	↑
TNFSF6	FasL	↓
TNFSF7	CD27L/CD70	↑
TNFSF8	CD30L/CD153	↑
TNFSF9	4-1BBL	↑
TNFSF10	TRAIL/Apo2-L	↓
TNFSF11	RANKL	–
TNFSF12	TWEAK	↑
TNFSF13	APRIL	↑
TNFSF13B TNFSF20	BAFF/BLys	↑
TNFSF14	LIGHT	↑
TNFSF15	VEGI/TL1A	↑
TNFSF18	GITRL	↑

TNF, tumour necrosis factor; SF, superfamily; 4-1BBL, 4-BB ligand; APRIL, a proliferation-inducing ligand; BAFF/BLys, B cell activating factor belonging to the TNF family/B lymphocyte stimulator; CD, cluster of differentiation; FasL, Fas ligand; GITRL, glucocorticoid-induced TNF receptor family related protein ligand; LIGHT, lymphotoxin-like, exhibits inducible expression and competes with herpes simplex virus glycoprotein D for herpes virus entry mediator, a receptor expressed by T lymphocytes; LT, lymphotoxin; OX40L, OX40 ligand; RANKL, receptor activator of nuclear factor-κB ligand; TNF, tumour necrosis factor; TRAIL/Apo2-L, TNF-related apoptosis-inducing ligand; TWEAK, TNF-like weak inducer of apoptosis; VEGI/TL1A, vascular endothelial growth inhibitor/ TNF receptor-like 1A; ↑, stimulation; ↓, inhibition.

CD40L has been associated with the progression of RA. Peripheral blood T lymphocytes from RA patients express CD40L on their membrane. The level of expression correlates with disease activity and radiographic progression (Berner *et al.* 2000).

The primary function of Fas is to trigger death signals when bound to its ligand, Fas ligand (FasL). The Fas-FasL interaction plays a key role in activation-induced cell death of T lymphocytes and is crucial for lymphocyte homeostasis. The concentration of soluble FasL in the SF is elevated in patients with severe RA. In addition to pro-apoptotic properties, FasL can exert pro-inflammatory effects as well; this occurs especially in the presence of IL-1 and TNF-α. However, in rat models of experimental arthritis, local injection of soluble FasL into the affected joints suppressed disease activity, suggesting its therapeutic potential in RA (Li *et al.* 2004).

CD27L/CD70 is expressed at high levels on activated T cells in patients with autoimmune disorders, including RA. Treatment of mice with CIA with an anti-CD70 Ab results in reductions in inflammation and bone and cartilage destruction in joints (Oflazoglu *et al.* 2009).

CD30L/CD153 is expressed in the synovium of patients with RA and serum levels of this TNF superfamily member correlate with its SF concentrations (Berner *et al.* 2000, Oflazoglu *et al.* 2009).

TNF-related apoptosis-inducing ligand (TRAIL) appears to have paradoxical anti-inflammatory properties. Blocking of TRAIL with the soluble receptor DR5 has been shown to increase the susceptibility of mice to collagen-induced arthritis (Dinarello and Moldawer 2002).

Receptor activator of nuclear factor κB (RANK) ligand (RANKL) is expressed by FLS, and activated synovial T cells. The main function of RANKL is stimulation of bone resorption through induction of osteoclast differentiation and enhancement of their activity. RANKL expression is regulated by inflammatory cytokines. The interaction of RANKL with its receptor RANK is modulated by osteoprotegerin (OPG), a soluble decoy receptor, which is expressed by stromal cells in the RA synovium. In RA, there is a disproportionate increase in RANK relative to OPG, which facilitates RANKL-induced bone loss (McInnes and Schett 2007).

TNF-like weak inducer of apoptosis (TWEAK) mediates pleiotropic effects on a variety of cells. Activation of the TWEAK receptor leads to production of IL-6 and IL-8. TWEAK may also inhibit chondrogenesis and osteogenesis and promote osteoclastogenesis. TWEAK expression is significantly increased in RA FLS and synovial macrophages (van Kuijk *et al.* 2010).

A proliferation-inducing ligand (APRIL) and B cell activating factor belonging to the TNF family/B lymphocyte stimulator (BAFF/BLys) share two receptors. The main function of these two TNF superfamily members is providing signals leading to B-cell survival and activation. In knockout APRIL$^{-/-}$ mice, the incidence of CIA is significantly reduced and is accompanied by suppressed IL-17 production (Xiao *et al.* 2008).

Lymphotoxin-like, exhibits inducible expression and competes with herpes simplex virus glycoprotein D for herpes virus entry mediator, a receptor expressed by T lymphocytes (LIGHT) promotes RA-FLS proliferation and induces expression of pro-inflammatory cytokines by RA FLS. LIGHT is up-regulated in both SF and synovium of RA patients (Ishida *et al.* 2008).

TNF-like ligand 1A (TL1A) promotes TNF-α production by T cells and enhances Th17 differentiation and IL-17 production. TL1A concentrations are elevated in RA SF. Human chrondrocytes and FLS are capable of secreting TL1A upon TNF-α or IL-1β stimulation (Zhang *et al.* 2009).

Finally, glucocorticoid-induced TNF receptor family related protein ligand (GITRL) acts as a co-stimulator in immune cells. GITRL is expressed by a variety of cells. In RA patients, GITRL expression is significantly increased in peripheral blood CD4+CD25+ T cells and positively correlates with disease activity (Grewal 2009).

IL-1 SUPERFAMILY LIGANDS AND RA

The IL-1 superfamily consists of 11 members, all of whom now have both systematic and traditional names (Table 3). The most studied cytokines of this family are IL-1α, IL-1-β, IL-1Ra, and IL-18. IL-1α, IL-1β and IL-18 are agonists and IL-1Ra is a specific antagonist of IL-1α and IL-1-β, but not IL-18.

Table 3 Interleukin-1 family members: effects on inflammation

Systematic name	*Traditional name*	*Effects on inflammation*
IL-1F1	IL-1α	↑
IL-1F2	IL-1β	↑
IL-1F3	IL-1ra	↓
IL-1F4	IL-18	↑
IL-1F5	IL-1Hy1/FIL1δ/IL-1H3/IL -1RP3/IL-1L1/IL-1δ	↓
IL-1F6	FIL1ε	↑
IL-1F7	FIL1 ζ/IL-1H4/ IL-1RP1/IL-1H	↓
IL-1F8	FIL1η/IL-1H2	↑
IL-1F9	IL-1H1/IL-1RP2/ IL-1ε	↑
IL-1F10	IL-1Hy2/FKSG75	↓ ?
IL-1F11	IL-33	↑

IL, interleukin; ↑, stimulation; ↓, inhibition.

IL-1 is a prototypical pro-inflammatory cytokine and activates multiple genes involved in inflammation. Its significance as a contributor to joint inflammation and damage has been demonstrated in various rodent models of arthritis and in human RA. Despite the well-established role of IL-1 in RA, the therapeutic blocking of IL-1 by recombinant IL-1Ra has only limited efficacy with clinical response in approximately 40% of patients.

IL-18, a more recently discovered pro-inflammatory mediator, is secreted by a variety of cell types and is important in inflammation and host response to infection. Synovial IL-18 concentrations correlate with RA disease activity and response to therapy. IL-18 enhances the inflammatory response by promoting the release of other cytokines, such as TNF-α, granulocyte-macrophage colony-stimulating factor (GM-CSF) and interferon-γ (IFN-γ).

IL-1F6, IL-1F8, and IL-1F9 seem to induce signals similar to those induced by IL-1, but only at much higher concentrations. IL-1F6 increases IL-6 and IL-8 production in epithelial cells and is over-expressed in the dermal plaques of psoriasis patients. IL-1F8 stimulates pro-inflammatory cytokine production by FLS and articular chondrocytes. IL-1F8 serum levels are elevated in RA but do not correlate with inflammation. It seems that joint cells are not a major source of IL-1F8 protein in arthritis as the concentration of this cytokine in SF is less than in the serum (Magne *et al.* 2006).

IL-1F5 and IL-1F7 have non-specific anti-inflammatory properties. IL-1F5 induces IL-4 synthesis and inhibits IL-1β responses. IL-1F7 can interact with IL-18-binding protein (IL-18BP) and increase its inhibitory effect on IL-18 activity. There have been no studies investigating the role of either IL-1F5 or IL-1F7 in RA. IL-1F10 has been described as a low-affinity, nonagonistic ligand for IL-1RI. Its role in autoimmunity is unclear (Dinarello 2009).

IL-33 (or IL-1F11) is a ligand for the orphan IL-1 family receptor T1/ST2. IL-33 receptor binding induces intracellular signals similar to those provoked by IL-1. Treatment with a mAb against the IL-33 receptor reduced the severity of CIA and the production of IFN-γ by lymph node cells. IL-33 is significantly elevated in sera and SF of patients with RA and correlates with serum levels of IL-1β and IL-6. In addition, IL-33 concentrations are associated with increased disease activity (Matsuyama *et al.* 2010, Gabay and McInnes 2009).

IL-6 SUPERFAMILY

IL-6 is a pleiotropic cytokine with multiple functions that may be schematically divided into those involved in the regulation of haematopoiesis and those involved in the activation of both innate and acquired immune responses. Other cytokines of this family sharing the same cytokine receptor subunit glycoprotein (gp) 130 are IL-11, ciliary neurotrophic factor (CNTF), oncostatin M (OSM), leukaemia inhibitory factor (LIF), and cardiotropin-1 (CT-1). In RA, IL-6 and its superfamily members have been implicated in joint inflammation and destruction as well as systemic manifestations of the disease. IL-6 is a principal inductor of hepatic acute phase reactants. IL-6 is produced by FLS and serum concentrations of IL-6 correlate with disease activity measures. There has been concern regarding therapeutic blocking of IL-6 in RA as this cytokine is a suppressor of TNF-α, IL-1, and chemokine expression, and thus possesses anti-inflammatory properties. However, therapeutic disruption of IL-6 signalling by monoclonal antibodies (mAb) against the IL-6 receptor has been an effective approach leading to remission in about 30% of patients and successfully retarding radiological progression.

Expression of other IL-6 superfamily members has been demonstrated in RA synovial membranes. LIF, OSM, and IL-11 have also been detected in RA SF (Dinarello and Moldawer 2002).

IL-10 FAMILY

The IL-10 family comprises both anti-inflammatory and pro-inflammatory cytokines (Table 4). IL-10 is produced by many cell populations including Th2 cells and macrophages, the latter being the main source of this cytokine. IL-10 inhibits Th1, Th17 cells and suppresses expression of various pro-inflammatory

Table 4 Effects of IL-10 family members on inflammation

Name	*Effects on inflammation*
IL-10	↓
IL-19	↑
IL-20	↑
IL-22	↑

IL, interleukin; ↑, stimulation; ↓, inhibition.

mediators. Tregs exert their suppressive action on the immune response by using IL-10 and transforming growth factor β (TGF-β). Administration of IL-10 leads to reduction of disease activity and damage in murine CIA. In human RA, the role of IL-10 is complex, because in addition to anti-inflammatory properties it also increases antibody production and activates B cells and thus may enhance autoimmune responses.

Another member of the IL-10 family, IL-19, was found to be over-expressed in RA joints and it can act as an inductor of IL-6 and TNF-α synthesis in monocytes (Liao *et al.* 2002).

IL-20 induces FLS to produce pro-inflammatory cytokines including IL-6 and IL-8, and it promotes endothelial cell proliferation. The concentration of this cytokine is elevated in RA SF (Hsu *et al.* 2006).

IL-22 is produced by several cell types including Th17 cells and FLS. It is over-expressed in RA joints and may have a pro-inflammatory role in arthritis, promoting osteoclastogenesis and regulating antibody production (McInnes and Schett 2007).

IL-17 AND RELATED CYTOKINES

In recent years, IL-17 has drawn close attention as a potential key effector of joint inflammation in RA. A subpopulation of IL-17-producing T cells has been termed T helper cell type 17 (Th17). Th17 lymphocytes seem to be reciprocal with Th1, Th2 and Treg. For a long time, RA has been believed to be a Th1-driven disease where the cytokine balance is skewed towards Th1-produced cytokines. However, this paradigm has recently been drawn into question by data indicating that IFN-γ-deficient mice develop more severe arthritis and by the paucity of key Th1-cytokines, IFN-γ and IL-2, in RA SF. Moreover, IFN-γ has been shown to inhibit osteoclast differentiation, which is not consistent with the hypothesis that RA is a Th1-driven disease (McInnes and Schett 2007).

IL-17 contributes to inflammation through enhancing the production of other pro-inflammatory cytokines, such as IL-1 and TNF. It also acts synergistically with TNF while promoting pro-inflammatory cytokine synthesis by FLS. In addition, IL-17 is implicated in cartilage destruction (McInnes and Schett 2007).

IL-17 production is induced by IL-23, which shares a common subunit with IL-12 that is produced by macrophages and dendritic cells.

There have been conflicting data regarding the prevalence of Th17 in RA patients. However, a recent study performed on a homogeneous group of RA patients and accounting for immunomodulating therapy has clearly shown that Th17 cells were increased in the peripheral circulation and the SF of patients with both early and established RA, and correlated with disease activity (Leipe *et al.* 2010).

IL-2/IL-15 FAMILY

The IL-2/IL-15 family of cytokines includes IL-2, IL-4, IL-7, IL-9, IL-15 and IL-21 (Table 5).

Table 5 Effects of IL-2/IL-15 family members on inflammation

Name	*Effects on inflammation*
IL-4	↓
IL-7	↑
IL-9	↓
IL-15	↑
IL-21	↑

IL, interleukin; ↑, stimulation; ↓, inhibition.

IL-2 is primarily a T cell growth factor. In RA, IL-2 is spontaneously produced by PBMC but in RA SF the concentration of IL-2 is low. As the treatment with an anti-IL-2 mAb results only in modest improvement, IL-2 does not seem to contribute significantly to RA pathogenesis.

IL-4 is produced by Th2 cells and has anti-inflammatory properties. It inhibits Th17 differentiation and down-regulates production of IL-1β, TNF-α, and IFN-γ (Dinarello and Moldawer 2002).

IL-7 has numerous functions and influences different cells including T cells, dendritic cells and bone biology. IL-7 is required for developing mature T cells in the thymus. The cytokine is over-expressed in RA synovium and SF. It seems that blocking IL-7 may also have therapeutic potential (Churchman and Ponchel 2008).

Another cytokine of this family, IL-15 can induct proliferation of memory CD8+ T cells and activate them. Thus, it probably contributes to tolerance loss in autoimmune disorders. IL-15 can also stimulate IL-17 production by PBMC. In addition, IL-15 has been shown to stimulate monocyte/macrophages to produce TNF, IL-6, IL-8, and IL-12 and protect fibroblasts and endothelial cells from apoptosis (Gabay and McInnes 2009).

Finally, IL-21 has been shown to play an important role in the production of Th17 cells. IL-21 is potently induced by IL-6 and can be produced by Th17 cells themselves (Niu *et al.* 2010).

OTHER CYTOKINES

Granulocyte colony-stimulating factor (G-CSF) and granulocyte-macrophage colony-stimulating factor (GM-CSF) are involved in the regulation of haematopoiesis, but they can also have pro-inflammatory properties. G-CSF and GM-CSF administration can lead to exacerbation of rheumatoid arthritis (RA). Both cytokines are detected in the joints of patients with RA. In a mouse model of RA, blocking of G-CSF and GM-CSF results in a significant reduction in disease activity (Cornish *et al.* 2009).

Transforming growth factor β (TGFβ), basic fibroblast growth factor (bFGF), and platelet-derived growth factor (PDGF) appear to be crucial in synovial hyperplasia. Macrophage migration-inhibitory factor (MIF) has pro-inflammatory properties including stimulation of phagocytosis, cytokine synthesis, T-cell activation, and fibroblast proliferation (McInnes and Schett 2007).

Another recently described cytokine, IL-32, is produced by monocytes and epithelial cells. It is capable of inducing cytokine synthesis by macrophages and is found in synovial tissue (Joosten *et al.* 2006).

ADIPOKINES

Recently, regulatory peptides produced by adipose tissue cells have emerged as a novel class of immunoregulatory factors. Studies have demonstrated that adipocytes produce more than 100 pro-inflammatory, anti-inflammatory and immunomodulating peptides belonging to the cytokine, chemokine, complement, and growth factor families. These data allow adipocytes to be considered part of immune system (Schaffler *et al.* 2007).

Leptin has anti-inflammatory effects in models of acute inflammation and during activation of innate immune responses. On the contrary, in experimental models of autoimmune diseases, leptin stimulates T lymphocyte responses, thus exerting rather a pro-inflammatory role. In rheumatoid arthritis, leptin levels have been either elevated or unchanged, suggesting a rather complex role of this molecule in human autoimmune disease.

Adiponectin was shown to have anti-inflammatory properties in obesity and obesity-related disorders. In RA, adiponectin levels are elevated and it seems to have a pro-inflammatory role.

Visfatin induces monocyte production of pro-inflammatory cytokines. In RA, circulating visfatin levels are elevated; this adipokine may thus contribute to synovial inflammation.

Resistin has been detected in the plasma and the SF of RA patients, and the injection of resistin into mice joints induces an arthritis-like condition. Resistin is also associated with laboratory markers of inflammation in RA (Schaffler *et al.* 2007).

Summary Points

- RA is a chronic autoimmune inflammatory disorder of the joints. Cytokines play a crucial role in sustaining inflammation in RA.
- Cytokines in RA synovium form a complex regulatory network whose hierarchical interactions are not completely understood.
- Currently, TNF-α and IL-6 are considered to be key cytokines in RA synovial inflammation as evidenced by the successful application of therapeutic blocking of TNF-α and IL-6.
- RA seems to be a Th17-driven disease.
- There are a number of emerging cytokines that may be attractive therapeutic targets in the future.

Abbreviations

Ab	:	antibody (mAb: monoclonal antibody)
APRIL	:	a proliferation-inducing ligand
BAFF/BlyS	:	B cell activating factor belonging to the TNF family/B lymphocyte stimulator
CIA	:	collagen-induced arthritis
CNTF	:	ciliary neurotrophic factor
CSF	:	colony-stimulating factor
FasL	:	Fas ligand
FLS	:	fibroblast-like synoviocytes
GITRL	:	glucocorticoid-induced TNF receptor family related protein ligand
HLA	:	human leukocyte antigen
ICAM	:	intracellular cell adhesion molecule
IFN	:	Interferon
LIF	:	leukaemia inhibitory factor

LIGHT	:	lymphotoxin-like, exhibits inducible expression and competes with herpes simplex virus glycoprotein D for herpes virus entry mediator, a receptor expressed by T lymphocytes
LT	:	lymphotoxin
MCP	:	monocyte chemotactic protein 1
MIF	:	macrophage migration inhibitory factor
MIP	:	macrophage inflammatory protein
NKT	:	natural killer T cells
OPG	:	osteoprotegerin
OSM	:	Oncostatin M
PBMC	:	peripheral blood mononuclear cell
RANK	:	receptor activator of nuclear factor κB
SF	:	synovial fluid
TCR	:	T cell receptor
TGF	:	transforming growth factor
TNF	:	tumour necrosis factor
TRAIL	:	TNF-related apoptosis-inducing ligand
Treg	:	regulatory T cells
TWEAK	:	TNF-like weak inducer of apoptosis

Definition of Terms

Collagen-induced arthritis (CIA): An experimental model of arthritis produced by injecting mice with foreign collagen together with adjuvant. This model is frequently used to study RA pathogenesis and to assess efficacy of newer treatments.

Th1 cells (T helper cell type 1): a subset of lymphocytes that produce IL-2, IL-15, and IFN-γ and promote cellular immune response.

Th2 cells (T helper cell type 2): a subgroup of lymphocytes producing IL-4, IL-5, IL-6, IL-10, and IL-13 and involved in B-cell antibody response.

Th17 cells (T helper cell type 17) a recently described subtype of lymphocytes that produce cytokines IL-17, IL-21, and IL-22.

Treg (regulatory T cells): a subpopulation of T lymphocytes that are able to suppress an immune response.

Acknowledgements

We would like to thank Dr. Elena Eberle for her invaluable help in editing this manuscript.

References

Berner, B. and G. Wolf, K.M. Hummel, G.A. Muller, and M.A. Reuss-Borst. 2000. Increased expression of CD40 ligand (CD154) on CD4+ T cells as a marker of disease activity in rheumatoid arthritis. Ann. Rheum. Dis. 59: 190-195.

Churchman, S. and M.F. Ponchel. 2008. Interleukin-7 in rheumatoid arthritis. Rheumatology (Oxford) 47: 753-759.

Cornish, A.L. and I.K. Campbell, B.S. McKenzie, S. Chatfield, and I.P. Wicks. 2009. G-CSF and GM-CSF as therapeutic targets in rheumatoid arthritis. Nat. Rev. Rheumatol. 5: 554-559.

Dinarello, C.A. 2009. Immunological and inflammatory functions of the interleukin-1 family. Annu. Rev. Immunol. 27: 519-550.

Dinarello, C.A. and L.L. Moldawer. 2002. Proinflammatory and Anti-inflammatory Cytokines in Rheumatoid Arthritis. A Primer for Clinicians. Thousand Oaks, Amgen Inc.

Gabay, C.I. and B. McInnes. 2009. The biological and clinical importance of the 'new generation' cytokines in rheumatic diseases. Arthritis Res. Ther. 11: 230.

Grewal, I.S. 2009. Therapeutic Targets of the TNF Superfamily. New York, Springer Science+Business Media.

Hsu, H.C. and Y. Wu, and J.D. Mountz. 2006. Tumor necrosis factor ligand-receptor superfamily and arthritis. Curr. Dir. Autoimmun. 9: 37-54.

Hsu, Y.H. and H.H. Li, M.Y. Hsieh, M.F. Liu, K.Y. Huang, L.S. Chin, P.C. Chen, H.H. Cheng, and M.S. Chang. 2006. Function of interleukin-20 as a proinflammatory molecule in rheumatoid and experimental arthritis. Arthritis Rheum. 54: 2722-2733.

Ishida, S. and S. Yamane, T. Ochi, S. Nakano, T. Mori, T. Juji, N. Fukui, T. Itoh, and R. Suzuki. 2008. LIGHT induces cell proliferation and inflammatory responses of rheumatoid arthritis synovial fibroblasts via lymphotoxin beta receptor. J. Rheumatol. 35: 960-968.

Joosten, L.A. and M.G. Netea, S.H. Kim, D.Y. Yoon, B. Oppers-Walgreen, T.R. Radstake, P. Barrera, F.A. van de Loo, C.A. Dinarello, and W.B. van den Berg. 2006. IL-32, a proinflammatory cytokine in rheumatoid arthritis. Proc. Natl. Acad. Sci. USA 103: 3298-3303.

Leipe, J. and M. Grunke, C. Dechant, C. Reindl, U. Kerzendorf, H. Schulze-Koops, and A. Skapenko. 2010. Th17 cells in autoimmune arthritis. Arthritis Rheum. 10: 2876-2885.

Li, N.L. and H. Nie, Q.W. Yu, J.Y. Zhang, A.L. Ma, B.H. Shen, L. Wang, J. Bai, X.H. Chen, T. Zhou, and D.Q. Zhang. 2004. Role of soluble Fas ligand in autoimmune diseases. World J. Gastroenterol. 10: 3151-3156.

Liao, Y.C. and W.G. Liang, F.W. Chen, J.H. Hsu, J.J. Yang, and M.S. Chang. 2002. IL-19 induces production of IL-6 and TNF-alpha and results in cell apoptosis through TNF-alpha. J. Immunol. 169: 4288-4297.

Magne, D. and G. Palmer, J.L. Barton, F. Mezin, D. Talabot-Ayer, S. Bas, T. Duffy, M. Noger, P.A. Guerne, M.J. Nicklin, and C. Gabay. 2006. The new IL-1 family member IL-1F8 stimulates production of inflammatory mediators by synovial fibroblasts and articular chondrocytes. Arthritis Res. Ther. 8: R80.

Matsuyama, Y. and H. Okazaki, H. Tamemoto, H. Kimura, Y. Kamata, K. Nagatani, T. Nagashima, M. Hayakawa, M. Iwamoto, T. Yoshio, S. Tominaga, and S. Minota. 2010.

Increased levels of interleukin 33 in sera and synovial fluid from patients with active rheumatoid arthritis. J. Rheumatol. 37: 18-25.

McInnes, I. and B.G. Schett. 2007. Cytokines in the pathogenesis of rheumatoid arthritis. Nat. Rev. Immunol. 7: 429-442.

Niu, X. and D. He, X. Zhang, T. Yue, N. Li, J.Z. Zhang, C. Dong, and G. Chen. 2010. IL-21 regulates Th17 cells in rheumatoid arthritis. Hum. Immunol. 71: 334-341.

Oflazoglu, E. and T.E. Boursalian, W. Zeng, A.C. Edwards, S. Duniho, J.A. McEarchern, C.L. Law, H.P. Gerber, and I.S. Grewal. 2009. Blocking of CD27-CD70 pathway by anti-CD70 antibody ameliorates joint disease in murine collagen-induced arthritis. J. Immunol. 183: 3770-3777.

Schaffler, A. and J. Scholmerich, and B. Salzberger. 2007. Adipose tissue as an immunological organ: Toll-like receptors, C1q/TNFs and CTRPs. Trends Immunol. 28: 393-399.

van Kuijk, A.W. and C.A. Wijbrandts, M. Vinkenoog, T.S. Zheng, K.A. Reedquist, and P.P. Tak. 2010. TWEAK and its receptor Fn14 in the synovium of patients with rheumatoid arthritis compared to psoriatic arthritis and its response to tumour necrosis factor blockade. Ann. Rheum. Dis. 69: 301-304.

Vandenbroeck, K. and S. Cunningham, A. Goris, I. Alloza, S. Heggarty, C. Graham, A. Bell, and M. Rooney. 2003. Polymorphisms in the interferon-gamma/interleukin-26 gene region contribute to sex bias in susceptibility to rheumatoid arthritis. Arthritis Rheum. 48: 2773-2778.

Xiao, Y. and S. Motomura, and E.R. Podack. 2008. APRIL (TNFSF13) regulates collagen-induced arthritis, IL-17 production and Th2 response. Eur. J. Immunol. 38: 3450-3458.

Zhang, J. and X. Wang, H. Fahmi, S. Wojcik, J. Fikes, Y. Yu, J. Wu, and H. Luo. 2009. Role of TL1A in the pathogenesis of rheumatoid arthritis. J. Immunol. 183: 5350-5357.

Index

D

E

F

About the Editor

Victor R. Preedy BSc, PhD, DSc, FIBiol, FRCPath, FRSPH is Professor of Nutritional Biochemistry, King's College London, Professor of Clinical Biochemistry, Kings College Hospital and Director of the Genomics Centre, King's College London. Presently he is a member of the Kings College London School of Medicine.

Professor Preedy graduated in 1947 with an Honours Degree in Biology and Physiology with Pharmacology. He gained his University of London PhD in 1981 when he was based at the Hospital for Tropical Disease and The London School of Hygiene and Tropical Medicine. In 1992, he received his Membership of the Royal College of Pathologists and in 1993 he gained his second doctoral degree, i.e. DSc, for his outstanding contribution to protein metabolism in health and disease.

Professor Preedy was elected as a Fellow to the Institute of Biology in 1995 and to the Royal Colllege of Pathologists in 2000. Since then he has been elected as a Fellow to the Royal Society for the Promotion of Health (2004) and The Royal Institute of Public Health (2004). In 2009, Professor Preedy became a Fellow of the Royal Society for Public Health.

In his career Professor Preedy has carried out research at the National Heart Hospital (part of Imperial College London) and the MRC Centre at Northwick Part Hospital. He has collaborated with research groups in Finland, Japan, Australia, USA and Germany. He is a leading expert on the pathology of disease and has lectured nationally and internationally. He has published over 570 articles, which includes over 165 peer-reviewed manuscripts based on original research, 90 reviews and 20 books.

Ross J. Hunter MD BSc MRC Path trained in medical sciences at King's College London (Times University ranking 11th in UK). He spent a further year at Imperial College London (Times University ranking 3rd in UK) and was awarded his BSc in Cardiovascular medicine in 1998. Since returning to his medical training at King's College School of Medicine, he has remained an honorary research fellow at The Department of Nutritional Sciences, researching the effect of different nutritional states and alcoholism on the cardiovascular system. He was awarded his bachelor of medicine & surgery

(MBBS) with distinction in 2001. He trained in general medicine in London and Brighton and was made a member of the Royal College of Physicians (UK) in 2005. He trained as a Registrar in the London Deanery from 2005-2008. Since 2008 he has been a research fellow at the Department of Cardiology & Electrophysiology at St Bartholomew's Hospital London, conducting clinical research and clinical trials in cardiology and electrophysiology. He has published over 60 scientific articles of various kinds.

Color Plate Section

Chapter 1

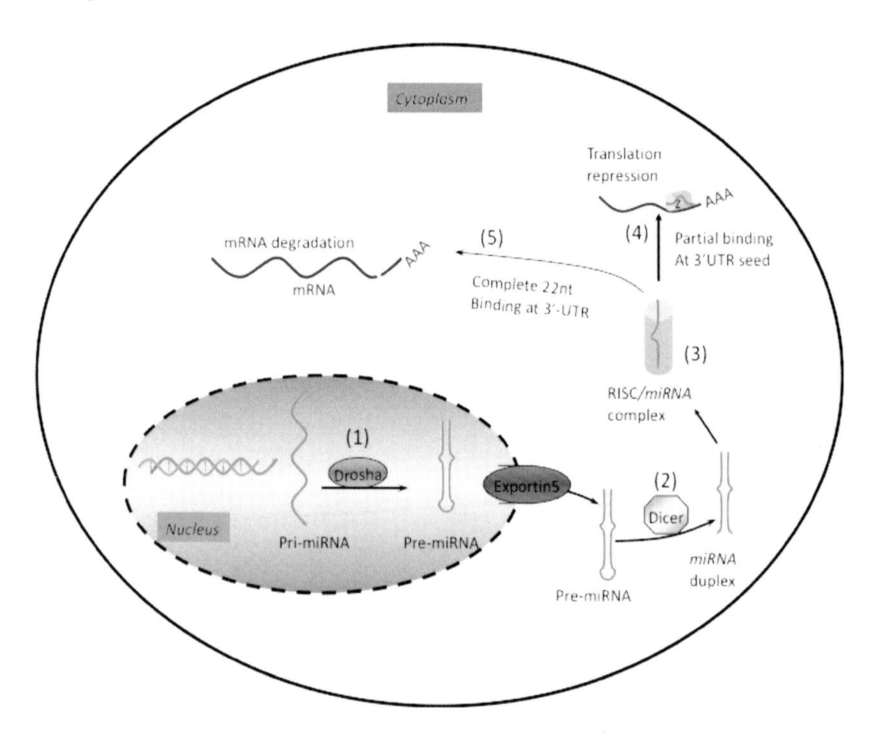

Fig. 1 miRNA biogenesis and function in a cell. Primary miRNAs (pri-miRNAs) are transcribed from independent genes by the RNA polymerase II. (1) The pri-miRNAs are further cleaved into 50- to 80-base premature miRNAs (pre-miRNAs) by Drosha. (2) The pre-miRNAs are translocated into cytoplasm by the GTP-dependent factor exportin-5, where the nuclease Dicer processes further into the miRNA duplexes. (3) Functional mature miRNAs are loaded with the RNA-induced silencing complex (RISC), and (4) represses translation of the target mRNAs by partial binding to the 3'UTR (7-8 mer) or (5) degrades mRNA by complete complementary binding to target (22nt).

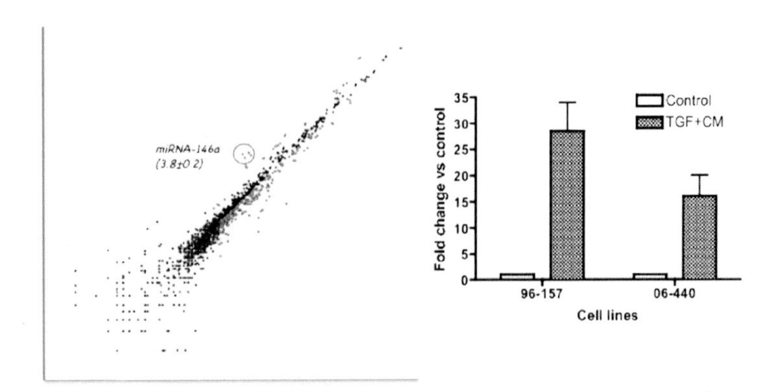

Fig. 3 MicroRNA-146a was up-regulated in HBECs in response to the cytokine stimulation. HBECs were treated with TGF-ß1 (2 ng/ml) plus cytomix (2.5 ng/ml IL-1ß + 5 ng/ml TNF-α + 5 ng/ml IFN-γ) for 24 h. MicroRNA expression was assessed by miRNA microarray and further confirmed by real time RT-PCR. (A) miRNA expression by microRNA microarray. Plotted data shown is one representative strain from 3 HBEC strains. Four dots in the circle indicate up-regulated miRNA-146a, an average of 3.8±0.2 fold increase in all 3 strains tested. Vertical axis: cells treated with TGF-ß1 plus cytomix. Horizontal axis: cells cultured in medium only. (B) Quantitative assay of miRNA-146a expression by real time RT-PCR. Data presented was from two strains of the HBECs (96-157 and 06-440). Vertical axis: fold change of miRNA-146a expression compared to control. Horizontal axis: cell strains. Control: medium only. TCM: TGF-ß1 plus cytomix.

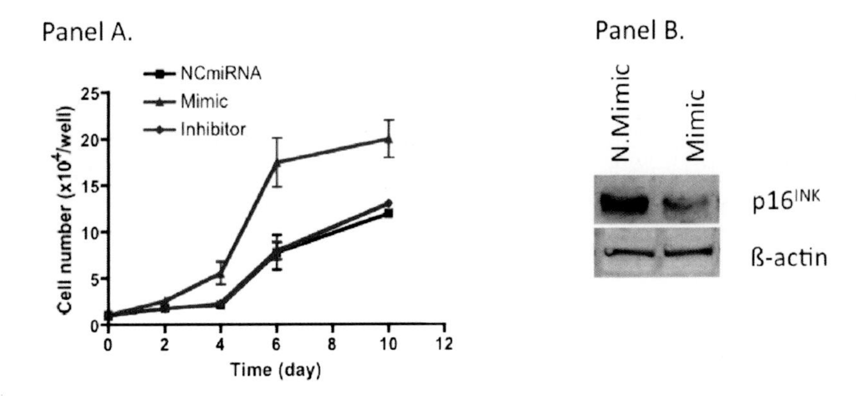

Fig. 4 miR-146a promotes HBEC proliferation and partially inhibits p16^{INK4A}. (A) Effect on cell growth. A negative control miRNA or a miRNA-146a mimic were transfected into HBECs using Lipofectamine 2000. Cells were then allowed to grow in complete medium for 10 d and medium was changed every 2 d. Cell number was counted by a Coulter Counter. Vertical axis: cell number per well. Horizontal axis: time (day). (B) Effect on p16^{INK4A}. HBECs were transfected with either a negative control miRNA (NC miRNA) or a mimic of miRNA-146a. After 4 d culture in complete medium, cell lysate was immunoblotted for p16^{INK4A} and ß-actin (as loading control).

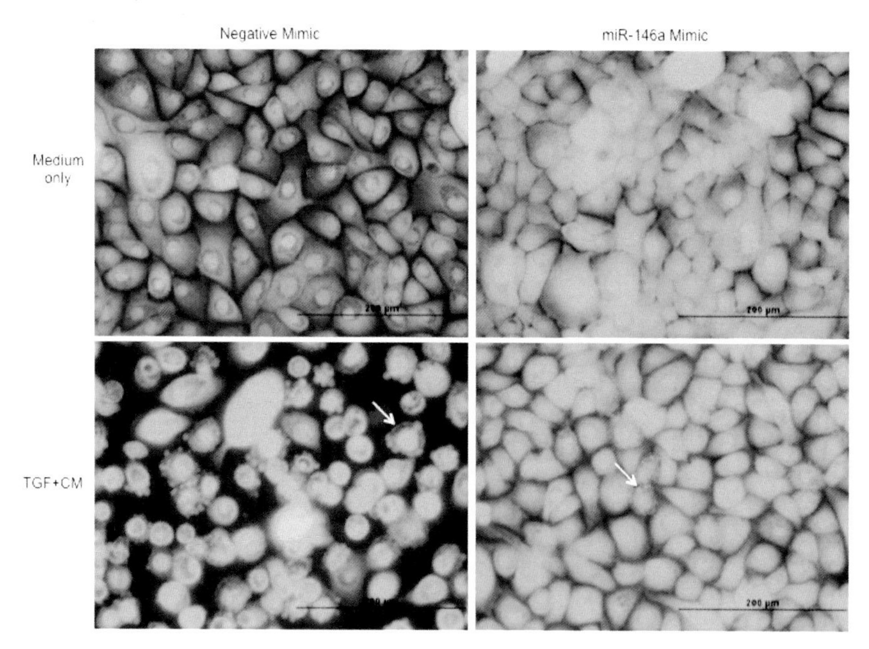

Fig. 5 Introduction of a miR-146a mimic protected HBECs from apoptosis. A negative control miRNA and a miR-146a mimic (100nM for both) were transfected into HBECs using Lipofectamine 2000. Cells were then treated with or without TGF-ß1 (2 ng/ml) plus cytomix (2 ng/ml IL-1ß, 4 ng/ml TNF-α and 4 ng/ml IFN-γ). After 48 h, cell morphology and viability was assessed by LIVE/DEAD staining. Arrows indicate examples of apoptotic cells with intact membrane but cell shrinkage with blebs, membrane asymmetry and condensed nuclei.

Chapter 4

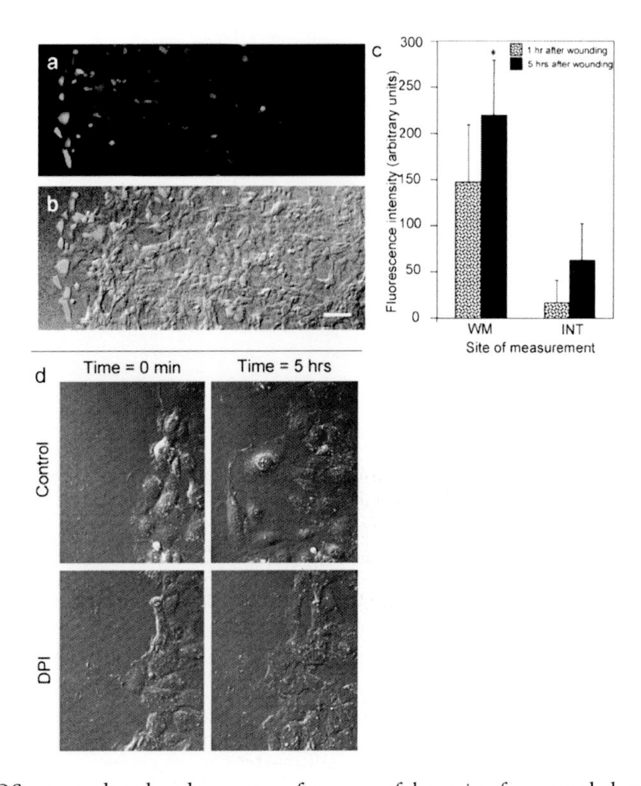

Fig. 1 ROS are produced and necessary for successful repair of a wounded endothelium. Increased production of ROS at the wound margin. a and b, CM-DCF-DA fluorescence (a) and overlay of the fluorescence and Differential Interference Contrast (DIC) Light Microscopy images of the same field (b). Bar = 100 μm. c, Measurement of CM-DCF-DA fluorescence in mouse aortic endothelial cells, 1 and 5 hr after wounding, at the wound margin (WM) and distant from the wound, in the intact monolayer (INT). Data correspond to mean ± SEM. *$P<0.05$ for WM vs INT at 5 hr. ROS are required for endothelial wound healing. d, DIC images of cells at the beginning (0 min) and end (5 hr) of representative experiments for each condition (control and 10 μm DPI). Bar = 50 μm. (With permission from Circ. Res. 2000: 86: 549-557.)

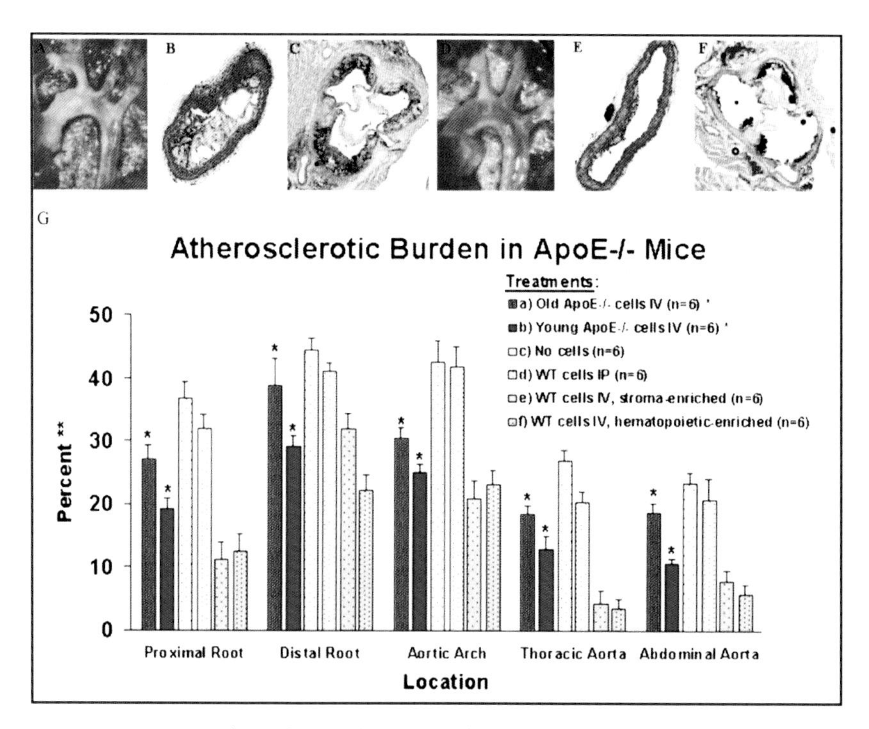

Fig. 3 Bone marrow-derived EPCs from young but not aged mice suppress atherosclerosis. Atherosclerosis assessment in mice receiving no cell treatment (**A-C**) or EPCs from age-matched wild-type mice (**D-F**). The relative efficacy of EPCs from young and aged apoE-/- mice as well as wild-type mice in different parts of the aortic wall is depicted in **G**. (With permission from Circulation 2003: 108: 457-463.)

Chapter 12

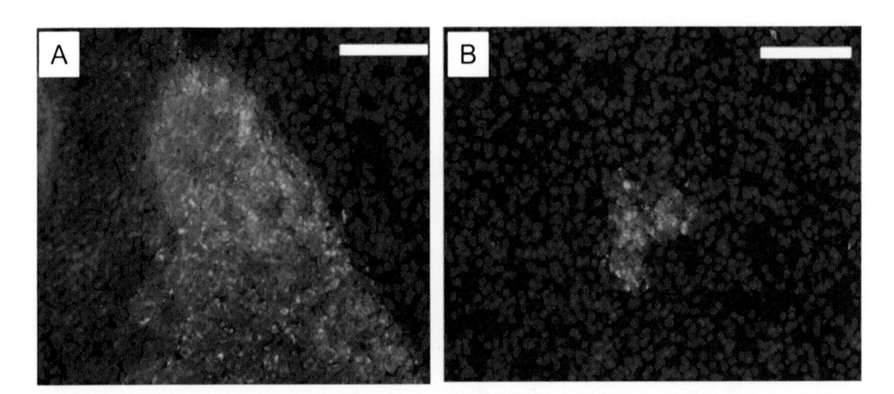

Fig. 1 Selective engraftment of IFN-β-over-expressing hUCMSC to lung metastasized MDA-231 breast carcinoma xenografts in SCID mice. MDA-231 lung metastatic breast carcinoma-bearing SCID mice were treated systemically via IV injection with IFN-β-hUCMSC (5×10^5 cells) labeled with SP-DiI red fluorescent dye, three times with one week interval. One week after the last treatment, lungs were dissected and subjected to immunohistochemical analysis. MDA-231 cells in the lung were identified by anti-human mitochondrial antibodies conjugated with a green fluorophore. Sections were counterstained with Hoechst 33342 nuclear stain (blue). These pictures clearly indicate that systemically administered IFN-β-hUCMSC selectively engrafted within a microtumor or in close proximity to the lung tumor (green). (Scale bar = 100 μm). Figure is modified from the original figure by Rachakatla *et al.* (2008).

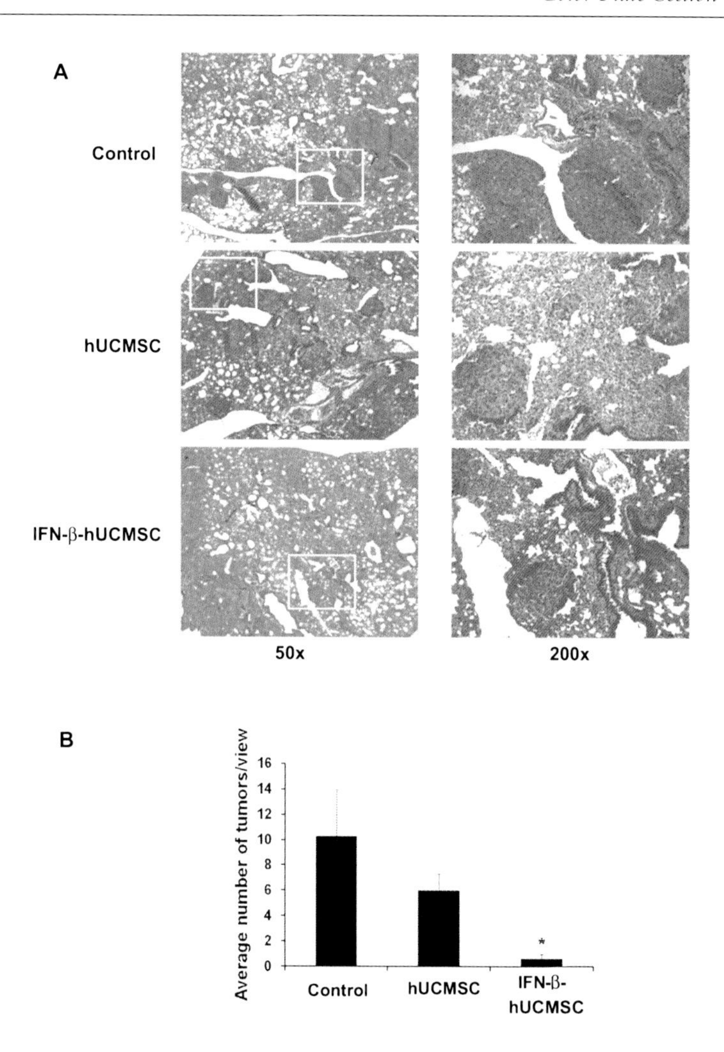

Fig. 3 Systemic administration of IFN-β-over-expressing hUCMSC significantly attenuated the growth of H358 orthotopic lung adenocarcinoma xenografts in SCID mice. H358 orthotopic lung adenocarcinoma xenografts were treated systemically via IV injection with IFN-β-hUCMSC (4×10^5 cells) four times with 5 d interval. Two weeks after the last treatment, lungs were dissected and fixed in 10% formalin. Paraffin-embedded lung sections were stained by H&E staining. Morphologies of lung tumors in three different treatments are presented in the upper panels (A). Multiple sizes of tumors, from as small as a cluster of several cancer cells to relatively large tumors, were counted under the microscope at 50x magnification. White squares in the 50x pictures indicate the area magnified in the 200x pictures. The average number of tumors in three view areas was expressed in the bar graph. *, $p < 0.05$ as compared to the level of PBS-treated control. Figure is modified from the original figure by Matsuzuka *et al.* (2010).

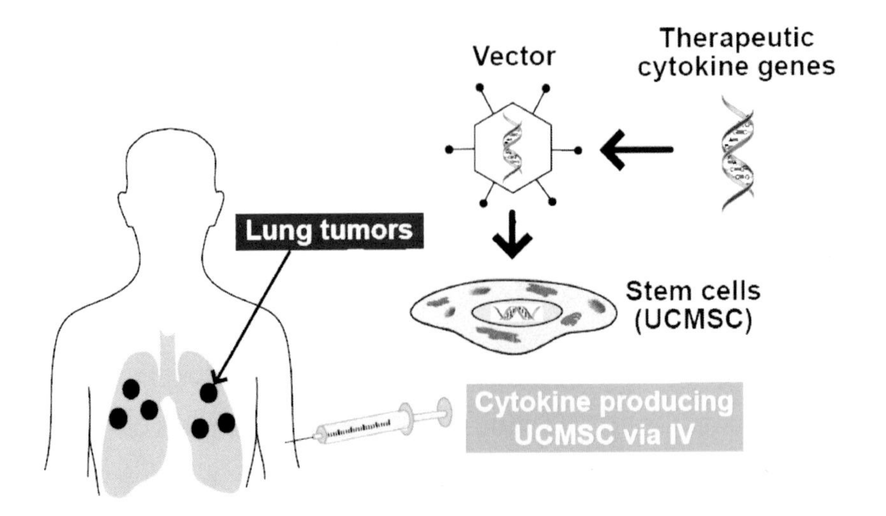

Fig. 5 Schematic illustration of cellular vehicle-based targeted gene therapy. Systemically administered stem cells transfected with therapeutic cytokine genes specifically migrate to cancer site and produce cytokine effectively. This procedure overcomes cytokine short half life and cytotoxicity to normal tissues.

Chapter 14

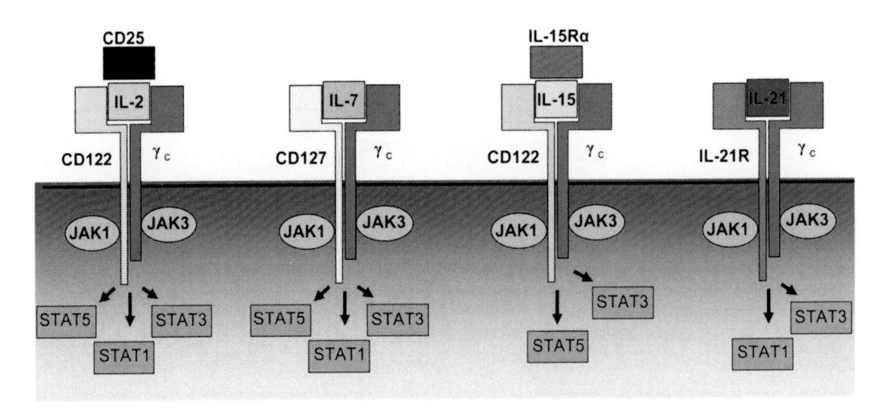

Fig. 1 Cytokine receptors on T cells. Common gamma chain cytokines (IL-2, IL-7, IL-15, IL-21) being used in adoptive T cell immunotherapy of cancer both *ex vivo* and *in vivo*, and their shared receptors, the IL-2 Rγ (γ c). Predominant STAT signaling for each cytokine is shown.

Chapter 15

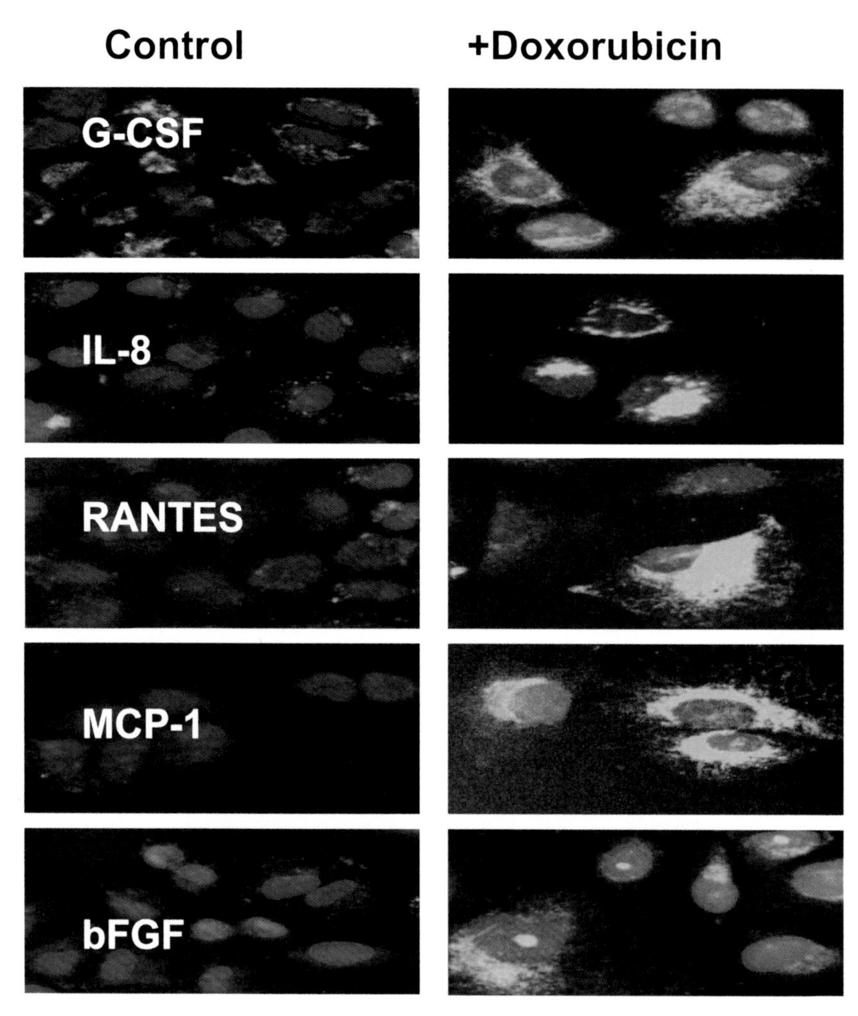

Fig. 1 Treating H460 NSCLC cells with doxorubicin increases multiple cytokine production. Intracellular accumulation of cytokines in doxorubicin-treated and untreated cells was analyzed. H460 cells were treated with doxorubicin (0.125 μg/ml) for 16 hr and were then incubated with monensin (2 μM) for 3 hr. Cells were incubated with primary Abs against CXCL8, CCL2, CCL5, G-CSF, bFGF or VEGF and then with secondary Alexa Fluor 488- or Alexa 680-conjugated Abs. Cell nuclei were stained with Hoechst 33342 and cell images were acquired using the Cellomics ArrayScan HCS Reader and analyzed using the Target Activation BioApplication Software Module. Images of immunofluorescently stained cells for CXCL8, CCL2, CCL5, G-CSF or bFGF (Alexa488) are presented.

Chapter 17

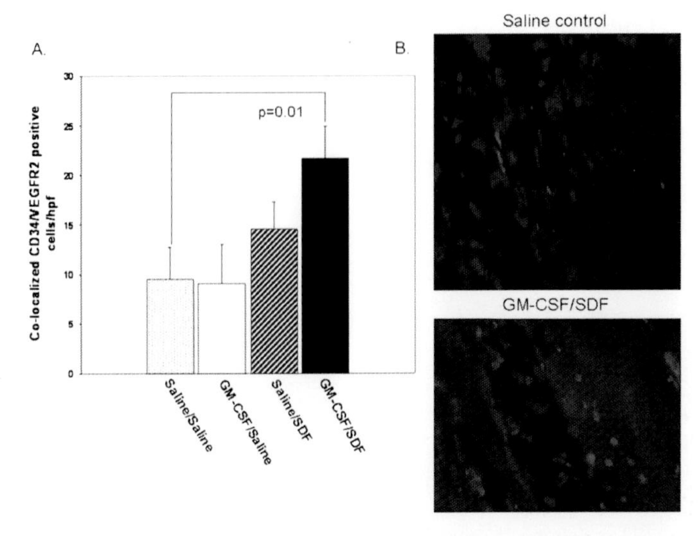

Fig. 3 Effect of stromal cell-derived factor 1α (SDF) and granulocyte-macrophage colony-stimulating factor (GM-CSF) on EPC migration in a rat model of ischemic cardiomyopathy. (a) Graph of the mean number of EPC markers CD34+/VEGFR2+ coexpressing cells per high-power field in hearts from the four groups. (b) Representative immunostains of saline control and SDF/GM-CSF-treated hearts for CD34 (green), VEGFR2 (red; merged = orange), and DAPI nuclei (blue) (reproduced from Woo *et al.*).

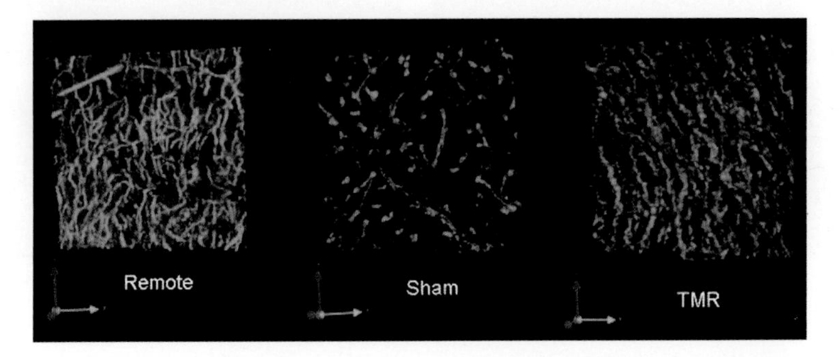

Fig. 4 Effect of transmyocardial laser revascularization (TMR) in a porcine model of ischemic heart failure. Three-dimensional microvascular lectin angiogram demonstrates enhanced myocardial perfusion after therapy with TMR. Representative angiograms from remote myocardium and ischemic myocardium from animals in sham and transmyocardial laser revascularization groups are presented (z-series, 25× oil magnification, Zeiss LSM-510 Meta Confocal Microscope; Carl Zeiss, Inc, Thornwood, NY). There was no difference in remote myocardial perfusion between sham and TMR groups. Perfusion of the ischemic myocardium was significantly increased after TMR. Orientation is presented in lower left corners. Red indicates y-axis; green indicates x-axis; blue indicates z-axis (reproduced from Atluri *et al.*).

Chapter 20

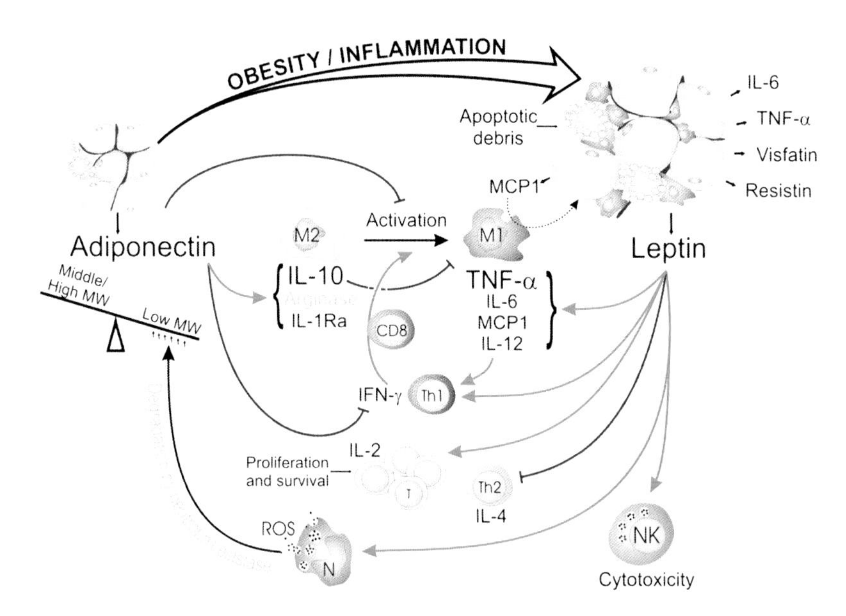

Fig. 1 Interaction of key pro- and anti-inflammatory cytokines, with particular emphasis on the effects of leptin and adiponectin. In the lean state, adipose tissue is in non-inflammatory state characterized by abundant concentrations of adiponectin, which has several anti-inflammatory effects. Also alternatively activated (M2) macrophages of wound-healing phenotype prevail and contribute to the immunological homeostasis by production of anti-inflammatory cytokines such as IL-10. With increasing obesity, changes in adipose tissue cause a shift towards pro-inflammatory secretions such as leptin and in time the adipose tissue macrophage population is increased and transformed to classically activated, pro-inflammatory phenotype (M1) characterized by pro-inflammatory cytokine production. Leptin promotes NK cytotoxicity, neutrophil chemotaxis and activation, T-cell differentiation to Th1, proliferation and survival of naïve T cells and production of pro-inflammatory cytokines such as TNF-α by monocytes/macrophages. IFN-γ producing Th1 cells promote the transition from M2 to M1 cells and may activate cytotoxic T cells (CD8+), which may also contribute to M1 activation and recruitment. Neutrophil action (production of neutrophil elastase) may result in degradation of middle and high molecular weight adiponectins, which may result in reduced anti-inflammatory action by this adipokine. Green arrows indicate promoting and red lines with blunt end inhibitory action. Th1/Th2, T-helper cell type 1/type 2; NK, natural killer cell; N, neutrophil; ROS, reactive oxygen species; M1, classically activated, pro-inflammatory, macrophage; M2, alternatively activated, wound healing/anti-inflammatory macrophage; Low/Middle/High MW, Low/Middle/High molecular weight adiponectin; CD8, CD8+ cytotoxic T-cells. (Figure produced by Pirkka Kirjavainen.)

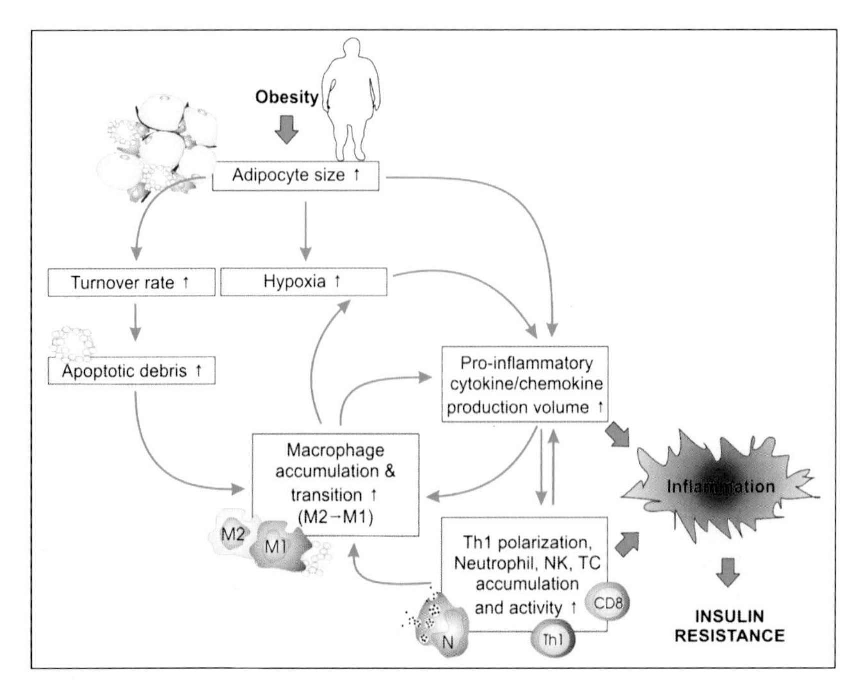

Fig. 2 Potential key events in the formation of a vicious cycle maintaining and exacerbating chronic inflammation in obesity, which interferes with insulin signaling resulting in reduced insulin sensitivity. In obesity, the size of adipocytes is increased. This causes (1) increased turnover and thus apoptosis rate and need for clearance by macrophages, (2) hypoxia, and (3) increased production volume of potentially pro-inflammatory cytokines such as leptin. The cytokine milieu together with other signals then skew the adipose tissue macrophage population to exhibit pro-inflammatory phenotype (M1), result in Th1 polarization and recruitment and activation of cytotoxic T-cells (CD8+), natural killer cells and neutrophils, which further exacerbate the inflammation. These events also result in further infiltration of monocytes/macrophages to the adipose tissue, causing further hypoxia and increase in pro-inflammatory cytokine/chemokine production; thus, a vicious cycle may be formed. Green arrows indicate promoting action. Th1, T-helper cell type 1; NK, natural killer cell; N, neutrophil; M1, classically activated, pro-inflammatory, macrophage; M2, alternatively activated, wound healing/anti-inflammatory macrophage; CD8, CD8+ cytotoxic T-cells. (Figure produced by Pirkka Kirjavainen.)

Chapter 24

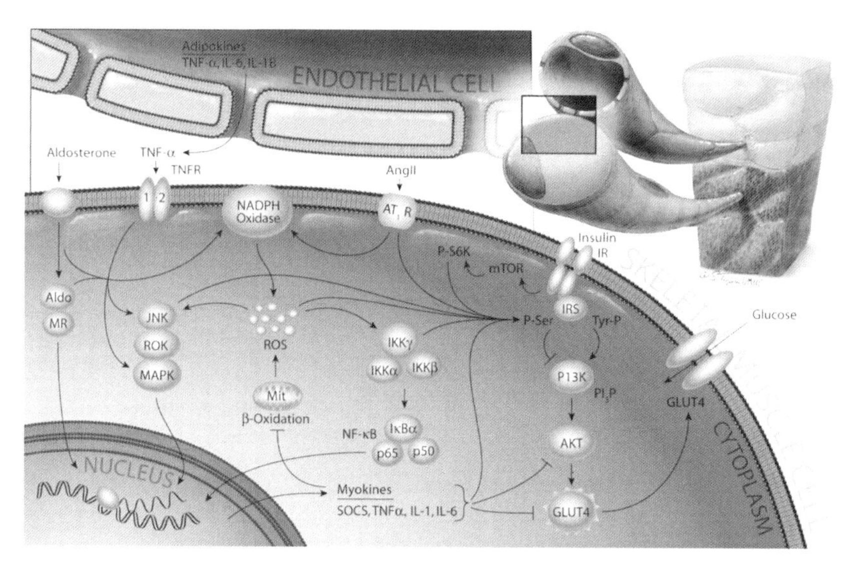

Fig. 1 Visceral adipose tissue is an important contributor of circulating cytokines (e.g., adipokines) that, in concert with local production of cytokines in skeletal muscle (e.g., myokines), contribute to inflammatory pathways that react with oxidative stress and activation of the RAAS to impair insulin metabolic signaling in skeletal muscle tissue. The consequent impairment in insulin-mediated glucose utilization in skeletal muscle is an important contributor to skeletal muscle insulin resistance as well as systemic insulin resistance. (Original artwork by Stacy Turpin.) (→ enhancement; inhibition)

Chapter 26

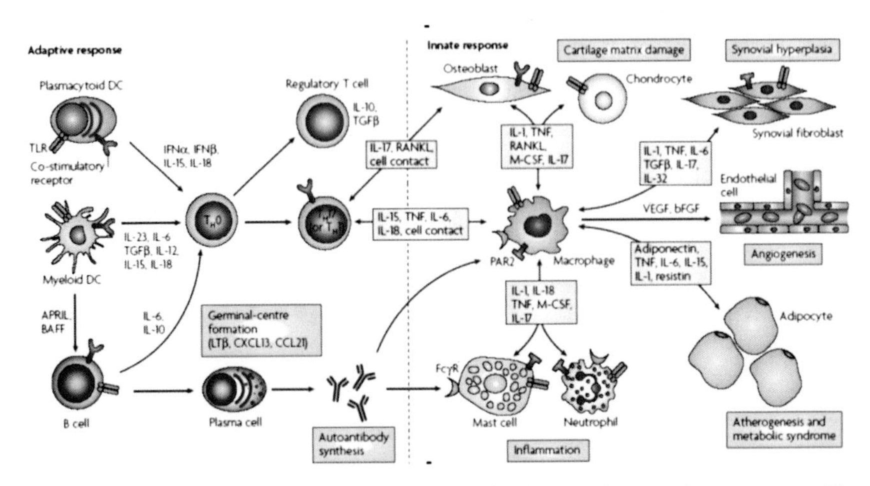

Fig. 2 An overview of the cytokine-mediated regulation of synovial interactions. The component cells of the inflamed rheumatoid synovial membrane are depicted in innate and adaptive predominant compartments of the inflammatory response. Pivotal cytokine pathways are depicted in which activation of dendritic cells (DCs), T cells, B cells and macrophages underpin the dysregulated expression of cytokines that in turn drive activation of effector cells, including neutrophils, mast cells, endothelial cells and synovial fibroblasts. The clinical manifestations of such effects are highlighted. Reprinted with permission of Nature Publishing Group from McInnes and Schett (2007). APRIL, a proliferation-inducing ligand; BAFF, B-cell activating factor; bFGF, basic fibroblast growth factor; CCL21, CC-chemokine ligand 21; CXCL13, CXC-chemokine ligand 13; FcγR, Fc receptor for IgG; IFN, interferon; IL, interleukin; LTβ, lymphotoxin-β; M-CSF, macrophage colony-stimulating factor; PAR2, protease-activated receptor 2; RANKL, receptor activator of nuclear factor-κB (RANK) ligand; TGFβ, transforming growth factor-β; TH, T helper; TLR, Toll-like receptor; TNF, tumour-necrosis factor; VEGF, vascular endothelial growth factor.